A

Comprehensive Introduction

to

DIFFERENTIAL GEOMETRY

VOLUME FOUR

A Comprehensive Introduction to DIFFERENTIAL GEOMETRY

VOLUME FOUR

Second Edition

MICHAEL SPIVAK

Publish or Perish, Inc.

Cloth Set, Vols. 3,4,5 ISBN 0-914098-82-9
Cloth Set, Vols, 1-5 ISBN 0-914098-83-7
Library of Congress Card Catalog Number 78-71771

PUBLISH OR PERISH, INC.
WILMINGTON, DELAWARE

In Japan distributed exclusively by
KINOKUNIYA COMPANY
TOKYO, JAPAN

Printed in the United States of America

CONTENTS

[Although the Chapters are not divided into sections, except for the major subdivisions of Chapter 7, the list for each Chapter gives some indication which topics are treated, and on which pages.]

Contents of Volume III

Contents of Volume V

A

Comprehensive Introduction

to

DIFFERENTIAL GEOMETRY

VOLUME FOUR

Chapter 7. Higher Dimensions and Codimensions

The aim of this chapter is, roughly speaking, to see whether and how the results of the previous chapters generalize; instead of surfaces in $\mathbb{R}^3$, we will be considering higher dimensional manifolds, of higher codimensions, imbedded or immersed in more general Riemannian manifolds. Even at the risk of making the chapter somewhat disorganized, I have tried to make it pretty complete, so that the reader does not have to sit gnawing his thumbs wondering whether a generalization does not appear because it is trivial or because it is false, or because it is unknown. It should be mentioned, however, that a few diddly topics, like the Dupin indicatrix, aren't considered at all. In addition, a few points are taken up in later chapters, and the bibliography for appropriate sections should also be consulted. Finally, the most notable omission of all is the generalization of the Gauss-Bonnet Theorem, which occupies the place of honor in the last chapter of the book.

A. The Geometry of Constant Curvature Manifolds

Although our aim in this chapter is to obtain results of the greatest possible generality, many of the theorems will not hold, or even make sense, unless the ambient manifold has constant curvature K_0. It will be necessary for us to be as familiar with the properties of these Riemannian manifolds as we are with the case of Euclidean space $(K_0 = 0)$. We will consider only the simply-connected complete n-dimensional Riemannian manifolds $(M, \langle\ ,\ \rangle)$ of constant curvature K_0; by Problem 1-5, the manifold $(M, \langle\ ,\ \rangle)$ is then uniquely determined up to isometry by K_0.

For $K_0 > 0$, the manifold $(M, \langle\ ,\ \rangle)$ is just the n-sphere $S^n(K_0)$ of radius $1/\sqrt{K_0}$ in $\mathbb{R}^{n+1}$,

$$S^n(K_0) = \left\{p \in \mathbb{R}^{n+1};\ \langle p,p\rangle = \frac{1}{K_0}\right\},$$

with the Riemannian metric induced from the ordinary metric $\langle\ ,\ \rangle$ of $\mathbb{R}^{n+1}$. For simplicity, we usually consider only the case $K_0 = 1$, setting $S^n = S^n(1)$. It is clear that every orthogonal map $A \in O(n+1)$ takes S^n to itself and is an isometry. Moreover, $O(n+1)$ is precisely the set of isometries of S^n, since a suitable $A \in O(n+1)$ takes any orthonormal frame $X_1,\dots,X_n \in S^n{}_p$ at any point $p \in S^n$ to any other orthonormal frame $Y_1,\dots,Y_n \in S^n{}_q$ at any point $q \in S^n$, and an isometry of S^n is determined by its action on $S^n{}_p$ (Problem 1-5).

For $K_0 < 0$, we can obtain an analogous submanifold of $\mathbb{R}^{n+1}$ by considering a non-positive definite Riemannian metric on $\mathbb{R}^{n+1}$. Denoting the components of a point $a \in \mathbb{R}^{n+1}$ by $a^0, a^1,\dots,a^n$, we consider first the non-degenerate inner product $\langle\,,\rangle$ on $\mathbb{R}^{n+1}$ defined by

$$\langle a,b\rangle = -a^0b^0 + a^1b^1 + \cdots + a^nb^n .$$

This is called the <u>Lorentzian</u> inner product on $\mathbb{R}^{n+1}$, and the group $O^1(n+1)$ of all linear transformations $f\colon \mathbb{R}^{n+1} \to \mathbb{R}^{n+1}$ which preserve $\langle\,,\rangle$ is called the <u>Lorentz group</u> [actually (Problem 1), any map $f\colon \mathbb{R}^{n+1} \to \mathbb{R}^{n+1}$ preserving $\langle\,,\rangle$ is automatically linear]. By means of the standard identification of $\mathbb{R}^{n+1}{}_p$ with $\mathbb{R}^{n+1}$, we obtain a non-degenerate inner product $\langle\,,\rangle_p$ on each $\mathbb{R}^{n+1}{}_p$, and thus a non-positive definite Riemannian metric on $\mathbb{R}^{n+1}$, which we denote also simply by $\langle\,,\rangle$. In terms of the standard coordinate system

$x^0, x^1, \ldots, x^n$ on $\mathbb{R}^{n+1}$ we have

$$\langle\, ,\rangle = -\, dx^0 \otimes dx^0 + dx^1 \otimes dx^1 + \cdots + dx^n \otimes dx^n .$$

The isometries of $(\mathbb{R}^{n+1}, \langle\, ,\rangle)$ are (Problem 2) precisely the maps of the form

$$p \longmapsto A(p) + q \qquad A \in O^1(n+1), \quad q \in \mathbb{R}^{n+1} .$$

Now for $K_0 < 0$ consider the quadric hypersurface

$$\left\{ p \in \mathbb{R}^{n+1} : \langle p, p\rangle = \frac{1}{K_0} \right\} .$$

This consists of 2 components, each homeomorphic to $\mathbb{R}^n$; we will pick one of

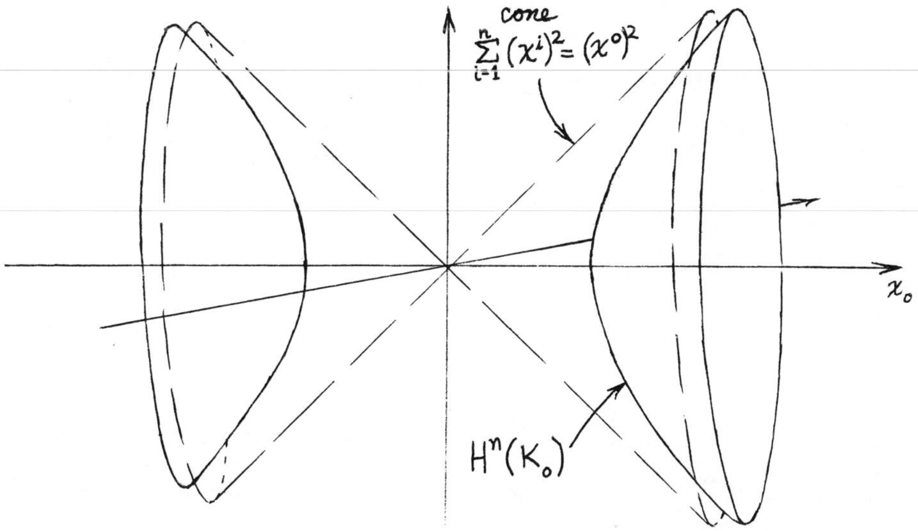

them, say the one consisting of points with $p^0 > 0$, and define

$$H^n(K_0) = \left\{ p \in \mathbb{R}^{n+1} : p^0 > 0 \text{ and } \langle p,p \rangle = \frac{1}{K_0} \right\} .$$

For simplicity, we usually consider only the case $K_0 = -1$, setting $H^n(-1) = H^n$, the "n-dimensional hyperbolic space." To find the tangent space $H^n{}_p$, we proceed precisely as in the case of S^n. Any curve c in H^n satisfies

$$\langle c(t),c(t) \rangle = 0 \quad \text{for all } t \quad \Rightarrow \quad \langle c'(t),c(t) \rangle = 0 ,$$

so $H^n{}_p$ contains only vectors v_p with $\langle v,p \rangle = 0$. Moreover, $\{v : \langle v,p \rangle = 0\}$ is the kernel of the non-zero linear functional $v \mapsto \langle v,p \rangle$, so it has dimension exactly $n-1$. Thus

$$H^n{}_p = \{v_p : \langle v,p \rangle = 0\} \qquad \text{for } p \in H^n, \text{ i.e., } \langle p,p \rangle = -1 .$$

We now claim that the induced Riemannian metric on H^n is positive definite. To show this, it is convenient to consider the index of a bilinear function $B\colon V \times V \to \mathbb{R}$ on a vector space V, which is defined to be the largest dimension of any subspace $W \subset V$ on which B is negative definite [that is, $B(w,w) < 0$ for all $0 \neq w \in W$]. The bilinear function

$$(a,b) \mapsto \langle a,b \rangle = -a^0b^0 + a^1b^1 + \cdots + a^nb^n$$

on $\mathbb{R}^{n+1}$ clearly has index ≥ 1, for it is negative definite on the subspace $U^- = \{(a^0,0,\ldots,0)\}$. Moreover, $\langle\;,\;\rangle$ is positive definite on the subspace $U^+ = \{(0,a^1,\ldots,a^n)\}$. If $\langle\;,\;\rangle$ were negative definite on a subspace W of dimension ≥ 2, then $\langle\;,\;\rangle$ would be negative definite on the non-zero subspace

$W \cap U^+$, which is clearly impossible. So $\langle\ ,\ \rangle$ has index 1. Naturally, each $\langle\ ,\ \rangle_p$ also has index 1. Now consider $\langle\ ,\ \rangle_p$ on $H^n{}_p$. If $v_p \in H^n{}_p$, then v is linearly independent of p, and we already have $\langle p,p\rangle < 0$, so we cannot have $\langle v,v\rangle < 0$, as $\langle\ ,\ \rangle$ has index 1. Nor can we even have $\langle v,v\rangle = 0$, for then we would have

$$\langle p+v,\ p+v\rangle = \langle p,p\rangle + 2\langle p,v\rangle + \langle v,v\rangle = \langle p,p\rangle < 0\ ,$$

which is also impossible. Thus $\langle\ ,\ \rangle_p$ is positive definite on $H^n{}_p$, and H^n is an ordinary Riemannian manifold. (In the above picture this is quite clear, since all tangent lines have greater slope than the generators of the cone $\Sigma(x^i)^2 = (x^0)^2$, and a vector v along one of these generators satisfies $\langle v,v\rangle = 0$.)

Naturally every element of $O^1(n+1)$ which keeps $\{p \in \mathbb{R}^{n+1}\colon p^0 > 0\}$ fixed will give an isometry of H^n onto itself. We also claim that all isometries of H^n arise in this way. To prove this, we just note that if $(v_1)_p,\ldots,(v_n)_p \in H^n{}_p$ is orthonormal, and similarly for $(w_1)_q,\ldots,(w_n)_q \in H^n{}_q$, so that

$$\langle p,p\rangle = \langle q,q\rangle = -1$$
$$\langle v_i,p\rangle = 0 = \langle w_i,q\rangle$$
$$\langle v_i,v_j\rangle = \langle w_i,w_j\rangle = \delta_{ij}\ ,$$

then the linear transformation taking

$$p \mapsto q \qquad \text{and} \qquad v_i \mapsto w_i$$

is clearly in $O^1(n+1)$. Since there are thus isometries of H^n taking any orthonormal basis at any point to any orthonormal basis at any other point,

H^n must have constant curvature. We can compute that $H^n(K_0)$ has constant curvature K_0 in a manner exactly analogous to a computation of the curvature of $S^n(K_0)$, by using theorems 1-1, 1-6, and 1-9; the only difference is that we must allow the ambient manifold in theorems 1-1 and 1-6 to have a non-positive definite Riemannian metric, and the "unit" normal field ν in 1-9 will actually satisfy $\langle\nu,\nu\rangle = -1$. The manifold $H^n(K_0)$ is (geodesically) complete. Because we are dealing with an indefinite metric on $\mathbb{R}^{n+1}$, this does not simply follow from the fact that $H^n(K_0)$ is a closed subset of $\mathbb{R}^{n+1}$. However, it is an easy exercise to prove completeness using the fact that there are isometries taking any orthonormal basis to any other. We also mention that the geodesics of H^n are (Problem 3) precisely the intersections $H^n \cap P$ where P is a plane in $\mathbb{R}^{n+1}$ through 0; more generally, the totally geodesic submanifolds of H^n are $H^n \cap P$ where P is a vector subspace of $\mathbb{R}^{n+1}$.

In the past we have given several other models for H^n, and for S^n minus a point. For example, we have described the metric of a space of constant curvature K_0 in terms of normal coordinates, in Addendum 1 to Chapter II.7. In Addendum 2 to that Chapter we found the most general isothermal coordinate systems on the manifolds of constant curvature, after first determining the expression for the metric on S^n in the coordinate system defined by "stereographic projection". To define this map, we considered S^n as the sphere of radius 1 around the point $(0,\dots,0,1)$, so that S^n is tangent to $\mathbb{R}^n = \mathbb{R}^n \times \{0\} \subset \mathbb{R}^{n+1}$. Letting $*$ be the "north pole" $* = (0,\dots,0,2) \in S^n$, the stereographic projection

$$\sigma\colon S^n - \{*\} \longrightarrow \mathbb{R}^n$$

is defined geometrically as follows: for any $p \neq *$ in S^n, we let $\sigma(p)$ be the point where the line between p and $*$ intersects $\mathbb{R}^n$. It is easy to

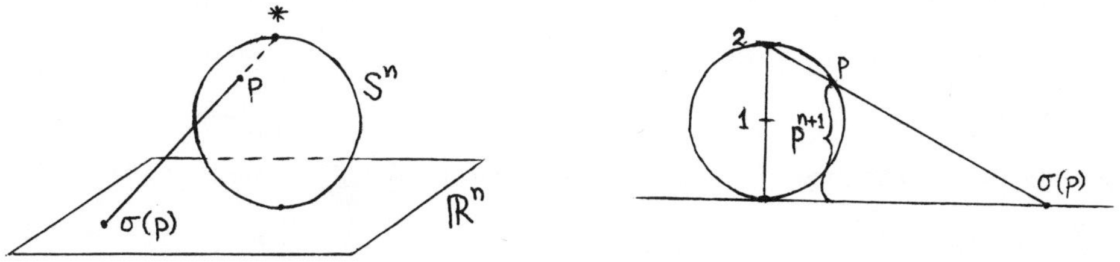

check (see the figure on the right) that

$$(1) \qquad \sigma(p) = \left(\frac{2p^1}{2 - p^{n+1}}, \ldots, \frac{2p^n}{2 - p^{n+1}}\right)$$

and that $f = \sigma^{-1}$ is given by

$$(2) \qquad \sigma^{-1}(y) = f(y) = \left(\frac{y^1}{1 + \frac{1}{4}\Sigma(y^i)^2}, \ldots, \frac{y^n}{1 + \frac{1}{4}\Sigma(y^i)^2}, \frac{\frac{1}{2}\Sigma(y^i)^2}{1 + \frac{1}{4}\Sigma(y^i)^2}\right).$$

If $y^1, \ldots, y^n$ denotes the standard coordinate system on $\mathbb{R}^n$, then the $y^i \circ \sigma$ give a coordinate system on $S^n - \{*\}$. We can compute the metric $\langle\ ,\ \rangle$ in terms of this coordinate system by computing

$$\begin{aligned} f^* \sum_{i=1}^{n+1} dx^i \otimes dx^i &= \sum_{i=1}^{n+1} df^i \otimes df^i \\ &= \sum_{i=1}^{n+1} \sum_{j,k=1}^{n} \frac{\partial f^i}{\partial y^j} \frac{\partial f^i}{\partial y^k} dy^j \otimes dy^k, \end{aligned}$$

by means of equation (2). However we can save ourselves a lot of computational work by first proving geometrically that σ is conformal. It clearly suffices to consider the case $n = 2$. Notice first that if $L \subset \mathbb{R}^2$ is a straight line, then the lines through $*$ and points of L form a plane with a horizontal line through $*$ deleted, so $\sigma^{-1}(L)$ is $\Sigma - \{*\}$ for some circle $\Sigma \subset S^2$. Now given two linearly independent vectors $X_1, X_2 \in S^2_p$, consider the straight lines L_1, L_2 through $\sigma(p)$ pointing in the directions of $\sigma_*(X_1)$ and $\sigma_*(X_2)$. Their inverse images under σ are $\Sigma_1 - \{*\}$ and $\Sigma_2 - \{*\}$ for

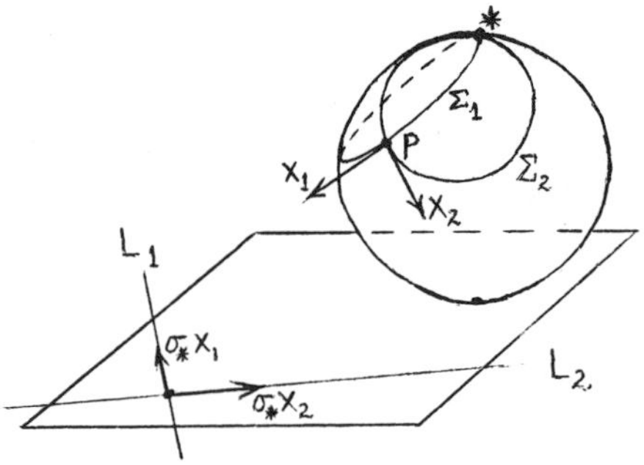

two circles $\Sigma_1, \Sigma_2 \subset S^2$ containing $*$. The angle between X_1 and X_2 is the angle of intersection of Σ_1 and Σ_2 at p, which is the same as the angle of intersection of Σ_1 and Σ_2 at $*$. But the tangent lines to Σ_1 and Σ_2 at $*$ are parallel to L_1 and L_2, respectively. So the angle of intersection at $*$ is the same as the angle between $\sigma_* X_1$ and $\sigma_* X_2$. Thus, σ is conformal.

Now for any point $y \in \mathbb{R}^n$, let $c\colon [0,2\pi] \to \mathbb{R}^n$ be a curve, parameterized proportionally to arclength, which goes once around a circle centered at 0 and passing through y; thus c' always has squared length $|y|^2$. Formula (2)

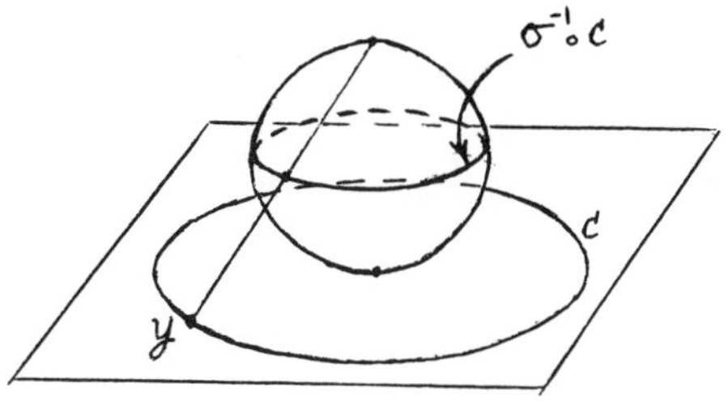

shows that $(\sigma^{-1}\circ c)'$ always has squared length

$$\sum_{i=1}^{n}\left[\frac{y^i}{1+\frac{1}{4}\Sigma(y^i)^2}\right]^2 = \frac{|y|^2}{\left[1+\frac{1}{4}\Sigma(y^i)^2\right]^2} .$$

This shows that in the conformal coordinate system $\{x^i = y^i\circ\sigma\}$ on $S^n - \{*\}$, the metric $\langle\ ,\ \rangle$ has the form

$$(3) \qquad \langle\ ,\ \rangle = \sum_{i=1}^{n} \frac{dx^i \otimes dx^i}{\left[1+\frac{1}{4}\Sigma(x^i)^2\right]^2} .$$

If we were dealing with a sphere of curvature K_0, the factor 1/4 would be replaced by $K_0/4$.

For later use, we mention one further property of the stereographic projection: it takes spheres in S^n to spheres and hyperplanes of $\mathbb{R}^n$, and visa-versa. Indeed, a sphere $\Sigma \subset S^n$ is the intersection of S^n with some hyperplane,

$$\Sigma = \left\{ p \in S^n \colon \sum_{i=1}^{n+1} \alpha_i p^i = \beta \right\} ,$$

and then

$$\begin{aligned} y \in \sigma(\Sigma) &\iff \sigma^{-1}(y) \in \Sigma \\ &\iff \sum_{i=1}^{n} \alpha_i \frac{y^i}{1+\frac{1}{4}\Sigma(y^i)^2} + \frac{1}{2}\alpha_{n+1} \frac{\Sigma(y^i)^2}{1+\frac{1}{4}\Sigma(y^i)^2} = \beta , \qquad \text{by (2)} . \end{aligned}$$

This is always a sphere or hyperplane in $\mathbb{R}^n$, and the converse works similarly.

Now for $K_0 < 0$, in particular for $K_0 = -1$, we can just formally replace the factor 1/4 in (3) by -1/4. In Addendum 2 to Chapter II.7 we showed that this metric does indeed have $K_0 = -1$. In fact, this metric was simply one possible choice for the conformal metrics of constant curvature $K_0 = -1$.

We have already pointed out that, in order to have a connected manifold, we must consider the metric

$$\langle\ ,\ \rangle = \frac{\sum_{i=1}^{n} dx^i \otimes dx^i}{\left[1-\frac{1}{4}\Sigma(x^j)^2\right]^2}$$

only on the open ball of radius 2,

$$B^n = B^n(2) = \{x \in \mathbb{R}^n \colon \Sigma(x^i)^2 < 4\} ,$$

but that $\langle\ ,\ \rangle$ is already complete on B^n (see p. II.339).

Thus $(B^n, \langle\ ,\ \rangle)$ must be isometric to the space $H^n \subset (\mathbb{R}^{n+1}, \langle\ ,\ \rangle)$; a method for constructing an explicit isometry between

$(B^n, <\ ,\ >)$ and H^n will be suggested later.

The model $(B^n, <\ ,\ >)$ will often be very useful, and we will examine it in great detail, determining, in particular, precisely what the isometries of $(B^n, <\ ,\ >)$ onto itself look like. In order to do this, however, we first need to generalize a few results from previous chapters.

First of all, Dupin's Theorem (4-10) on triply orthogonal systems of surfaces generalizes immediately to a theorem on n-orthogonal systems of hypersurfaces in $\mathbb{R}^n$. We will also need to generalize Theorem 2-2, concerning all-umbilic surfaces in $\mathbb{R}^3$. For a hypersurface $M \subset \mathbb{R}^{n+1}$ we locally have a unit normal field $\nu\colon M \to S^n \subset \mathbb{R}^{n+1}$, and a map $d\nu\colon M_p \to M_p$ (Theorem 1-8); we call $p \in M$ an **umbilic** if $d\nu\colon M_p \to M_p$ is multiplication by a constant.

1. LEMMA. For $n \geq 2$, let $M \subset \mathbb{R}^{n+1}$ be a connected hypersurface with all points umbilics. Then M is part of a hyperplane or an n-dimensional sphere.

Remark. Later on we will have much more general results.

Proof. As in the proof of Theorem 2-2, it suffices to prove this locally. Choose an adapted orthonormal moving frame $X_1, \ldots, X_n, X_{n+1} = \nu$ on M. By hypothesis, there is a function λ on M such that

$$(1) \qquad \nabla'_X X_{n+1} = -\lambda X \qquad X \text{ tangent to } M .$$

In terms of the dual and connection forms we have

$$\psi^j_{n+1}(X) = \langle \nabla'_X X_{n+1}, X_j \rangle = -\lambda \langle X, X_j \rangle ,$$

and thus

$$\psi_j^{n+1} = -\psi_{n+1}^j = \lambda\theta^j .$$

Taking the exterior derivative of this equation, we obtain

$$d\lambda \wedge \theta^j + \lambda\, d\theta^j = d\psi_j^{n+1} = -\sum_i \psi_i^{n+1} \wedge \omega_j^i \qquad \text{(p. III.30)}$$
$$= -\lambda \sum_i \theta^i \wedge \omega_j^i ,$$

while

$$d\theta^j = -\sum_i \omega_i^j \wedge \theta^i .$$

So we find that

$$d\lambda \wedge \theta^j = 0 \qquad j = 1,\ldots,n .$$

This implies that $d\lambda = 0$, so λ is constant.

The remainder of the argument can be carried out as in the proof of Theorem 2-2, by considering an immersion $f\colon U \longrightarrow M$ for $U \subset \mathbb{R}^n$ open. Here is an alternative (essentially equivalent) argument. If $\lambda = 0$, then all $\psi_j^{n+1} = 0$, so the second fundamental form $s = 0$; thus M is totally geodesic (Propositions 1-16 and 1-17), so M lies in a hyperplane. So we assume $\lambda \neq 0$. Let V be the vector field on $\mathbb{R}^{n+1}$ defined by

$$V(p) = p_p \in \mathbb{R}^{n+1}{}_p .$$

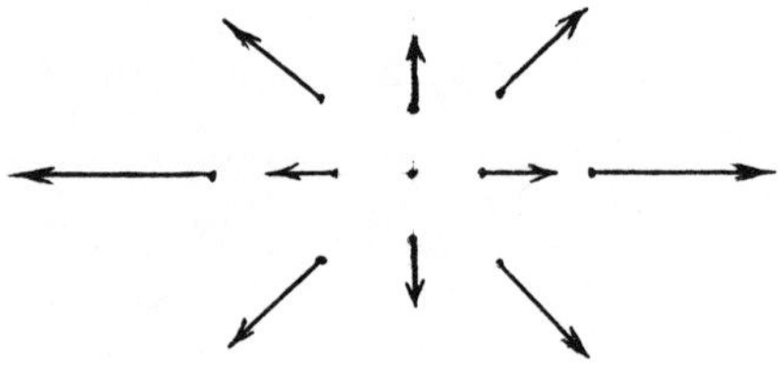

If $x^1,\ldots,x^{n+1}$ is the standard coordinate system on $\mathbb{R}^{n+1}$, then

$$V = \Sigma\, x^i \frac{\partial}{\partial x^i} ,$$

and we easily see that $\nabla'_X V = X$ for all tangent vectors X of $\mathbb{R}^{n+1}$. Thus equation (1) can be written

$$\nabla'_X(X_{n+1} + \lambda V) = 0 .$$

Thus the vector field $X_{n+1} + \lambda V$ is parallel along M. Identifying tangent vectors of $\mathbb{R}^{n+1}$ with elements of $\mathbb{R}^{n+1}$, this means that $X_{n+1} + \lambda V$ is a constant vector v_0 on M, so we have

$$X_{n+1}(p) + \lambda p = v_0 \in \mathbb{R}^{n+1} .$$

Thus

$$p = \frac{v_0 - X_{n+1}(p)}{\lambda}$$

for all $p \in M$, which means that M lies in the sphere of radius $1/\lambda$ around the point v_0/λ. ■

Using Lemma 1, and the generalization of Dupin's Theorem, it is now a straightforward matter to generalize Liouville's Theorem (4-12) to $\mathbb{R}^n$: every conformal map of an open subset of $\mathbb{R}^n$ onto an open subset of $\mathbb{R}^n$ is the restriction of a composition of similarities and inversions. The discussion on pp. III.304-310 also generalizes, so every such composition can be represented as the composition of at most one similarity and one inversion. In addition, these conformal maps take hyperplanes and spheres to hyperplanes and spheres.

With this information we are now in a good position to consider the isometries of $(B^n, \langle\ ,\ \rangle)$. Since $\langle\ ,\ \rangle$ is conformally equivalent to the usual metric $\Sigma\, dx^i \otimes dx^i$ on B^n, we see immediately that

(1) Every isometry $f\colon (B^n, \langle\ ,\ \rangle) \to (B^n, \langle\ ,\ \rangle)$ onto itself is a conformal map of B^n onto itself (as a subset of $\mathbb{R}^n$ with the usual metric).

We next claim

(2) If $f\colon B^n \to B^n$ is a conformal map of B^n <u>onto</u> itself and $f_*\colon B^n{}_p \to B^n{}_p$ is a multiple of the identity for some $p \in B^n$, then f is the identity (or possibly minus the identity, if $p = 0$).

To prove this, consider the sphere S = boundary B^n. Then S is taken into itself by f (more precisely, by the composition of similarities and inversions of which f is the restriction). If P is a hyperplane through p, then $f(P)$ is a hyperplane or sphere tangent to P at p (since $f_*\colon B^n{}_p \to B^n{}_p$ is a multiple of the identity). But also the angle at which P cuts S must equal the angle at which $f(P)$ cuts $f(S) = S$. It follows easily that $f(P) = P$. Consequently, f cannot be an inversion or the composition of one inversion and one similarity, for the inversion must be through a point $p_* \notin B^n$, and then $f(P)$ could not be a plane. So f must be a similarity, and the desired result follows easily.

Now consider any conformal map $f\colon B^n \to B^n$ of B^n <u>onto</u> itself, and let $p \in B^n$ be a point $\neq 0$. If $X_1,\dots,X_n \in B^n{}_p$ is an orthonormal basis with respect to $< , >_p$, then there is some $\lambda > 0$ with

$$<f_*(X_i),f_*(X_j)>_{f(p)} = \lambda\cdot\delta_{ij} ,$$

so $\{f_*(X_i)/\sqrt{\lambda}\}$ is an orthonormal basis for $B^n{}_{f(p)}$. Consequently, there is an isometry $g: (B^n, < , >) \longrightarrow (B^n, < , >)$ with

$$g_*(X_i) = \frac{1}{\sqrt{\lambda}} f_*(X_i) .$$

Then $g^{-1}\circ f: B^n \longrightarrow B^n$ is a conformal map of B^n onto itself (by (1)), and $(g^{-1}\circ f)_*: B^n{}_p \longrightarrow B^n{}_p$ is a multiple of the identity. So $g = f$ by (2). Thus

(3) Every conformal map $f: B^n \longrightarrow B^n$ of B^n onto itself is an isometry of $(B^n, < , >)$ onto itself.

We can now deduce some further information about $(B^n, < , >)$. We know (p.III.37) that the d-dimensional totally geodesic submanifolds through $0 \in B^n$ are just $B^n \cap P$, where P is a d-dimensional plane through 0 in $\mathbb{R}^n$. Now any totally geodesic submanifold is the image of $B^n \cap P$ under some isometry $f: B^n \longrightarrow B^n$. The map f is conformal by (1), so

(4) Every totally geodesic submanifold of B^n is the intersection of B^n with a plane or sphere which intersects S = boundary B^n orthogonally.

Conversely, suppose that Σ is a plane or sphere which intersects S orthogonally, and let $p \in \Sigma \cap B^n$. There is a totally geodesic submanifold of B^n tangent to Σ at p. By (4), this submanifold must intersect S

orthogonally. So it must be precisely $\Sigma \cap B^n$. Thus

(5) The intersection with B^n of a plane or sphere which intersects S orthogonally is a totally geodesic submanifold.

Next consider a geodesic sphere Σ around $0 \in B^n$ (that is, let Σ be the set of points at fixed $\langle\ ,\ \rangle$ distance from 0). By symmetry of $\langle\ ,\ \rangle$, the set Σ is an ordinary (hyper) sphere. Now any geodesic sphere is the image of Σ under some isometry $f\colon B^n \longrightarrow B^n$. Since this isometry is a conformal map we see that

(6) Every geodesic sphere of $(B^n, \langle\ ,\ \rangle)$ is an ordinary hypersphere completely contained in B^n.

Now we will work on proving the converse of (6). Suppose we have an ordinary hypersphere Σ completely contained in B^n. We claim first of all that there is a hypersphere Σ' which is orthogonal to both Σ and S. To prove

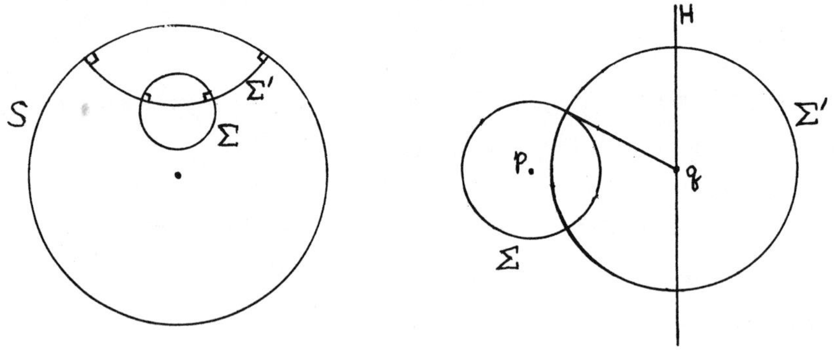

this we note that by means of an inversion through a point of S, we can reduce the problem to that of finding a hypersphere Σ' orthogonal to a hyperplane H and a hypersphere Σ lying completely on one side of it. If Σ has center p and $q \in H$ is the point closest to p, then we just choose Σ' to be a hypersphere around q whose radius has the length of a tangent from q to Σ. Now that we have the hypersphere Σ' orthogonal to both Σ and S, we consider the intersection $\bar{p}$ of Σ' and the line L between 0 and the center of Σ. Let $f\colon B^n \longrightarrow B^n$ be an isometry taking $\bar{p}$ to 0. We know by

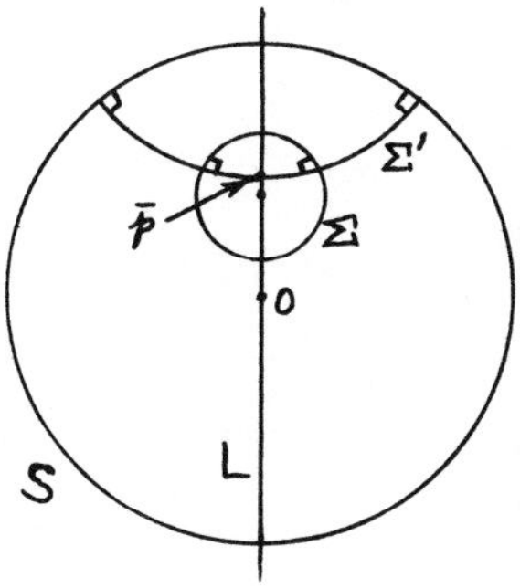

(5) that $B^n \cap \Sigma'$ is a totally geodesic hypersurface. Therefore f must take Σ' to a hyperplane H through 0. Moreover, f takes the geodesic L to another geodesic through 0, i.e., to a straight line L' through 0, but not lying in H. The image hypersphere $f(\Sigma)$ must be perpendicular to both H and L', which can happen only when $f(\Sigma)$ has center 0. So $f(\Sigma)$ is a geodesic sphere, which implies that Σ is also:

(7) Every ordinary hypersphere completely contained in B^n is a geodesic sphere.

All of this information, by the way, was obtained only for the case $n \geq 3$, since we made use of Liouville's Theorem. The case $n = 2$ is sometimes analysed by explicit computation, making use of the identification of $\mathbb{R}^2$ with $\mathbb{C}$ (see Problems 4,5,6), but we can also use the information which we already have for $n \geq 3$. To do this we consider B^2 as a totally geodesic surface in

B^3. An isometry $f\colon B^2 \to B^2$ of B^2 onto itself clearly extends to an isometry $\tilde{f}\colon B^3 \to B^3$ of B^3 onto itself. Since $\tilde{f}$ is conformal, f is also. Moreover, since $\tilde{f}$ is a composition of at most one similarity and inversion, it is not hard to see that the same must be true of f (this information is not redundant in the 2-dimensional case). Conversely, if $f\colon B^2 \to B^2$ is a conformal map of B^2 onto itself which happens to be a composition of at most one similarity and inversion, then f can easily be extended to a similar conformal map $\tilde{f}\colon B^3 \to B^3$ of B^3 onto itself. Since $\tilde{f}$ is an isometry, so is f. It now follows, exactly as before, that the geodesics of B^2 are portions of lines or circles intersecting S orthogonally, while the geodesic circles are the ordinary circles completely contained in B^2.

In Addendum 2 to Chapter II.7 we also described a complete manifold of constant curvature $K_0 = -1$ by means of the metric

$$\sum_{i=1}^{n} \frac{dx^i \otimes dx^i}{(x^n)^2}$$

on the upper half-space $\mathbb{H}^n = \{a \in \mathbb{R}^n\colon a^n > 0\}$. It is easy to describe the isometry between $\mathbb{H}^n$ and B^n. In fact, since the metric on each of them is conformally equivalent to the usual metric on $\mathbb{R}^n$, the isometry $f\colon B^n \to \mathbb{H}^n$ must be a conformal map. If we take an inversion I about a point $*$ on the boundary sphere S of B, then $I(B)$ will be an open half-space, and it is only necessary to compose I with an appropriate similarity. We now easily see that the isometries of $\mathbb{H}^n$ onto itself are precisely the conformal maps taking $\mathbb{H}^n$ onto itself, that the totally geodesic submanifolds of $\mathbb{H}^n$ are $\Sigma \cap \mathbb{H}^n$ for planes and spheres Σ intersecting $\mathbb{R}^{n-1}$ orthogonally, and that the geodesic spheres of $\mathbb{H}^n$ are the ordinary spheres completely contained in

$\mathbb{H}^n$. It will prove extremely useful to be able to shuttle back and forth between B^n and $\mathbb{H}^n$.

We have now given intrinsic characterizations of the sets $\Sigma \cap B^n$ [or $\Sigma \cap \mathbb{H}^n$] when Σ is a hyperplane or hypersphere intersecting S [or $\mathbb{R}^{n-1}$], either orthogonally, or else not at all. We also want to give intrinsic characterizations when Σ intersects non-orthogonally. There are two different cases to consider, the first of which is related to a certain limiting construction which played an essential role in the earliest investigations of non-Euclidean geometry. Take a ray L, with initial point p, in a non-Euclidean space. For each q on L, consider the sphere with center q that passes through p. As $q \to \infty$, this sphere approaches a surface. In

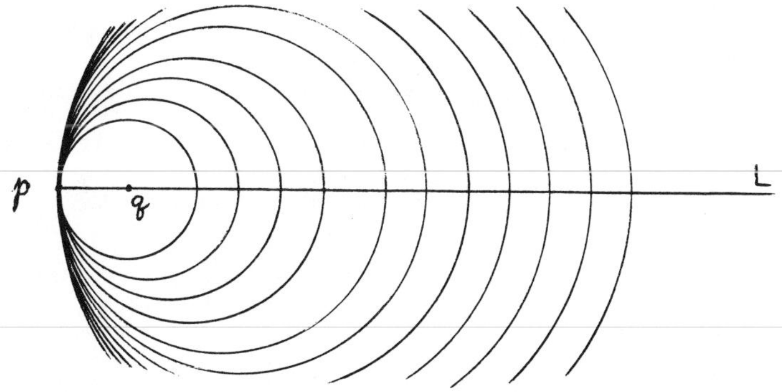

the Euclidean case, this surface is just the plane through p perpendicular to L; in the non-Euclidean case, the limiting set is called a "limit sphere" or horosphere. It is easy to see that the horospheres of $(B^n, \langle\ ,\ \rangle)$ are precisely $B^n \cap \Sigma$ where Σ is a hypersphere completely inside B except for one point. (First consider the horospheres determined by a ray starting at 0, as in the figure below, and note that there are isometries of B^n taking any horosphere to any other.) The early non-Euclidean geometers had their minds

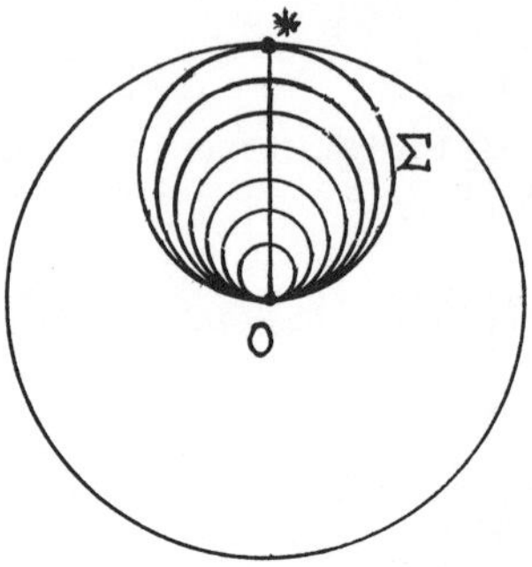

blown when they proved that the laws of Euclidean geometry hold on the horosphere; in other words, the horosphere is flat. The easiest way for us to see this is to consider an isometry $f\colon B^n \longrightarrow \mathbb{H}^n$ which involves an inversion around the unique point $* \in \Sigma \cap S$. The image $f(\Sigma)$ is then a hyperplane parallel to $\mathbb{R}^{n-1}$. But the metric induced on this hyperplane is a constant multiple of $\Sigma\, dx^i \otimes dx^i$, so this hyperplane (which is a horosphere of $\mathbb{H}^n$) is flat; all other horospheres are isometric images of this one, so they are also flat.

To describe the other sets $\Sigma \cap B^n$ and $\Sigma \cap \mathbb{H}^n$, we first do a short computation in $\mathbb{H}^2$. Consider a semi-circle intersecting $\mathbb{R}^1$ orthogonally, parameterized by

$$c(\theta) = (a + r\cos\theta,\ r\sin\theta)\ .$$

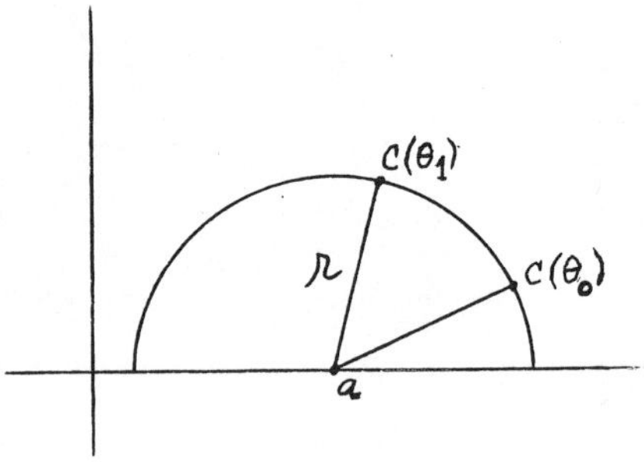

This curve is a geodesic, apart from its parameterization. Its length from the point $c(\theta_0)$ to $c(\theta_1)$ is

$$\int_{\theta_0}^{\theta_1} |c'(\theta)|\,d\theta = \int_{\theta_0}^{\theta_1} \left|(-r\sin\theta)\frac{\partial}{\partial x^1} + (r\cos\theta)\frac{\partial}{\partial x^2}\right| d\theta$$

$$= \int_{\theta_0}^{\theta_1} \sqrt{\frac{(-r\sin\theta)^2 + (r\cos\theta)^2}{(r\cos\theta)^2}}\, d\theta$$

$$= \int_{\theta_0}^{\theta_1} \frac{1}{\cos\theta}\, d\theta\ .$$

Notice that this is independent of r. It follows that for a geodesic L which is a straight line perpendicular to $\mathbb{R}$, the set of points at a fixed distance d from L is a pair of straight lines making equal angles with L.

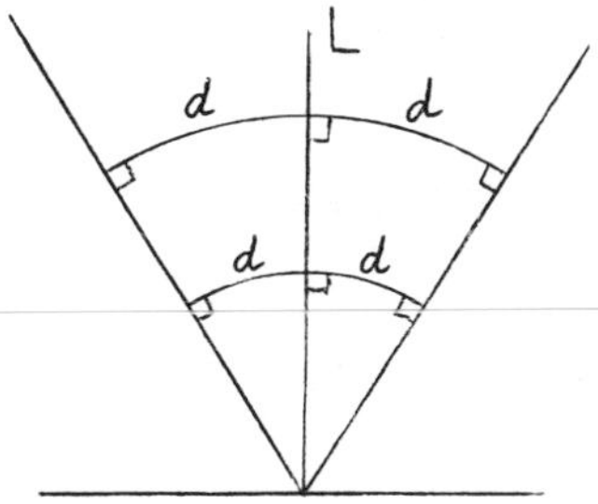

set of points at distance d from L

Similarly, if P is a totally geodesic hypersurface in $\mathbb{H}^n$ consisting of a hyperplane perpendicular to $\mathbb{R}^{n-1}$, then the set of points at a fixed distance d from P is a pair of hyperplanes P_1, P_2 making equal angles with P. For

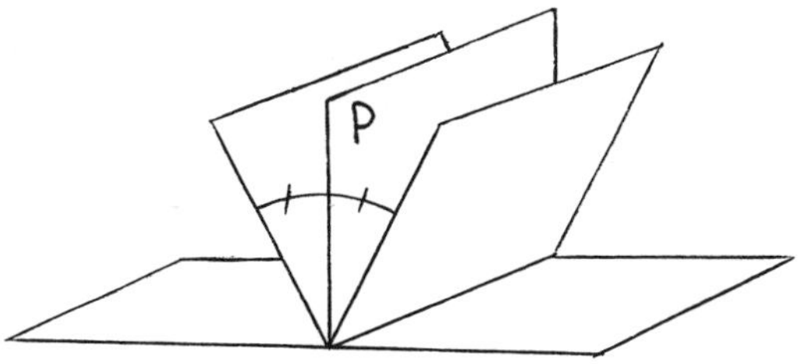

the isometry $f\colon B^n \to \mathbb{H}^n$, involving an inversion about the point $* \in S$, the set $f^{-1}(P)$ is $\Sigma \cap B^n$ for some hyperplane or hypersphere Σ with $* \in \Sigma \cap S$;

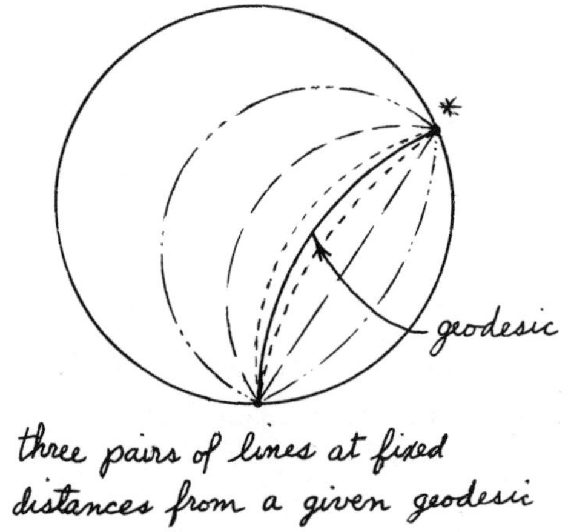

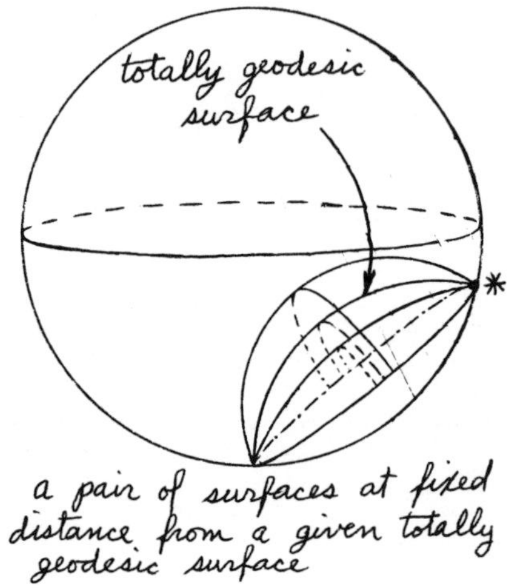

and the sets $f^{-1}(P_i)$ are sets of the same sort. We thus see that for hyperplanes or hyperspheres Σ which intersect S [or $\mathbb{R}^{n-1}$] in more than one point, but not orthogonally, the set $\Sigma \cap B^n$ [or $\Sigma \cap \mathbb{H}^n$] is one component of the set of points at a fixed distance from a totally geodesic hypersurface; these sets are thus called <u>equidistant hypersurfaces</u>.

By the way, we can also describe the geodesic spheres, horospheres, and equidistant hypersurfaces for

$$H^n = \{p \in \mathbb{R}^{n+1} : p^0 > 0 \text{ and } \langle p,p\rangle = -1\} .$$

They are all of the form $H^n \cap P$ for some hyperplane P. The parabolas, which occur when P is parallel to a generator of the cone $\Sigma(x^i)^2 = (x^0)^2$, are horo-

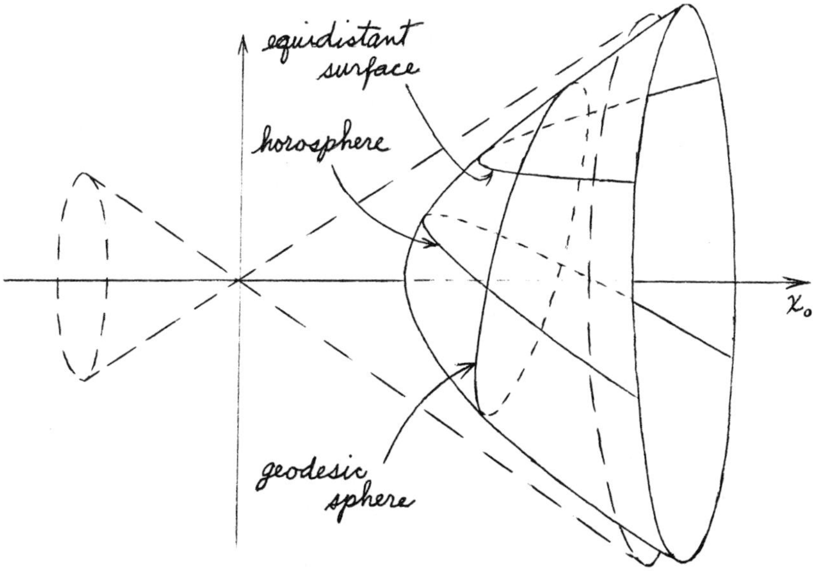

spheres. Ellipses, which occur when P makes a larger angle with the x^0-axis, are geodesic spheres. Hyperbolas, which occur when P makes a smaller angle, are equidistant hypersurfaces. Although these assertions should look pictorially reasonable, we are not yet in a position to prove them (see p. 115). For the moment we merely want to note that we would not obtain any new hypersurfaces by considering the sets $H^n \cap Q$ where Q is another quadric hypersurface of the form

$$Q = \{p \in \mathbb{R}^{n+1} : \langle p - p_0, p - p_0 \rangle = c\} ;$$

for it is easy to see that $H^n \cap Q$ is always of the form $H^n \cap P$ for some hyperplane P. (Analogously, the intersection of two ordinary spheres in $\mathbb{R}^{n+1}$ is also the intersection of one sphere with a hyperplane.)

Much of this discussion evolved from the existence of conformal maps from S^n or H^n to $\mathbb{R}^n$. Another kind of map will play an important role. A homeomorphism $\phi: M_1 \to M_2$ into M_2 is called a geodesic mapping if for every geodesic

γ of M_1, the composition $\phi \circ \gamma$ is a reparameterization of a geodesic of M_2. Notice that a geodesic mapping $\phi: M_1 \to M_2$ clearly also takes totally geodesic submanifolds of M_1 to totally geodesic submanifolds of M_2.

As usual, we first consider S^n. We define the central projection ϕ of S^n to be the map which takes a point p in the open northern hemisphere S^{n+} of S^n to the intersection of $\mathbb{R}^n = \mathbb{R}^n \times \{1\} \subset \mathbb{R}^{n+1}$ with the straight line through p and the origin $0 \in \mathbb{R}^{n+1}$. It is clear that $\phi: S^{n+} \to \mathbb{R}^n$ is a

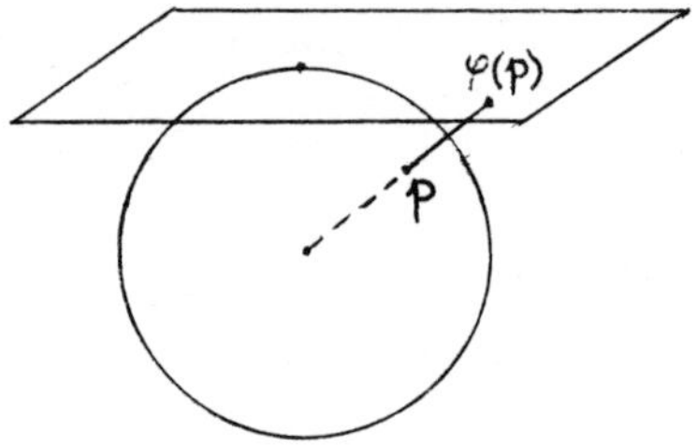

geodesic mapping, since the geodesics of S^n are intersections of S^n with planes through the center of S^n. An exactly analogous construction works for $H^n \subset (\mathbb{R}^n, \langle,\rangle)$, except now we obtain a map defined on all of H^n. We define $\phi: H^n \to \mathbb{R}^n$ to be the map which takes $p \in H^n$ to the intersection of $\mathbb{R}^n = \{(1,a^1,\ldots,a^n) \in \mathbb{R}^{n+1}\}$ with the straight line through p and $0 \in \mathbb{R}^{n+1}$.

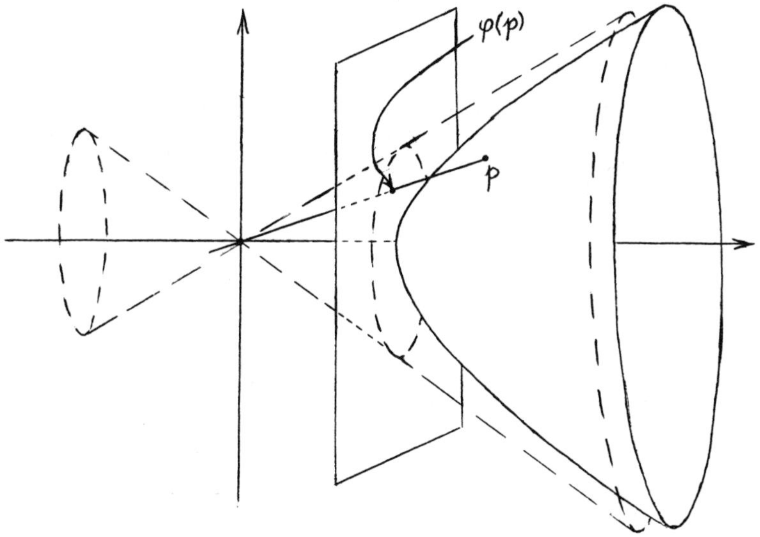

In this case, the image of H^n is the open ball in $\mathbb{R}^n$ bounded by the intersection of $\mathbb{R}^n$ with the cone $\Sigma(x^i)^2 = (x^0)^2$; thus $\phi(H^n)$ is the open ball $B^n(1)$ of radius 1.

We can also construct a geodesic mapping by using the model $(B^n, \langle\ ,\ \rangle) = (B^n(2), \langle\ ,\ \rangle)$. To do this, we regard S^n as the unit sphere tangent to $\mathbb{R}^n = \mathbb{R}^n \times \{0\}$ at 0. Then the stereographic projection σ from the north pole

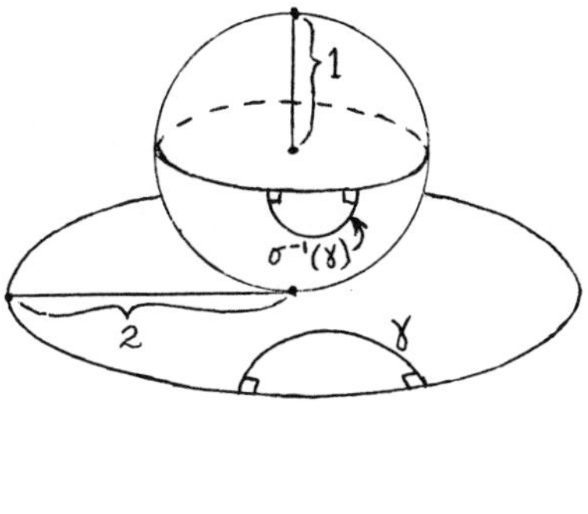

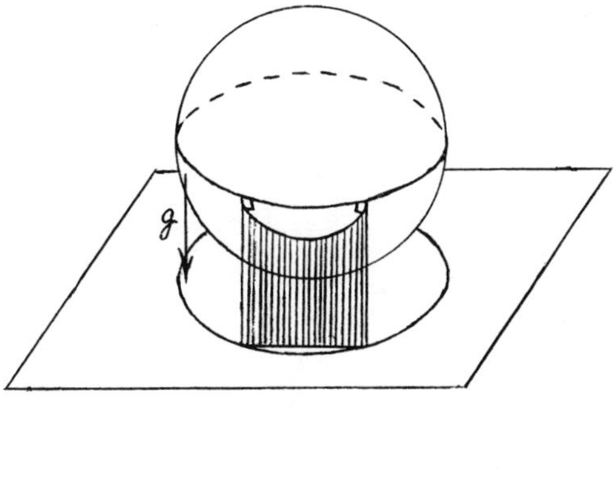

of S^n takes the open southern hemisphere of S^n diffeomorphically onto $B^n(2)$. A geodesic γ in $(B^n, \langle\ ,\ \rangle)$ is a straight line or circle intersecting S = boundary B^n orthogonally. It follows from the properties of stereographic projection that $\sigma^{-1}(\gamma)$ is a semi-circle intersecting the equator of S^n orthogonally. Now let $g: S^n \to \mathbb{R}^n$ be the orthogonal projection $g(x^1,\ldots,x^{n+1}) = (x^1,\ldots,x^n)$. Then g takes circles intersecting the equator of S^n orthogonally onto straight line segments of $\mathbb{R}^n$. So $\phi = g\circ\sigma^{-1}: B^n(2) \to B^n(1)$ is a geodesic mapping. Using these geodesic mappings $\phi: H^n \to B^n(1)$ and $\phi: B^n(2) \to B^n(1)$, it is not hard (Problems 9,10) to describe an isometry between H^n and $B^n(2)$.

Naturally, the geodesic mappings $\phi: S^{n+} \to \mathbb{R}^n$ and $\phi: H^n \to B^n(1)$, together with the standard coordinate system on $\mathbb{R}^n$, lead to new coordinate systems for S^{n+} and H^n. In particular, the unit ball $B^n(1)$, together with the metric induced by the metric on H^n, is called the "projective model" of H^n. We could calculate the form of the metric in these coordinate systems, and describe the geodesic spheres, horospheres, and equidistant hypersurfaces of H^n in the projective model, but we will never need to know any of this information. For us, the only important result will be the existence of the geodesic maps $\phi: S^{n+} \to \mathbb{R}^n$ and $\phi: H^n \to B^n(1)$. This is not surprising in view of the following

2. THEOREM (BELTRAMI). If M is a connected Riemannian n-manifold such that every point has a neighborhood that can be mapped geodesically to $\mathbb{R}^n$, then M has constant curvature.

Proof. The case $n \geq 3$ follows immediately from Theorem 1-18; it is the case

$n = 2$ which causes all the trouble. Note first that in the case of a surface, Lemma II.7-18 implies that the curvature K satisfies

$$R_{hijk} = K(g_{hj}g_{ik} - g_{hk}g_{ij})$$
$$\Downarrow$$

(1) $$R^{\ell}{}_{ijk} = \sum_{h=1}^{2} g^{\ell h} R_{hijk} = K(\delta^{\ell}_{j} g_{ik} - \delta^{\ell}_{k} g_{ij}) .$$

We will use the mapping in the hypothesis of the theorem to identify our neighborhood in M with an open set in $\mathbb{R}^2$, on which we use the standard coordinate system (x^1,x^2). Thus the metric $\Sigma g_{ij}\, dx^i \otimes dx^j$ has the same geodesics as the metric $\Sigma\, \delta_{ij}\, dx^i \otimes dx^j$. Since the Christoffel symbols for the latter metric are all zero, Proposition II.6-18 shows that the Christoffel symbols for g_{ij} satisfy

$$\Gamma^{i}_{jk} = \delta^{i}_{j}\omega_k + \delta^{i}_{k}\omega_j$$

for certain functions ω_i. Hence we have

(2) $$\Gamma^{2}_{11} = \Gamma^{1}_{22} = 0\ , \qquad \Gamma^{1}_{11} = 2\Gamma^{2}_{12}\ , \qquad \Gamma^{2}_{22} = 2\Gamma^{1}_{21}\ .$$

From equation (1), and the formula (**) for $R^{\ell}{}_{ijk}$ on p. II.188, we find that

(3a) $$Kg_{11} = (\Gamma^{2}_{12})^2 - \frac{\partial}{\partial x^1}\Gamma^{2}_{12}$$ (3b) $$Kg_{12} = \Gamma^{1}_{21}\Gamma^{2}_{12} - \frac{\partial}{\partial x^2}\Gamma^{2}_{12}$$

(3c) $$Kg_{22} = (\Gamma^{1}_{12})^2 - \frac{\partial}{\partial x^2}\Gamma^{1}_{12}$$ (3d) $$Kg_{21} = \Gamma^{2}_{12}\Gamma^{1}_{12} - \frac{\partial}{\partial x^1}\Gamma^{1}_{12}\ .$$

Notice that equations (3b) and (3d) imply that

(4) $$\frac{\partial}{\partial x^2}\Gamma^2_{12} = \frac{\partial}{\partial x^1}\Gamma^1_{12} .$$

We also have

$$\begin{aligned} 2g_{12}\Gamma^2_{12} &= g_{12}\Gamma^1_{11} && \text{by (2)} \\ &= g_{12}\Gamma^1_{11} + g_{22}\Gamma^2_{11} && \text{by (2)} \\ &= [11,2] \\ &= \frac{1}{2}\left(\frac{\partial g_{12}}{\partial x^1} + \frac{\partial g_{12}}{\partial x^1} - \frac{\partial g_{11}}{\partial x^2}\right) \\ &= \frac{\partial g_{12}}{\partial x^1} - \frac{1}{2}\frac{\partial g_{11}}{\partial x^2} . \end{aligned}$$

Subtracting this equation from

$$g_{11}\Gamma^1_{12} + g_{12}\Gamma^2_{12} = [12,1] = \frac{1}{2}\frac{\partial g_{11}}{\partial x^2} ,$$

we obtain

$$g_{11}\Gamma^1_{12} - g_{12}\Gamma^2_{12} = \frac{\partial g_{11}}{\partial x^2} - \frac{\partial g_{12}}{\partial x^1} .$$

Multiplying by K, and using (3a) and (3b), we obtain

(5) $$K\left(\frac{\partial g_{11}}{\partial x^2} - \frac{\partial g_{12}}{\partial x^1}\right) = \Gamma^2_{12}\frac{\partial}{\partial x^2}\Gamma^2_{12} - \Gamma^1_{12}\frac{\partial}{\partial x^1}\Gamma^2_{12} .$$

Now differentiate (3a) with respect to x^2, and subtract the result of differentiating (3b) with respect to x^1. We obtain

$$g_{11}\frac{\partial K}{\partial x^2} - g_{12}\frac{\partial K}{\partial x^1} + K\left(\frac{\partial g_{11}}{\partial x^2} - \frac{\partial g_{12}}{\partial x^1}\right) = 2\Gamma_{12}^2\frac{\partial}{\partial x^2}\Gamma_{12}^2 - \Gamma_{21}^1\frac{\partial}{\partial x^1}\Gamma_{12}^2 - \Gamma_{12}^2\frac{\partial}{\partial x^1}\Gamma_{21}^1 .$$

Using (5) we have

$$g_{11}\frac{\partial K}{\partial x^2} - g_{12}\frac{\partial K}{\partial x^1} = \Gamma_{12}^2\frac{\partial}{\partial x^2}\Gamma_{12}^2 - \Gamma_{12}^2\frac{\partial}{\partial x^1}\Gamma_{21}^1 .$$

Hence by (4) we have

$$g_{11}\frac{\partial K}{\partial x^2} - g_{12}\frac{\partial K}{\partial x^1} = 0 .$$

Similarly,

$$-g_{12}\frac{\partial K}{\partial x^2} - g_{22}\frac{\partial K}{\partial x^1} = 0 .$$

Since the determinant $g_{11}g_{22} - (g_{12})^2 \neq 0$, this implies that $\partial K/\partial x^1 = \partial K/\partial x^2 = 0$. ■

B. Curves in a Riemannian Manifold

Before investigating general submanifolds of a Riemannian manifold, we will consider the special case of 1-dimensional submanifolds, which works out quite differently than all other cases. Our aim is not to obtain any startling theorems about curves in Riemannian manifolds, but merely to show briefly how the Serret-Frenet formulas of Chapter II.1 generalize; along the way we will derive a few results which are needed to discuss higher dimensions.

Consider a Riemannian manifold $(N, \langle\ ,\ \rangle)$, and an arc-length parameterized curve $c\colon [a,b] \longrightarrow N$. We use N for the ambient manifold to conform with the notation to be used in the general case of a submanifold $M \subset N$. For consistency of notation, we also use ∇' for the covariant derivative in N, even though there will be no occasion to consider the covariant derivative ∇ in the 1-dimensional manifold $c([a,b])$. We will let $\mathbf{v}_1 = c'$ denote the unit tangent vector of c. Since $\langle \mathbf{v}_1, \mathbf{v}_1 \rangle = 1$ we have

$$0 = \frac{d}{ds}\langle \mathbf{v}_1(s), \mathbf{v}_1(s)\rangle = 2\left\langle \mathbf{v}_1(s), \frac{D'\mathbf{v}_1(s)}{ds}\right\rangle .$$

We define the first "curvature function" κ_1 of c by

$$\kappa_1(s) = \left|\frac{D'\mathbf{v}_1(s)}{ds}\right| ,$$

and if $\kappa_1(s) \neq 0$ for all s we set

$$\mathbf{v}_2(s) = \kappa_1(s)^{-1} \cdot \frac{D'\mathbf{v}_1(s)}{ds}$$

so that $\mathbf{v}_2$ is a unit vector field along c which is everywhere perpendicular to $\mathbf{v}_1$. We then have the "Frenet formula"

$$(F_1) \qquad \frac{D'\mathbf{v}_1(s)}{ds} = \kappa_1(s)\mathbf{v}_2(s) .$$

Now

$$\langle \mathbf{v}_2, \mathbf{v}_2 \rangle = 1 \quad \Rightarrow \quad \left\langle \mathbf{v}_2(s), \frac{D'\mathbf{v}_2(s)}{ds} \right\rangle = 0 .$$

Moreover,

$$\langle \mathbf{v}_1, \mathbf{v}_2 \rangle = 0 \quad \Rightarrow \quad 0 = \left\langle \frac{D'\mathbf{v}_1(s)}{ds}, \mathbf{v}_2(s) \right\rangle + \left\langle \mathbf{v}_1(s), \frac{D'\mathbf{v}_2(s)}{ds} \right\rangle$$

$$= \kappa_1(s) + \left\langle \mathbf{v}_1(s), \frac{D'\mathbf{v}_2(s)}{ds} \right\rangle \qquad \text{by (1).}$$

This implies that

$$\frac{D'\mathbf{v}_2(s)}{ds} = -\kappa_1(s)\mathbf{v}_1(s) + \text{vector perpendicular to } \mathbf{v}_1(s) \text{ and } \mathbf{v}_2(s) .$$

We define the second "curvature function" κ_2 by

$$\kappa_2(s) = \left| \frac{D'\mathbf{v}_2(s)}{ds} + \kappa_1(s)\mathbf{v}_1(s) \right| ,$$

and if $\kappa_2(s) \neq 0$ for all s, we set

$$\mathbf{v}_3(s) = \kappa_2(s)^{-1}\left[\frac{D'\mathbf{v}_2(s)}{ds} + \kappa_1(s)\mathbf{v}_1(s) \right] ,$$

so that $\mathbf{v}_3$ is a unit vector field along c which is everywhere perpendicular to $\mathbf{v}_1$ and $\mathbf{v}_2$. We then have

$$(F_2) \qquad \frac{D'\mathbf{v}_2(s)}{ds} = -\kappa_1(s)\mathbf{v}_1(s) + \kappa_2(s)\mathbf{v}_3(s) .$$

Now suppose, inductively, that for $j \leq m = \dim N$ we have orthonormal vector fields $\mathbf{v}_1, \ldots, \mathbf{v}_j$ along c and nowhere zero curvature functions $\kappa_1, \ldots, \kappa_{j-1}$

such that

$$(F_1) \qquad \frac{D'\mathbf{v}_1(s)}{ds} = \kappa_1(s)\mathbf{v}_2(s)$$

$$(F_2) \qquad \frac{D'\mathbf{v}_2(s)}{ds} = -\kappa_1(s)\mathbf{v}_1(s) + \kappa_2(s)\mathbf{v}_3(s)$$

$$\vdots$$

$$(F_{j-1}) \qquad \frac{D'\mathbf{v}_{j-1}(s)}{ds} = -\kappa_{j-2}(s)\mathbf{v}_{j-2}(s) + \kappa_{j-1}(s)\mathbf{v}_j(s) \ .$$

Then

$$\langle \mathbf{v}_j, \mathbf{v}_j \rangle = 1 \implies \left\langle \frac{D'\mathbf{v}_j(s)}{ds}, \mathbf{v}_j(s) \right\rangle = 0 \ ,$$

while for $i < j$ we have

$$\langle \mathbf{v}_j, \mathbf{v}_i \rangle = 0 \implies \left\langle \mathbf{v}_i(s), \frac{D'\mathbf{v}_j(s)}{ds} \right\rangle = -\left\langle \frac{D'\mathbf{v}_i(s)}{ds}, \mathbf{v}_j \right\rangle$$
$$= \begin{cases} 0 & i \neq j-1 \\ -\kappa_{j-1}(s) & i = j-1 \ . \end{cases}$$

Hence

$$(*) \qquad \frac{D'\mathbf{v}_j(s)}{ds} = -\kappa_{j-1}(s)\mathbf{v}_{j-1}(s)$$
$$+ \text{ vector perpendicular to } \mathbf{v}_1(s), \ldots, \mathbf{v}_j(s) \ .$$

If $j < m$ we set

$$\kappa_j(s) = \left| \frac{D'\mathbf{v}_j(s)}{ds} + \kappa_{j-1}(s)\mathbf{v}_{j-1}(s) \right| \ ,$$

and if $\kappa_j(s) \neq 0$ for all s we set

$$\mathbf{v}_{j+1}(s) = \kappa_j(s)^{-1} \cdot \left[\frac{D'\mathbf{v}_j(s)}{ds} + \kappa_{j-1}(s)\mathbf{v}_{j-1}(s) \right] .$$

We then have

$$(F_j) \qquad \frac{D'\mathbf{v}_j(s)}{ds} = -\kappa_{j-1}(s)\mathbf{v}_{j-1}(s) + \kappa_j(s)\mathbf{v}_{j+1}(s) .$$

If $j = m$, then only the zero vector is perpendicular to $\mathbf{v}_1(s),\ldots,\mathbf{v}_m(s)$, so equation (*) becomes

$$(F_m) \qquad \frac{D'\mathbf{v}_m(s)}{ds} = -\kappa_{m-1}(s)\mathbf{v}_{m-1}(s) .$$

It is easy to see that we have equations (F_1) to (F_{j-1}) with nowhere zero functions $\kappa_1,\ldots,\kappa_{j-1}$ if and only if

$$c'(s), \quad \frac{D'c'(s)}{ds}, \quad \ldots, \quad \frac{D'^{(j-1)}c'(s)}{ds^{j-1}}$$

are everywhere linearly independent; the vector fields $\mathbf{v}_1,\ldots,\mathbf{v}_j$ along c are then precisely the result of applying the Gram-Schmidt orthonormalization process to these vectors. If $D'^{(j)}c'(s)/ds^j$ is everywhere linearly dependent on $c'(s),\ldots,D'^{(j-1)}c'(s)/ds^{j-1}$, then the function κ_j will be everywhere 0, and we cannot define $\mathbf{v}_{j+1}$, but we can write instead

$$(F_j') \qquad \frac{D'\mathbf{v}_j(s)}{ds} = -\kappa_{j-1}(s)\mathbf{v}_{j-1}(s) .$$

[Note, in particular, that (F_m') is just (F_m).] As in the theory of curves in

$\mathbb{R}^3$, we consider only intervals where a set of equations $(F_1),\ldots,(F_{j-1}),(F'_j)$ holds for some $j \le m$. In other words, we assume that $\kappa_1,\ldots,\kappa_{j-1}$ are nowhere zero, while κ_j is identically zero. We call $\mathbf{v}_1,\ldots,\mathbf{v}_j$ the "Frenet frame" for c. The subspace of $N_{c(s)}$ spanned by $\mathbf{v}_1(s)$ and $\mathbf{v}_i(s)$ is sometimes called the $(i-1)$st <u>osculating plane</u> of c at s.

Notice that once we have $c'(s),\ldots,D'^{(m-2)}c'(s)/ds^{m-2}$ linearly independent, so that $\mathbf{v}_1,\ldots,\mathbf{v}_{m-1}$ are defined, then there are only two possible choices for each $\mathbf{v}_m(s)$. Having made a choice of $\mathbf{v}_m(a)$, there is then a unique continuous way of choosing $\mathbf{v}_m(s)$ for all $s \in [a,b]$. We still have equations (F_1) to (F_m), but now the function κ_{m-1} might take on negative values, whereas all other κ_j, being non-zero norms, are everywhere positive. The particularly interesting situation occurs when N is oriented. Then we define $\mathbf{v}_m(s)$ to be the unique unit vector in $N_{c(s)}$ orthogonal to $\mathbf{v}_1(s),\ldots,\mathbf{v}_{m-1}(s)$ such that $(\mathbf{v}_1(s),\ldots,\mathbf{v}_m(s))$ is positively oriented [equivalently, we can define $\mathbf{v}_m(s) = \mathbf{v}_1(s) \times \cdots \times \mathbf{v}_{m-1}(s)$, where the cross-product is determined by the metric and the orientation (see Problem 11)]. For curves in $\mathbb{R}^3$ this is precisely how we defined the binormal $\mathbf{b} = \mathbf{v}_3$, and obtained the torsion $\tau = \kappa_2$ which could be positive, negative, or zero. When we apply this procedure to arc-length parameterized curves c in an oriented 2-dimensional Riemannian manifold, we obtain an everywhere defined curvature κ_1, whose values may be positive, negative, or zero. Clearly κ_1 is just the geodesic curvature κ_g defined previously.

In the next theorem we will, for simplicity, ignore these refinements and consider only curves with $c'(s),\ldots,D'^{(m-1)}c'(s)/ds^{m-1}$ everywhere linearly independent. The reader may sort out for himself the details which have to be changed when N is oriented and we allow κ_{m-1} to take on non-positive values.

3. THEOREM. (1) Let c, $\bar{c}$: $[a,b] \to N^m$ be arc-length parameterized curves with nowhere zero curvature functions $\kappa_1,\dots,\kappa_{m-1}$ and $\bar{\kappa}_1,\dots,\bar{\kappa}_{m-1}$, respectively, and Frenet frames $\mathbf{v}_1,\dots,\mathbf{v}_m$ and $\bar{\mathbf{v}}_1,\dots,\bar{\mathbf{v}}_m$, respectively. Suppose that $\kappa_i = \bar{\kappa}_i$ for $1 \le i \le m-1$, and that

$$c(a) = \bar{c}(a) \qquad \text{and} \qquad \mathbf{v}_i(a) = \bar{\mathbf{v}}_i(a) \qquad \text{for } i = 1,\dots,m\ .$$

Then $c = \bar{c}$.

(2) Let N^m be complete, let $\kappa_1,\dots,\kappa_{m-1}$: $[a,b] \to \mathbb{R}$ be everywhere positive continuous functions, and let $\mathring{\mathbf{v}}_1,\dots,\mathring{\mathbf{v}}_m$ be an orthonormal basis for some N_p. Then there is an arc-length parameterized curve c: $[a,b] \to N$ with $c(a) = p$ whose curvature functions are $\kappa_1,\dots,\kappa_{n-1}$ and whose Frenet frame $\mathbf{v}_1,\dots,\mathbf{v}_m$ satisfies $\mathbf{v}_i(a) = \mathring{\mathbf{v}}_i$ for $1 \le i \le m$.

Proof. To prove (1) it clearly suffices (by a least upper bound argument) to show that $c(s) = \bar{c}(s)$ for s sufficiently close to a. So we might as well assume that $M = \mathbb{R}^m$, with some metric $\sum g_{ij}\, dx^i \otimes dx^j$, where $x^1,\dots,x^m$ is the standard coordinate system for $\mathbb{R}^m$. Let $v_i(s) \in \mathbb{R}^m$ be the vector representing $\mathbf{v}_i(s)$ when we identify tangent vectors of $\mathbb{R}^m$ with elements of $\mathbb{R}^m$ in the usual way. We thus have $m+1$ functions

$$c,\ v_1,\dots,v_m\colon [a,b] \to \mathbb{R}^m\ .$$

We will also let $Dv_i(s)/ds$ be the vector representing $D\mathbf{v}_i(s)/ds$ when we identify tangent vectors of $\mathbb{R}^m$ with elements of $\mathbb{R}^m$. The formula on p. II.249 shows that $Dv_j(s)/ds$ can be written in terms of

$$c(s),\ c'(s),\ v_j(s),\ v_j'(s)\ , \qquad \text{i.e.,} \qquad c(s),\ v_1(s),\ v_j(s),\ v_j'(s).$$

Each Frenet equation (F_j) then gives us an equation

$$(E_j) \qquad v_j'(s) = F_j(c(s), v_1(s), \ldots, v_m(s)) \; .$$

We also have the equation

$$(E_0) \qquad c'(s) = v_1(s) \; .$$

So the equations $(E_0), (E_1), \ldots, (E_m)$ give us an equation

$$(*) \qquad \alpha'(s) = F(\alpha(s))$$

for the function $\alpha = (c, v_1, \ldots, v_m)$. The function F depends only on $\kappa_1, \ldots, \kappa_{m-1}$ (and the Christoffel symbols).

For the function $\bar{\alpha} = (\bar{c}, \bar{v}_1, \ldots, \bar{v}_m)$ there is a similar equation

$$\bar{\alpha}'(s) = \bar{F}(\bar{\alpha}(s)) \; .$$

Moreover, since $\bar{\kappa}_i = \kappa_i$ for all i, the function $\bar{F}$ is exactly the same as F. Now by hypothesis, the functions α and $\bar{\alpha}$ are equal at a, so by uniqueness of solutions of $(*)$ we have $\alpha = \bar{\alpha}$. Hence $c = \bar{c}$.

To prove (2) we first show that the desired curve c can be defined on some interval $[a, a+\varepsilon]$. Again we may clearly assume that $M = \mathbb{R}^m$. Now we can solve equation $(*)$ on some interval $[a, a+\varepsilon]$, with any given initial conditions. This gives us a curve $c\colon [a, a+\varepsilon] \longrightarrow M$ and vector fields $c' = \mathbf{v}_1, \ldots, \mathbf{v}_m$ along c satisfying the Frenet formulas (F_1) to (F_m), with

$$c(a) = p \quad \text{and} \quad \mathbf{v}_i(a) = \mathring{\mathbf{v}}_i\,, \quad 1 \le i \le m\,.$$

We have

$$\langle \mathbf{v}_i, \mathbf{v}_j \rangle'(s) = \left\langle \frac{D\mathbf{v}_i(s)}{ds}, \mathbf{v}_j(s) \right\rangle + \left\langle \mathbf{v}_i(s), \frac{D\mathbf{v}_j(s)}{ds} \right\rangle;$$

using the formulas (F_1) to (F_m), we find that this is zero. Since $\{\mathbf{v}_i(a)\} = \{\mathring{\mathbf{v}}_i\}$ is orthonormal, $\{\mathbf{v}_i\}$ must therefore be orthonormal everywhere. So $\mathbf{v}_1, \ldots, \mathbf{v}_m$ is the Frenet frame of c, and the κ_i are its curvatures.

In order to extend c to all of $[a,b]$, we first consider the equation (*) once again. If we choose our initial point p to lie in some compact set, then there will be $\varepsilon > 0$ with the property that for any orthonormal $\{\mathring{\mathbf{v}}_i\}$ at any such p, there is a curve $c\colon [a, a+\varepsilon] \to N$ with curvature functions κ_i on $[a, a+\varepsilon]$, whose Frenet frame $\mathbf{v}_1, \ldots, \mathbf{v}_m$ satisfies $\mathbf{v}_i(a) = \mathring{\mathbf{v}}_i$. The size of ε will depend on bounds for F, and hence only on bounds for the κ_i, as well as bounds for the Christoffel symbols Γ_{ij}^k. It is thus clear that for every point $q \in M$ there is $\delta(q) > 0$ with the following property:

(**) If $d(p,q) < \delta(q)$ and $\{\mathring{\mathbf{v}}_i\}$ is an orthonormal basis for M_p, then for any a' with $a < a' < b$ there is a curve $c\colon [a', \min(a' + \delta(q), b)] \to M$ with curvature functions κ_i on this interval, whose Frenet frame $\mathbf{v}_1, \ldots, \mathbf{v}_m$ satisfies $\mathbf{v}_i(a') = \mathring{\mathbf{v}}_i$.

Now by a least upper bound argument it clearly suffices to show that the curve $c\colon [a, a+\varepsilon) \to M$, with Frenet frame $\mathbf{v}_1, \ldots, \mathbf{v}_m$, can always be extended to the closed interval $[a, a+\varepsilon]$. The curve c is parameterized by arclength (since $\mathbf{v}_1 = c'$ has length 1), so for all $a < a' < a+\varepsilon$ we have $d(c(a), c(a')) \leq$ length of c on $[a,a'] = a' - a < \varepsilon$. So the image of c on $[a, a+\varepsilon)$ lies in some compact subset K of the complete manifold M. Thus there is $\delta > 0$ which will serve as $\delta(q)$ in (**) for all $q \in K$. Now choose $a < a' < a+\varepsilon$, so that $(a+\varepsilon) - a' < \delta$, and find the curve $\bar{c}$ with curvature functions κ_i, whose Frenet frame $\bar{\mathbf{v}}_1,\ldots,\bar{\mathbf{v}}_m$ satisfies $\bar{\mathbf{v}}_i(a') = \mathbf{v}_i(a')$. The curve $\bar{c}$ is defined at least on $[a',\varepsilon]$ by (**), and c followed by $\bar{c}$ is an extension of c at least as far as $c+\varepsilon$. ■

4. COROLLARY. Let N be a complete m-dimensional Riemannian manifold of constant curvature K_0. Let $c, \bar{c}\colon [a,b] \to N$ be arclength parameterized curves with nowhere zero curvature functions $\kappa_1,\ldots,\kappa_{m-1}$ and $\bar{\kappa}_1,\ldots,\bar{\kappa}_{m-1}$, respectively. If $\kappa_i = \bar{\kappa}_i$ for $1 \leq i \leq m-1$, then there is a unique isometry $A\colon N \to N$ such that $\bar{c} = A \circ c$.

Proof. Left to the reader. ■

We also want to consider curves with $\kappa_1,\ldots,\kappa_{j-1}$ nowhere zero, but κ_j identically zero, for some $j \leq m-1$. In Chapter II.1 we found that curves with $\kappa = 0$ are straight lines, while curves with $\tau = 0$ lie in a plane. The generalization for curves in $\mathbb{R}^m$ is the following.

5. THEOREM. Let $c\colon [a,b] \to \mathbb{R}^m$ be an arclength parameterized curve with $\kappa_1,\ldots,\kappa_{j-1}$ nowhere zero, and κ_j everywhere zero. Then c lies in some j-dimensional plane in $\mathbb{R}^m$.

Proof. Let $\mathbf{v}_1,\ldots,\mathbf{v}_j$ be the Frenet frame for c, and let $\Delta(s) \subset \mathbb{R}^m{}_{c(s)}$ be the j-dimensional subspace of $\mathbb{R}^m{}_{c(s)}$ spanned by $\mathbf{v}_1(s),\ldots,\mathbf{v}_j(s)$. We claim that all $\Delta(s)$ are parallel (considered as j-dimensional planes in $\mathbb{R}^m$). To prove this, we note that since $D'\mathbf{v}_i(s)/ds$ is just $\mathbf{v}_i'(s)$ in $\mathbb{R}^m$, the Frenet equations $(F_1),\ldots,(F_{j-1}),(F_j')$ show that each $\mathbf{v}_i'(s)$ is a linear combination of certain of the $\mathbf{v}_i(s)$,

$$\mathbf{v}_i'(s) = \sum_{l=1}^{j} a_{li}(s)\mathbf{v}_l(s) .$$

So if $\mathbf{w}$ is a parallel vector field on $\mathbb{R}^m$ (that is, $\mathbf{w}(p) = w_p$ for some $w \in \mathbb{R}^m$), then

$$(*) \quad \frac{d}{ds}\langle \mathbf{v}_i(s),\mathbf{w}(c(s))\rangle = \langle \mathbf{v}_i'(s),\mathbf{w}(c(s))\rangle = \sum_{l=1}^{j} a_{li}(s)\langle \mathbf{v}_l(s),\mathbf{w}(c(s))\rangle .$$

By uniqueness of solutions of the system (*), we see that if all $\langle \mathbf{v}_i(a),\mathbf{w}(c(a))\rangle = 0$, then all $\langle \mathbf{v}_i(s),\mathbf{w}(c(s))\rangle = 0$ for all s. In other words, $\Delta(s)$ is always orthogonal to the same vectors as $\Delta(a)$. Hence $\Delta(s)$ is parallel to $\Delta(a)$. Our result now follows from

6. LEMMA. Let $c\colon [a,b] \to \mathbb{R}^m$ be an immersed curve, and for each s let $\Delta(s) \subset \mathbb{R}^m{}_{c(s)}$ be a j-dimensional subspace of $\mathbb{R}^m{}_{c(s)}$ with $c'(s) \in \Delta(s)$. Suppose that all $\Delta(s)$ are parallel. Then c is a curve in some j-dimensional plane $P \subset \mathbb{R}^m$, and P is just $\exp(\Delta(s))$ for any s.

Proof. Let $P = \Delta(a)$, considered as a j-dimensional plane in $\mathbb{R}^m$. Without loss of generality we may assume that P is parallel to the $(x^1,\ldots,x^j)$-plane.

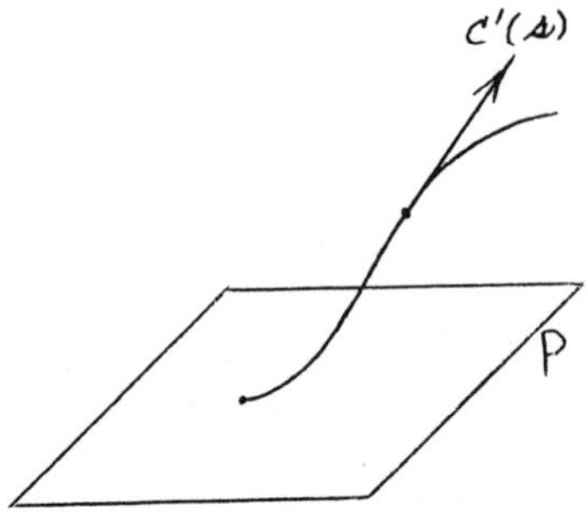

If c does not lie entirely in P, then by the mean value theorem some tangent vector $c'(s)$ has a non-zero k^{th} component for some $k > j$. But this is impossible, since $c'(s) \in \Delta(s)$ and $\Delta(s)$ is parallel to $P = \Delta(a)$. So c lies in P. Since each $\Delta(s)$ is parallel to $P = \Delta(a)$ and also contains the point $c(s) \in P$, each $\Delta(s)$ must equal P when $\Delta(s)$ is considered as a j-dimensional plane in $\mathbb{R}^m$. In other words, $P = \exp(\Delta(s))$. ∎

As soon as we try to replace $\mathbb{R}^m$ in Theorem 5 with a manifold N of constant curvature, we find that the proof of Lemma 6 doesn't generalize at all. However, the result is still true, and we will give two proofs, exploiting two different descriptions of constant curvature manifolds. First consider a curve $c: [a,b] \to N$ in any Riemannian manifold N, and suppose that for each s we have a j-dimensional subspace $\Delta(s) \subset N_{c(s)}$, so that Δ is a "distribution along c." Let $\tau_s: N_{c(s)} \to N_{c(s)}$ be the parallel translation along c from $c(a)$ to $c(s)$. We say that Δ is parallel along c if $\tau_s(\Delta(a)) = \Delta(s)$ for all s. Suppose that Δ is parallel along c and that V is a smooth vector field along c belonging to Δ (that is, $V(s) \in \Delta(s)$ for all s). Proposition II.6-3 immediately shows that

$$\frac{D'V(s)}{ds} \in \Delta(s) \qquad \text{for all } s ,$$

so that $D'V(s)/ds$ also belongs to Δ. We need the converse assertion also.

<u>7. PRE-LEMMA</u>. Let $c: [a,b] \to N$ be a curve in a Riemannian manifold N, and let Δ be a smooth j-dimensional distribution along c. Suppose that $D'V(s)/ds$ belongs to Δ whenever V is a smooth vector field belonging to Δ. Then Δ is parallel along c.

<u>Proof</u>. Choose everywhere linearly independent smooth vector fields $V_1,\ldots,V_j$ along c belonging to Δ. By hypothesis, we can write

$$\frac{D'V_i(s)}{ds} = \sum_{\iota=1}^{j} a_{\iota i}(s)V_\iota(s) \tag{1}$$

for certain smooth functions $a_{\iota i}$. We claim that there are functions $b_{\lambda i}$, with arbitrary initial conditions $b_{\lambda i}(a)$, such that

$$\frac{D'}{ds}\sum_{\lambda=1}^{j} b_{\lambda i}(s)V_\lambda(s) = 0\ , \qquad i = 1,\ldots,j\ . \tag{2}$$

In fact, equation (2) is equivalent to

$$\begin{aligned} 0 &= \sum_{\lambda=1}^{j} b_{\lambda i}{}'(s)V_\lambda(s) + \sum_{\lambda=1}^{j} b_{\lambda i}(s)\frac{D'V_\lambda(s)}{ds} \\ &= \sum_{\iota=1}^{j} b_{\iota i}{}'(s)V_\iota(s) + \sum_{\lambda,\iota=1}^{j} b_{\lambda i}(s)a_{\iota\lambda}(s)V_\iota(s) \qquad \text{by (1)}\ , \end{aligned}$$

and hence to

$$b_{\iota i}{}'(s) = \sum_{\lambda=1}^{j} a_{\iota\lambda}(s)b_{\lambda i}(s) \qquad i = 1,\ldots,j\ . \tag{3}$$

Since (3) is a linear equation, we can solve it on the whole interval $[a,b]$, with arbitrary initial conditions. Choose the initial conditions $b_{\lambda i}(a) = \delta_{\lambda i}$, and set

$$W_i(s) = \sum_{\lambda=1}^{j} b_{\lambda i}(s) V_\lambda(s) .$$

Then the vector fields W_i along c are parallel, by (2), and linearly independent at a, hence linearly independent everywhere. So the $W_i(s)$ span $\Delta(s)$ for all s, which shows that Δ is parallel along c. ∎

8. LEMMA. Let N be a manifold of constant curvature K_0. Let $c\colon [a,b] \longrightarrow N$ be an immersed curve, and let Δ be a j-dimensional distribution along c such that $c'(s) \in \Delta(s)$ for all s. Suppose that Δ is parallel along c. Then c lies in some j-dimensional totally geodesic submanifold $P \subset N$, and $\exp(\Delta(s)) \subset P$ for all s.

First Proof. It is easy to see that this result is essentially a local one, so without loss of generality we take N to be the complete simply-connected manifold of constant curvature K_0. Consider first the case $K_0 > 0$, so that $N = S^m(K_0) \subset \mathbb{R}^{m+1}$. We can then consider $V(s) \subset \mathbb{R}^{m+1}{}_{c(s)}$. We denote the covariant derivatives in N and $\mathbb{R}^{m+1}$ by ∇' and $\boldsymbol{\nabla}'$, respectively, and we will let $\boldsymbol{\nu}$ be a unit normal field on $N = S^m(K_0)$.

Let V be a vector field along c which belongs to Δ, so that $D'V/ds$ also belongs to Δ, since Δ is parallel along c. If $\mathbf{D}'/ds$ denotes the covariant derivative along c in $\mathbb{R}^{n+1}$, then Corollary 1-2 gives

$$\mathbf{T}\left(\frac{D'V(s)}{ds}\right) = \frac{D'V(s)}{ds} \in \Delta(s) .$$

Thus we see that

(1) $V(s) \in \Delta(s)$ for all s $\Longrightarrow$ $\dfrac{D'V(s)}{ds} \in \Delta(s) + \mathbb{R}\cdot\boldsymbol{\nu}(c(s))$ for all s .

On the other hand, we also have

(2) $\dfrac{D'\boldsymbol{\nu}(c(s))}{ds} = \nabla'_{c'(s)}\boldsymbol{\nu} = \text{constant}\cdot c'(s)$,

since all points of N are umbilics

$\in \Delta(s)$, by assumption on Δ .

Now let

$$\boldsymbol{\Delta}(s) = \Delta(s) \oplus \mathbb{R}\cdot\boldsymbol{\nu}(c(s)) \in \mathbb{R}^{m+1}{}_{c(s)} .$$

From (1) and (2) we see that for a vector field W along c in $\mathbb{R}^{m+1}$ we have

$$W(s) \in \boldsymbol{\Delta}(s) \text{ for all } s \Longrightarrow \frac{D'W(s)}{ds} \in \boldsymbol{\Delta}(s) \text{ for all } s .$$

By our prelemmanary remark we see that $\boldsymbol{\Delta}$ is parallel along c in $\mathbb{R}^{m+1}$. So Lemma 6 shows that c lies in some $(j+1)$-dimensional plane $P \subset \mathbb{R}^{m+1}$, and $P = \exp(\boldsymbol{\Delta}(s))$ for all s. Since $\boldsymbol{\nu}(c(s)) \in \boldsymbol{\Delta}(s)$, the plane P must pass through the origin $0 \in \mathbb{R}^{n+1}$. Hence c is contained in $P \cap S^m(K_0)$, which is a j-dimensional totally geodesic subspace of $S^m(K_0)$. Clearly, we also have $\exp(\Delta(s)) \subset P \cap S^m(K_0)$ for all s.

The case $K_0 < 0$ can be proved similarly, taking N to be $H^m(K_0)$, considered as a subset of $\mathbb{R}^{m+1}$ with the Lorentzian metric.

Second Proof. As the result is essentially local, we can assume that we have a geodesic mapping $\phi: N \longrightarrow \mathbb{R}^m$. If $\bar{\nabla}$ denotes covariant differentiation on $\mathbb{R}^m$, then Proposition II.6-18 shows that there is a 1-form ω on $\mathbb{R}^m$ with

$$(1) \qquad \bar{\nabla}_{\phi_* X}\phi_* Y - \phi_*(\nabla'_X Y) = \omega(\phi_* X)\cdot\phi_* Y + \omega(\phi_* Y)\cdot\phi_* X .$$

Let

$$\gamma(s) = \phi(c(s))$$
$$\bar{\Delta}(s) = \phi_*\Delta(s) \subset \mathbb{R}^m{}_{\gamma(s)} .$$

Then we have

$$(2) \qquad \gamma'(s) = \phi_* c'(s) \in \phi_*\Delta(s) = \bar{\Delta}(s) .$$

If V is a vector field along c with $V(s) \in \Delta(s)$ for all s, and hence $D'V(s)/ds \in \Delta(s)$ for all s, then equation (1) implies that

$$\frac{\bar{D}\phi_* V(s)}{ds} = \phi_*\left(\frac{D'V(s)}{ds}\right) + \omega(\gamma'(s))\cdot\phi_* V(s) + \omega(\phi_* V(s))\cdot\gamma'(s)$$
$$\in \bar{\Delta}(s) , \qquad \text{by (2)} .$$

In other words, if W is a vector field along γ with $W(s) \in \bar{\Delta}(s)$ for all s, then also $\bar{D}W(s)/ds \in \bar{\Delta}(s)$ for all s. Once again, this implies that $\bar{\Delta}$ is parallel along γ. So Lemma 6 implies that γ lies in some j-dimensional plane $P \subset \mathbb{R}^m$. Then c lies in $\phi^{-1}(P)$, which is a totally geodesic j-dimensional submanifold of N. Clearly, we also have $\exp(\Delta(s)) \subset \phi^{-1}(P)$ for all s. ∎

9. THEOREM. Let N be a manifold of constant curvature K_0. Let $c\colon [a,b] \longrightarrow N$ be an arclength parameterized curve with $\kappa_1,\ldots,\kappa_{j-1}$ nowhere zero, and κ_j everywhere zero. Then c lies in some j-dimensional totally geodesic submanifold of N.

Proof. Let $\mathbf{v}_1,\ldots,\mathbf{v}_j$ be the Frenet frame for c, and let $\Delta(s) \subset M_{c(s)}$ be the subspace spanned by $\mathbf{v}_1(s),\ldots,\mathbf{v}_j(s)$. The argument in the proof of Theorem 5 shows generally that Δ is parallel along c. So our result follows from Lemma 8. ∎

10. COROLLARY. Let N^m be a complete manifold of constant curvature K_0. Let $c, \bar{c}\colon [a,b] \longrightarrow N$ be arclength parameterized curves with $\kappa_1,\ldots,\kappa_{j-1}$ and $\bar{\kappa}_1,\ldots,\bar{\kappa}_{j-1}$ nowhere zero, and κ_j and $\bar{\kappa}_j$ everywhere zero. If $\kappa_i = \bar{\kappa}_i$ for $1 \leq i \leq j-1$, then there is an isometry $A\colon N \longrightarrow N$ such that $\bar{c} = A \circ c$. The group of such isometries is isomorphic to the orthogonal group $O(m-j-1)$.

Proof. Left to the reader. ∎

For later use, we note a consequence of Lemma 8 for higher dimensional submanifolds M of N.

11. COROLLARY. Let N be a manifold of constant curvature K_0. Let M be a connected submanifold immersed in N, and let Δ be a j-dimensional distribution along M such that $M_p \subset \Delta(p)$ for all $p \in M$. Suppose that Δ is parallel along every curve c in M. Then M lies in some j-dimensional totally geodesic submanifold $P \subset N$, and $\exp(\Delta(p)) \subset P$ for all $p \in M$.

Proof. Choose a point $p_0 \in M$, and let P be the largest j-dimensional totally geodesic submanifold of N with $P \supset \exp(\Delta(p_0))$. For any $p \in M$, choose a curve $c\colon [0,1] \longrightarrow M$ with $c(0) = p_0$ and $c(1) = p$. Lemma 8, applied to the distribution $s \mapsto \Delta(c(s))$ along c, implies that c lies in some j-dimensional totally geodesic submanifold $P' \subset M$, and $\exp(\Delta(c(s))) \subset P'$ for all s. Applying this for $s = 0$, we see that $P' \subset P$. Hence $p \in P$, and also $\exp(\Delta(p)) = \exp(\Delta(c(1))) \subset P' \subset P$. ■

C. The Fundamental Equations for Submanifolds

In Chapter 1, we considered a submanifold M^n of a Riemannian manifold $(N^m, \langle\ ,\ \rangle)$ with $i\colon M \longrightarrow N$ the inclusion map. For each $p \in M$, we have $N_p = M_p \oplus M_p{}^{\perp}$, and we used this decomposition to define two projections, $\mathbf{T}\colon N_p \longrightarrow M_p$ and $\perp\colon N_p \longrightarrow M_p{}^{\perp}$. For vector fields X and Y tangent along M we wrote

$$\nabla'_{X_p} Y = \mathbf{T}(\nabla'_{X_p} Y) + \perp(\nabla'_{X_p} Y)$$

where ∇' is the covariant differentiation in N, and we showed that $\mathbf{T}(\nabla'_{X_p} Y) = \nabla_{X_p} Y$, where ∇ is the covariant differentiation in M determined by the metric $i^*\langle\ ,\ \rangle$, while $\perp(\nabla'_{X_p} Y) = s(X_p, Y_p)$ is symmetric in X_p and Y_p (and independent of the extension Y of Y_p). This gave us

The Gauss Formulas	$\nabla'_{X_p} Y = \nabla_{X_p} Y + s(X_p, Y_p)$

and we then derived

Gauss' Equation
$\langle R'(X,Y)Z,W\rangle = \langle R(X,Y)Z,W\rangle$
$\qquad + \langle s(X,Z),s(Y,W)\rangle - \langle s(Y,Z),s(X,W)\rangle$

for all tangent vectors $X, Y, Z, W \in M_p$. (For convenience we will often use $X,Y,Z,\ldots$ without subscripts to denote vectors as well as vector fields.)

For a hypersurface with a unit normal field ν, we showed that $\nabla'_{X_p}\nu \in M_p$, and we determined this vector explicitly by

The Weingarten Equations: $\langle \nabla'_{X_p}\nu, Y_p\rangle = -\langle \nu, s(X_p,Y_p)\rangle$.

Defining a tensor II by $s(X_p,Y_p) = II(X_p,Y_p)\cdot\nu(p)$, we then derived the

Codazzi-Mainardi Equations:

$$\langle R'(X,Y)Z,\nu\rangle = \langle\nabla_X II)(Y,Z) - (\nabla_Y II)(X,Z) .$$

Now we want to consider a submanifold of arbitrary codimension. We define the **normal bundle** Nor M **of** M **in** N to be

$$\text{Nor } M = \bigcup_{p\in M} M_p{}^{\perp} ,$$

and we define the projection map

$$\varpi\colon \text{Nor } M \longrightarrow M$$

to be the one which takes all vectors in $M_p{}^{\perp}$ to p. Thus (compare p.I.465)

ϖ: Nor $M \longrightarrow M$ is a vector bundle whose fibre $\varpi^{-1}(p)$ over p is $M_p{}^{\perp}$. A section ξ of E is a map with $\xi(p) \in M_p{}^{\perp}$ for all p, in other words, a normal vector field along M.

Unlike the case of a hypersurface, it is no longer true that $\nabla'_{X_p}\xi \in M_p$, even if ξ always has length 1, so we will look at the general decomposition

$$\nabla'_{X_p}\xi = \mathbf{T}(\nabla'_{X_p}\xi) + \boldsymbol{\perp}(\nabla'_{X_p}\xi) .$$

The tangential component is just as nice as in the case of hypersurfaces:

12. PROPOSITION. If ξ is a section of the normal bundle of M, and $X_p \in M_p$, then the vector $\mathbf{T}(\nabla'_{X_p}\xi) \in M_p$ satisfies

$$\langle \mathbf{T}(\nabla'_{X_p}\xi), Y_p\rangle = \langle \nabla'_{X_p}\xi, Y_p\rangle = -\langle \xi(p), s(X_p,Y_p)\rangle , \qquad \text{for all } Y_p \in M_p .$$

Consequently, $\mathbf{T}(\nabla'_{X_p}\xi)$ depends only on X_p and $\xi(p)$.

Proof. If Y is a vector field tangent along M which extends Y_p, then $\langle \xi, Y\rangle = 0$, so

$$0 = X_p(\langle \xi, Y\rangle) = \langle \nabla'_{X_p}\xi, Y_p\rangle + \langle \xi(p), \nabla'_{X_p}Y\rangle$$
$$= \langle \nabla'_{X_p}\xi, Y_p\rangle + \langle \xi(p), s(X_p,Y_p)\rangle ,$$

since $\xi(p) \in M_p{}^{\perp}$, by assumption. ∎

For any vector $\xi_p \in M_p{}^{\perp}$, we will define $A_{\xi_p} : M_p \longrightarrow M_p$ as follows. For each $X_p \in M_p$, we let $A_{\xi_p}(X_p) \in M_p$ be the unique vector satisfying

$$<A_{\xi_p}(X_p), Y_p> = <s(X_p, Y_p), \xi_p> \qquad \text{for all } Y_p \in M_p .$$

By Proposition 12, we also have

$$A_{\xi_p}(X_p) = -\mathbf{T}(\nabla'_{X_p}\xi) ,$$

where ξ is any normal vector field extending ξ_p. When M is a hypersurface in $\mathbb{R}^{n+1}$ with unit normal vector field ν, the map $A_{\nu_p} : M_p \longrightarrow M_p$ is the same as $-d\nu : M_p \longrightarrow M_p$.

For the normal component $\perp(\nabla'_{X_p}\xi)$ we will simply introduce a new symbol, just as we did for $\perp(\nabla'_{X_p}Y)$. For a section ξ of the normal bundle of M, and for $X_p \in M_p$, we define

$$D_{X_p}\xi = \perp(\nabla'_{X_p}\xi) \in M_p{}^{\perp} .$$

Unlike the case of $\perp(\nabla'_{X_p}Y)$, the value of $\perp(\nabla'_{X_p}\xi)$ depends on the values of ξ in a neighborhood of p, not just on $\xi(p)$.

<u>13. PROPOSITION</u>. The map $(X_p, \xi) \longmapsto D_{X_p}\xi$ is a connection on the normal bundle Nor M; that is (compare p.II.242 and also p. II.397),

(1) $D_{X_p+Y_p}\xi = D_{X_p}\xi + D_{Y_p}\xi$,

(2) $D_{X_p}(\xi + \eta) = D_{X_p}\xi + D_{X_p}\eta$,

(3) $D_{aX_p}\xi = aD_{X_p}\xi$ for all $a \in \mathbb{R}$,

(4) $D_{X_p} f\cdot\xi = f(p)\cdot D_{X_p}\xi + X_p(f)\cdot\xi(p)$ for all C^∞ functions f ,

(5) If X is a C^∞ vector field and ξ is a C^∞ section of the normal bundle, then $p \longmapsto D_{X_p}\xi$ is also C^∞.

Moreover, D is compatible with the metric $\langle\ ,\ \rangle$ on the normal bundle:

$$X_p(\langle\xi,\eta\rangle) = \langle D_{X_p}\xi,\eta\rangle + \langle\xi,D_{X_p}\eta\rangle .$$

Proof. All properties follow immediately from the corresponding properties for ∇'. ■

We will call D the normal connection for the imbedding $M \subset N$. With the notation we have just introduced, we may now write the decomposition $\nabla'_{X_p}\xi = \mathbf{T}(\nabla'_{X_p}\xi) + \perp(\nabla'_{X_p}\xi)$ as:

The Weingarten Equations. $\nabla'_{X_p}\xi = -A_{\xi_p}(X_p) + D_{X_p}\xi$.

In the case of a hypersurface, we used the second fundamental form s and a unit normal vector field ν to define a real-valued second fundamental form II. In the general case, we choose $\nu_{n+1},\ldots,\nu_m$ to be everywhere orthonormal sections of E defined in a neighborhood of a point and we define $m-n$ real-valued second fundamental forms II^r by

$$II^r(X_p,Y_p) = \langle \nabla'_{X_p} Y_p, \nu_r(p)\rangle = \langle s(X_p,Y_p), \nu_r(p)\rangle \qquad r = m+1,\ldots,n \,.$$

We thus have

$$s(X_p,Y_p) = \sum_r II^r(X_p,Y_p)\cdot \nu_r(p) \,.$$

Notice that the set $\{II^r\}$ depends on the choice of the $\{\nu_r\}$; there are many possible choices, unlike the case of a hypersurface, where the choice of the single unit normal field ν was essentially unique. Using the II^r instead of s, we can write

<u>Gauss' Equation.</u>

$$\langle R'(X,Y)Z,W\rangle = \langle R(X,Y)Z,W\rangle + \sum_r \{II^r(X,Z)II^r(Y,W) - II^r(Y,Z)II^r(X,W)\} \,.$$

Since the tensors II^r give us an explicit expression for s, they also essentially give us an expression for the $A_{\nu_r(p)}$, for the equation

$$\langle A_{\nu_r(p)}(X_p), Y_p\rangle = \langle \nu_r(p), s(X_p,Y_p)\rangle = II^r(X_p,Y_p)$$

determines $A_{\nu_r(p)}(X_p)$. We also want quantities by means of which we can express D. So we introduce certain 1-forms, the <u>normal fundamental forms</u> β_r^s, by

$$\beta_r^s(X_p) = \langle \nabla'_{X_p} \nu_r, \nu_s\rangle = \langle D_{X_p} \nu_r, \nu_s\rangle \,.$$

Then

$$D_{X_p} \nu_r = \sum_s \beta_r^s(X_p) \cdot \nu_s ,$$

and $D_{X_p} \xi$ can be computed for any $\xi = \sum_r a^r \nu_r$ by using Proposition 13. Notice that since $\langle \nu_r, \nu_s \rangle = 1$ or 0, we have $\beta_r^s = -\beta_s^r$. In particular, for hypersurfaces we have the single 1-form $\beta_{n+1}^{n+1} = 0$.

14. THEOREM. Let M be a submanifold of N, with corresponding s and D. Then for all vector fields X, Y, Z tangent along M we have

$$\begin{aligned} \perp(R'(X,Y)Z) = {} & [D_X(s(Y,Z)) - s(\nabla_X Y,Z) - s(Y,\nabla_X Z)] \\ & - [D_Y(s(X,Z)) - s(\nabla_Y X,Z) - s(X,\nabla_Y Z)] . \end{aligned}$$

If $\nu_{n+1}, \ldots, \nu_m$ are everywhere orthonormal sections of $\operatorname{Nor} M$, with corresponding II^r and β_r^s, then for all vectors $X, Y, Z \in M_p$ we have

The Codazzi-Mainardi Equations.

$$\begin{aligned} \langle R'(X,Y)Z, \nu^r(p) \rangle = {} & (\nabla_X II^r)(Y,Z) - (\nabla_Y II^r)(X,Z) \\ & + \sum_s II^s(Y,Z)\beta_s^r(X) - II^s(X,Z)\beta_s^r(Y) . \end{aligned}$$

Proof. The first equation is precisely equation (3) in the proof of Theorem 1-11. Now Proposition 12 gives

$$\begin{aligned} (1) \qquad D_X(s(Y,Z)) &= D_X\Big(\sum_s II^s(Y,Z)\nu_s\Big) \\ &= \sum_s X(II^s(Y,Z)) \cdot \nu_s + \sum_s II^s(Y,Z) D_X \nu_s . \end{aligned}$$

Moreover,

(2) $s(\nabla_X Y,Z) + s(Y,\nabla_X Z) = \sum_s II^s(\nabla_X Y,Z)\cdot\nu_s + \sum_s II^s(Y,\nabla_X Z)\cdot\nu_s$.

Then (1) and (2) give

(3) $$D_X(s(Y,Z)) - s(\nabla_X Y,Z) - s(Y,\nabla_X Z)$$
$$= \sum_s [X(II^s(Y,Z)) - II^s(\nabla_X Y,Z) - II^s(Y,\nabla_X Z)]\cdot\nu_s + \sum_s II^s(Y,Z)D_X\nu_s$$
$$= \sum_s (\nabla_X II^s)(Y,Z)\cdot\nu_s + \sum_s II^s(Y,Z)D_X\nu_s ,$$ by Corollary II.6-5*.

Hence

(4) $$\langle D_X(s(Y,Z)) - s(\nabla_X Y,Z) - s(Y,\nabla_X Z),\nu_r\rangle$$
$$= (\nabla_X II^r)(Y,Z) + \sum_s II^s(Y,Z)\beta_s^r(X) .$$

Naturally there is a similar equation with X and Y interchanged. Substituting into the first part of the Theorem, we obtain the Codazzi-Mainardi equations. ■

The Codazzi-Mainardi equations which we have derived here are obviously a lot less satisfying than they were in the case of a hypersurface, since the set $\{II^r\}$ is not unique. As a matter of fact, the nicest form of the Codazzi-Mainardi equation is obtained by looking a little harder at the expression

$$D_X(s(Y,Z)) - s(\nabla_X Y,Z) - s(Y,\nabla_X Z)$$

which appears in the first part of Theorem 14. A quick check shows that this expression is linear in X, Y, and Z <u>over the</u> C^∞ <u>functions</u>. So the value

*As on p.III.16, we really need this Corollary for tensors of type $\binom{k}{0}$.

of this expression at p depends only on X_p, Y_p, Z_p. To obtain an explicit description of this function of three vectors, we consider s as a section of the bundle $\mathrm{Hom}(TM \times TM, \mathrm{Nor}\ M)$ whose fibre at p is the vector space of all bilinear maps $M_p \times M_p \longrightarrow M_p^{\perp}$. Now, using the connection ∇ in TM and the connection D in Nor M, a connection $\tilde{\nabla}$ in the bundle $\mathrm{Hom}(TM \times TM, \mathrm{Nor}\ M)$ can be defined in the following natural way. Given

$$\begin{cases} \text{a section } \psi \text{ of } \mathrm{Hom}(TM \times TM, \mathrm{Nor}\ M)\ , \\ \text{a vector } X_p \in M_p\ , \end{cases}$$

we want to have a bilinear map

$$\tilde{\nabla}_{X_p} \psi \colon M_p \times M_p \longrightarrow M_p^{\perp}\ ,$$

so we want to define

$$(\tilde{\nabla}_{X_p} \psi)(Y_p, Z_p)\ , \qquad \text{for } Y_p, Z_p \in M_p\ .$$

Let c be a curve in M with $c'(0) = X_p$, and let

τ_h = the parallel translation in TM along c from $c(0)$ to $c(h)$ determined by ∇,

ρ_h = the parallel translation in Nor M along c from $c(0)$ to $c(h)$ determined by D.

Then we define

$$(\tilde{\nabla}_{X_p} \psi)(Y_p, Z_p) = \lim_{h \to 0} \frac{1}{h}\Big[\rho_h^{-1}\big(\psi(c(h))(\tau_h Y_p, \tau_h Z_p)\big) - \psi(p)(Y_p, Z_p)\Big]\ .$$

[Notice that if Nor M were the trivial bundle $M \times \mathbb{R}$, making ψ essentially a tensor of type $\binom{2}{0}$, and D were the flat connection, with parallel translation the same along all curves, taking (p,a) to (q,a) for all $p, q \in M$ and $a \in \mathbb{R}$, then this definition would reduce to the definition of $\nabla_{X_p} \psi$ already given (p. II.252). On the other hand, if Nor M were TM, with the connection ∇, then this definition would reduce to the definition of $\nabla_{X_p} \psi$ when ψ is a tensor of type $\binom{2}{1}$.] Now it is easy to see that if Y and Z are vector fields, then

$$(\tilde{\nabla}_{X_p} \psi)(Y_p, Z_p) = D_{X_p}(\psi(Y,Z)) - \psi(\nabla_{X_p} Y, Z_p) - \psi(Y_p, \nabla_{X_p} Z)$$

[Corollary II.6-5 is the special case when Nor M is the trivial bundle]. We can therefore also express the Codazzi-Mainardi equations in an intrinsic form:

<u>15. COROLLARY</u>. Let M be a submanifold of N. Then for all vectors $X, Y, Z \in M_p$ we have

<u>The Codazzi-Mainardi Equations</u>

$$\perp(R'(X,Y)Z) = (\tilde{\nabla}_X s)(Y,Z) - (\tilde{\nabla}_Y s)(X,Z)$$

where $\tilde{\nabla}$ is the covariant differentiation on $\mathrm{Hom}(TM \times TM, \text{Nor } M)$ determined by the covariant differentiations ∇ on TM and D on the normal bundle Nor M.

The Gauss and Codazzi-Mainardi equations tell us what $\langle R'(X,Y)Z,W\rangle$ is when all four vectors are in M_p, or when three are in M_p and one is in $M_p^{\perp}$ (which one doesn't matter, because the symmetry properties of R' allow us to express all possibilities in terms of the one where $W \in M_p^{\perp}$). We can just as well ask what $\langle R'(X,Y)Z,W\rangle$ is when two vectors are in M_p and two are in $M_p^{\perp}$. The answer is known (though not very well known) as Ricci's equations, or sometimes as the Ricci-Kühne equations. We need one more definition. Given $X, Y \in M_p$, and an orthonormal basis $U_1,\dots,U_n$ of M_p, we set

$$II^r{*}II^s(X,Y) = \sum_{i=1}^{n} II^r(X,U_i)\cdot II^s(Y,U_i) .$$

It is easy to check that $II^r{*}II^s$ does not depend on the choice of the orthonormal basis $U_1,\dots,U_n$. Classically, $II^r{*}II^s$ would be written as a contraction involving the components of II^r, II^s and the metric $\langle\ ,\ \rangle^*$ on T^*M (compare pp.III.190-191).

<u>16. THEOREM</u>. Let M be a submanifold of N, with corresponding s, A, and D. Then for all vector <u>fields</u> X and Y tangent along M, and all <u>sections</u> ξ of the normal bundle $\text{Nor}\, M$ we have

$$\perp(R'(X,Y)\xi) = s(A_\xi(X),Y) - s(A_\xi(Y),X) + [D_X(D_Y\xi) - D_Y(D_X\xi) - D_{[X,Y]}\xi] .$$

If $\nu_{n+1},\dots,\nu_m$ are everywhere orthonormal sections of $\text{Nor}\, M$, with corresponding II^r and β_r^s, then for all vectors $X, Y \in M_p$ we have

The Ricci Equations

$$\langle R'(X,Y)\nu_r(p),\nu_s(p)\rangle = II^r * II^s(X,Y) - II^r * II^s(Y,X) + (\nabla_X\beta_r^s)(Y) - (\nabla_Y\beta_r^s)(X) + \sum_w \beta_w^s(X)\beta_r^w(Y) - \beta_w^s(Y)\beta_r^w(X) .$$

(Notice that these equations are trivial if M is a hypersurface.)

Proof. The Weingarten equations and the Gauss formulas give

$$\begin{aligned}\nabla'_X(\nabla'_Y\xi) &= -\nabla'_X(A_\xi(Y)) + \nabla'_X(D_Y\xi)\\ &= -\nabla_X(A_\xi(Y)) - s(X,A_\xi(Y)) - A_{(D_Y\xi)}(X) + D_X(D_Y\xi) ,\end{aligned}$$

and hence

$$(1) \qquad \perp(\nabla'_X(\nabla'_Y\xi)) = -s(X,A_\xi(Y)) + D_X(D_Y\xi) ,$$

$$(1') \qquad \perp(\nabla'_Y(\nabla'_X\xi)) = -s(Y,A_\xi(X)) + D_Y(D_X\xi) .$$

Also,

$$\nabla'_{[X,Y]}\xi = A_\xi([X,Y]) + D_{[X,Y]}\xi ,$$

so

$$(2) \qquad \perp(\nabla'_{[X,Y]}\xi) = D_{[X,Y]}\xi .$$

Equations (1), (1'), and (2) give the first part of the theorem.

Now if $U_1,\ldots,U_n$ is an orthonormal basis for M_p, then

$$A_{\nu_r(p)}(X_p) = \sum_{i=1}^{n} \langle A_{\nu_r(p)}(X_p),U_i\rangle\cdot U_i$$

$$= \sum_{i=1}^{n} \langle \nu_r(p),s(X_p,U_i)\rangle\cdot U_i$$

$$= \sum_{i=1}^{n} II^r(X_p,U_i)\cdot U_i ,$$

so

$$(3) \qquad \langle s(A_{\nu_r(p)}(X_p),Y_p),\nu_s(p)\rangle = II^s(A_{\nu_r(p)}(X_p),Y_p)$$

$$= II^s(\sum_{i=1}^{n} II^r(X_p,U_i)\cdot U_i,Y_p)$$

$$= \sum_{i=1}^{n} II^r(X_p,U_i)\cdot II^s(Y_p,U_i)$$

$$= II^r{*}II^s(X_p,Y_p) ,$$

$$(3') \qquad \langle s(A_{\nu_r(p)}(Y_p),X_p),\nu_s(p)\rangle = II^r{*}II^s(Y_p,X_p) .$$

We also have

$$D_X(D_Y\nu_r) = \sum_w D_X(\beta_r^w(Y)\cdot\nu_w)$$

$$= \sum_w X(\beta_r^w(Y))\cdot\nu_w + \sum_w \beta_r^w(Y)\cdot D_X\nu_w$$

$$= \sum_w X(\beta_r^w(Y))\cdot\nu_w + \sum_w \beta_r^w(Y)\cdot\sum_v \beta_w^v(X)\nu_v ,$$

so

$$(4) \qquad \langle D_X(D_Y\nu_r),\nu_s\rangle = X(\beta^s_r(Y)) + \sum_w \beta^s_w(X)\beta^w_r(Y) ,$$

$$(4') \qquad \langle D_Y(D_X\nu_r),\nu_s\rangle = Y(\beta^s_r(X)) + \sum_w \beta^s_w(Y)\beta^w_r(X) .$$

Also,

$$(5) \qquad \langle D_{[X,Y]}\nu_r,\nu_s\rangle = \langle D_{\nabla_X Y}\nu_r,\nu_s\rangle - \langle D_{\nabla_Y X}\nu_r,\nu_s\rangle$$

$$= \beta^s_r(\nabla_X Y) - \beta^s_r(\nabla_Y X) .$$

From (4), (4') and (5) we get

$$(6) \qquad \langle D_X(D_Y\nu_r) - D_Y(D_X\nu_r) - D_{[X,Y]}\nu_r,\nu_s\rangle$$

$$= [X(\beta^s_r(Y)) - \beta^s_r(\nabla_X Y)] - [Y(\beta^s_r(X)) - \beta^s_r(\nabla_Y X)]$$
$$+ \sum_w \beta^s_w(X)\beta^w_r(Y) - \beta^s_w(Y)\beta^w_r(X)$$

$$= (\nabla_X\beta^s_r)(Y) - (\nabla_Y\beta^s_r)(X) + \sum_w \beta^s_w(Y)\beta^w_r(X) - \beta^s_w(Y)\beta^w_r(X) ,$$

by Corollary II.6-5.

Substituting (3), (3'), (6) into the first part of the theorem, we obtain the Ricci equations. ■

The expression $D_X(D_Y\xi) - D_Y(D_X\xi) - D_{[X,Y]}\xi$ which appears in the first part of Theorem 16 can be treated just like the expressions which arose in Theorem 14. In fact, for any connection D in any vector bundle $\varpi\colon E \longrightarrow M$, the map

$$(X,Y,\xi) \longmapsto D_X(D_Y\xi) - D_Y(D_X\xi) - D_{[X,Y]}\xi$$

(for vector fields X, Y on M and sections ξ of E) is linear in X, Y, and ξ over the C^{∞} functions. Consequently, its value at $p \in M$ depends only on X_p, Y_p, $\xi(p)$: we already know this for the two vector fields X, Y (Theorem I.4-2), and the proof for the section ξ is essentially the same. We therefore have a well-defined map

$$R_D = R_D(p)\colon M_p \times M_p \times \varpi^{-1}(p) \longrightarrow \varpi^{-1}(p) ,$$

the curvature of the connection D, given by

$$R_D(p)(X_p, Y_p)\xi_p = D_X(D_Y\xi)(p) - D_Y(D_X\xi)(p) - D_{[X,Y]}\xi(p) ,$$

for any vector fields X and Y extending X_p and Y_p, and any section ξ of E with $\xi(p) = \xi_p$. Thus we can state a more intrinsic form of the Ricci equations:

17. COROLLARY. Let M be a submanifold of N, with corresponding s and A. Then for all vectors $X, Y \in M_p$ and $\xi \in M_p{}^{\perp}$ we have

The Ricci Equations

$$\perp(R'(X,Y)\xi) = R_D(X,Y)\xi + s(A_\xi(X),Y) - s(A_\xi(Y),X)$$

where R_D is the curvature of the connection D in $\text{Nor}\, M$.

The Gauss, Codazzi-Mainardi, and Ricci equations are the only general equations which we have for submanifolds of a Riemannian manifold. It would not be reasonable to expect an interesting formula for $\langle R'(X,Y)Z,W\rangle$ when <u>three</u> of the vectors are in $M_p{}^{\perp}$. For if $X, Y, Z \in M_p{}^{\perp}$, then $R'(X,Y,Z)$ has nothing to do with M at all, and $\langle R'(X,Y)Z,W\rangle$ would depend only on the position of M_p. The classical reason for resting content with these three equations was somewhat different. In Chapter 2 we saw (at least in a special case) that the Gauss and Codazzi-Mainardi equations were precisely the integrability conditions for the Gauss formulas. It turned out that the integrability conditions for the Weingarten equations reduced to the Codazzi-Mainardi equations, but this was only because we happened to be dealing with a hypersurface. In general, the integrability conditions for the Weingarten equations lead to two sets of equations; one set reduces to the Codazzi-Mainardi equations, while the other set is precisely the Ricci equations [notice that our proof of Theorem 16 essentially investigated integrability conditions also, for we compared $\nabla'_X(\nabla'_Y\xi)$ with $\nabla'_Y(\nabla'_X\xi)$]. It was therefore clear to classical differential geometers that the II^r and β_r^s determine an n-dimensional submanifold of $\mathbb{R}^m$ up to Euclidean motion, and that any set of II^r and β_r^s comes from some submanifold if the three fundamental equations are satisfied. In order to derive these results without writing everything out in very classical terms, we will first see what our fundamental equations say in terms of moving frames.

Consider an adapted orthonormal moving frame $X_1,\ldots,X_n,X_{n+1},\ldots,X_m$ on M. As in Chapter 1, we let θ^i, ω^i_j, and Ω^i_j be the dual forms, connection forms, and curvature forms on M for the frame $X_1,\ldots,X_n$, and we let ϕ^α, ψ^α_β and Ψ^α_β be the forms on N for the frame $X_1,\ldots,X_m$. Then on TM we have

$$\phi^i = \theta^i\,, \qquad \phi^r = 0\,.$$

By looking at the first structural equations, we found that

$$\psi^i_j = \omega^i_j \ ,$$

and that there are unique functions s^r_{ij} on M satisfying

$$\psi^r_j = \sum_i s^r_{ij}\theta^i \ , \qquad s^r_{ij} = s^r_{ji} \ .$$

These functions are related to s by the equation (p. III.28)

$$s(X_j,X_k) = \sum_r s^r_{jk}X_r \ .$$

So if we choose the orthonormal vectors $X_{n+1},\ldots,X_m$ to be our $\nu_{n+1},\ldots,\nu_m$, then the II^r are given simply by

(a) $$II^r(X_j,X_k) = s^r_{jk} \quad \Longrightarrow \quad \psi^r_j(X) = II^r(X,X_j) \ ,$$

while the normal forms β^s_r are simply

(b) $$\beta^s_r = \psi^s_r \quad \text{on } TM \ .$$

Since the map A is determined by

$$\begin{aligned} \langle A_{\nu_r}(X_i),X_j\rangle &= \langle \nu_r, s(X_i,X_j)\rangle \\ &= s^r_{ij} \ , \end{aligned}$$

we also have the explicit formula

$$A_{\nu_r}(X_i) = \sum_j s^r_{ij}X_j \ .$$

More important, we have

$$\text{(c)} \qquad II^r * II^s(X_i,X_j) = \sum_{k=1}^{n} II^r(X_i,X_k)II^s(X_j,X_k)$$

$$= \sum_{k=1}^{n} s^r_{ik}s^s_{jk} .$$

Now let us look at the second structural equation

$$d\psi^\alpha_\beta = -\sum_\gamma \psi^\alpha_\gamma \wedge \psi^\gamma_\beta + \Psi^\alpha_\beta .$$

If we restrict to TM, and choose various ranges for the indices, we obtain the following three equations (for the first we also use the structural equation on M, as in Chapter 1):

$$\text{(A)} \qquad \Psi^i_j = \Omega^i_j - \sum_r \psi^r_i \wedge \psi^r_j .$$

$$\text{(B)} \qquad d\psi^r_j = -\sum_i \psi^r_i \wedge \omega^i_j - \sum_w \psi^r_w \wedge \psi^w_j + \Psi^r_j .$$

$$\text{(C)} \qquad d\psi^s_r = \sum_i \psi^s_i \wedge \psi^r_i - \sum_w \psi^s_w \wedge \psi^w_r + \Psi^s_r .$$

Using equation (a) we see immediately that equation (A) is precisely equivalent to Gauss' equation (in the form given on p. 51). For equations (B) and (C) we recall that for a 1-form η we have (p. I.292)

$$\text{(d)} \qquad d\eta(X_k,X_\ell) = X_k(\eta(X_\ell)) - X_\ell(\eta(X_k)) - \eta([X_k,X_\ell]) .$$

We also have

(e) $$[X_k,X_\ell] = \nabla_{X_k} X_\ell - \nabla_{X_\ell} X_k = \sum_i \omega_\ell^i(X_k)X_i - \omega_k^i(X_\ell)X_i$$

and (Corollary II.6-5)

(f) $$(\nabla_{X_k} II^r)(X_\ell,X_j) = X_k(II^r(X_\ell,X_j)) - II^r(\nabla_{X_k} X_\ell,X_j) - II^r(X_\ell,\nabla_{X_k} X_j)$$

(g) $$(\nabla_{X_k} \beta_r^s)(X_\ell) = X_k(\beta_r^s(X_\ell)) - \beta_r^s(\nabla_{X_k} X_\ell) .$$

When we apply equations (B) and (C) to (X_k,X_ℓ), and use equations (a)-(g), we find that (B) and (C) are equivalent to the Codazzi-Mainardi equations in Theorem 14, and the Ricci equations in Theorem 16, respectively. Equations (A), (B), (C), involving differential forms, are much more convenient for considering questions connected with integrability conditions.

<u>18. THEOREM</u>. (1) Let M, $\bar{M} \subset \mathbb{R}^m$ be two connected n-manifolds imbedded in $\mathbb{R}^m$, let $\nu_{n+1},\ldots,\nu_m$ be everywhere orthonormal sections for the normal bundle of M, and let $\bar{\nu}_{n+1},\ldots,\bar{\nu}_m$ be everywhere orthonormal sections for the normal bundle of $\bar{M}$. Let I, II^r, β_r^s be the first, second, and normal fundamental forms for M (defined with respect to the $\{\nu^r\}$), and define $\bar{I}$, $\overline{II}^r$, $\bar{\beta}_r^s$ similarly. Let $\phi\colon M \to \bar{M}$ be a diffeomorphism which preserves all the fundamental forms:

$$\phi^*\bar{I} = I , \qquad \phi^*\overline{II}^r = II^r , \qquad \phi^*\bar{\beta}_r^s = \beta_r^s .$$

Then there is a Euclidean motion A such that $\phi = A|M$ and $A_*(\nu_r) = \bar{\nu}_r$ for $r = n+1,\ldots,m$.

(2) Let $(M, \langle\langle\ ,\ \rangle\rangle)$ be an n-dimensional Riemannian manifold with curvature tensor R. For $r, s = n+1,\ldots,m$, let S^r be symmetric tensors on M, covariant of order 2, and let b_r^s be 1-forms on M with $b_r^s = -b_s^r$. Suppose that the S^r and b_r^s satisfy

(1) Gauss' Equation

$$0 = \langle\langle R(X,Y)Z,W\rangle\rangle - \sum_r S^r(Y,Z)S^r(X,W) + S^r(X,Z)S^r(Y,W)$$

(2) The Codazzi-Mainardi Equations

$$0 = (\nabla_X S^r)(Y,Z) - (\nabla_Y S^r)(X,Z) + \sum_s \{S^s(Y,Z)b_s^r(X) - S^s(X,Z)b_s^r(Y)\}$$

(3) The Ricci Equations

$$0 = S^r{*}S^s(X,Y) - S^r{*}S^s(Y,X) + (\nabla_X b_r^s)(Y) - (\nabla_Y b_r^s)(X)$$

$$+ \sum_w \{b_w^s(X)b_r^w(Y) - b_w^s(Y)b_r^w(X)\}\ .$$

Then for every point of M there is a neighborhood U and isometric imbedding $f: U \to \mathbb{R}^m$ such that there are everywhere orthonormal sections $\nu_{n+1},\ldots,\nu_m$ of the normal bundle of $f(U)$ in $\mathbb{R}^m$ for which the corresponding forms II^r and β_r^s on $f(U)$ satisfy

$$S^r = f^*II^r\ , \qquad b_r^s = f^*\beta_r^s\ .$$

Proof. We will consider the proof of (2) first, since the proof of (1) will come along for free. Since we are trying to prove a local result, we might as well assume that M is $\mathbb{R}^n$. Let $X_1,\ldots,X_n$ be an orthonormal moving frame for $\ll, \gg$ on $\mathbb{R}^n$, with dual forms θ^i, and connection forms ω^i_j. Define 1-forms ψ^α_β $(1 \leq \alpha, \beta \leq m)$ on $\mathbb{R}^n$ as follows:

$$\psi^i_j = \omega^i_j \qquad 1 \leq i, j \leq n$$

$$\psi^r_j(X) = s^r(X, X_j) \qquad 1 \leq j \leq n < r \leq m$$

$$\psi^r_s = b^r_s \qquad n < r, s \leq m .$$

Then the forms ψ^α_β satisfy two crucial equations:

$$(*) \qquad \sum_{j=1}^{n} \psi^r_j \wedge \theta^j = 0 \qquad r = n+1, \ldots, m$$

$$(**) \qquad d\psi^\alpha_\beta = -\sum_{\gamma=1}^{m} \psi^\alpha_\gamma \wedge \psi^\gamma_\beta \qquad \alpha, \beta = 1, \ldots, m .$$

Equation (*) follows directly from the definition of ψ^r_j, and symmetry of s^r. Equation (**) follows from the Gauss, Codazzi-Mainardi, and Ricci equations in the hypothesis. This should be clear from our verifications, previous to the statement of the theorem, that equations (A), (B), and (C) are equivalent to Gauss' equation on p. 51 , and to the Codazzi-Mainardi equations in Theorem 14 and 16, respectively.

Now suppose for the moment that we have an immersion $f\colon (\mathbb{R}^n, \ll\ ,\ \gg) \to \mathbb{R}^m$, and orthonormal sections $\nu_{n+1}, \ldots, \nu_m$ of the normal bundle of $f(\mathbb{R}^n)$. Identifying tangent vectors of $\mathbb{R}^m$ with elements of $\mathbb{R}^m$, as usual, we thus have a map

$$v = (v_1, \ldots, v_m) = (f_*(X_1), \ldots, f_*(X_n), \nu_{n+1}, \ldots, \nu_m)\colon \mathbb{R}^n \to \mathbb{R}^{m^2} .$$

If f is an isometry and $S^r = f^*II^r$ and $b_r^s = f^*\beta_r^s$, then the components v_α^β of the functions v_α will satisfy

$$(1) \qquad dv_\alpha^\beta = \sum_{\gamma=1}^m v_\gamma^\beta \cdot \psi_\alpha^\gamma .$$

So we will first show that a map $v = (v_1, \ldots, v_m)\colon \mathbb{R}^n \to \mathbb{R}^{m^2}$ satisfying (1) can be found. The idea of the proof is to look for the graph $\Gamma \subset \mathbb{R}^n \times \mathbb{R}^{m^2}$ of v (compare p. II.291). Let $\pi_1\colon \mathbb{R}^n \times \mathbb{R}^{m^2} \to \mathbb{R}^n$ and $\pi_2\colon \mathbb{R}^n \times \mathbb{R}^{m^2} \to \mathbb{R}^{m^2}$ be the projections on the first and second factors, and let $\{x_\alpha^\beta\}$ be the standard coordinate system on $\mathbb{R}^{m^2}$. It is easy to see that if $v\colon \mathbb{R}^n \to \mathbb{R}^{m^2}$ satisfying (1) exists, then its graph $\Gamma \subset \mathbb{R}^n \times \mathbb{R}^{m^2}$ is a submanifold on which the m^2 linearly independent 1-forms

$$(2) \qquad d(x_\alpha^\beta \circ \pi_2) - \sum_{\gamma=1}^m (x_\gamma^\beta \circ \pi_2) \cdot \pi_1{}^*\psi_\alpha^\gamma$$

all vanish. Conversely, if all these 1-form vanish on an n-dimensional manifold $\Gamma \subset \mathbb{R}^n \times \mathbb{R}^{m^2}$, then Γ is the graph of the desired function v. So by the Frobenius integrability theorem (I.7-14), we just have to show that the exterior derivative of each form (2) is in the ideal $\mathcal{I}$ generated by these forms. Now

$$d\left(\sum_{\gamma=1}^m (x_\gamma^\beta \circ \pi_2) \cdot \pi_1{}^*\psi_\alpha^\gamma\right) = \sum_{\gamma=1}^m d(x_\alpha^\beta \circ \pi_2) \wedge \pi_1{}^*\psi_\alpha^\gamma$$
$$+ \sum_{\lambda=1}^m (x_\lambda^\beta \circ \pi_2) \cdot \pi_1{}^*d\psi_\alpha^\lambda$$

$$= \sum_{\gamma=1}^{m} d(x_{\alpha}^{\beta}\circ\pi_2) \wedge \pi_1{}^*\psi_{\alpha}^{\gamma}$$

$$- \sum_{\lambda=1}^{m} (x_{\lambda}^{\beta}\circ\pi_2)\cdot\left(\sum_{\gamma=1}^{m} \pi_1{}^*\psi_{\gamma}^{\lambda} \wedge \pi_1{}^*\psi_{\alpha}^{\gamma}\right) \qquad \text{by } (**)$$

$$= \sum_{\gamma=1}^{m} \left(d(x_{\alpha}^{\beta}\circ\pi_2) - \sum_{\lambda=1}^{m} (x_{\lambda}^{\beta}\circ\pi_2)\cdot\pi_1{}^*\psi_{\gamma}^{\lambda}\right) \wedge \pi_1{}^*\psi_{\alpha}^{\gamma} ,$$

which is indeed in the ideal $\mathscr{I}$ generated by the forms (2). Thus we see that there is a function $v\colon \mathbb{R}^n \to \mathbb{R}^{m^2}$ satisfying (1). In fact, we can choose $v(0)$ to be any linearly independent set of vectors in $\mathbb{R}^{m^2}$ (just choose the integral submanifold of $\mathscr{I} = 0$ which passes through this set of vectors); in particular, we can choose $v(0)$ to be a set of orthonormal vectors.

We next note that the functions $v_1,\ldots,v_m$ satisfy

$$d(\langle v_\alpha,v_\beta\rangle) = \langle dv_\alpha,v_\beta\rangle + \langle v_\alpha,dv_\beta\rangle \qquad (\langle\ ,\ \rangle = \text{ordinary inner product in } \mathbb{R}^m)$$

$$= \sum_{\gamma=1}^{m} v_{\beta}^{\gamma}\cdot dv_{\alpha}^{\gamma} + \sum_{\gamma=1}^{m} v_{\alpha}^{\gamma}\cdot dv_{\beta}^{\gamma}$$

$$= \sum_{\gamma,\lambda=1}^{m} v_{\beta}^{\gamma}v_{\lambda}^{\gamma}\psi_{\alpha}^{\lambda} + \sum_{\gamma,\lambda=1}^{m} v_{\alpha}^{\gamma}v_{\lambda}^{\gamma}\psi_{\beta}^{\lambda} \qquad \text{by (1)}$$

$$= \sum_{\lambda=1}^{m} \langle v_\beta,v_\lambda\rangle\psi_{\alpha}^{\lambda} + \sum_{\lambda=1}^{m} \langle v_\alpha,v_\lambda\rangle\psi_{\beta}^{\lambda} .$$

In particular, for any curve c in $\mathbb{R}^m$, the functions

$$f_{\alpha\beta}(t) = \langle v_\alpha(c(t)),v_\beta(c(t))\rangle$$

satisfy the differential equation

$$f_{\alpha\beta}'(t) = \sum_{\lambda=1}^{m} \psi_\alpha^\lambda(c'(t)) f_{\beta\lambda}(t) + \sum_{\lambda=1}^{m} \psi_\beta^\lambda(c'(t)) f_{\alpha\lambda}(t) .$$

Since $\psi_\beta^\alpha = -\psi_\alpha^\beta$, this same equation is satisfied by the functions $f_{\alpha\beta}(t) = \delta_{\alpha\beta}$. So by uniqueness of solutions with a given initial condition, we conclude that $v_1,\ldots,v_m$ are orthonormal everywhere.

Now we want to show that there is actually a function $f\colon \mathbb{R}^n \to \mathbb{R}^m$ such that

$$f_*(X_i) = v_i \qquad i = 1,\ldots,n .$$

For the component functions $f^1,\ldots,f^m$ of f we want

$$df^\alpha(X_i(p)) = v_i^\alpha(p) \quad \text{i.e.,} \quad df^\alpha(p) = \sum_{j=1}^{n} v_j^\alpha(p)\cdot\theta^j(p) .$$

To prove that f exists, we look for its graph $\Gamma \subset \mathbb{R}^n \times \mathbb{R}^m$. We let $\pi_1\colon \mathbb{R}^n\times\mathbb{R}^m \to \mathbb{R}^n$ and $\pi_2\colon \mathbb{R}^n\times\mathbb{R}^m \to \mathbb{R}^m$ be the projections on the first and second factors, and we let $\{x^\alpha\}$ be the standard coordinate system on $\mathbb{R}^m$. The Γ we are seeking is a submanifold of $\mathbb{R}^n\times\mathbb{R}^m$ on which the 1-forms

$$(3) \qquad d(x^\alpha\circ\pi_2) - \sum_{j=1}^{n} (v_j^\alpha\circ\pi_1)\cdot\pi_1{}^*\theta^j$$

all vanish. Now

$$d\left(\sum_{j=1}^{n} (v_j^\alpha\circ\pi_1)\cdot\pi_1{}^*\theta^j\right) = \sum_{j=1}^{n} \pi_1{}^*dv_j^\alpha \wedge \pi_1{}^*\theta^j$$
$$- \sum_{i=1}^{n} (v_i^\alpha\circ\pi_1)\cdot \sum_{j=1}^{n} \pi_1{}^*\omega_j^i \wedge \pi_1{}^*\theta^j$$

$$= \sum_{j=1}^{n} \sum_{\gamma=1}^{m} (v_\gamma^\alpha \circ \pi_1) \cdot \pi_1{}^*\psi_j^\gamma \wedge \pi_1{}^*\theta^j$$

$$- \sum_{i,j=1}^{n} (v_i^\alpha \circ \pi_1) \cdot \pi_1{}^*\omega_j^i \wedge \pi_1{}^*\theta^j \qquad \text{by (1)}$$

$$= \sum_{j=1}^{n} \sum_{r=n+1}^{m} (v_r^\alpha \circ \pi_1) \cdot \pi_1{}^*\psi_j^r \wedge \pi_1{}^*\theta^j$$

$$= 0 \qquad \text{by } (*).$$

So the Frobenius integrability theorem proves that there is $f\colon \mathbb{R}^n \to \mathbb{R}^m$ with $f_*(X_i) = v_i$ $(i = 1,\ldots,n)$. Then f is an isometry, since $X_1,\ldots,X_n$ and $v_1,\ldots,v_n$ are both orthonormal sets. This proves part (2).

As for part (1), we first note that the required Euclidean motion A, if it exists, is unique [for if $p \in M$, then $A_*|M_p$ must be ϕ_{*p}, and $A_*(\nu_r(p))$ must be $\bar{\nu}_r(p)$]. Therefore it suffices to prove existence of A locally. By the usual argument, it suffices to show that $M = \bar{M}$ if for some $p \in M$ we have $p = \phi(p)$ and $M_p = \bar{M}_p$, and ϕ_{*p} = identity and $\nu_r(p) = \bar{\nu}_r$. But this follows immediately from the fact that there is just one function $v\colon \mathbb{R}^n \to \mathbb{R}^{m^2}$ satisfying (1), with a given value of $v(0)$. ■

This classical formulation of Theorem 18 was given in order to emphasize the extra problems which arise when the codimension is greater than 1. But it is a very unsatisfactory way of bringing the normal bundle into the picture. After all, everywhere orthonormal sections $\nu_{m+1},\ldots,\nu_n$ of the normal bundle can usually be found only in a neighborhood of each point. So the first part of Theorem 18 really makes sense only locally, even though it is supposed to be global. We have similar problems in the second part, where we would like to obtain a global result when M is simply connected. These defects are easily rectified, since we also have invariant statements of our fundamental equations. We need one simple bit of terminology. Let M and $\bar{M}$ be C^∞ manifolds, and let $\varpi\colon E \to M$ and $\bar{\varpi}\colon \bar{E} \to \bar{M}$ be C^∞ vector bundles of the same dimension over M and $\bar{M}$, respectively. If $\phi\colon M \to \bar{M}$ is a diffeomorphism, a C^∞ map $\tilde{\phi}\colon E \to \bar{E}$ is called a <u>bundle isomorphism covering</u> ϕ if:

(1) the diagram

$$\begin{array}{ccc} E & \xrightarrow{\tilde{\phi}} & \bar{E} \\ {\scriptstyle\varpi}\downarrow & & \downarrow{\scriptstyle\bar{\varpi}} \\ M & \xrightarrow{\phi} & \bar{M} \end{array}$$

commutes, so that $\tilde{\phi}$ takes $\varpi^{-1}(p)$ to $\bar{\varpi}^{-1}(\phi(p))$,

(2) each map

$$\tilde{\phi}|\varpi^{-1}(p)\colon \varpi^{-1}(p) \to \bar{\varpi}^{-1}(\phi(p))$$

is a vector space isomorphism.

It then follows easily that $\tilde{\phi}$ is a diffeomorphism. Notice that if ξ is a section of E, then we have a section $\tilde{\phi}(\xi)$ of $\bar{E}$ defined by

$$\tilde{\phi}(\xi)(q) = \tilde{\phi}(\xi(\phi^{-1}(q))) \qquad q \in \bar{M}\,.$$

<u>19. THEOREM</u>. (1) Let M be a connected submanifold of $\mathbb{R}^m$ with normal bundle $\varpi\colon \text{Nor}\, M \to M$, and corresponding second fundamental form s and normal connection D. Similarly, let $\bar{M}$ be a connected submanifold with normal bundle $\bar{\varpi}\colon \text{Nor}\, \bar{M} \to \bar{M}$ and corresponding $\bar{s}$ and $\bar{D}$. Let $\phi\colon M \to \bar{M}$ be an

isometry. Suppose that there is a bundle isomorphism $\tilde{\phi}$: Nor $M \longrightarrow$ Nor $\bar{M}$ covering ϕ such that $\tilde{\phi}$ preserves inner products, second fundamental forms, and normal connections:

$$\langle\tilde{\phi}(\xi),\tilde{\phi}(\eta)\rangle = \langle\xi,\eta\rangle \qquad \text{for all } \xi, \eta \in M_p^{\perp}$$

$$\tilde{\phi}(s(X,Y)) = \bar{s}(\phi_*X,\phi_*Y) \qquad \text{for all } X, Y \in M_p$$

$$\tilde{\phi}(D_X\xi) = \bar{D}_{\phi_*X}(\tilde{\phi}(\xi)) \qquad \text{for all } X \in M_p \text{ and all sections } \xi \text{ of Nor } M.$$

Then there is a Euclidean motion A such that $\phi = A|M$ and $\tilde{\phi} = A_*|\text{Nor } M$.

(2) Let $(M, \langle\!\langle\ ,\ \rangle\!\rangle)$ be a simply-connected n-dimensional Riemannian manifold, with covariant differentiation ∇ and curvature tensor R. Let $\varpi: E \longrightarrow M$ be an $(m-n)$-dimensional vector bundle over M with a Riemannian metric $\{\ ,\ \}$, let δ be a connection on E compatible with $\{\ ,\ \}$, with curvature tensor R_δ, and let σ be a symmetric section of the bundle $\text{Hom}(TM \times TM, E)$. Denote by $\tilde{\nabla}$ the connection on $\text{Hom}(TM \times TM, E)$ determined by ∇ and δ; and for $X \in M_p$ and $\xi \in \varpi^{-1}(p)$, let $A_\xi(X) \in M_p$ be the unique vector satisfying

$$\langle\!\langle A_\xi(X),Y\rangle\!\rangle = \{\xi,\sigma(X,Y)\} \qquad \text{for all } Y \in M_p .$$

Suppose that σ and δ satisfy

(1) <u>Gauss' Equation</u>

$$\langle\!\langle R(X,Y)Z,W\rangle\!\rangle = \{\sigma(Y,Z),\sigma(X,W)\} - \{\sigma(X,Z),\sigma(Y,W)\}$$

(2) <u>The Codazzi-Mainardi equations</u>

$$(\tilde{\nabla}_X\sigma)(Y,Z) = (\tilde{\nabla}_Y\sigma)(X,Z)$$

(3) The Ricci equations

$$R_\delta(X,Y)\xi = \sigma(A_\xi(Y),X) - \sigma(A_\xi(X),Y) .$$

Then there is an isometric immersion $f: M \to \mathbb{R}^m$ and a bundle isomorphism $\tilde{f}: E \to \{\text{normal bundle of } f(M)\}$ covering f such that

$$\langle\tilde{f}(\xi),\tilde{f}(\eta)\rangle = \{\xi,\eta\} \qquad \text{for all } \xi, \eta \in \varpi^{-1}(p)$$

$$\tilde{f}(\sigma(X,Y)) = s(f_*X,f_*Y) \qquad \text{for all } X, Y \in M_p$$

$$\tilde{f}(\delta_X\xi) = D_{\phi_*X}(\tilde{f}(\xi)) \qquad \text{for all } X \in M_p \text{ and all sections } \xi \text{ of } E ,$$

s and D being the second fundamental form and normal connection for $f(M)$.

Proof. It's just one big translation job locally. Then simple connectivity is used to prove the global result, as in Problem 2-3. ■

Naturally, Theorem 19 simplifies considerably for the case of hypersurfaces, when Nor M is 1-dimensional. In part (1) we can dispense with the normal connection D; in part (2) we can dispense with δ (and the Ricci equations). When we deal with oriented hypersurfaces, we can ignore Nor M completely, since the orientation of M determines a unit normal field ν, and thus a second fundamental form II. In part (1), we simply need that ϕ is an isometry with $\phi^*\overline{II} = II$; then the Euclidean motion A of the conclusion is actually a proper Euclidean motion with $A_*\nu = \bar{\nu}$. In part (2), we simply supply our Riemannian manifold $(M, \langle\langle\ ,\ \rangle\rangle)$ with a symmetric tensor S, covariant of order 2, satisfying

$$\langle\langle R(X,Y)Z,W\rangle\rangle = S(Y,Z)\cdot S(X,W) - S(X,Z)\cdot S(Y,Z) ,$$

$$(\nabla_X S)(Y,Z) = (\nabla_Y S)(X,Z) .$$

In the general case, Theorem 19 is much less satisfying, for it does not tell us when a map $\phi\colon M \to \bar{M}$ between two submanifolds of $\mathbb{R}^m$ is the restriction of a Euclidean motion; it only gives us information about maps $\phi\colon M \to \bar{M}$ together with bundle isomorphisms $\tilde{\phi}\colon \text{Nor } M \to \text{Nor } \bar{M}$ covering ϕ (as a slight compensation, we find a Euclidean motion A preserving this additional structure). In the theory of curves we have a situation more closely resembling the case of hypersurfaces, since certain functions $\kappa_1,\ldots,\kappa_{m-1}$ determine ("general") curves parameterized by arclength. For curves in $\mathbb{R}^m$, the results of part B are actually a special case of Theorem 19, for the vector fields $\mathbf{v}_2,\ldots,\mathbf{v}_m$ along c give a trivialization of the normal bundle of c; it is not hard to see (Problem 12) that if we choose $\nu_r = \mathbf{v}_r$, then the corresponding II^r and β_r^s are all expressible in terms of $\kappa_1,\ldots,\kappa_{m-1}$. For higher dimensional submanifolds of higher codimension there is also a theory which determines "general" submanifolds, up to Euclidean motions, by means of tensors on the submanifold. Although this theory is certainly more appealing geometrically, it is rather elaborate, and is presented separately in Addendum 4.

For the present, we merely wish to generalize Theorem 19 by replacing $\mathbb{R}^m$ with a more general ambient space. Now the first part of Theorem 19 simply isn't true for an arbitrary ambient space $(N, \langle\ ,\ \rangle)$, which may not have any isometries onto itself except the identity. For example, we can easily construct a metric on $\mathbb{R}^{n+1}$ such that the hypersurfaces $\mathbb{R}^n \times \{0\}$ and $\mathbb{R}^n \times \{1\}$ are isometric and have vanishing second fundamental forms, without there being any non-trivial isometry of $\mathbb{R}^{n+1}$ onto itself, and in particular none taking $\mathbb{R}^n \times \{0\}$ to $\mathbb{R}^n \times \{1\}$. It should be mentioned, however, that one can at least prove the following (Problem 12):

Suppose that M and $\bar{M}$ are connected submanifolds of $(N, \langle\ ,\ \rangle)$ and that there is a point $p \in M \cap \bar{M}$ with $M_p = \bar{M}_p$. Let $\phi: M \to \bar{M}$ be an isometry with $\phi(p) = p$ and $\phi_{*p}: M_p \to \bar{M}_p$ the identity. Suppose that there is a bundle isomorphism $\tilde{\phi}$: Nor $M \to$ Nor $\bar{M}$ covering ϕ which preserves inner products, second fundamental forms, and normal connections, and such that $\tilde{\phi}$ is the identity map on $M_p{}^{\perp}$. Then $M = \bar{M}$ and ϕ is the identity.

We encounter difficulties of another sort when we try to generalize the second part of Theorem 19 for an arbitrary ambient manifold $(N, \langle\ ,\ \rangle)$. Now we don't even know what conditions to place on δ and σ, since the Codazzi-Mainardi and Ricci equations for δ and σ involve terms $R'(X,Y)Z$ which we cannot evaluate unless we already have the imbedding of M into N.

These difficulties do not arise when N is a complete simply-connected manifold of constant curvature K_0. The Euclidean motions of part (1) will be replaced by the isometries $A: N \to N$; such isometries can be found taking any orthonormal frame at one point of N to any orthonormal frame at any other point (Problem 1-5). Moreover, the Codazzi-Mainardi and Ricci equations for a submanifold $M \subset N$ are exactly the same as in the Euclidean case, since

$$R'(X,Y)Z = K_0[\langle Y,Z\rangle X - \langle X,Z\rangle Y] \qquad \text{(p.III.16)}$$

$$\Downarrow$$

$$\perp(R'(X,Y)Z) = 0 \qquad \text{for } X, Y, Z \text{ tangent to } M$$

$$R'(X,Y)\xi = 0 \qquad \text{and } \xi \text{ normal to } M.$$

Gauss' equation, on the other hand, becomes (Corollary 1-12)

$$K_0[\langle X,W\rangle\cdot\langle Y,Z\rangle - \langle X,Z\rangle\cdot\langle Y,W\rangle]$$
$$= \langle R(X,Y)Z,W\rangle + \langle s(X,Z),s(Y,W)\rangle - \langle s(Y,Z),s(X,W)\rangle .$$

For an adapted orthonormal moving frame on M, equations (B) and (C) have $\Psi^r_j = \Psi^s_r = 0$, while equation (A) becomes

$$\text{(A')} \qquad K_0[\theta^i \wedge \theta^j] = \Omega^i_j - \sum_r \psi^r_i \wedge \psi^r_j ,$$

since

$$\Psi^i_j(X,Y) = \langle R'(X,Y)X_j, X_i\rangle = K_0[\langle X,X_i\rangle\cdot\langle Y,X_j\rangle - \langle X,X_j\rangle\cdot\langle Y,X_i\rangle]$$
$$= K_0[\theta^i(X)\theta^j(Y) - \theta^i(Y)\theta^j(X)] .$$

One fairly straightforward method of generalizing Theorem 19 is to regard N as $S^m(K_0) \subset \mathbb{R}^{m+1}$, or as $H^m(K_0) \subset \mathbb{R}^{m+1}$ with the Lorentzian metric. The details of this pleasant proof are left to Problems 13, 14, partly because it uses a result from the next section, but mainly because we want to provide a (rather unpleasant) proof which involves only the intrinsic description of N as a manifold of constant curvature K_0. The proof of Theorem 19 itself will not generalize at all, because it involves the natural identification of $T\mathbb{R}^m$ with $\mathbb{R}^m \times \mathbb{R}^m$, an identification that essentially depends on the fact that $\mathbb{R}^m$ is flat. For a general manifold $(N, \langle\ ,\ \rangle)$ of constant curvature, we will have to consider the tangent bundle TN, and work with Ehresmann connections.

Consider a principal bundle $\pi\colon P \to M$ with group G. For each $X \in \mathfrak{g}$ = Lie algebra of G, we have defined (p. II.352) the <u>fundamental vector field</u> on P corresponding to X; we will change notation slightly and denote this vector field by $\boldsymbol{\sigma}(X)$ [so that σ can still be used as in the statement of Theorem 19]. Recall that an Ehresmann connection on P is a $\mathfrak{g}$-valued 1-form on P. We will use a bold face Greek letter, like $\boldsymbol{\omega}$, for such connections. Here our aim is to avoid confusing $\boldsymbol{\omega}$ with the (closely related) connection forms ω^i_j of a moving frame on a manifold. Recall that a frame u for M_p is an ordered basis $u = (u_1,\ldots,u_n)$ of M_p, and that we have a principal bundle $F(TM) \to M$ with group $GL(n,\mathbb{R})$, where $F(TM)$ is the set of all frames at all $p \in M$. An Ehresmann connection $\boldsymbol{\omega} = (\boldsymbol{\omega}^i_j)$ on $F(TM)$ is a $\mathfrak{gl}(n,\mathbb{R})$-valued 1-form on $F(TM)$. For any moving frame $s = (X_1,\ldots,X_n)\colon U \to F(TM)$ on an open set $U \subset M$ we thus have the matrix of 1-forms $\omega = s^*\boldsymbol{\omega}$, and the assignment of $\omega = s^*\boldsymbol{\omega}$ to $s = (X_1,\ldots,X_n)$ is a (Cartan) connection on M; then by defining $\nabla_X X_j = \sum \omega^i_j(X)X_i$, we obtain a covariant differentiation operator ∇ on M. Conversely, every Cartan connection on M comes from a unique Ehresmann connection $\boldsymbol{\omega}$ in this way, and thus every covariant differentiation operator ∇ comes from a unique $\boldsymbol{\omega}$. More generally, given a k-dimensional vector bundle $\varpi\colon E \to M$, we let $F(E)$ denote the set of all ordered bases u of $\varpi^{-1}(p)$, for all $p \in M$. Then $F(E) \to M$ is a principal bundle with group $GL(k,\mathbb{R})$, and Ehresmann connections $\boldsymbol{\omega}$ on $F(E)$ correspond to covariant differentiation operators $(X,\xi) \mapsto \nabla_X\xi$ on sections ξ of E. In the special case of $F(TM)$ we also have the $\mathbb{R}^n$-valued dual form $\boldsymbol{\theta} = (\boldsymbol{\theta}^1,\ldots,\boldsymbol{\theta}^n)$; for any moving frame $s = (X_1,\ldots,X_n)$, the forms $\theta^i = s^*\boldsymbol{\theta}^i$ are just the dual forms for this moving frame.

When we have a vector bundle $\varpi\colon E \to M$ with a Riemannian metric $\{\ ,\ \}$, it is often more convenient to consider the bundle $O(E) \subset F(E)$ consisting of orthonormal frames. Notice that for $X \in \mathfrak{o}(k) \subset \mathfrak{gl}(k,\mathbb{R})$, the vector field $\sigma(X)$ on $O(E)$ is just the restriction of the vector field $\sigma(X)$ which is defined on $F(E)$. Now suppose we have a covariant differentiation $(X,\xi) \mapsto \nabla_X\xi$ which is compatible with the metric $\{\ ,\ \}$, so that

$$X(\{\xi,\eta\}) = \{\nabla_X\xi,\eta\} + \{\xi,\nabla_X\eta\}\ .$$

If $\xi_1,\ldots,\xi_k$ are local sections of E with $\{\xi_i,\xi_j\} = \delta_{ij}$, and we define the 1-forms ω^i_j $(1 \leq i, j \leq k)$ by

$$\nabla_X \xi_j = \sum_{i=1}^{k} \omega^i_j(X)\cdot\xi_i ,$$

then we have

$$\omega^i_j = -\,\omega^j_i . \tag{1}$$

Let $\boldsymbol{\omega} = (\boldsymbol{\omega}^i_j)$ be the Ehresmann connection on $F(E)$ corresponding to the covariant differentiation ∇. We claim that $\boldsymbol{\omega}|O(E)$ actually takes values in $\mathfrak{o}(k)$ = {skew symmetric $k\times k$ matrices}. In fact, if Y is a vertical vector at some $u \in O(E)$, then $Y = \boldsymbol{\sigma}(X)(u)$ for some $X \in \mathfrak{o}(k)$, and

$$\boldsymbol{\omega}(Y) = \boldsymbol{\omega}(\boldsymbol{\sigma}(X)(u)) = X \in \mathfrak{o}(k) .$$

On the other hand, every non-vertical vector at a frame $u \in O(E)$ at $p \in M$ is of the form $s_*(Z)$ for some local orthonormal section $s = (\xi_1,\ldots,\xi_k)$ and some tangent vector $Z \in M_p$, and then

$$\boldsymbol{\omega}(s_*(Z)) = s^*\boldsymbol{\omega}(Z) = \omega(Z) ,$$

so the claim follows from equation (1). Thus we see that $\boldsymbol{\omega}|O(E)$ is an Ehresmann connection on the principal bundle $O(E)$. [Conversely, an Ehresmann connection on $O(E)$ clearly extends in a natural way to an Ehresmann connection on $F(E)$ whose corresponding ∇ is compatible with the metric $\{\ ,\ \}$.] It is clear that at any point $u \in O(E)$, the horizontal subspace for the connection $\boldsymbol{\omega}|O(E)$ is exactly the same as the horizontal subspace at $u \in F(E)$ for the connection $\boldsymbol{\omega}$. So for $Y_1, Y_2 \in O(E)_u$, the covariant differential $D(\boldsymbol{\omega}|O(E))$ has value

$$\begin{aligned} D(\boldsymbol{\omega}|O(E))(Y_1,Y_2) &= d(\boldsymbol{\omega}|O(E))(hY_1,hY_2) \\ &= d\boldsymbol{\omega}(hY_1,hY_2) \\ &= \boldsymbol{\Omega}(Y_1,Y_2) . \end{aligned} \qquad \left\{ \begin{array}{l} hY_i = \text{horizontal} \\ \text{component of } Y_i \\ \text{in either } F(E)^i \\ \text{or } O(E). \end{array} \right.$$

In other words, the curvature form of $\boldsymbol{\omega}|O(E)$ is just the restriction to $O(E)$ of the curvature form $\boldsymbol{\Omega}$ of $\boldsymbol{\omega}$. So no confusion will arise if we simply use $\boldsymbol{\omega}$ for the restriction $\boldsymbol{\omega}|O(E)$, and $\boldsymbol{\Omega}$ for the curvature form of $\boldsymbol{\omega}|O(E)$.

We will apply these considerations, in particular, to the case where $E = TM$ is the tangent bundle of a Riemannian manifold $(M^n, \langle\!\langle\ ,\ \rangle\!\rangle)$ and $\boldsymbol{\omega}$ the Ehresmann connection corresponding to the Levi-Civita connection for $\langle\!\langle\ ,\ \rangle\!\rangle$. Thus we have $\mathfrak{o}(n)$-valued forms $\boldsymbol{\theta}^i$, $\boldsymbol{\omega}^i_j$, $\boldsymbol{\Omega}^i_j$ on $O(TM)$. For another Riemannian manifold $(N^m, \langle\ ,\ \rangle)$ we have, similarly, $\mathfrak{o}(m)$-valued forms $\boldsymbol{\phi}^\alpha$, $\boldsymbol{\psi}^\alpha_\beta$, $\boldsymbol{\Psi}^\alpha_\beta$ on $O(TN)$.

Now suppose we are given a Riemannian manifold $(M^n, \langle\!\langle\ ,\ \rangle\!\rangle)$, and an $(m-n)$-dimensional vector bundle $\varpi\colon E \to M$ with a Riemannian metric $\{\ ,\ \}$. Let $O(TM,E)$ denote the set of all pairs (u,v) where $u \in O(TM)$ and $v \in O(E)$ lie over the same point $p \in M$. Then $O(TM,E)$ is a principal bundle, whose group G is the set of all $m \times m$ matrices of the form

$$\begin{pmatrix} A & 0 \\ 0 & B \end{pmatrix} \qquad A \in O(n) \text{ and } B \in O(m-n) .$$

These bundles will come into our proof of the generalization of Theorem 19 in the following way. If we succeed in finding an isometric immersion $f\colon M \to N$, covered by an inner product preserving bundle isomorphism

$\tilde{f}$: E $\longrightarrow$ {normal bundle of f(M)}, then we will also have a "principal bundle isomorphism" from O(TM,E) into O(TN)|f(M). Instead of looking for f directly, we will look for this principal bundle isomorphism. Since the graph of this principal bundle isomorphism is a subset of O(TM,E) × O(TN), we will look for the graph as an integral submanifold of a certain distribution on O(TM,E) × O(TN); if (u,v) ε O(TM,E) and w ε O(TN), then the integral manifold through ((u,v),w) will turn out to be the graph of the principal bundle isomorphism determined by $(f_*,\tilde{f})$, where f: M $\longrightarrow$ N is an isometry with $f_*(u_i) = w_i$, and $\tilde{f}(v_r) = w_r$. Thus, although our set-up is now rather complicated, it is very natural to look for maps from O(TM,E) to O(TN), because they involve precisely the right amount of leeway which we expect in the choice of the isometric imbedding f: M $\longrightarrow$ N.

<u>20. THEOREM</u>. The results of Theorem 19 hold when $\mathbb{R}^m$ is replaced by a complete connected Riemannian manifold (N, < , >) of constant curvature K_0, and the following modifications are made:

(1) The map A in part (1) is replaced by an isometry A: N $\longrightarrow$ N.

(2) Gauss' Equation in the hypothesis of part (2) is stated as:

$$K_0[\langle\langle X,W\rangle\rangle\langle\langle Y,Z\rangle\rangle - \langle\langle X,Z\rangle\rangle\langle\langle Y,W\rangle\rangle]$$
$$= \langle\langle R(X,Y)Z,W\rangle\rangle + \{\sigma(X,Z),\sigma(Y,W)\} - \{\sigma(Y,Z),\sigma(X,W)\} .$$

<u>Proof</u>. Again we will begin by considering the second part of the theorem. So we are given (M, ≪ , ≫) and the bundle ϖ: E $\longrightarrow$ M, with metric { , }, a connection δ compatible with { , }, and a symmetric section σ of the bundle Hom(TM × TM,E). Then δ gives us a connection form $\bar{\psi}$ on O(E) which is

$\mathfrak{o}(m-n)$-valued; we will denote its components by $\bar{\psi}^r_s$ for $r, s = n+1,\dots,m$. Similarly, $(\bar{\Psi}^r_s)$ will be the curvature form on $O(E)$.

Now we have obvious maps

$$\begin{array}{ccc} & O(TM,E) & \\ {}^{\lambda_1}\swarrow & & \searrow^{\lambda_2} \\ O(TM) & & O(E) \end{array} .$$

For convenience we will denote

$$\lambda_1^*(\theta^i),\quad \lambda_1^*(\omega^i_j),\quad \lambda_1^*(\Omega^i_j) \qquad \text{simply by} \quad \theta^i,\ \omega^i_j,\ \Omega^i_j$$

$$\lambda_2^*(\bar{\psi}^r_s),\quad \lambda_2^*(\bar{\Psi}^r_s) \qquad \text{simply by} \quad \bar{\psi}^r_s,\ \bar{\Psi}^r_s .$$

We define functions $s^r_{ij}\colon O(TM,E) \to \mathbb{R}$ as follows. An element of $O(TM,E)$ is a pair (u,v), where u and v are orthonormal frames, of TM and E, respectively, at the same point p. Then $\sigma(u_i,u_j)$ can be written uniquely as

$$\sigma(u_i,u_j) = \sum_r s^r_{ij}((u,v))\cdot v_r .$$

We now define forms $\bar{\psi}^r_i$ directly on $O(TM,E)$ by

$$\bar{\psi}^r_i = \sum_j s^r_{ij}\,\theta^j \qquad \Big(= \sum_j s^r_{ij}\lambda_1^*(\theta^j)\Big) .$$

The symmetry of σ implies that $s^r_{ij} = s^r_{ji}$, and thus that

$$(1) \qquad \sum_i \bar{\psi}^r_i \wedge \theta^i = 0 .$$

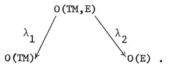

Now we claim that on the bundle $O(TM,E)$ we have

$$(2)\qquad K_0[\boldsymbol{\theta}^i \wedge \boldsymbol{\theta}^j] = \boldsymbol{\Omega}^i_j - \sum_r \overline{\boldsymbol{\psi}}^r_i \wedge \overline{\boldsymbol{\psi}}^r_j$$

$$(3)\qquad d\overline{\boldsymbol{\psi}}^r_j = -\sum_i \overline{\boldsymbol{\psi}}^r_i \wedge \boldsymbol{\omega}^i_j - \sum_s \overline{\boldsymbol{\psi}}^r_s \wedge \overline{\boldsymbol{\psi}}^s_j$$

$$(4)\qquad d\overline{\boldsymbol{\psi}}^r_s = \sum_i \overline{\boldsymbol{\psi}}^r_i \wedge \overline{\boldsymbol{\psi}}^s_i - \sum_w \overline{\boldsymbol{\psi}}^r_w \wedge \overline{\boldsymbol{\psi}}^w_s .$$

The proof is in two steps. First, consider a section $(X_1,\dots,X_n,\nu_{n+1},\dots,\nu_m) = \xi\colon U \longrightarrow O(TM,E)$ on $U \subset M$. Denote $\xi^*(\boldsymbol{\theta}^i)$ by θ^i, etc., and $\xi^*(\overline{\boldsymbol{\psi}}^r_i)$ by $\bar{\psi}^r_i$. Then the θ^i, ω^i_j, and Ω^i_j are the forms for the moving frame $X_1,\dots,X_n$. When we apply ξ^* to equations (2)-(4), we obtain equations on M, of which the first, for example, reads

$$(2')\qquad K_0[\theta^i \wedge \theta^j] = \Omega^i_j - \sum_r \bar{\psi}^r_i \wedge \bar{\psi}^r_j .$$

When we take into account the fact that

$$\bar{\psi}^r_i = \xi^*\overline{\boldsymbol{\psi}}^r_i = \sum_k (\mathbf{s}^r_{ik}\circ\xi)\cdot\xi^*\boldsymbol{\theta}^k = \sum_k \langle\sigma(X_i,X_k),\nu_r\rangle\theta^k ,$$

we find, by a straightforward calculation, that equation (2') is equivalent to Gauss' equation. As in the proof of Theorem 18, we can even avoid the calculation by realizing that it will be essentially the same as the calculation which shows that equation (A') is equivalent to Gauss' equation. In a similar manner, we see that true equations result from applying ξ^* to (3) and (4). This means that equations (2)-(4) hold for tangent vectors which are not vertical.

So we just have to prove that (2)-(4) hold when applied to a pair of vectors of which at least one is vertical.

The 1-forms θ^i on O(TM) are zero on any vertical vector, while the 2-forms Ω^i_j are zero on any pair of vectors of which at least one is vertical. Since the vertical vectors of O(TM,E) are precisely the vectors Y for which $\lambda_{1*}(Y)$ is vertical in O(TM) and $\lambda_{2*}(Y)$ is vertical in O(E), we see that the forms θ^i and Ω^i_j on O(TM,E) have the same property as the forms θ^i and Ω^i_j on O(TM). Analogous statements holds for the forms Ψ^r_s on O(TM,E). Moreover, the forms $\overline{\psi}^r_i$ are clearly 0 on vertical vectors of O(TM,E). It is thus clear that (2) holds when applied to a pair of vectors one of which is vertical. To treat equation (4), we note that the structural equation for O(E) gives

$$d\overline{\psi}^r_s = -\sum_w \overline{\psi}^r_w \wedge \overline{\psi}^w_s + \Psi^r_s .$$

Since Ψ^r_s is zero on a pair of vectors one of which is vertical, while ψ^r_i and ψ^s_i are zero on vertical vectors, this clearly gives the result for equation (4). Thus we are left with equation (3). If <u>both</u> our vectors Y_1, Y_2 are vertical, then the right side of (3) is zero; on the other hand, if $\tilde{Y}_1, \tilde{Y}_2$ are vertical vector fields extending Y_1, Y_2, then the left side is

$$\begin{aligned} d\overline{\psi}^r_j(Y_1,Y_2) &= Y_1(\overline{\psi}^r_j(\tilde{Y}_2)) - Y_2(\overline{\psi}^r_j(\tilde{Y}_1)) - \overline{\psi}^r_j([\tilde{Y}_1,\tilde{Y}_2]) \\ &= 0 , \end{aligned}$$

since $\overline{\psi}^r_j$ is zero on vertical vectors. This leaves us with equation (3) in the case where just one vector is vertical.

Every vertical vector at a frame (u,v) at p is $\sigma(\Gamma)(u,v)$ for some

Γ in the Lie algebra $\mathfrak{g}$ of the group G of $O(TM,E)$. We want to show that (3) holds when applied to

$$Y_1 = \boldsymbol{\sigma}(\Gamma)(u,v) \ , \qquad Y_2 = \xi_*(X_p) \ ,$$

for some $X_p \in M_p$ and some local section $(X_1,\ldots,X_n,\nu_{n+1},\ldots,\nu_n) = \xi\colon U \longrightarrow O(TM,E)$. We extend Y_2 to a vector field $\tilde{Y}_2$ as follows. First extend X_p to a vector field X on M. Then $\xi_*(X)$ is a vector field defined at just one point in each fibre. We extend $\xi_*(X)$ to $\tilde{Y}_2$ by making it invariant under R_{a*} for all $a \in G$. This means, in particular, that for the Lie derivative we have

$$0 = L_{\boldsymbol{\sigma}(\Gamma)}\tilde{Y}_2 = [\boldsymbol{\sigma}(\Gamma),\tilde{Y}_2] \ .$$

Equation (3), applied to Y_1, Y_2 thus becomes

$$(3') \qquad Y_1(\overline{\boldsymbol{\psi}}{}^r_j(\tilde{Y}_2)) = \sum_i \overline{\boldsymbol{\psi}}{}^r_i(Y_2)\,\boldsymbol{\omega}^i_j(Y_1) - \sum_s \overline{\boldsymbol{\psi}}{}^r_s(Y_1)\overline{\boldsymbol{\psi}}{}^s_j(Y_2) \ .$$

To prove equation (3'), we need information about $R_a{}^*\overline{\boldsymbol{\psi}}{}^r_j$, for $a \in G$. We write a as

$$a = \begin{pmatrix} A & 0 \\ 0 & B \end{pmatrix} , \qquad A \in O(n), \quad B \in O(m-n) \ .$$

Let us note first that the definition of $\mathbf{s}^r_{ij}$ gives

$$\sigma((u\cdot A)_i,(u\cdot A)_j) = \sum_r \mathbf{s}^r_{ij}((u,v)\cdot a)\cdot(v\cdot B)_r \ .$$

From this we easily find that

$$s^r_{ij}((u,v)\cdot a) = \sum_{k,\ell=1}^{n} \sum_{\rho=n+1}^{m} A^k_i A^\ell_j \, s^\rho_{k\ell}(u,v)(B^{-1})^r_\rho \,.$$

We can now compute

$$R_a{}^*\overline{\boldsymbol{\psi}}^r_i = \sum_j (s^r_{ij} \circ R_a)\cdot R_a{}^*\boldsymbol{\theta}^i \,,$$

using the equation (Proposition II.8-12)

$$R_A{}^*\boldsymbol{\theta}^i = \sum_k (A^{-1})^i_k \boldsymbol{\theta}^k \,;$$

we find that the $(m-n)\times n$ matrix $[\overline{\boldsymbol{\psi}}] = (\boldsymbol{\psi}^r_i)$ satisfies

(*) $$R_a{}^*[\overline{\boldsymbol{\psi}}] = B^{-1}[\overline{\boldsymbol{\psi}}]A \,.$$

Now suppose we write $\Gamma \in \mathfrak{g}$ as

$$\Gamma = \begin{pmatrix} \Gamma_1 & 0 \\ 0 & \Gamma_2 \end{pmatrix} \qquad \text{for } \Gamma_1 \in \mathfrak{o}(n), \quad \Gamma_2 \in \mathfrak{o}(m-n) \,.$$

An integral curve of $Y_1 = \boldsymbol{\sigma}(\Gamma)(u,v)$ is given by

$$t \longmapsto (u,v)\cdot \exp t\Gamma = R_{\exp t\Gamma}(u,v) \,.$$

So for the left side of (3') we find that

$$Y_1(\overline{\boldsymbol{\psi}}^r_j(\tilde{Y}_2)) = \lim_{h\to 0} \frac{1}{h}\Big[\overline{\boldsymbol{\psi}}^r_j(\tilde{Y}_2(R_{\exp h\Gamma}(u,v))) - \overline{\boldsymbol{\psi}}^r_j(Y_2)\Big]$$

$$= \lim_{h\to 0} \frac{1}{h}\Big[\overline{\boldsymbol{\psi}}^r_j(R_{\exp h\Gamma *}Y_2) - \overline{\boldsymbol{\psi}}^r_j(Y_2)\Big]$$

(from the way $\tilde{Y}_2$ was defined)

$$= \lim_{h\to 0} \frac{1}{h}\left[\left((R_{\exp\, h\Gamma})^* \overline{\boldsymbol{\psi}}{}^r_j\right)(Y_2) \;-\; \overline{\boldsymbol{\psi}}{}^r_j(Y_2)\right]$$

$$= \binom{r}{j} \text{ component of } \lim_{h\to 0} \frac{1}{h}\left[(\{\exp h\Gamma_2\}^{-1}\cdot[\overline{\boldsymbol{\psi}}]\cdot \exp h\Gamma_1)(Y_2) \;-\; [\overline{\boldsymbol{\psi}}](Y_2)\right]$$

by equation (*)

$$= \binom{r}{j} \text{ component of } \{-\,\Gamma_2\cdot[\overline{\boldsymbol{\psi}}(Y_2)] \;+\; [\overline{\boldsymbol{\psi}}(Y_2)]\Gamma_1\}$$

$$= -\sum_s (\Gamma_2)^r_s \overline{\boldsymbol{\psi}}{}^s_j(Y_2) \;+\; \sum_i \overline{\boldsymbol{\psi}}{}^r_i(Y_2)\cdot(\Gamma_1)^i_j$$

$$= -\sum_s \overline{\boldsymbol{\psi}}{}^r_s(\boldsymbol{\sigma}(\Gamma))\overline{\boldsymbol{\psi}}{}^s_j(Y_2) \;+\; \sum_i \overline{\boldsymbol{\psi}}{}^r_i(Y_2)\,\boldsymbol{\omega}^i_j(\boldsymbol{\sigma}(\Gamma_1))$$

$$= -\sum_s \overline{\boldsymbol{\psi}}{}^r_s(Y_1)\overline{\boldsymbol{\psi}}{}^s_j(Y_2) \;+\; \sum_i \overline{\boldsymbol{\psi}}{}^r_i(Y_2)\,\boldsymbol{\omega}^i_j(Y_1)\;,$$

which is precisely the right side of (3').

After all this work, we are ready to construct a distribution on $O(TM,E) \times O(N)$. We introduce the two projections

$$\begin{array}{ccc} O(TM,E) \times O(TN) & \xrightarrow{\pi_2} & O(TN) \\ \Big\downarrow \pi_1 & & \\ O(TM,E) & & \end{array} \quad .$$

Define $\Delta_{((u,v),w)}$ to be the set of all tangent vectors at $((u,v),w)$ on which the following 1-forms all vanish:

(a) $\pi_2{}^*\boldsymbol{\phi}^i \;-\; \pi_1{}^*\boldsymbol{\theta}^i$

(b) $\pi_2{}^*\boldsymbol{\phi}^r$

(c) $\pi_2{}^*\boldsymbol{\psi}^i_j \;-\; \pi_1{}^*\boldsymbol{\omega}^i_j$

(d) $\pi_2{}^*\psi^r_s - \pi_1{}^*\overline{\psi}^r_s$

(e) $\pi_2{}^*\psi^r_j - \pi_1{}^*\overline{\psi}^r_j$.

Since the forms ϕ^α, ψ^α_β $(\alpha < \beta)$ are a basis for the dual space of the tangent space $O(TN)_w$, it is clear that each $\Delta_{((u,v),w)}$ has dimension

$$\dim O(TM,E) \times O(TN) - \dim O(TN) = \dim O(TM,E) ,$$

and that $\pi_{1*}\colon \Delta_{((u,v),w)} \longrightarrow O(TM,E)_{(u,v)}$ is an isomorphism. We thus obtain a distribution Δ.

For any $a \in G$, we have the map $R_a\colon O(TM,E) \longrightarrow O(TM,E)$. Since $a \in G \subset O(m)$ = group of the bundle $O(TN)$, we also have maps, which we will denote by the same letter, $R_a\colon O(TN) \longrightarrow O(TN)$. These maps then give us maps

$$R_a\colon O(TM,E) \times O(TN) \longrightarrow O(TM,E) \times O(TN) .$$

We claim that $R_{a*}\Delta = \Delta$. To prove this, it suffices to show that $R_a{}^*\eta$ is a linear combination of the forms (a)-(e) whenever η is any of the forms in (a)-(e). If

$$a = \begin{pmatrix} A & 0 \\ 0 & B \end{pmatrix} ,$$

then Proposition II.8-12 shows that the $\mathbb{R}^m$-valued form $\pi_2{}^*\phi$ satisfies

$$R_a{}^*\pi_2{}^*\phi = \pi_2{}^*R_a{}^*\phi = \pi_2{}^*a^{-1}\phi = \pi_2{}^*\begin{pmatrix} A^{-1} & 0 \\ 0 & B^{-1} \end{pmatrix}\phi ,$$

while

$$R_a{}^*\pi_1{}^*\boldsymbol{\theta} = \pi_1{}^*R_a{}^*\boldsymbol{\theta} = \pi_1{}^*A^{-1}\boldsymbol{\theta} .$$

From this we see that $R_a{}^*\eta$ has the required property when η is one of the forms in (a) or (b). We also have, by Proposition II.8-11,

$$R_a{}^*\pi_2{}^*\boldsymbol{\psi} = \pi_2{}^*R_a{}^*\boldsymbol{\psi} = \pi_2{}^*a^{-1}\boldsymbol{\psi}a = \pi_2{}^*\begin{pmatrix} A^{-1} & 0 \\ 0 & B^{-1} \end{pmatrix}\boldsymbol{\psi}\begin{pmatrix} A & 0 \\ 0 & B \end{pmatrix} ;$$

for the same reason we have

$$R_a{}^*\pi_1{}^*\boldsymbol{\omega} = \pi_1{}^*A^{-1}\boldsymbol{\omega}A ,$$

$$R_a{}^*\pi_1{}^*\overline{\boldsymbol{\psi}} = \pi_1{}^*B^{-1}\overline{\boldsymbol{\psi}}B , \qquad \overline{\boldsymbol{\psi}} = (\overline{\boldsymbol{\psi}}{}^r_s) ,$$

while (*) gives

$$R_a{}^*\pi_1{}^*[\overline{\boldsymbol{\psi}}] = \pi_1{}^*B^{-1}[\overline{\boldsymbol{\psi}}]A \qquad [\overline{\boldsymbol{\psi}}] = (\overline{\boldsymbol{\psi}}{}^r_i) .$$

From these equations we see that $R_a{}^*\eta$ has the required property when η is one of the forms in (c)-(e).

Now we claim that our distribution Δ is <u>integrable</u>. According to Proposition I.7-14, we just have to show that the differential of all the forms in (a)-(e) are in the ideal $\mathscr{I}$ generated by these forms. Now we have, for example,

$$d(\pi_2{}^*\phi^i - \pi_1{}^*\theta^i) = \pi_2{}^*(-\sum_j \psi^i_j \wedge \phi^j) - \pi_1{}^*(-\sum_j \omega^i_j \wedge \theta^j)$$

$$+ \pi_2{}^*(-\sum_r \psi^i_r \wedge \phi^r) .$$

The last term is in $\mathscr{I}$, since the forms (b) are. The first 2 terms can be written

$$-\sum_j \pi_2{}^*\psi^i_j \wedge (\pi_2{}^*\phi^j - \pi_1{}^*\theta^j) - \sum_j(\pi_2{}^*\psi^i_j - \pi_1{}^*\omega^i_j) \wedge \pi_1{}^*\theta^j ,$$

which is in $\mathscr{I}$. For the exterior differential

$$d(\pi_2{}^*\phi^r) = \pi_2{}^*(-\sum_i \psi^r_i \wedge \phi^i) + \pi_2{}^*(-\sum_s \psi^r_s \wedge \phi^s) ,$$

we note that the second term is in $\mathscr{I}$, while the first can be written

$$-\sum_i(\pi_2{}^*\psi^r_i - \pi_1{}^*\overline{\psi}^r_i) \wedge \pi_2{}^*\phi^i - \sum_i \pi_1{}^*\overline{\psi}^r_i \wedge (\pi_2{}^*\phi^i - \pi_1{}^*\theta^i)$$

$$- \sum_i \pi_1{}^*(\overline{\psi}^r_i \wedge \theta^i) ;$$

the first two terms are in $\mathscr{I}$, while the third is zero by equation (1). We will briefly outline the check for (c), and leave the others for the reader. We have

$$d(\pi_2{}^*\psi^i_j - \pi_1{}^*\omega^i_j) = \pi_2{}^*(-\sum_k \psi^i_k \wedge \psi^k_j) - \pi_1{}^*(-\sum_k \omega^i_k \wedge \omega^k_j)$$

$$+ \pi_2{}^*(-\sum_r \psi^i_r \wedge \psi^r_j) + \pi_2{}^*(\Psi^i_j) - \pi_1{}^*(\Omega^i_j).$$

Since N has constant curvature K_0, we have $\pi_2{}^*(\Psi^i_j) = K_0\pi_2{}^*(\phi^i \wedge \phi^j)$, while equation (2) tells us how to get rid of $\pi_1{}^*(\Omega^i_j)$. We obtain, in particular, the term

$$K_0[\pi_2{}^*\phi^i \wedge \pi_2{}^*\phi^j - \pi_1{}^*\theta^i \wedge \pi_1{}^*\theta^j]$$

$$= K_0[(\pi_2{}^*\phi^i - \pi_1{}^*\theta^i) \wedge \pi_2{}^*\phi^j$$

$$+ \pi_1{}^*\theta^i \wedge (\pi_2{}^*\phi^j - \pi_1{}^*\theta^j)] ,$$

which is in $\mathcal{A}$. The other terms are easily paired off and treated as above.

Now consider an integral manifold Γ of the distribution Δ. Since $\pi_{1*}\colon \Delta_{((u,v),w)} \to O(TM,E)_{(u,v)}$ is always an isomorphism, the map $\pi_1\colon \Gamma \to O(TM,E)$ is a diffeomorphism in a neighborhood of any point. Replacing M by a sufficiently small open subset of M if necessary, we may assume that $\pi_1\colon \Gamma \to O(TM,E)$ is a diffeomorphism. Then Γ is the graph of a function $g\colon O(TM,E) \to O(TN)$, given explicitly by

$$g = \pi_2 \circ (\pi_1|\Gamma)^{-1} .$$

Because $R_{a*}\Delta = \Delta$ for all $a \in G$, it is easy to see that g takes fibres of $O(TM,E)$ to fibres of $O(TN)$, so that there is a diffeomorphism $f\colon M \to N$ for which the following diagram commutes.

$$\begin{array}{ccc} O(TM,E) & \xrightarrow{\;g\;} & O(TN) \\ \pi \downarrow & & \downarrow \pi_N \\ M & \xrightarrow{\;f\;} & N \end{array}$$

Now suppose we have a tangent vector $X_p \in M_p$, a frame $(u,v) \in \pi^{-1}(p)$, and a tangent vector $Y \in O(TM,E)_{(u,v)}$ with $\pi_* Y = X_p$. Then, by definition of θ^i, we have

$$\theta^i(Y) = i^{\underline{th}} \text{ component of } X_p \text{ with respect to the frame } u .$$

Now

$$f_* X_p = f_* \pi_* Y = \pi_{N*} g_* Y ,$$

so we likewise have

$$\begin{aligned} i^{\underline{th}} &\text{ component of } f_* X_p \text{ with respect to the frame } g((u,v)) \\ &= \phi^i(g_* Y) \\ &= \phi^i(\pi_{2*}(\pi_1 | \Gamma)^{-1}{}_*(Y)) \\ &= \pi_2{}^* \phi^i((\pi_1 | \Gamma)^{-1}{}_*(Y)) \\ &= \pi_1{}^* \theta^i((\pi_1 | \Gamma)^{-1}{}_*(Y)) \qquad \text{since the forms (a) are zero on } \Gamma \\ &= \theta^i(Y) . \end{aligned}$$

Similarly, since the forms (b) vanish on Γ, we find that

$$r^{\underline{th}} \text{ component of } f_* X_p \text{ with respect to the frame } g((u,v)) = 0 .$$

This shows us that g is of the form

$$g((u,v)) = (f_*(u), \tilde{f}(v)) ,$$

for some bundle isomorphism $\tilde{f}\colon E \longrightarrow \{\text{normal bundle of } f(M) \text{ in } N\}$ covering f. The map f is an isometry, since f_* takes orthonormal frames to orthonormal frames, and the map $\tilde{f}$ is inner product preserving, for the same reason.

The proof that $\tilde{f}$ makes σ correspond to s and δ correspond to D is similar to the above arguments, using the fact that the forms (c)-(e) vanish on Γ.

We have thus proved the existence part of the theorem locally. Simple-connectivity is then used to prove the global result, in the standard way. The uniqueness part of the theorem is handled just like the uniqueness part of Theorem 18. ■

For ease of reference, we want to have an explicit statement of Theorem 20 in the case of hypersurfaces. We will assume that the ambient space N is orientable. In this case we claim that a diffeomorphism $\phi\colon M \to \bar{M}$ between immersed hypersurfaces is always covered by an inner product preserving bundle isomorphism $\tilde{\phi}\colon \text{Nor } M \to \text{Nor } \bar{M}$. To construct $\tilde{\phi}$ we first choose a particular orientation for N. Then for $p \in M$ we choose an ordered basis $X_1,\ldots,X_n \in M_p$ and a unit normal $\nu_p \in M_p{}^{\perp}$ such that $(X_1,\ldots,X_n,\nu_p)$ is positively oriented in N_p. Then there is a unique unit normal $\bar{\nu}_{\phi(p)} = M_{\phi(p)}{}^{\perp}$ such that $(\phi_*X_1,\ldots,\phi_*X_n,\bar{\nu}_{\phi(p)})$ is positively oriented in $N_{\phi(p)}$. We let $\tilde{\phi}\colon M_p{}^{\perp} \to \bar{M}_{\phi(p)}{}^{\perp}$ be the linear map taking ν_p to $\bar{\nu}_{\phi(p)}$; it is clear that this map is well-defined. If $-1\colon \text{Nor } M \to \text{Nor } M$ is the bundle equivalence taking $X \in M_p{}^{\perp}$ to $-X \in M_p{}^{\perp}$, then $\tilde{\tilde{\phi}} = \tilde{\phi}\circ -1\colon \text{Nor } M \to \text{Nor } \bar{M}$ is another inner product preserving bundle isomorphism covering ϕ, and these are clearly the only such. When N is oriented, the hypersurfaces $M, \bar{M}$ are also oriented, and $\phi\colon M \to \bar{M}$ is orientation preserving, things are even simpler, for there are unit normal fields ν and $\bar{\nu}$ on M and $\bar{M}$, determined by their orientations (and the orientation of N), and the obvious $\tilde{\phi}$ to consider is the one taking ν to $\bar{\nu}$.

21. THEOREM. Let $(N^{n+1}, \langle\ ,\ \rangle)$ be an orientable complete connected Riemannian manifold of constant curvature K_0.

(1) Let M and $\bar{M}$ be connected hypersurfaces of N, let $\phi\colon M \to \bar{M}$ be an isometry, and let $\tilde{\phi}\colon \text{Nor } M \to \text{Nor } \bar{M}$ be one of the two inner product preserving bundle isomorphisms covering ϕ. Suppose that either

$$\bar{s}(\phi_* X, \phi_* Y) = \tilde{\phi}(s(X,Y))$$

for all tangent vectors X, Y at all points of M, or

$$\bar{s}(\phi_* X, \phi_* Y) = -\tilde{\phi}(s(X,Y))$$

for all X, Y. Then ϕ is the restriction of an isometry $A\colon N \to N$ with $A_* = \tilde{\phi}$ or $-\tilde{\phi}$ on $\text{Nor } M$.

(1') Choose an orientation for N, and let M and $\bar{M}$ be connected oriented hypersurfaces of N, with unit normal fields ν and $\bar{\nu}$ determined by their orientations (and the orientation of N), with corresponding second fundamental forms II and $\overline{II}$. Suppose that $\phi\colon M \to \bar{M}$ is an orientation preserving isometry with $\phi^*\overline{II} = II$. Then ϕ is the restriction of an orientation preserving isometry $A\colon N \to N$ with $A_*\nu = \bar{\nu}$.

(2) Let $(M, \langle\langle\ ,\ \rangle\rangle)$ be a simply-connected n-dimensional Riemannian manifold, with covariant differentiation ∇ and curvature tensor R, and let S be a symmetric tensor on M, covariant of order 2. Suppose that S satisfies

(1) Gauss' Equation

$$K_0[\langle\langle X,W\rangle\rangle\cdot\langle\langle Y,Z\rangle\rangle - \langle\langle X,Z\rangle\rangle\cdot\langle\langle Y,W\rangle\rangle]$$

$$= \langle\langle R(X,Y)Z,W\rangle\rangle + S(X,Z)\cdot S(Y,W) - S(Y,Z)\cdot S(X,W)$$

(2) The Codazzi-Mainardi Equations

$$(\nabla_X S)(Y,Z) = (\nabla_Y S)(X,Z) .$$

Then there is an isometric immersion $f: M \longrightarrow N$ such that $S = f^*II$, where II is the second fundamental form on $f(M)$ for some unit normal field ν.

One concluding remark is in order. In Chapter 2 we showed that the Gauss and Codazzi-Mainardi equations for a surface in $\mathbb{R}^3$ are equivalent to the equations of structure of $O(3)$, and that the Fundamental Theorem of Surface Theory reduces to Theorems I.10-17, 18 about Lie groups. It is not hard (Problem 15) to show, similarly, that the Gauss, Codazzi-Mainardi, and Ricci equations for a submanifold of $\mathbb{R}^m$ are equivalent to the equations of structure of $O(m)$, and that Theorem 19 reduces to theorems about Lie groups. The Lie group $O(m)$ makes its appearance here because of the fact that the group of Euclidean motions of $\mathbb{R}^m$ is a semi-direct product $\mathbb{R}^m \times O(m)$. For a general Riemannian manifold $(N, \langle\ ,\ \rangle)$ of constant curvature K_0, the group of isometries cannot be factored in this way. This is why our proof of Theorem 20 involved the bundle $O(N)$ of orthonormal frames of N. As a matter of fact, the bundle $O(N)$ is the group of isometries of N (as a set), since an isometry is determined by knowing which orthonormal frame $u \in O(N)$ is the image of some fixed orthonormal frame u_0. Thus we ought to be able to interpret the Gauss, Codazzi-Mainardi, and Ricci equations for a submanifold of N as the equations of structure of $O(N)$, equipped with the appropriate group structure, and Theorem 20 should reduce to Theorems I.10-17 and I.10-18. However, we forebear to enter any further into such considerations.

D. First Consequences

We begin by considering hypersurfaces $M^n \subset \mathbb{R}^{n+1}$. In a neighborhood of any point of M there is a unit normal field $\nu: M^n \to S^n \subset \mathbb{R}^{n+1}$, unique up to sign, and hence a single second fundamental form $II: M_p \times M_p \to \mathbb{R}$ (which is defined only up to sign). We also have the map $d\nu: M_p \to M_p$ with

$$\begin{aligned} II(X_p,Y_p) &= \langle s(X_p,Y_p),\nu(p)\rangle \\ &= \langle \nabla'_{X_p} Y,\nu(p)\rangle \\ &= -\langle \nabla'_{X_p} \nu,Y_p\rangle \\ &= \langle -d\nu(X_p),Y_p\rangle . \end{aligned}$$

Thus $-d\nu: M_p \to M_p$ is a symmetric linear transformation. As in Chapter 2, we define the <u>principal directions</u> at p to be the unit eigenvectors $X_p \in M_p$ for $-d\nu: M_p \to M_p$, and we define the <u>principal curvatures</u> to be the corresponding eigenvalues. Equivalently, the principal curvatures are the eigenvalues of the symmetric matrix $(II(X_i,X_j)) = (\langle s(X_i,X_j),\nu(p)\rangle)$ for $X_1,\ldots,X_n$ an orthonormal basis of M_p.

Various kinds of curvatures can be defined in terms of the k_i. Since the ordering of the k_i is arbitrary, we obviously want to consider only combinations of the k_i which are invariant under all permutations of the indices $1,\ldots,n$. It is well-known that any polynomial function of n variables $t_1,\ldots,t_n$ which is invariant under all permutations of $1,\ldots,n$ can be written as a polynomial in the "elementary symmetric functions" $\sigma_1,\ldots,\sigma_n$ defined by

$$\sigma_1(t_1,\ldots,t_n) = \sum_{i=1}^{n} t_i \,, \qquad \sigma_2(t_1,\ldots,t_n) = \sum_{i<j} t_i t_j \,,$$

$$\sigma_3(t_1,\ldots,t_n) = \sum_{i<j<k} t_i t_j t_k \qquad \cdots$$

$$\sigma_n(t_1,\ldots,t_n) = t_1 t_2 \cdots t_n \,.$$

These functions are the coefficients, up to sign, of the various powers of x in the polynomial

$$\begin{aligned} P_{t_1,\ldots,t_n}(x) &= (x - t_1)(x - t_2)\cdots(x - t_n) \\ &= x^n - \sigma_1(t_1,\ldots,t_n)x^{n-1} + \cdots + (-1)^n\sigma_n(t_1,\ldots,t_n) \,. \end{aligned}$$

[Recall also that if $\sigma_i(t_1,\ldots,t_n) = \sigma_i(u_1,\ldots,u_n)$ for all i, then the polynomials $P_{t_1,\ldots,t_n}(x)$ and $P_{u_1,\ldots,u_n}(x)$ are equal, and thus the set of their roots, $\{t_1,\ldots,t_n\}$ and $\{u_1,\ldots,u_n\}$, are also equal, counting multiplicities.] We define the (elementary symmetric) curvatures $K_1(p),\ldots,K_n(p)$ by

$$\binom{n}{j}K_j(p) = \sigma_j(k_1,\ldots,k_n) \,,$$

where the k_i are the principal curvatures at p; the binomial coefficient $\binom{n}{j}$ is inserted for sentimental reasons. In particular,

$$H(p) = K_1(p) = \frac{k_1 + \cdots + k_n}{n} \qquad \text{is called the mean curvature} \,,$$

$$K(p) = K_n(p) = k_1 \cdots k_n \qquad \text{is called the Gaussian curvature} \,.$$

Notice that $K_j(p)$ is independent of the choice of ν for j even, while

$K_j(p)$ is only defined up to sign for j odd.

In the case of surfaces, we found that the Gaussian curvature $K = k_1 \cdot k_2$ is an invariant under isometry. In general, we have

22. PROPOSITION. For hypersurfaces of $\mathbb{R}^{n+1}$, the set of $\binom{n}{2}$ numbers $\{k_i k_j : i < j\}$ is invariant under isometry: If $f: M \to \bar{M}$ is an isometry between two hypersurfaces $M, \bar{M} \subset \mathbb{R}^{n+1}$, and $k_1, \ldots, k_n$ are the principal curvatures of M at p, while $\bar{k}_1, \ldots, \bar{k}_n$ are the principal curvatures of $\bar{M}$ at $f(p)$, then the sets $\{k_i k_j : i < j\}$ and $\{\bar{k}_i \bar{k}_j : i < j\}$ are equal, counting multiplicities.

Proof. For $X, Y \in M_p$, let $\tilde{R}(X,Y)$ denote the map $M_p \times M_p \to \mathbb{R}$ defined by

$$\tilde{R}(X,Y)(Z,W) = \langle R(X,Y)Z, W\rangle \ .$$

The symmetry properties of R show that the map $\tilde{R}(X,Y)$ is skew-symmetric, so that $\tilde{R}(X,Y) \in \Omega^2(M_p)$. Now the inner product $\langle \ , \ \rangle_p$ on M_p gives us a map $X \mapsto X^*$ from M_p to $M_p{}^*$, defined by $X^*(Y) = \langle X,Y\rangle$. Choose a basis $X_1, \ldots, X_n$ for M_p and consider the map from $\Omega^2(M_p)$ to $\Omega^2(M_p)$ given by

$$X_i{}^* \wedge X_j{}^* \mapsto \tilde{R}(X_i, X_j) \ ;$$

this makes sense since the $X_i{}^* \wedge X_j{}^*$ for $i < j$ are a basis for $\Omega^2(M_p)$ and since $\tilde{R}(X_j, X_i) = -\tilde{R}(X_i, X_j)$. We see immediately that under this map

$$(\Sigma\, a_i X_i)^* \wedge (\Sigma\, b_j X_j)^* \mapsto \tilde{R}(\Sigma\, a_i X_i, \Sigma\, b_j X_j) \ ,$$

so we can describe our map, without any choice of basis, as

$$(1) \qquad X^* \wedge Y^* \longmapsto \tilde{R}(X,Y) \qquad \text{from } \Omega^2(M_p) \longrightarrow \Omega^2(M_p) .$$

Now the vector space $\Omega^2(M_p)$ has dimension $\binom{n}{2}$, so this map has $\binom{n}{2}$ eigenvalues (counting multiplicities). But if $X_1,\dots,X_n$ are principal vectors at p, with corresponding eigenvalues $k_1,\dots,k_n$, then Gauss' equation tells us that

$$\tilde{R}(X_i,X_j) = -\, k_i k_j\, X_i{}^* \wedge X_j{}^* .$$

So the set $\{-\, k_i k_j : i < j\}$ is the set of eigenvalues of the map (1). Since (1) is defined in terms of the curvature tensor R and the metric $\langle\ ,\ \rangle$, this proves that $\{-\, k_i k_j : i < j\}$ is invariant under isometry. ■

23. COROLLARY (THEOREMA EGREGIUM). For hypersurfaces in $\mathbb{R}^{n+1}$, the Gaussian curvature K is invariant under isometry if n is even, and invariant up to sign if n is odd.

Proof. Observe that

$$K^{n-1} = \Big(\prod_{i=1}^{n} k_i\Big)^{n-1} = \prod_{i<j} k_i k_j . \quad ■$$

There is another way of reaching this result, which will provide us with an explicit formula for K in terms of R and $\langle\ ,\ \rangle$, a formula which will be extremely important in Chapter 13. First we will do a little linear algebra. Let V be a vector space of even dimension n, and let $f: V \longrightarrow V$ be a linear transformation having matrix $A = (a_{ij})$ with respect to a basis $v_1,\dots,v_n$.

We propose to find $\det f = \det A$ _in terms of the determinants of all_ 2×2 _submatrices of_ A. We will let

$$D(i_1,i_2;j_1,j_2) = a_{i_1j_1}\, a_{i_2j_2} - a_{i_1j_2}\, a_{i_2j_1} ,$$

so that if $i_1 < i_2$ and $j_1 < j_2$, then $D(i_1,i_2;j_1,j_2)$ is the determinant of the 2×2 submatrix of A obtained by selecting rows i_1 and i_2, and columns j_1 and j_2. Recall that $\det f$ can be defined as follows. The linear transformation f gives us a map $f^*\colon \Omega^k(V) \longrightarrow \Omega^k(V)$ defined by

$$f^*(T)(v_1,\dots,v_k) = T(f(v_1),\dots,f(v_k)) , \qquad \text{all } T \in \Omega^k(V) .$$

In particular, we have the map $f^*\colon \Omega^n(V) \longrightarrow \Omega^n(V)$. Since $\Omega^n(V)$ is 1-dimensional, this map must be multiplication by a constant; and this constant is, in fact, just $\det f$. Now our map f^* also satisfies

$$f^*(\phi_1 \wedge \cdots \wedge \phi_k) = f^*(\phi_1) \wedge \cdots \wedge f^*(\phi_k) \qquad \text{all } \phi_i \in \Omega^1(V) .$$

In particular, let the ϕ_i be the dual basis to the v_i. Then

$$f(v_i) = \sum_{j=1}^n a_{ji}v_j \implies f^*(\phi_i) = \sum_{j=1}^n a_{ij}\phi_j .$$

So

$$\begin{aligned} f^*(\phi_1 \wedge \cdots \wedge \phi_n) &= [f^*(\phi_1) \wedge f^*(\phi_2)] \wedge \cdots \\ &= (\sum_{j=1}^n a_{1j}\phi_j \wedge \sum_{k=1}^n a_{2k}\phi_k) \wedge \cdots \end{aligned}$$

$$= \Big(\sum_{j<k} [a_{1j}a_{2k} - a_{1k}a_{2j}]\phi_j \wedge \phi_k\Big) \wedge \cdots$$

$$= \left(\frac{1}{2} \sum_{j,k} D(1,2;j,k)\phi_j \wedge \phi_k\right) \wedge \cdots .$$

From this we see that

$$\det f = \frac{1}{2^{n/2}} \sum_{j_1,\ldots,j_n} D(1,2;j_1,j_2)\cdots D(n-1,n;j_{n-1},j_n)\varepsilon^{j_1\cdots j_n} ,$$

where

$$\varepsilon^{j_1\cdots j_n} = \begin{cases} 1 & j_1,\ldots,j_n \text{ is an even permutation of } 1,\ldots,n \\ -1 & j_1,\ldots,j_n \text{ is an odd permutation of } 1,\ldots,n \\ 0 & j_1,\ldots,j_n \text{ are not all distinct.} \end{cases}$$

We can clearly also write

$$\det f = \frac{1}{2^{n/2}n!} \sum_{\substack{i_1,\ldots,i_n \\ j_1,\ldots,j_n}} D(i_1,i_2;j_1,j_2)\cdots D(i_{n-1},i_n;j_{n-1},j_n)\varepsilon^{i_1\cdots i_n}\varepsilon^{j_1\cdots j_n} .$$

Now we apply this formula to evaluate

$$K(p) = \det\ -d\nu\colon M_p \longrightarrow M_p$$

in terms of a basis $X_1,\ldots,X_n$ of M_p. Using Fact 0 from Chapter 2, we have

$$K = \frac{1}{\det(\langle X_i,X_j\rangle)} \cdot \det(II(X_i,X_j)) .$$

For the determinants of the 2×2 submatrices of the matrix $(II(X_i,X_j))$ we have, by Gauss' equation,

$$D(i_1,i_2;j_1,j_2) = \langle R(X_{i_2},X_{i_1})X_{j_1},X_{j_2}\rangle .$$

So

$$K(p) = \frac{1}{2^{n/2}n!} \sum_{\substack{i_1,\ldots,i_n \\ j_1,\ldots,j_n}} \langle R(X_{i_2},X_{i_1})X_{j_1},X_{j_2}\rangle \cdots$$

$$\cdots \langle R(X_{i_n},X_{i_{n-1}})X_{j_{n-1}},X_{j_n}\rangle \cdot \frac{\varepsilon^{i_1\cdots i_n}\varepsilon^{j_1\cdots j_n}}{\det(\langle X_i,X_j\rangle)} .$$

If we have a coordinate system $x^1,\ldots,x^n$ on M, and let $X_i = \partial/\partial x^i$, then

$$\langle R(X_{i_2},X_{i_1})X_{j_1},X_{j_2}\rangle = \left\langle R\left(\frac{\partial}{\partial x^{i_2}},\frac{\partial}{\partial x^{i_1}}\right)\frac{\partial}{\partial x^{j_1}},\frac{\partial}{\partial x^{j_2}}\right\rangle$$

$$= R_{j_2 j_1 i_2 i_1} \qquad \text{(see p. II.191)}$$

$$= R_{i_1 i_2 j_1 j_2} .$$

So we can write

$$K = \frac{1}{2^{n/2}n!} \sum_{\substack{i_1,\ldots,i_n \\ j_1,\ldots,j_n}} R_{i_1 i_2 j_1 j_2} \cdots R_{i_{n-1} i_n j_{n-1} j_n} \cdot \frac{\varepsilon^{i_1\cdots i_n}}{\sqrt{\det(g_{ij})}} \cdot \frac{\varepsilon^{j_1\cdots j_n}}{\sqrt{\det(g_{ij})}} .$$

The symbol $\varepsilon^{i_1\cdots i_n}/\sqrt{\det(g_{ij})}$ which appears in this formula has the following natural interpretation. We have a map

$$\underbrace{M_p{}^* \times \cdots \times M_p{}^*}_{n \text{ times}} \xrightarrow{\wedge} \Omega^n(M_p)$$

given by

$$(\phi_1,\ldots,\phi_n) \mapsto \phi_1 \wedge \cdots \wedge \phi_n \ .$$

In particular,

$$\left(dx^{i_1}(p),\ldots,dx^{i_n}(p)\right) \mapsto \varepsilon^{i_1\cdots i_n}\cdot\left(dx^1(p)\wedge\cdots\wedge dx^n(p)\right) \ .$$

Now the metric $\langle\ ,\ \rangle_p$ on M_p determines (compare p. I.422) two elements of norm 1 in the 1-dimensional vector space $\Omega^n(M_p)$, namely

$$\pm\sqrt{\det(g_{ij}(p))}\cdot dx^1(p)\wedge\cdots\wedge dx^n(p) \ .$$

If we choose an orientation for M, then we have a way of choosing between these two elements (choose the $+$ sign if and only if $x^1,\ldots,x^n$ is a positively oriented coordinate system), and we therefore have a map $\Omega^n(M_p) \to \mathbb{R}$ defined by taking this element to 1. The composition

$$\boldsymbol{\mathcal{E}} : \underbrace{M_p{}^* \times \cdots \times M_p{}^*}_{n \text{ times}} \xrightarrow{\wedge} \Omega^n(M_p) \to \mathbb{R}$$

is then a contravariant vector field of order n, and its components in the $x^1,\ldots,x^n$ coordinate system are precisely $\varepsilon^{i_1\cdots i_n}/\sqrt{\det(g_{ij})}$. If we use $\mathcal{R}$ for the tensor

$$\mathcal{R}(X,Y,Z,W) = \langle R(X,Y)Z,W\rangle \ ,$$

we can then write our formula for K as

$$K = \frac{1}{2^{n/2} n!} \cdot \text{contraction of } (\underbrace{\mathcal{R} \otimes \cdots \otimes \mathcal{R}}_{n/2 \text{ times}} \otimes \mathcal{E} \otimes \mathcal{E}) .$$

A different choice of orientation for M changes $\mathcal{E}$ to $-\mathcal{E}$, but doesn't change K.

Proposition 22 also shows that K_2 is invariant under isometry, since

$$\binom{n}{2} K_2(p) = \sigma_2(k_1,\ldots,k_n) = \sum_{i<j} k_i k_j = \sigma_1(\{k_i k_j : i < j\}) .$$

The other elementary symmetric functions of $\{k_i k_j : i < j\}$ are also invariant under isometry, but in general these functions do not have very nice expressions in terms of $k_1,\ldots,k_n$. More interesting is the fact that K_r is invariant under isometry whenever r is even; this follows from the algebraic fact (Problem 16) that the coefficients of <u>even</u> powers of λ in the characteristic polynomial $\chi(\lambda)$ of A can always be expressed in terms of the determinants of the 2×2 submatrices of A.

Now consider a hypersurface M^n of a general Riemannian manifold $(N^{n+1}, \langle\ ,\ \rangle)$. We still have a unit normal field ν on M, and corresponding second fundamental form II with

$$\begin{aligned} II(X_p,Y_p) &= \langle s(X_p,Y_p),\nu(p)\rangle \\ &= \langle \nabla'_{X_p} Y,\nu(p)\rangle = -\langle \nabla'_{X_p} \nu, Y_p\rangle \\ &= \langle A_\nu(X_p),Y_p\rangle . \end{aligned}$$

We can define the principal directions at p to be the unit eigenvectors for the self-adjoint map $A_\nu: M_p \to M_p$, and the principal curvatures to be the corresponding eigenvalues. Equivalently, the principal curvatures are the eigenvalues of the symmetric matrix $(II(X_i,X_j))$ for $X_1,\ldots,X_n$ an orthonormal basis of M_p. We no longer expect the Theorema Egregium to be true in general -- even for surfaces, Gauss' equation for the Gaussian curvature involves not only the metric induced on the surface, but also the curvature of N, which varies from point to point. We do obtain a generalization of the Theorema Egregium in the one case where we would expect it:

24. PROPOSITION. Let N^{n+1} be a Riemannian manifold of constant curvature K_0. Then for hypersurfaces in N, the set $\{k_ik_j: i < j\}$ of products of principal curvatures is invariant under isometry. Consequently, the Gaussian curvature K_n is invariant under isometry if n is even, and invariant up to sign if n is odd.

Proof. Exactly like the proof of Proposition 22, except that Gauss' equation gives

$$\tilde{R}(X_i,X_j) = -(k_ik_j + K_0)X_i^* \wedge X_j^* ,$$

so the set $\{-k_ik_j - K_0: i < j\}$ is the set of eigenvalues of the map $X^* \wedge Y^* \mapsto \tilde{R}(X,Y)$. ■

When we consider submanifolds $M \subset N$ of higher codimension, the definitions given previously no longer make sense. However, if we choose any normal vector $\xi \in M_p^\perp$, then we have the map $A_\xi: M_p \to M_p$, satisfying

$$\langle s(X,Y),\xi\rangle = \langle A_\xi(X),Y\rangle \qquad X,\ Y \in M_p\ ,$$

so we can define the *principal directions* and *principal curvatures for* ξ to be the unit eigenvectors and corresponding eigenvalues for A_ξ; equivalently, the principal curvatures are the eigenvalues of the symmetric matrix $(\langle s(X_i,X_j),\xi\rangle)$ for $X_1,\ldots,X_n$ an orthonormal basis of M_p. We can then define the *(elementary symmetric) curvatures* $K_{1;\xi},\ldots,K_{n;\xi}$ by

$$\binom{n}{j}K_{j;\xi} = \sigma_j(k_1,\ldots,k_n)\ ,$$

where the k_i are the principal curvatures for ξ. We thus have maps

$$M_p{}^\perp \longrightarrow \mathbb{R} \qquad \text{given by} \qquad \xi \longmapsto K_{j;\xi}\ .$$

The one interesting (and also very important) case arises for the map

$$M_p{}^\perp \longrightarrow \mathbb{R} \qquad \text{given by} \qquad \xi \longmapsto H_\xi = K_{1;\xi}\ .$$

This map is *linear*, since $A_{\xi+\xi'} = A_\xi + A_{\xi'}$ and since trace is a linear function of matrices. Therefore there is a unique vector $\eta(p) \in M_p{}^\perp$ such that

$$\langle\eta(p),\xi\rangle = H_\xi = \frac{\operatorname{trace}(\langle s(X_i,X_j),\xi\rangle)}{n} \qquad X_1,\ldots,X_n \in M_p \ \text{ orthonormal}$$
$$\text{for all } \xi \in M_p{}^\perp\ .$$

This vector $\eta(p)$ is called the *mean curvature normal* at p. In the case of a hypersurface, $\eta(p) = H(p)\cdot\nu(p)$, where ν is the unit normal (changing ν to $-\nu$ changes H to $-H$, so $H\cdot\nu$ is well-defined). In general, if

$\nu_{n+1},\ldots,\nu_m \in M_p{}^\perp$ is an orthonormal basis, then clearly

$$\eta(p) = \sum_{r=n+1}^{m} H_{\nu_r}\cdot\nu_r .$$

If, moreover, $X_1,\ldots,X_n$ are vector fields tangent to M with $X_1(p),\ldots,X_n(p)$ an orthonormal basis for M_p, then

$$\begin{aligned} H_\xi &= \frac{1}{n}\,\mathrm{trace}(\langle s(X_i(p),X_j(p)),\xi\rangle) \\ &= \frac{1}{n}\sum_{i=1}^{n}\langle s(X_i(p),X_i(p)),\xi\rangle \\ &= \frac{1}{n}\sum_{i=1}^{n}\langle \nabla'_{X_i(p)}X_i,\xi\rangle . \end{aligned}$$

Consequently,

$$\begin{aligned} \eta(p) &= \sum_{r=n+1}^{m} H_{\nu_r}\cdot\nu_r \\ &= \frac{1}{n}\sum_{r=n+1}^{m}\sum_{i=1}^{n}\langle \nabla'_{X_i(p)}X_i,\nu_r\rangle\cdot\nu_r , \end{aligned}$$

whence

$$\boxed{\eta(p) = \frac{1}{n}\perp\left(\sum_{i=1}^{n}\nabla'_{X_i(p)}X_i\right), \qquad X_1(p),\ldots,X_n(p) \text{ orthonormal} .}$$

The mean curvature H for a hypersurface, and the mean curvature normal field η in general, will play an important role in Chapter 9.

Even though principal directions and curvatures cannot be defined for submanifolds $M \subset N$ of higher codimension, one definition still makes sense. A point $p \in M$ is called an umbilic if the principal curvatures for ξ are all equal, for every $\xi \in M_p{}^\perp$. In other words, each map $A_\xi \colon M_p \longrightarrow M_p$ must be some multiple of the identity, so for each ξ there must be a λ with

$$A_\xi(X) = \lambda X \implies \langle s(X,Y),\xi\rangle = \lambda \cdot \langle X,Y\rangle \qquad \text{for all } X, Y \in M_p .$$

It clearly suffices to have

$$A_{\nu_r}(X) = \lambda_r X$$

for a basis $\nu_{n+1},\dots,\nu_m$ of $M_p{}^\perp$.

If p is an umbilic and we choose an orthonormal basis $\nu_{n+1},\dots,\nu_m$ of $M_p{}^\perp$ and constants $\lambda_{n+1},\dots,\lambda_m$ with

$$\langle s(X,Y),\nu_r\rangle = \lambda_r\langle X,Y\rangle \qquad \text{for all } X, Y \in M_p ,$$

then

$$s(X,Y) = \sum_{r=n+1}^{m} \langle s(X,Y),\nu_r\rangle\nu_r = \langle X,Y\rangle \cdot \left(\sum_{r=n+1}^{m} \lambda_r\nu_r\right) .$$

This means that for every $\xi \in M_p{}^\perp$, and every orthonormal basis $X_1,\dots,X_n$ of M_p, we have

$$\langle s(X_i,X_j),\xi\rangle = \delta_{ij}\langle \sum_r \lambda_r\nu_r,\xi\rangle ,$$

so

$$\frac{1}{n}\operatorname{trace}(\langle s(X_i,X_j),\xi\rangle) = \frac{1}{n}\langle\sum_r \lambda_r\nu_r,\xi\rangle\operatorname{trace}(\delta_{ij})$$

$$= \langle\sum_r \lambda_r\nu_r,\xi\rangle .$$

It follows that $\sum_r \lambda_r\nu_r$ is precisely $\eta(p)$, so we have

$$s(X,Y) = \langle X,Y\rangle\eta(p) , \qquad \text{at an umbilic } p .$$

When $s\colon M_p \times M_p \longrightarrow M_p{}^{\perp}$ is not the zero map we can set

$$\eta(p) = \sum_{r=m+1}^{n} \lambda_r\nu_r = \lambda\nu_*$$

for a unique non-zero $\lambda \in \mathbb{R}$ and unit vector ν_*, and for all $X, Y \in M_p$ we have

$$(*) \qquad \begin{cases} \langle s(X,Y),\nu_*\rangle = \lambda\cdot\langle X,Y\rangle \\ \langle s(X,Y),\nu\rangle = 0 \qquad \text{for } \langle\nu,\nu_*\rangle = 0 . \end{cases}$$

<u>25. LEMMA</u>. Let $(N^m, \langle\ ,\ \rangle)$ be a space of constant curvature K_0, and for $n \geq 2$ let M^n be a connected immersed submanifold with all points umbilics. Then either $s = 0$ everywhere, so that M is totally geodesic (by 1-16 and 1-17), or else $\lambda \neq 0$ is constant and M lies in some $(n+1)$-dimensional totally geodesic submanifold.

<u>Proof</u>. Suppose that $s(p) \neq 0$, so that $\lambda(p) \neq 0$. In a neighborhood of p we choose an adapted orthonormal moving frame $X_1,\ldots,X_n,X_{n+1},\ldots,X_m$ on M with $X_{n+1} = \nu_*$ at each point. Then for $1 \leq i \leq n$, and X tangent to M

we have, by (*),

$$\psi_i^r(X) = \langle \nabla'_X X_i, X_r \rangle = \langle s(X,X_i), X_r \rangle = \begin{cases} \lambda\langle X,X_i\rangle & r = n+1 \\ 0 & r > n+1 , \end{cases}$$

which means that on TM we have

$$(1) \qquad \psi_i^{n+1} = \lambda\theta^i$$

$$(2) \qquad \psi_i^r = 0 , \qquad r > n+1 .$$

From equation (1) and the Codazzi-Mainardi equations we find that on TM we have

$$\begin{aligned} d\lambda \wedge \theta^i + \lambda\, d\theta^i = d\psi_i^{n+1} &= - \sum_\alpha \psi_\alpha^{n+1} \wedge \psi_i^\alpha \\ &= - \sum_{k=1}^n \lambda\theta^k \wedge \omega_i^k , \end{aligned}$$

while the first structural equation gives

$$d\theta^i = - \sum_{k=1}^n \omega_k^i \wedge \theta^k = - \sum_{k=1}^n \theta^k \wedge \omega_i^k .$$

So we find that

$$d\lambda \wedge \theta^i = 0 , \qquad 1 \le i \le n .$$

Since $n \ge 2$, this implies that $d\lambda = 0$, so that λ is constant in the neighborhood. This argument shows in general that $\{q \in M: \lambda(q) = \lambda(p)\}$ is open. But this set is also closed, and hence all of M. Thus λ is constant.

Now note that equation (2) gives

$$0 = d\psi_i^r = -\sum_\alpha \psi_\alpha^r \wedge \psi_i^\alpha = -\psi_{n+1}^r \wedge \lambda\theta^i$$

$$\Longrightarrow \psi_{n+1}^r = 0 \quad \text{on } TM, \text{ for } \quad r > n+1 .$$

Therefore

$$(3) \qquad \nabla'_X \nu_* = \nabla'_X X_{n+1} = \sum_{j=1}^{n} \psi_{n+1}^j(X)\cdot X_j = -\lambda \sum_{j=1}^{n} \langle X, X_j\rangle \cdot X_j \qquad \text{by (1)}$$

$$= -\lambda X .$$

We also have

$$(4) \qquad \nabla'_X X_i = \sum_{k=1}^{n} \psi_i^k(X) X_k + \psi_i^{n+1}(X)\cdot \nu_* \qquad i = 1,\ldots,n .$$

Let Δ be the $(n+1)$-dimensional distribution on M with $\Delta(p) = M_p + \mathbb{R}\cdot\nu_*(p)$. Equations (3) and (4) and Pre-Lemma 7 show that Δ is parallel along every curve c lying in M. So Corollary 11 implies that M lies in an $(n+1)$-dimensional totally geodesic subspace of N. ∎

For the case $K_0 = 0$, we can immediately characterize the all-umbilic submanifolds:

26. THEOREM. For $n \geq 2$, let $M^n \subset \mathbb{R}^m$ be a connected immersed submanifold of $\mathbb{R}^m$ with all points umbilics. Then either M lies in some n-dimensional plane or else M lies in some n-dimensional sphere in some $(n+1)$-dimensional plane.

Proof. We just have to show that if $\lambda \neq 0$ in Lemma 25, then M lies in a sphere of radius $1/\lambda$. We simply repeat the proof from Lemma 1: Let V be the vector field on $\mathbb{R}^m$ defined by

$$V(p) = p_p \in \mathbb{R}^m{}_p \ .$$

Then $\nabla'_X V = X$ for all tangent vectors X of $\mathbb{R}^m$, so we can write equation (3) in Lemma 25 as

$$\nabla'_X (X_{n+1} + \lambda V) = 0 \ .$$

Thus the vector field $X_{n+1} + \lambda V$ is parallel along M. Identifying tangent vectors of $\mathbb{R}^m$ with elements of $\mathbb{R}^m$, this means that $X_{n+1} + \lambda V$ is a constant vector v_0 on M, so we have

$$X_{n+1}(p) + \lambda \cdot p = v_0 \in \mathbb{R}^m \ .$$

Thus

$$p = \frac{v_0 - X_{n+1}(p)}{\lambda}$$

for all $p \in M$, which means that M lies in the sphere of radius $1/\lambda$ around the point v_0/λ. ■

This proof, which depends so strongly on the special properties of $\mathbb{R}^m$, breaks down completely when we replace $\mathbb{R}^m$ by a complete simply connected manifold of constant curvature $K_0 \neq 0$. Again we have to exploit different descriptions of these manifolds. First we consider the case $K_0 > 0$.

27. THEOREM. Let $S \subset \mathbb{R}^{m+1}$ be an m-sphere. For $n \geq 2$, let M^n be a connected immersed submanifold of S with all points of M umbilics. Then M is part of an n-sphere.

Proof. We have $M \subset S \subset \mathbb{R}^{m+1}$, with corresponding covariant differentiations ∇, ∇', $\boldsymbol{\nabla}'$. Given $X_p, Y_p \in M_p$, extend them to vector fields X, Y in $\mathbb{R}^m$ tangent to M along M, and tangent to S along S. If $\xi \in M_p{}^\perp \subset S_p$, then

$$<\boldsymbol{\nabla}'_{X_p} Y,\xi> = <\nabla'_{X_p} Y,\xi> ,$$

since $\nabla'_{X_p} Y$ is the component of $\boldsymbol{\nabla}'_{X_p} Y$ tangent to S; so we have

$$(1) \qquad <\boldsymbol{\nabla}'_{X_p} Y,\xi> = \lambda<X_p,Y_p> \qquad \text{for some } \lambda ,$$

since p is an umbilic. On the other hand, if $\boldsymbol{\nu} \in S_p{}^\perp \subset \mathbb{R}^{m+1}{}_p$ is the unit normal, then

$$(2) \qquad <\boldsymbol{\nabla}'_{X_p} Y,\boldsymbol{\nu}> = \frac{1}{r}<X,Y> , \qquad r = \text{radius of } S ,$$

since all points of S are umbilics in $\mathbb{R}^{m+1}$. Equations (1) and (2) show that all points of M are umbilics when M is considered as a submanifold of $\mathbb{R}^{m+1}$. Thus the desired result follows immediately from Theorem 26. ■

Notice that, as predicted by Lemma 25, an n-sphere $\Sigma \subset S$ is either a totally geodesic submanifold of S (when the radius of Σ equals the radius of S), or else is contained in some $(n+1)$-dimensional totally geodesic submanifold Σ' of S. In the latter case, Σ is a geodesic sphere in Σ'; thus we have a complete analogy with Theorem 26.

In order to use the same scheme for investigating all-umbilic submanifolds of H^n, we would first have to consider the all-umbilic submanifolds of $\mathbb{R}^{n+1}$ with the Lorentzian metric; these are the planes $P \subset \mathbb{R}^{n+1}$ of various dimensions, and the quadrics

$$Q = \{p \in P: \langle p - p_0, p - p_0 \rangle = c\} \subset P .$$

Then the all-umbilic submanifolds of H^n must be of the form $H^n \cap P$ or $H^n \cap Q$, and we already noted that the latter submanifolds are contained among the former. However, we merely mentioned, but did not prove, the characterization of the sets $H^n \cap P$. So we will use a different method for the case $K_0 < 0$. We have already used the projective model of H^n, in the second proof of Lemma 8. Now we will use the conformal model. We appeal to a classical result about conformally equivalent manifolds.

<u>28. PROPOSITION</u>. Let $f: N \longrightarrow \bar{N}$ be a conformal equivalence, and $M \subset N$ a submanifold of N with an umbilic $p \in M$. Then $f(p)$ is an umbilic of $f(M) \subset \bar{N}$ (but the λ for $f(p)$ need not be the λ for p).

<u>Proof</u>. Since the result is purely local, we can assume that the underlying spaces of N and $\bar{N}$ are both $\mathbb{R}^m$, that f is the identity with $p = f(p) = 0$, and that $M_p = f(M)_{f(p)}$ is the $(x^1,\ldots,x^n)$-plane $\subset \mathbb{R}^m{}_0$. The metrics for N and $\bar{N}$ have components $g_{\alpha\beta}$ and $\bar{g}_{\alpha\beta}$ satisfying

$$\bar{g}_{\alpha\beta} = e^{2\sigma} g_{\alpha\beta}$$

for some function σ. Then $\bar{g}^{\alpha\beta} = e^{-2\sigma} g^{\alpha\beta}$, and straightforward calculations show that the corresponding Christoffel symbols satisfy the following equations, in which subscripts on σ denote partial derivatives:

$$\overline{[\alpha\beta,\gamma]} = e^{2\sigma}([\alpha\beta,\gamma] + g_{\alpha\gamma}\sigma_\beta + g_{\beta\gamma}\sigma_\alpha - g_{\alpha\beta}\sigma_\gamma)$$

$$\bar{\Gamma}^\gamma_{\alpha\beta} = \Gamma^\gamma_{\alpha\beta} + \delta^\gamma_\alpha\sigma_\beta + \delta^\gamma_\beta\sigma_\alpha - g_{\alpha\beta}\sum_{\mu=1}^m g^{\gamma\mu}\sigma_\mu \ .$$

In particular, for $i, j \le n$ and $r > n$ we have

$$\text{(1)} \qquad \bar{\Gamma}^r_{ij} = \Gamma^r_{ij} - g_{ij} \cdot \sum_{\mu=1}^m g^{r\mu}\sigma_\mu \ .$$

The hypothesis that $p = 0$ is an umbilic point for M means that for each $r > n$ there is a constant λ_r with

$$\Gamma^r_{ij}(0) = \lambda_r g_{ij}(0) \qquad i \le i,\ j \le n \ .$$

Then equation (1) gives

$$\bar{\Gamma}^r_{ij}(0) = [\lambda_r - \sum_{\mu=1}^m g^{r\mu}(0)\sigma_\mu(0)]\cdot g_{ij}(0) \ ,$$

which shows that $f(p) = 0$ is an umbilic for $f(M)$. ∎

<u>29. THEOREM</u>. For $n \ge 2$, let M^n be a connected immersed submanifold of $H^m(K_0)$ with all points of M umbilics. Then either M is totally geodesic, or else M is either a geodesic sphere, a horosphere, or an equidistant hypersurface in some $(n+1)$-dimensional totally geodesic submanifold of $H^m(K_0)$.

Proof. Immediate from Lemma 25, Theorem 26, Proposition 28, and our discussion of $(B^m, \langle\ ,\ \rangle)$ in section A. ■

Proposition 28 could just as well be used to prove Theorem 27. Conversely, if we apply the method used in proving Theorem 27 with the results of Theorem 29, then it is not hard to work backwards and verify the description of geodesic spheres, horospheres, and equidistant hypersurfaces in H^n which was given on p. 23. A particular consequence of Theorem 29 is also noteworthy: Any n-sphere contained in H^n, and any n-sphere which intersects $\mathbb{R}^{m-1}$ non-orthogonally, lies in some (n+1)-sphere or (n+1)-plane which intersects $\mathbb{R}^{m-1}$ orthogonally. Presumably one could also hack this result out by

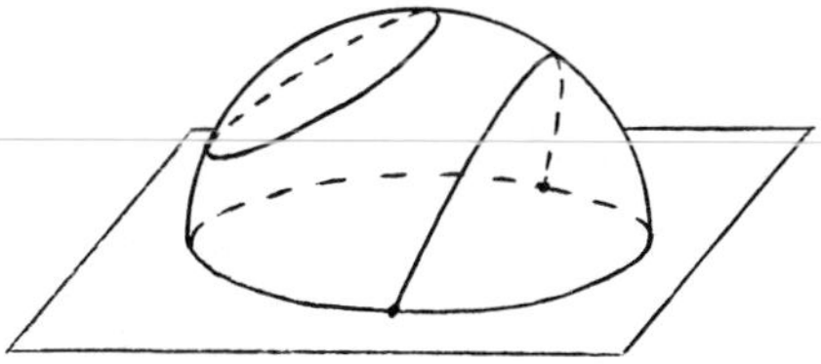

elementary geometry.

For an orthonormal frame $X_1,\dots,X_n$ on an all umbilic hypersurface $M^{m-1} \subset H^m(K_0)$ $(m \geq 3)$ with (constant) λ we have

$$\tilde{R}(X_i,X_j) = -(\lambda^2 + K_0)X_i{}^* \wedge X_j{}^* \qquad \text{(compare p. 104)},$$

which implies that

$$\langle R(X_i,X_j)X_j,X_i\rangle = \tilde{R}(X_i,X_j)(X_j,X_i)$$
$$= \lambda^2 + K_0 ,$$

so that M has constant curvature $\lambda^2 + K_0$. Any two all umbilic hypersurfaces with the same λ are related by an isometry of $H^m(K_0)$, by the first part of Theorem 21. Moreover, there exists a hypersurface with any given $\lambda \geq 0$ (for $\lambda < 0$ we just have the same hypersurface with the opposite choice of unit normal field). In fact, if $(M, \langle\!\langle\ ,\ \rangle\!\rangle)$ is a simply connected $(m-1)$-dimensional manifold of constant curvature $\lambda^2 + K_0$, and we define the tensor S on M by $S(X,Y) = \lambda\langle\!\langle X,Y\rangle\!\rangle$, then M, together with $\langle\!\langle\ ,\ \rangle\!\rangle$ and S, satisfies Gauss' equation and the Codazzi-Mainardi equations, so by the second part of Theorem 21 there is an isometry of M into $H^m(K_0)$ with second fundamental form II satisfying $\text{II} = \lambda \cdot \text{I}$.

It is not hard to determine how the various λ are attached to the various types of all-umbilic hypersurfaces of $H^m(K_0)$. For simplicity, consider $(B^m, \langle\ ,\ \rangle)$, with constant curvature $K_0 = -1$. We know that the horospheres have constant curvature $0 = \lambda^2 - 1 \Rightarrow \lambda = 1$, while the totally geodesic hypersurfaces have constant curvature $-1 = \lambda^2 - 1 \Rightarrow \lambda = 0$. We can take a family of

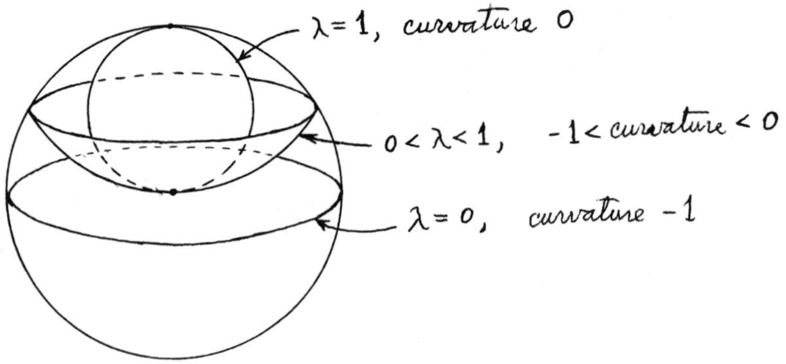

all-umbilic hypersurfaces passing continuously from a totally geodesic hypersurface to a horosphere, with all members of the family distinct up to isometry of B^m. The intermediate hypersurfaces will be equidistant hypersurfaces, and include all such hypersurfaces (up to isometry of B^m). The corresponding λ's must vary monotonically from 0 to 1. This shows that equidistant hypersurfaces, and only equidistant hypersurfaces, have $0 < \lambda < 1$. So all $\lambda > 1$ must occur for the geodesic spheres. If λ_r is the λ for the geodesic sphere of radius r around 0, then $r \to \lambda_r$ must be a monotonic function of r. Clearly $\lambda_r \to \infty$ as $r \to 0$, and $\lambda_r \to 1$ as $r \to \infty$.

We have now generalized essentially the material in Chapter 2 which preceeds the discussion of the third fundamental form. The facts about higher fundamental forms in general will be left to the Problems. The next generalization on our agenda is then the following.

30. PROPOSITION. If M^n is a compact submanifold immersed in $\mathbb{R}^m$, then there is a point $p \in M$ and a normal $\xi \in M_p{}^\perp$ for which the map $A_\xi\colon M_p \to M_p$,

$$\langle A_\xi(X),Y\rangle = \langle s(X,Y),\xi\rangle , \qquad X, Y \in M_p ,$$

is positive definite, $\langle A_\xi(X),X\rangle > 0$ for $X \neq 0$. So if M is a compact hypersurface, with unit normal field ν, then there is a point $p \in M$ for which $-d\nu\colon M_p \to M_p$ is either positive or negative definite (depending on the choice of ν). In particular, the Gaussian curvature $K_n(p)$ is non-zero, and in fact $K_n(p) > 0$ for n even.

Proof. As in the proof of Proposition 2-8, let p be a point of M furthest from 0. Then the line from 0 to p is normal to M at p, and we choose ξ to be the unit vector in M_p pointing in this direction. The rest of the argument is left as an exercise for the reader. ■

31. COROLLARY. There are no compact submanifolds M^n immersed in $\mathbb{R}^m$ with mean curvature normal $\eta = 0$. In particular, there are no immersed hypersurfaces in $\mathbb{R}^m$ with mean curvature $H = 0$.

Proof. If ξ is given by Proposition 30, then

$$\langle\eta(p),\xi\rangle = H_\xi = \text{trace } A_\xi ,$$

and trace $A_\xi > 0$ since A_ξ is positive definite. ■

Suppose we replace $\mathbb{R}^m$ in Proposition 30 by the space $(B^m, \langle\ ,\ \rangle)$ of constant curvature $K_0 < 0$. If $p \in M$ is a point furthest from 0, then M is contained in the geodesic sphere around 0 which passes through p. All principal curvatures of this sphere are equal to some $\lambda > \sqrt{-K_0}$ (compare

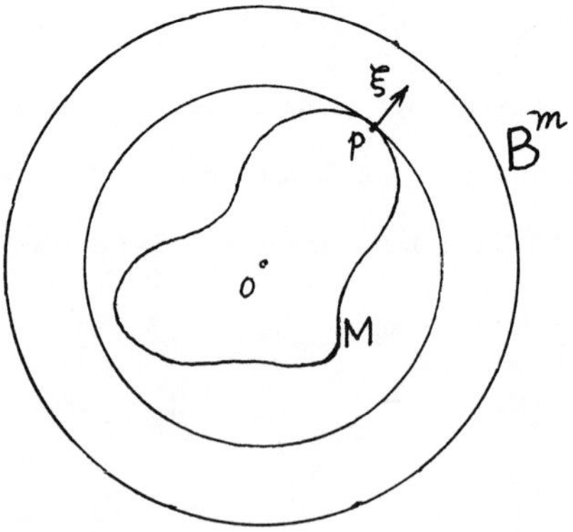

p. III.91). The geodesic from 0 to p is normal to M at p, and if we choose ξ to be the unit normal in $M_p^{\perp}$ pointing in this direction, then we will have

$$\langle A_\xi(X), X\rangle \geq \lambda > \sqrt{-K_0}\ .$$

For a hypersphere M, and a correctly chosen unit normal field ν, we thus find that all principal curvatures $k_1, \ldots, k_n$ are $\geq \lambda > \sqrt{-K_0}$. Hence

$$K_n(p) = \prod_{i=1}^{n} k_i \geq \lambda^n > (\sqrt{-K_0})^n\ .$$

In particular, for n even this holds for either choice of ν. We also see that there are no compact immersed submanifolds of N with mean curvature normal $\eta = 0$, and hence no compact immersed hypersurfaces of N with mean curvature $H = 0$.

Now let us replace $\mathbb{R}^m$ by a sphere S of radius $1/\sqrt{K_0}$, for $K_0 > 0$, and suppose moreover, that M <u>is contained in an open hemisphere of</u> S, say the hemisphere centered around the point x. By choosing a point $p \in M$ furthest from x, and ξ a unit normal in M_p pointing along the geodesic from x to p, we find that

$$\langle A_\xi(X), X\rangle \geq \lambda$$

for some $\lambda > 0$. But there is obviously no positive lower bound for all λ's. For hypersurfaces M we find that $K_n(p) \neq 0$, and $K_n(p) > 0$ for n even, but again there is no positive lower bound for K_n. Similarly, we find that Corollary 31 generalizes to compact submanifolds of an open hemisphere.

Naturally, our results break down if we replace the hemisphere by the whole sphere, for the equatorial $(m-1)$-sphere has second fundamental form $s = 0$. You might think that this is the only exception, but there are actually many other possibilities. In fact, we easily compute that for $p, q \geq 1$ with $p+q = m-1$, the hypersurface

$$M = \left\{(x_1,\dots,x_{p+1},y_1,\dots,y_{q+1}) \in \mathbb{R}^{m+1} : \Sigma {x_k}^2 = \frac{p}{m-1} \text{ and } \Sigma {y_k}^2 = \frac{q}{m-1}\right\} \subset S^m$$

has mean curvature $H = 0$ in the unit sphere S^m, so there is certainly no point $p \in M$ where $d\nu\colon M_p \to M_p$ is definite.

To complete our generalization of the material in Chapter 2, we want to discuss the relationship between positive curvature and convexity of hypersurfaces. For a hypersurface $M^n \subset \mathbb{R}^{n+1}$, the proper analogue of positivity of the Gaussian curvature at p is the condition that all sectional curvatures at p are positive; equivalently, all principal curvatures should have the same sign, or yet again, the map $d\nu\colon M_p \to M_p$ should be (positive or negative) definite. It is easy to see that definiteness of $d\nu\colon M_p \to M_p$ implies that M locally lies on one side of the tangent hyperplane of M at p. If $d\nu\colon M_p \to M_p$ is merely semi-definite (that is, $\langle d\nu(X),X\rangle \geq 0$ for all X, or $\langle d\nu(X),X\rangle \leq 0$ for all X), then no conclusion can be drawn. But if $d\nu\colon M_p \to M_p$ is <u>not</u> semi-definite, then M locally lies on both sides of its tangent hyperplane at p. Propositions 2-9 and 2-10 clearly generalize to hypersurfaces in $\mathbb{R}^{n+1}$; we will not bother to write down all the details, but will henceforth use the word "convex" for a hypersurface in either of its two equivalent meanings.

32. PROPOSITION.

(1) If M is a convex hypersurface in $\mathbb{R}^{n+1}$, then $d\nu\colon M_p \to M_p$ is semi-definite for all $p \in M$.

(2) Let M be a compact connected n-manifold, and $f\colon M \to \mathbb{R}^{n+1}$ an immersion with normal map n such that $dn\colon M_p \to M_p$ is definite for all $p \in M$. Then

(i) The manifold M is orientable, and the normal map $n\colon M \to S^n \subset \mathbb{R}^{n+1}$ is a diffeomorphism,

(ii) The map $f\colon M \to \mathbb{R}^{n+1}$ is an imbedding, and $f(M)$ is convex.

Proof. This generalization of Hadamard's Theorem (2-11) is proved in exactly the same way as the original. ∎

The most significant part of this result is the fact that the immersion f must be an imbedding. In fact, the definiteness of $d\nu$ implies that M is locally convex, and there are general arguments to show that a locally convex set in $\mathbb{R}^m$ is actually convex, which implies the theorem for an imbedded hypersurface M. On the other hand, we have already mentioned in Chapter 2 that for $n = 2$ Hadamard's Theorem holds even under the weakened assumption that $K(p) \geq 0$ for all $p \in M$. Here the result is not clear even for imbedded $M \subset \mathbb{R}^3$, since the condition $K \geq 0$ does not imply local convexity for arbitrary (non-compact) M. For example, the graph of $(x,y) \mapsto x^3(1+y^2)$ has

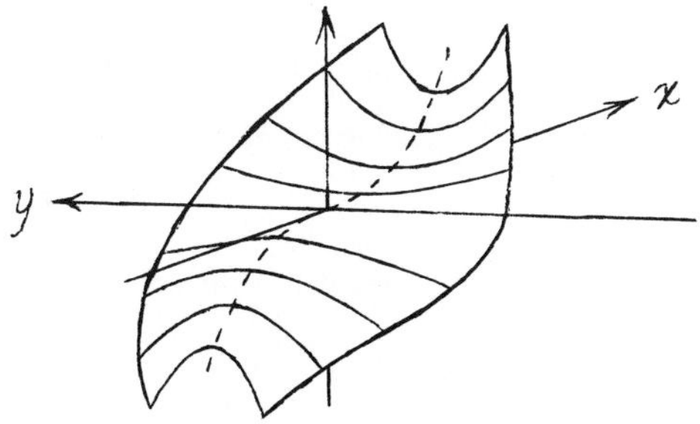

$K \geq 0$ in a neighborhood of $0 \in \mathbb{R}^3$ (by an easy calculation), but is clearly not locally convex. The extension of Hadamard's Theorem for $K \geq 0$ (and $n = 2$) was originally proved by Chern and Lashof [1], using a little Morse theory. Sacksteder [1] then gave a proof for all n under the weakened assumption that $dn\colon M_p \to M_p$ is semi-definite for all $p \in M$; in fact, compactness of M can be replaced by completeness, provided that there is at least one point $p \in M$ where at least one sectional curvature is non-zero (without this last condition, M might be a generalized cylinder). Sacksteder's proof is more "elementary," and, as one might guess, much harder. (For the case where all $dn\colon M_p \to M_p$ are definite, but M is merely complete and immersed, there is an earlier proof by Stoker [1].) Do Carmo and Lima [1] gave a simple proof of a result even more general than Sacksteder's when M is compact: If $f\colon M^n \to \mathbb{R}^m$ is an immersion with all maps $A_\xi\colon M_p \to M_p$ semi-definite (for $\xi \in M_p{}^\perp$) for all $p \in M$, and all maps A_ξ definite for at least one $p \in M$ [for $m = n+1$ this latter condition follows from Proposition 30], then $f(M)$ is contained in some $(n+1)$-dimensional plane in $\mathbb{R}^m$, and f is an imbedding of M as a convex set. In Do Carmo and Lima [2], they also give a simple argument which reproves Sacksteder's result for complete M (but which does not recapture all of the additional information obtained in the course of Sacksteder's analysis).

We can also consider convex sets in spaces of constant curvature K_0. For $H^m(K_0)$, the definition is precisely the same as for $\mathbb{R}^m$: a set $A \subset H^m(K_0)$ is convex if A contains the segment of the unique geodesic between p and q whenever $p, q \in A$. For $K_0 > 0$, we consider only an open hemisphere of $S^m(K_0)$, so that there is a unique geodesic between any two points, and the same definition can be used. Since geodesic mappings preserve convexity, we see immediately that Proposition 2-10 generalizes when we replace the tangent plane of M at p by the totally geodesic hypersurface $\exp(M_p)$. Again we will use "convex" for hypersurfaces in either of its two equivalent meanings. It also looks as if we should be able to use geodesic mappings to generalize Proposition 32 to hypersurfaces of $H^{n+1}(K_0)$ and $S^{n+1}(K_0)$. The details of this program turn out to be a little sticky, and since the arguments have been covered in a recent paper, Do Carmo and Warner [1], we will merely quote their results:

<u>33. THEOREM (DO CARMO-WARNER).</u>

(1) If M is a convex hypersurface in $H^{n+1}(K_0)$ for $K_0 < 0$, or a convex hypersurface in a hemisphere of $S^{n+1}(K_0)$ for $K_0 > 0$, then all sectional curvatures of M are $\geq K_0$. Moreover, if ϕ is a geodesic mapping from $H^{n+1}(K_0)$, or a hemisphere of $S^{n+1}(K_0)$, to $\mathbb{R}^{n+1}$, then all sectional curvatures of M are $> K_0$ at p if and only if all sectional curvatures of $\phi(M)$ are > 0 at $\phi(p)$.

(2) Let M be a compact connected n-manifold, and $f\colon M \to S^{n+1}(K_0)$ an immersion, for $K_0 > 0$, such that all sectional curvatures are $\geq K_0$. Then M is orientable, the immersion f is an imbedding, and either $f(M)$ is totally geodesic, or else $f(M)$ is contained in some open hemisphere and is convex.

(3) Let M be a compact connected n-manifold, and $f\colon M \to H^{n+1}(K_0)$ an immersion, for $K_0 < 0$, such that all sectional curvatures are $\geq K_0$. Then M is orientable, the immersion f is an imbedding, and $f(M)$ is convex.

In part (2) of this result, compactness of M is really equivalent to completeness, by Corollary 8-22. In part (3), compactness does not follow from completeness, and if we try to deal with complete M in $H^{n+1}(K_0)$ we run into the problem that the image $\phi(H^{n+1}(K_0))$ of the geodesic map $\phi\colon H^{n+1}(K_0) \to \mathbb{R}^{n+1}$ is an open ball, and hence $\phi \circ f(M)$ need not be complete. As a matter of fact, part (3) is false if M is merely assumed complete. Even if all sectional curvatures of an immersion $f\colon M \to H^{n+1}(K_0)$ are $> K_0$, it does not follow that f is an imbedding. To see this, we consider an immersed, but not imbedded, surface in $\mathbb{R}^3$ with everywhere positive curvature. Such a surface

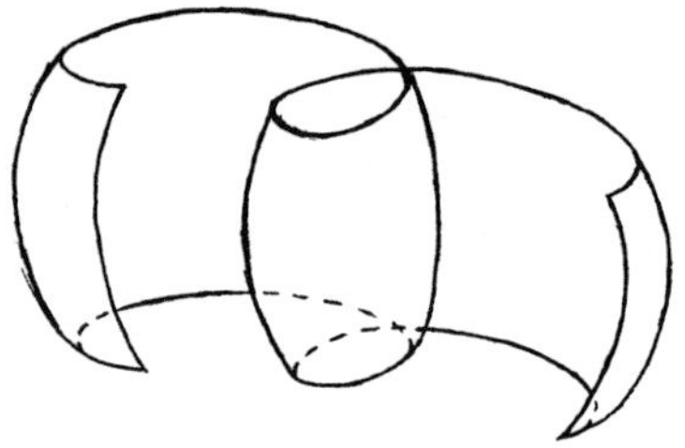

cannot be complete in $\mathbb{R}^3$, but its intersection with the projective model of H^3 may very well be complete in H^3, even though its (extrinsic) curvature is > -1, by part (1) of Theorem 33. Similarly, if $M \subset \mathbb{R}^3$ is the non-convex surface pictured on p. 121, with non-negative curvature near 0, then the intersection of M with the projective model of H^3 can be a complete imbedded surface with extrinsic curvature ≥ -1 everywhere, but it will not be convex in H^3. As a concluding remark, we point out that a complete convex hypersurface in $\mathbb{R}^{n+1}$ is of very restricted topological type; it is homeomorphic

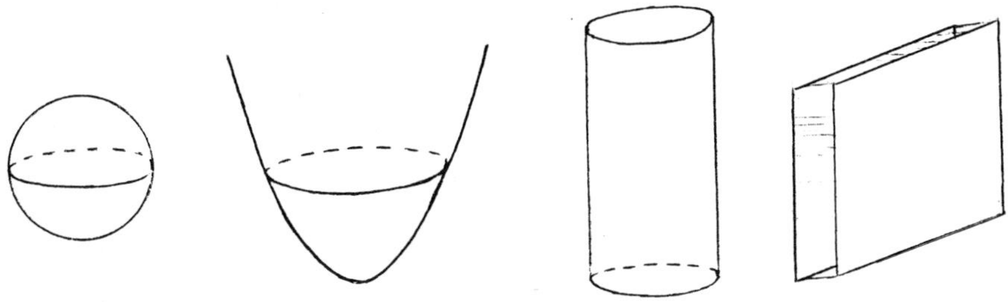

to S^n (if it is compact) or to $\mathbb{R}^n$ or $S^1 \times \mathbb{R}^{n-1}$ otherwise. On the other hand, there are complete convex hypersurfaces of H^{n+1} which are homeomorphic to $\mathbb{R}^n$ with any number of holes, as shown below for the projective model of H^3.

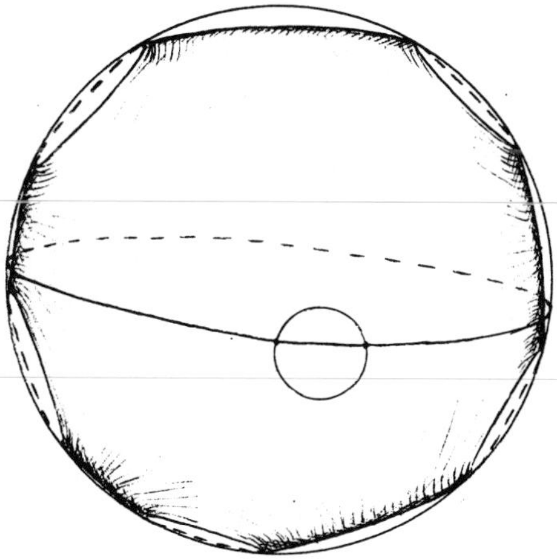

E. Further Results

This section is devoted to generalizations of certain material in Chapters 3 and 4. The first thing we want to consider are ruled surfaces in $\mathbb{R}^m$, given by

$$f(s,t) = c(s) + t\delta(s)$$

for two curves c, δ in $\mathbb{R}^m$. When $m = 3$ we found that the surface is flat precisely when c', δ, δ' are everywhere linearly dependent, by using the Gaussian curvature $k_1 \cdot k_2$, the product of the principal curvatures. For $m > 3$, we have to compute the curvature of the surface f from an intrinsic formula. We can assume that $|\delta| = 1$, and hence $\langle \delta, \delta' \rangle = 0$. Then

$$\left.\begin{array}{l} f_1 = c' + t\delta' \\ f_2 = \delta \end{array}\right\} \Rightarrow \left\{\begin{array}{l} E = \langle f_1, f_1 \rangle = \langle c', c' \rangle + 2t\langle c', \delta' \rangle + t^2\langle \delta', \delta' \rangle \\ F = \langle f_1, f_2 \rangle = \langle c', \delta \rangle \\ G = 1 \; . \end{array}\right.$$

Most of the terms in the formula on p. II.109 vanish, and we end up with

$$\begin{aligned} 4(EG - F^2)^2 K &= G \cdot \left(\frac{\partial E}{\partial t}\right)^2 - 2(EG - F^2)\frac{\partial^2 E}{\partial t^2} \\ &= [2\langle c', \delta' \rangle + 2t\langle \delta', \delta' \rangle]^2 \\ &\quad - 2[\langle c', c' \rangle + 2t\langle c', \delta' \rangle + t^2\langle \delta', \delta' \rangle - \langle c', \delta \rangle^2] \cdot 2\langle \delta', \delta' \rangle \; . \end{aligned}$$

The coefficients of t and t^2 vanish, and we find that

$$K = 0 \iff 0 = \langle c', \delta' \rangle^2 - \langle c', c' \rangle \cdot \langle \delta', \delta' \rangle + \langle c', \delta \rangle^2 \cdot \langle \delta', \delta' \rangle \; .$$

This condition is automatic when $\delta' = 0$. At points where $\delta' \neq 0$, we can write

$$K = 0 \iff \langle c', c' \rangle = \left\langle c', \frac{\delta'}{|\delta'|} \right\rangle^2 + \langle c', \delta \rangle^2 \; .$$

Since $\delta, \delta'/|\delta'|$ are orthonormal, this happens precisely when c' is a linear combination of δ, δ'. So in all cases,

$$K = 0 \iff c', \delta, \delta' \text{ are linearly dependent} .$$

We can now repeat the analysis on pp. III.351-353 and see that flat ruled surfaces in $\mathbb{R}^m$ are "in general" cylinders, cones, or tangents to a curve.

It should be pointed out that there are plenty of non-ruled flat surfaces in $\mathbb{R}^m$ for $m > 3$. For example, the torus

$$S^1 \times S^1 \subset \mathbb{R}^2 \times \mathbb{R}^2 = \mathbb{R}^4 ,$$

with the product metric, is flat.

We can also define ruled surfaces in an arbitrary Riemannian manifold $(N^m, \langle\ ,\ \rangle)$. They are the surfaces which can be parameterized as

$$f(s,t) = \exp_{c(s)}(tV(s)) ,$$

where V is a unit vector field along c.

We want to consider, in particular, the case where N has constant curvature K_0, and try to describe the ruled surfaces in N which also have constant curvature K_0. First we consider the case $m = 3$. For a surface $M \subset N^3$ it is important to make a distinction which does not arise in the case of surfaces in $\mathbb{R}^3$. The surface M has an induced Riemannian metric, and thus an intrinsic curvature

$$K_{int}(p) = \langle R(X_p, Y_p)Y_p, X_p \rangle \quad \text{for orthonormal } X_p, Y_p \in M_p .$$

It also has an extrinsic Gaussian curvature $K_{ext}(p) = k_1 \cdot k_2$, the product of the principal curvatures at p. If N has constant curvature K_0, then Gauss' equation tells us that

$$(*) \qquad K_{int} = K_{ext} + K_0 .$$

Recall, by the way, that a surface M having constant curvature just means that the function K_{int} on M is constant, while the condition that a higher dimensional manifold have constant curvature is more involved.

The reason for considering the case $m = 3$ first is that in this case the hypothesis that M is ruled is essentially redundant:

34. PROPOSITION. Let N be a 3-dimensional manifold of constant curvature K_0, and let $M \subset N$ be a surface with constant intrinsic curvature $K_{int} = K_0$. If $p \in M$ is a point where the second fundamental form $s\colon M_p \times M_p \longrightarrow M_p{}^{\perp}$ is not 0, then p has a neighborhood which is a ruled surface.

Proof. Since M has

$$K_{int} = K_0 \implies K_{ext} = 0 \quad \text{by } (*) ,$$

one principal curvature, k_1, is always 0. Since s is non-zero at p, the other principal curvature, k_2, is non-zero in a neighborhood of p. Choose orthonormal vector fields X_1, X_2 on this neighborhood so that each $X_1(q)$ is a principal vector with principal curvature $k_1(q) = 0$, and $X_2(q)$ is a principal vector with principal curvature $k_2(q) \neq 0$. Now the

Codazzi-Mainardi equations for N are exactly the same as for $\mathbb{R}^3$, so the proof of Proposition 5-4 goes through unchanged, leading to the conclusion that $\nabla'_{X_1}X_1 = 0$, which means that the integral curves of X_1 are geodesics in N. ∎

Naturally, this result does not hold when N has dimension > 3, so for a general manifold (N, < , >) of constant curvature K_0 we will now restrict our attention to ruled surfaces $M \subset N$. By Synge's inequality (Corollary 1-7) we always have $K_{int}(p) \leq K_0$. Moreover,

$$K_{int} = K_0 \text{ along a ruling } \gamma \text{ of } M \iff M_{\gamma(t)} \text{ is parallel along } \gamma .$$

But Lemma 8 shows that

$$M_{\gamma(t)} \text{ is parallel along } \gamma \iff M \text{ is tangent to a 2-dimensional totally geodesic submanifold of } N \text{ along } \gamma .$$

The interesting thing about this last condition is that it does not involve metrics, but only their geodesics. Hence

35. THEOREM. Let N be a manifold of constant curvature K_0 and let $\phi: N \longrightarrow \mathbb{R}^m$ be a geodesic mapping. Let $M \subset N$ be a ruled surface. Then M has constant intrinsic curvature $K_{int} = K_0$ if and only if the ruled surface $\phi(M) \subset \mathbb{R}^m$ is flat.

Proof. Immediate from the above equivalences. ∎

From Theorem 35 we see that the surfaces $M \subset N$ with $K_{int} = K_0$ are "in general" ϕ^{-1} of cones, cylinders, and tangent developables. As a local classification, this works equally well for $K_0 < 0$ and $K_0 > 0$. But the situation is quite different when we look for complete surfaces with $K_{int} = K_0$. In the sphere, any pair of geodesics intersect, so there cannot be "cylinders" as in $\mathbb{R}^m$ (this is reflected in the fact that the geodesic mapping from S^m to $\mathbb{R}^m$ is actually defined only on a hemisphere). Once one realizes this, it seems very hard for there to be many such surfaces. In fact, in the next section we will see that in S^3 the only complete surfaces with $K_{int} = K_0$ are the great 2-spheres. Now consider hyperbolic space H^m. We know that there is a geodesic mapping $\phi: H^m \longrightarrow B^m(1)$. Equivalently, there is a metric $\langle\ ,\ \rangle$ on $B^m(1)$ with constant curvature $K_0 < 0$, whose geodesics are just straight lines of $\mathbb{R}^m$ (with a different parameterization). A cone, cylinder, or tangent developable in $\mathbb{R}^m$ then intersects $B^m(1)$ in a surface with $K_{int} = K_0$ with the metric induced from $\langle\ ,\ \rangle$. The interesting thing is that we can take the vertex of our cone, or the generating curve for the tangent developable to lie outside of B. Then the intersection

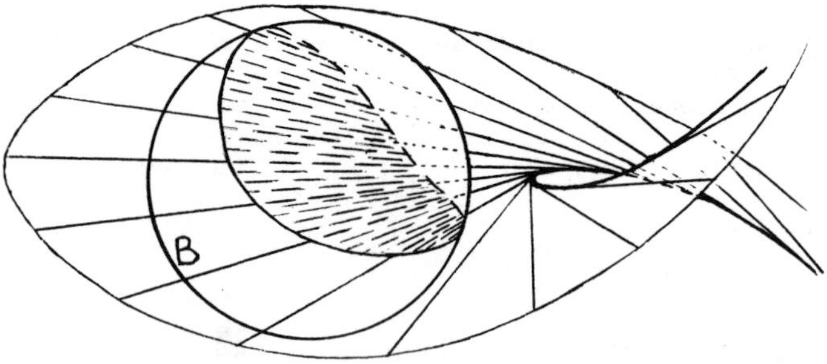

with B will be a complete flat surface, without singularities, of constant intrinsic curvature $K_{int} = K_0$. Thus there are many such surfaces, of far greater variety than in $\mathbb{R}^m$. In the next section we will see this in a startling way.

Now consider an oriented surface M in an arbitrary oriented 3-dimensional Riemannian manifold $(N, \langle\ ,\ \rangle)$, and an arclength parameterized curve c in M. We again define the Darboux frame of c on M to be the moving frame

$$\mathbf{t}(s) = c'(s)\ , \qquad \mathbf{u}(s)\ , \qquad \mathbf{v}(s) = \mathbf{t}(s) \times \mathbf{u}(s) = \nu(c(s))\ ,$$

where $\mathbf{u}(s) \in M_{c(s)}$ is a unit vector perpendicular to $\mathbf{t}(s)$ with $(\mathbf{t}(s), \mathbf{u}(s))$ positively oriented in M, and the unit normal field ν is chosen so that $(\mathbf{t}(s), \mathbf{u}(s), \mathbf{v}(s))$ is positively oriented in N. We still have

$$\begin{aligned} \mathbf{t}' &= \phantom{-\kappa_n\mathbf{t}} \quad \kappa_g \mathbf{u} + \kappa_n \mathbf{v} \\ \mathbf{u}' &= -\kappa_g \mathbf{t} \qquad\qquad + \tau_g \mathbf{v} \\ \mathbf{v}' &= -\kappa_n \mathbf{t} - \tau_g \mathbf{u} \qquad\qquad , \end{aligned}$$

for certain functions $\kappa_n, \kappa_g, \tau_g$. Everything in Chapter 4 up to and including Proposition 4-5 generalizes almost without any change (asymptotic directions on M are defined just as before, as unit vectors $X \in M_p$ with $II(X_p, X_p) = 0$; they exist only on regions where $K_{ext}(p) \le 0$). Moreover, Theorem 4-7 also generalizes, essentially without change. For reference, we merely state this generalization:

36. THEOREM (BELTRAMI-ENNEPER). Let M be a surface in an oriented 3-dimensional Riemannian manifold $(N, \langle\ ,\ \rangle)$. If c is an asymptotic curve in M with $c(0) = p$ and first curvature $\kappa_1(0) \neq 0$, then

$$|\kappa_2(0)| = \sqrt{-K_{ext}(p)}\ .$$

Moreover, if $K_{ext}(p) < 0$ and the two distinct asymptotic curves through p both have non-zero first curvature κ_1 at p, then their second curvatures κ_2 at p are negatives of each other.

The next result generalizes Theorem 4-8.

37. THEOREM. Let N^m be a manifold of constant curvature K_0, let c be an immersed curve in a hypersurface $M \subset N$, and let S be the ruled surface formed by the geodesics of N which are perpendicular to M along c. Then c is a line of curvature if and only if S has constant intrinsic curvature $K_{int} = K_0$.

Proof. Since the result is a local one, we can assume that there is a geodesic mapping $\phi\colon N \longrightarrow \mathbb{R}^m$. The surface S is $\{\exp_{c(s)} t\nu(c(s))\}$, where ν is a unit normal field of M. Hence, identifying tangent vectors of $\mathbb{R}^m$ with elements of $\mathbb{R}^m$ as usual, we have

$$\phi(S) = \{\phi(c(s)) + t\phi_*(\nu(c(s)))\}$$
$$= \{\gamma(s) + t\delta(s)\} \ , \quad \text{say.}$$

If $\overline{\nabla}$ denotes covariant differentiation in $\mathbb{R}^m$, then, as in the second proof of Lemma 8, we have

$$\overline{\nabla}_{\phi_* X}\phi_* Y - \phi_*(\nabla'_X Y) = \omega(\phi_* X)\cdot\phi_* Y + \omega(\phi_* Y)\cdot\phi_* X \ ,$$

for some 1-form ω on $\mathbb{R}^m$. Hence

$$\delta'(s) - \phi_*(\nabla'_{c'(s)}\nu) = \text{a linear combination of } \phi_*(c'(s)) \text{ and } \phi_*(\nu(c(s))) \ .$$

Consequently, we can write

$$(1) \qquad \delta'(s) = \phi_*(\nabla'_{c'(s)}\nu) + a\phi_*(c'(s)) + b\phi_*(\nu(c(s))) \ .$$

First suppose that c is a line of curvature, so that $\nabla'_{c'(s)}\nu$ is a multiple of $c'(s)$ for all s. Then equation (1) shows that we can write

$$\delta'(s) = \alpha\phi_*(c'(s)) + b\phi_*(\nu(c(s)))$$
$$= \alpha\gamma'(s) + b\delta(s) \ .$$

So γ', δ, δ' are always linearly independent, and the ruled surface $\phi(S) \subset \mathbb{R}^m$ is flat. Hence S has constant intrinsic curvature $K_{int} = K_0$ by Theorem 35.

Conversely, if S has constant intrinsic curvature $K_{int} = K_0$, then $\phi(S)$ is flat, so γ', δ, δ' are always linearly dependent. Then (1) shows that for each s there are numbers A, B, C, not all 0, with

$$(2) \qquad Ac'(s) + B\nu(c(s)) + C[\nabla'_{c'(s)}\nu + ac'(s) + b\nu(c(s))] = 0 \ .$$

Clearly $C \neq 0$. Taking the inner product of (2) with $\nu(c(s))$ gives

$$B + Cb = 0 ,$$

and hence (2) becomes

$$(A + Ca)c'(s) + C\nabla'_{c'(s)}\nu = 0 ,$$

which shows that c is a line of curvature. ■

In Eisenhardt {1 ; p. 213} this result is verified in a more direct way, by using special coordinates -- the Weierstrasse coordinates. I like the above proof because it has the strange feature that it uses geodesic mappings even though such mappings preserve neither perpendicularity nor lines of curvature. Once one realizes this, it becomes clear how to generalize the theorem vastly (Problem 19).

F. Complete Surfaces of Constant Curvature

In this section we will classify, so far as possible, the complete constant curvature surfaces in the complete simply-connected 3-dimensional manifolds of constant curvature. First consider a surface M in any 3-dimensional manifold $(N, \langle\ ,\ \rangle)$. By Theorem 4-17, for any point $p \in M$ we can find an imbedding $f\colon U \to M$ with $U \subset \mathbb{R}^2$ open and $p \in f(U)$ whose coordinate lines are the lines of curvature, or the asymptotic lines [if $K_{ext}(p) < 0$]. We want to see what the formulas in the Addendum to Chapter 4 become in these cases. As before, E, F, G are the components of $f^*\langle\ ,\ \rangle$ with respect to the standard coordinate system (s,t) on $\mathbb{R}^2$, while ℓ, m, n are the components of f^*II, where II is the second fundamental form of the hypersurface $M \subset N$ for some choice of a unit normal field ν on M. The formula in Problem 4-13 gives the intrinsic curvature K_{int}, so we see that

(A) When the parameter lines of $M^2 \subset N^3$ are orthogonal, we have

$$F = 0$$

$$K_{int} = -\frac{1}{2\sqrt{EG}}\left[\left(\frac{E_2}{\sqrt{EG}}\right)_2 + \left(\frac{G_1}{\sqrt{EG}}\right)_1\right].$$

We also know that the Codazzi-Mainardi equations for an ambient manifold of constant curvature are the same as in the Euclidean case, so

(B) When N^3 has constant curvature and the parameter lines of $M^2 \subset N^3$ are lines of curvature, we have

$$\ell = k_1 E, \quad n = k_2 G, \quad m = 0, \quad F = 0$$

$$\ell_2 = \frac{E_2}{2}\left(\frac{\ell}{E} + \frac{n}{G}\right)$$

$$n_1 = \frac{G_1}{2}\left(\frac{\ell}{E} + \frac{n}{G}\right).$$

(C) When N^3 has constant curvature and the parameter lines of $M^2 \subset N^3$ are asymptotic curves, we have

$$\ell = n = 0$$

$$m_1 = \frac{\left[\frac{1}{2}(EG - F^2)_1 + FE_2 - EG_1\right]}{EG - F^2}\cdot m$$

$$m_2 = \frac{\left[\frac{1}{2}(EG - F^2)_2 + FG_1 - GE_2\right]}{EG - F^2} \cdot m \; .$$

Recall, finally, that when N has constant curvature K_0, the intrinsic curvature K_{int} of M and the extrinsic curvature K_{ext} are related by

(*) $$K_{int} = K_{ext} + K_0 \; .$$

The first thing we are going to do is to see what the basic lemmas of Chapter 5 give in our more general situation. The main problem is keeping track of the times when the curvature K in the Euclidean case should be replaced by K_{int} and when it should be replaced by K_{ext}.

<u>38. LEMMA</u>. Let M be a surface immersed in a 3-manifold N of constant curvature, and let $p \in M$ be a non-umbilic point. Let $k_1 \geq k_2$ be the two principal curvatures on M and suppose that k_1 has a local maximum at p, and k_2 has a local minimum at p. Then $K_{int}(p) \leq 0$.

<u>Proof</u>. The proof is exactly the same as the proof of Lemma 5-1. ■

<u>39. THEOREM</u>. Let N be a 3-manifold of constant curvature. If M is a compact connected surface in N with constant extrinsic curvature $K_{ext} \geq 0$ and (constant) intrinsic curvature $K_{int} > 0$, then all points of M are umbilics.

<u>Proof</u>. First suppose that $K_{ext} > 0$. As in the proof of Theorem 5-2, let

$k_1 \geq k_2$ be the principal curvatures and let k_1 achieve its maximum at p. Then $k_2 = K_2/k_1$ has its minimum at p. If p were not an umbilic, then by Lemma 38 we would have $K_{int}(p) \leq 0$, contradicting the hypothesis. So $k_1(p) = k_2(p)$, and, reasoning as in the proof of Theorem 5-2, we see that all points are umbilics.

Next suppose that $K_{ext} = 0$. Suppose there is a non-umbilic point $p \in M$. Then $0 = k_1(p)\cdot k_2(p)$, but $k_1(p) \neq k_2(p)$, so either $k_1(p) > 0$ or $0 > k_2(p)$, say the first. Let $\bar{p}$ be the point where k_1 takes on its maximum $k_1(\bar{p}) > 0$. Then $k_1 > 0$ in a whole neighborhood of p, so $k_2 = 0$ in a whole neighborhood of $\bar{p}$, and hence k_2 has a local minimum at $\bar{p}$. Then Lemma 38 gives $K_{int}(\bar{p}) \leq 0$, a contradiction. ∎

<u>40. THEOREM</u>. Let N be a 3-manifold of constant curvature. Let M be a 2-dimensional immersed submanifold of N with constant extrinsic curvature $K_{ext} < 0$. Then for every point $p \in M$ there is a diffeomorphism

$$g\colon (-\varepsilon,\varepsilon) \times (-\varepsilon,\varepsilon) \longrightarrow M$$

$$g(0,0) = p$$

whose parameter curves are asymptotic curves <u>parameterized by arclength</u>.

<u>Proof</u>. The proof is exactly the same as the first proof of Lemma 5-10. ∎

<u>41. THEOREM</u>. Let N be a 3-manifold of constant curvature. Then there is no complete surface M immersed in N with constant extrinsic curvature $K_{ext} < 0$ and (constant) intrinsic curvature $K_{int} < 0$.

Proof. Suppose such a surface M existed. Using Theorem 40, we can repeat the first argument in the (first) proof of Theorem 5-12 verbatim and conclude that there is a Tschebyscheff net $f: \mathbb{R}^2 \to M$. If ω is the angle between the first and second parameter lines, then by Lemma 5-11 we have

$$\frac{\partial^2 \omega}{\partial s \partial t} = (-K_{int}) \sin \omega \qquad 0 < \omega < \pi ,$$

where $-K_{int}$ is a positive constant. Then part (B) of the proof of Theorem 5-12 shows that there is no such ω. ∎

Now we will begin putting these results together. Take N to be S^3, with constant curvature 1, and consider the possibilities for complete surfaces in S^3 with constant extrinsic curvature K_{ext}. Since equation (*) now becomes

$$K_{int} = K_{ext} + 1 ,$$

we see that $K_{ext} < -1 \Rightarrow K_{int} < 0$. So Theorem 41 shows that there are no complete surfaces immersed in S^3 with constant $K_{ext} < -1$. We also see that $K_{ext} \geq 0 \Rightarrow K_{int} > 0$, so Theorem 39 and Theorem 27 show that the only compact surfaces in S^3 with constant $K_{ext} \geq 0$ are spheres (Theorem 8-17 again shows that compactness can be replaced by completeness).

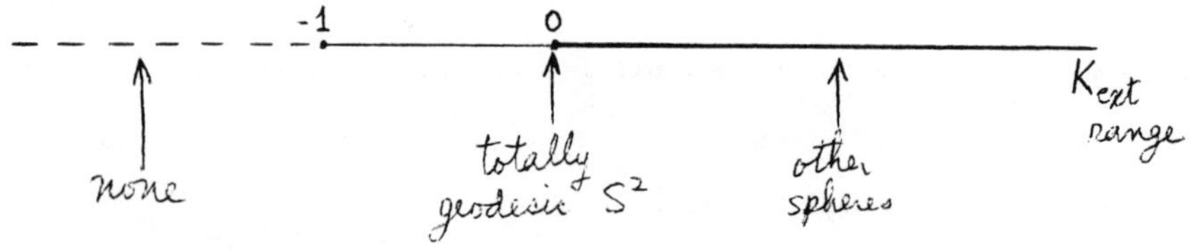

How about the range $-1 \leq K_{ext} < 0$? First of all we have

42. PROPOSITION. There are no complete surfaces M immersed in S^3 with constant K_{ext} satisfying $-1 < K_{ext} < 0$.

Proof. The intrinsic curvature of M would satisfy $K_{int} > 0$, so M would be compact, by Theorem 8-17. We can assume that M is orientable, for otherwise we can look at the orientable 2-fold covering of M, which will also be immersed, with the same K_{ext}. Then M must be homeomorphic to S^2, by the Gauss-Bonnet Theorem. Since $K_{ext} < 0$, at every point $p \in S^2$ the principle curvatures $k_1(p)$, $k_2(p)$ have opposite signs. By choosing the vectors pointing in the principal directions which correspond to the positive principal curvature, we would have a continuous choice of 1-dimensional subspaces of $S^2{}_p$. But this is impossible (Problem I.9-7). ∎

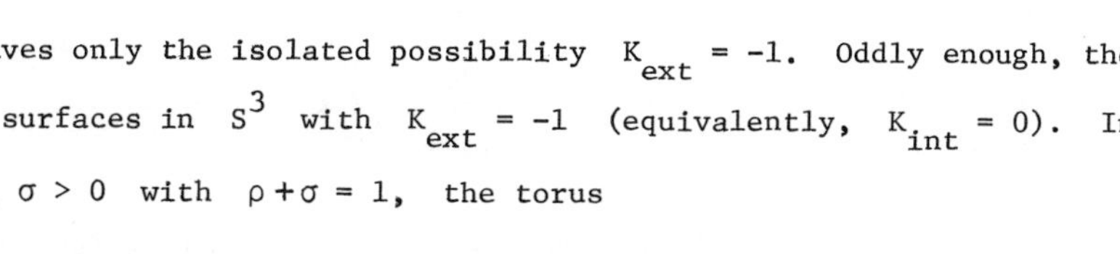

This leaves only the isolated possibility $K_{ext} = -1$. Oddly enough, there are complete surfaces in S^3 with $K_{ext} = -1$ (equivalently, $K_{int} = 0$). In fact, for $\rho, \sigma > 0$ with $\rho + \sigma = 1$, the torus

$$\left\{x \in \mathbb{R}^4 : x_1{}^2 + x_2{}^2 = \rho \text{ and } x_3{}^2 + x_4{}^2 = \sigma\right\} \subset S^3$$

is a (flat) product of two circles. Moreover, there is an infinite variety of other complete flat surfaces in S^3. Such surfaces can be classified, modulo a few sticky details, and we will essentially find the most general way to construct them. The classification actually works even for a piece of a flat surface, but we will deal only with complete surfaces, just to simplify some of the description; this classification is based on the work of Bianchi [1].

It will be necessary to first consider some of the geometry which is special to the manifold S^3. For two points $x, y \in S^3$, the distance $d(x,y)$ between x and y as elements of S^3 (<u>not</u> the Euclidean distance between x and y) is just the radian measure of the angle between x and y. Consequently, we have

(1) $\cos d(x,y) = \langle x,y\rangle$.

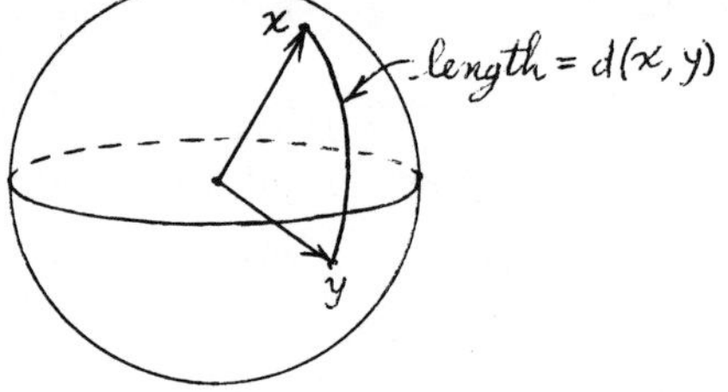

Now we ask whether there are any isometries $A \in O(4)$ of S^3 with the property that $d(x,A(x))$ is the same for all $x \in S^3$. Such isometries would be the analogues of the translations in $\mathbb{R}^n$; notice that S^2, for example, certainly has no isometries with this property, other than the identity, since every $A \in O(3)$ has a fixed point in S^2. If $A = (a_{ij})$, then

$$\langle x,Ax\rangle = \sum_{i,j=1}^{4} a_{ji}x_j x_i .$$

Taking into account equation (1) we see that we are looking for A with

$$\sum_{i,j=1}^{4} a_{ji}x_jx_i = \text{constant} \qquad \text{for all} \quad x \in S^4 .$$

This implies that

$$\sum_{i,j=1}^{4} a_{ji}x_jx_i = (\text{constant})\cdot \sum_{i=1}^{4} x_i^2 \qquad \text{for all} \quad x \in \mathbb{R}^4 .$$

Regarding this as a polynomial identity in the variables $x_1,\ldots,x_4$ we see that we must have

$$a_{11} = a_{22} = a_{33} = a_{44} \qquad a_{ij} + a_{ji} = 0 \qquad i \neq j .$$

Since A is also orthogonal we have

(2) $$0 = a_{11}a_{12} + a_{21}a_{22} + a_{31}a_{32} + a_{41}a_{42} = a_{31}a_{32} + a_{41}a_{42}$$

as well as

$$a_{11}^2 + a_{21}^2 + a_{31}^2 + a_{41}^2 = a_{12}^2 + a_{22}^2 + a_{32}^2 + a_{42}^2$$
$$\Downarrow$$

(3) $$a_{31}^2 + a_{41}^2 = a_{32}^2 + a_{42}^2 .$$

Equation (2) says that the vectors $(a_{31},a_{41}),\ (a_{32},a_{42}) \in \mathbb{R}^2$ are perpendicular, while equation (3) says that they have the same length. It follows that

$$\left.\begin{aligned} a_{32} &= a_{41} \\ a_{42} &= -a_{31} \end{aligned}\right\} \quad \text{or} \quad \left\{\begin{aligned} a_{32} &= -a_{41} \\ a_{42} &= +a_{31} \end{aligned}\right. .$$

We thus find two different kinds of A's with the desired property:

$$\begin{pmatrix} a & -b & -c & -d \\ b & a & -d & c \\ c & d & a & -b \\ d & -c & b & a \end{pmatrix} \quad \text{or} \quad \begin{pmatrix} a & b & c & d \\ -b & a & -d & c \\ -c & d & a & -b \\ -d & -c & b & a \end{pmatrix}, \qquad a^2+b^2+c^2+d^2 = 1 .$$

The existence of these "translations" in S^3 is directly related to the fact that S^3 is a group, the group of quaternions of norm 1. Recall that the quaternions are $\mathbb{R}^4$ with the structure of a non-commutative division algebra over $\mathbb{R}$ having unit $1 = (1,0,0,0)$ and elements

$$i = (0,1,0,0) , \qquad j = (0,0,1,0) , \qquad k = (0,0,0,1)$$

satisfying

$$\begin{matrix} i \cdot j = k = - j \cdot i \\ j \cdot k = i = - k \cdot j \\ k \cdot i = j = - i \cdot k \end{matrix} \qquad \text{and} \qquad i \cdot i = j \cdot j = k \cdot k = - 1 .$$

The norm $|x|$ of a quaternion x satisfies $|xy| = |x| \cdot |y|$, so the quaternions of norm 1 (i.e. S^3) are a non-commutative Lie group. It is easily checked that the two matrices given above are just left and right translation by the quaternion $a + bi + cj + dk \in S^3$. In particular, this shows that the usual Riemannian metric on S^3 is left and right invariant. Moreover, the map

$$a + bi + cj + dk \mapsto \begin{pmatrix} a & -b & -c & -d \\ b & a & -d & c \\ c & d & a & -b \\ d & -c & b & a \end{pmatrix}$$

is an isomorphism of S^3 into a subgroup of $O(4)$, namely the subgroup of all left translations by elements of S^3. It will be convenient to identify S^3 with a subgroup of $O(4)$ by this isomorphism.

We will need the first part of the following general result; the other parts are included for independent interest.

<u>43. THEOREM</u>. Let G be a Lie group with bi-invariant metric $\langle\ ,\ \rangle$. If X, Y, Z and W are left invariant vector fields on G, then

(1) $\nabla_X Y = \frac{1}{2}[X,Y]$

(2) $\langle [X,Y],Z\rangle = \langle X,[Y,Z]\rangle$

(3) $R(X,Y)Z = -\frac{1}{4}[[X,Y],Z]$

(4) $\langle R(X,Y)Z,W\rangle = -\frac{1}{4}\langle [X,Y],[Z,W]\rangle$.

<u>Proof</u>. The integral curves of X are left translates of 1-parameter subgroups (recall the second proof of Corollary I.10-8). Consequently, they are geodesics (Proposition I.10-21). This means that $\nabla_X X = 0$. So

$$0 = \nabla_{X+Y} X + Y = \nabla_X X + \nabla_X Y + \nabla_Y X + \nabla_Y Y = \nabla_X Y + \nabla_Y X .$$

But also

$$\nabla_X Y - \nabla_Y X = [X,Y] ,$$

which gives (1).

For (2) we note that

$$\begin{aligned} 0 = Y\langle X,Z\rangle &= \langle \nabla_Y X,Z\rangle + \langle X,\nabla_Y Z\rangle \\ &= \frac{1}{2}\langle [Y,X],Z\rangle + \frac{1}{2}\langle [X,[Y,Z]]\rangle . \end{aligned}$$

For (3) we have

$$\begin{aligned} R(X,Y)Z &= \nabla_X(\nabla_Y Z) - \nabla_Y(\nabla_X Z) - \nabla_{[X,Y]}Z \\ &= \tfrac{1}{4}[X,[Y,Z]] - \tfrac{1}{4}[Y,[X,Z]] - \tfrac{1}{4}[[X,Y],Z] , \end{aligned}$$

which gives the desired result when we apply the Jacobi identity.

Finally, (4) follows from (2) and (3). ■

Now we want to look at the Lie algebra $\mathcal{L}(S^3)$ of the group S^3. This is the tangent space of S^3 at $(1,0,0,0)$, and is therefore spanned by the vectors

$$\begin{aligned} X_1 &= (0,1,0,0) \\ X_2 &= (0,0,1,0) \\ X_3 &= (0,0,0,1) , \end{aligned}$$

regarded as tangent vectors at $(1,0,0,0)$. Notice that $X_1 = c'(0)$, where

$$\begin{aligned} c(t) &= (\cos t, \sin t, 0, 0) \in S^3 \\ &= \cos t + (\sin t)i \\ &= \begin{pmatrix} \cos t & -\sin t & 0 & 0 \\ \sin t & \cos t & 0 & 0 \\ 0 & 0 & \cos t & -\sin t \\ 0 & 0 & \sin t & \cos t \end{pmatrix} , \end{aligned}$$

under the identification of S^3 with a subgroup of $O(4)$. Thus X_1 can be identified with

$$c'(0) = \begin{pmatrix} 0 & -1 & 0 & 0 \\ 1 & 0 & 0 & 0 \\ 0 & 0 & 0 & -1 \\ 0 & 0 & 1 & 0 \end{pmatrix} \in \mathfrak{o}(4) .$$

Similarly X_2 and X_3 can be identified with

$$\begin{pmatrix} 0 & 0 & -1 & 0 \\ 0 & 0 & 0 & 1 \\ 1 & 0 & 0 & 0 \\ 0 & -1 & 0 & 0 \end{pmatrix}, \qquad \begin{pmatrix} 0 & 0 & 0 & -1 \\ 0 & 0 & -1 & 0 \\ 0 & 1 & 0 & 0 \\ 1 & 0 & 0 & 0 \end{pmatrix}.$$

A short calculation then shows that

(1) $$[X_1,X_2] = 2X_3 , \qquad [X_2,X_3] = 2X_1 , \qquad [X_3,X_1] = 2X_2 .$$

If we think of the X_i as vectors in $\mathbb{R}^3$, by simply ignoring their first components, then we have

$$X_1 \times X_2 = X_3 , \qquad X_2 \times X_3 = X_1 , \qquad X_3 \times X_1 = X_2 .$$

Equivalently, this relation holds when we define $\times$ in $S^3_{(1,0,0,0)}$ in terms of the usual inner product and the usual orientation for S^3. So if $\tilde{X}_i$ is the left invariant vector field on S^3 which extends X_i, then

(2) $$\tilde{X}_1 \times \tilde{X}_2 = \tilde{X}_3 , \qquad \tilde{X}_2 \times \tilde{X}_3 = \tilde{X}_1 , \qquad \tilde{X}_3 \times \tilde{X}_1 = \tilde{X}_2 ,$$

where $\times$ in each tangent space is defined in terms of the usual metric $\langle\ ,\ \rangle$ on S^3 and the usual orientation for S^3.

Now the theory of curves in S^3 can be given a special development, because we can express all tangent vectors in terms of the left invariant vector fields $\tilde{X}_i$. Suppose c is a curve in S^3 parameterized by arclength, and let the unit tangent vector $\mathbf{t} = \mathbf{v}_1$ of c be given by

$$(3) \qquad \mathbf{t}(s) = \sum_{i=1}^{3} f_i(s)\cdot\tilde{X}_i(c(s)) ,$$

where

$$(4) \qquad \sum_i f_i^2 = 1 \Rightarrow \sum_i f_i f_i' = 0 .$$

As usual, we denote the covariant derivative in our ambient manifold S^3 by ∇'. Then for any vector field $\sum_j h_j(s)\cdot\tilde{X}_j(c(s))$ along c we have

$$\begin{aligned}\frac{D'}{ds}\Big[\sum_j h_j(s)\cdot\tilde{X}_j(c(s))\Big] &= \sum_j h_j'(s)\cdot\tilde{X}_j(c(s)) + \sum_j h_j(s)\frac{D'}{ds}\tilde{X}_j(c(s))\\ &= \sum_j h_j'(s)\cdot\tilde{X}_j(c(s)) + \sum_j h_j(s)\sum_i f_i(s)\nabla'_{\tilde{X}_i}\tilde{X}_j(c(s)) .\end{aligned}$$

Using Theorem 43 to write $\nabla'_{\tilde{X}_i}\tilde{X}_j = \frac{1}{2}[\tilde{X}_i,\tilde{X}_j]$, and computing the brackets from (1), we get

$$\begin{aligned}(5) \qquad \frac{D'}{ds}\Big[\sum_j h_j(s)\cdot\tilde{X}_j(c(s))\Big] = &\sum_j h_j'\cdot\tilde{X}_j\\ &+ \big[(f_2h_3 - f_3h_2)\tilde{X}_1 + (f_3h_1 - f_1h_3)\tilde{X}_2\\ &\qquad + (f_1h_2 - f_2h_1)\tilde{X}_3\big]\end{aligned}$$

{all functions evaluated at s,
all $\tilde{X}_i$ at $c(s)$} .

In particular, we have

$$(6)\qquad \frac{D'\,\mathbf{t}\,(s)}{ds} = \sum_i f_i'\cdot\tilde{X}_i\ ;$$

hence the curvature κ $(= \kappa_1)$ is given by

$$(7)\qquad \kappa = \sqrt{\Sigma(f_i')^2}\ ,$$

and $\mathbf{n} = \mathbf{v}_2$ is given by

$$(8)\qquad \mathbf{n} = \frac{\Sigma\, f_i'\cdot\tilde{X}_i}{\kappa}\ .$$

Therefore $\mathbf{b} = \mathbf{v}_3$ is given by

$$(9)\qquad \mathbf{b} = \mathbf{t}\times\mathbf{n} = \frac{1}{\kappa}\cdot(\sum_i f_i\cdot\tilde{X}_i)\times(\sum_j f_j'\cdot\tilde{X}_j)$$

$$= \frac{1}{\kappa}\sum_{i,j} f_i f_j'(\tilde{X}_i\times\tilde{X}_j)$$

$$= \frac{1}{\kappa}[(f_2f_3' - f_3f_2')\tilde{X}_1 + (f_3f_1' - f_1f_3')\tilde{X}_2 + (f_1f_2' - f_2f_1')\tilde{X}_3]$$

by (2)

$$= \frac{1}{\kappa}\sum_i g_i\cdot\tilde{X}_i\ ,\quad \text{say.}$$

Now we have

$$\frac{D'\,\mathbf{b}(s)}{ds} = \frac{1}{\kappa}\frac{D'}{ds}(\sum_i g_i\cdot\tilde{X}_i) - \frac{\kappa'}{\kappa^2}\Sigma\, g_i\cdot\tilde{X}_i$$

$$= \frac{1}{\kappa}[(f_2g_3 - g_2f_3)\tilde{X}_1 + \cdots + \sum_i g_i'\tilde{X}_i] - \frac{\kappa'}{\kappa^2}\Sigma\, g_i\tilde{X}_i \qquad \text{by (5)}$$

$$= \frac{1}{\kappa}[(f_2g_3 - g_2f_3)\tilde{X}_1 + \cdots] + \sum_i(\frac{g_i}{\kappa})'\tilde{X}_i\ .$$

But

$$\begin{aligned} f_2g_3 - g_2f_3 &= f_2(f_1f_2' - f_2f_1') - f_3(f_3f_1' - f_1f_3') && \text{by (9)} \\ &= f_1(f_2f_2' + f_3f_3') - f_1'(f_2{}^2 + f_3{}^2) \\ &= f_1(-f_1f_1') - f_1'(1 - f_1{}^2) && \text{by (4)} \\ &= -f_1' , \end{aligned}$$

and similarly for the other terms. Hence we obtain

$$(10) \qquad \frac{D'\,\mathbf{b}(s)}{ds} = \frac{-\sum_i f_i'\cdot\tilde{X}_i}{\kappa} + \sum_i \left(\frac{g_i}{\kappa}\right)'\cdot\tilde{X}_i$$

$$= -\mathbf{n} + \sum_i \left(\frac{g_i}{\kappa}\right)'\cdot\tilde{X}_i \qquad \text{by (8).}$$

We therefore have the rather remarkable, and for us very important

<u>44. THEOREM</u>. If c is a curve in S^3 whose torsion τ $(=\kappa_2)$ satisfies $\tau = 1$ everywhere, then $\mathbf{b}$ is left invariant along c, that is,

$$\mathbf{b}(s) = L_{c(s)c(0)^{-1}*}\,\mathbf{b}(0) .$$

If c has torsion $\tau = -1$ everywhere, then $\mathbf{b}$ is right invariant along c.

<u>Proof</u>. The Serret-Frenet formulas give

$$\frac{D\,\mathbf{b}(s)}{ds} = -\tau\mathbf{n} .$$

So $\tau = 1$ implies that $(g_i/\kappa)' = 0$, and hence that g_i/κ is constant. But equation (9) shows that g_i/κ are the components of **b** with respect to the left-invariant vector fields $\tilde{X}_i$.

To deduce the second part of the theorem, consider the map $f(x) = x^{-1}$ of S^3 into itself. It reverses 1-parameter subgroups through (1,0,0,0), so $f_*\colon \mathcal{L}(S^3) \longrightarrow \mathcal{L}(S^3)$ is multiplication by -1. This shows that f is orientation reversing. It follows that the binormal of the curve $f \circ c$ is $-f_*\mathbf{b}$. Thus $f \circ c$ has $\tau = 1$ if and only if c has $\tau = -1$. ∎

Finally we are ready to consider connected immersed surfaces M in S^3 with $K_{ext} = -1$, and hence $K_{int} = 0$. We consider only oriented M; non-orientable surfaces may then be analysed by considering the 2-fold oriented covering of M. Since $K_{ext} < 0$, there are 2 distinct asymptotic directions at each point. The argument in the (first) proof of Theorem 5-12, in conjunction with Theorem 40, again shows that there is a Tsychebyscheff net $f\colon \mathbb{R}^2 \to M$. It is not hard to see that f is actually onto M (by essentially the argument used in the second proof of Theorem 5-12; for this part, it is not necessary that the ϕ_s be defined for all $s \in \mathbb{R}$, and simple connectivity is irrelevant). The metric $I_f = f^*\langle\ ,\ \rangle$ on $\mathbb{R}^2$ is then

$$I_f = f^*\langle\ ,\ \rangle = ds\otimes ds + \cos\omega[ds\otimes dt + dt\otimes ds] + dt\otimes dt\ ,$$

where ω is the oriented angle between the first and second parameter curves.

Now consider the curve $c(s) = (s,t)$ in $\mathbb{R}^2$, which is an arclength parameterized curve for the metric I_f. Its tangent vector $c'(s) = \partial/\partial s$ is a unit vector for the metric I_f. If D/ds temporarily denotes the covariant derivative determined by the metric I_f, then from the formula on p. II.249 we compute that

$$\frac{Dc'(s)}{ds} = \frac{\frac{\partial\omega}{\partial s}}{\sin\omega}\cdot\left[\cos\omega\cdot\frac{\partial}{\partial s} - \frac{\partial}{\partial t}\right].$$

If $\overline{\left(\frac{\partial}{\partial s}\right)}$ is the unique vector field with $\frac{\partial}{\partial s}, \overline{\left(\frac{\partial}{\partial s}\right)}$ orthonormal for the metric I_f and $\left(\frac{\partial}{\partial s}, \overline{\left(\frac{\partial}{\partial s}\right)}\right)$ positively oriented, then

$$\frac{\partial}{\partial t} = \cos\omega\cdot\frac{\partial}{\partial s} + \sin\omega\cdot\overline{\left(\frac{\partial}{\partial s}\right)}.$$

so we find that

$$\frac{Dc'(s)}{ds} = -\frac{\partial\omega}{\partial s}\cdot\overline{\left(\frac{\partial}{\partial s}\right)}.$$

Equivalently, if $\mathbf{t}$ denotes the (unit) tangent vector to the parameter curve $s \mapsto f(s,t)$ in M, and D/ds now denotes the covariant derivative in M, then

$$\frac{D\mathbf{t}}{ds} = -\frac{\partial\omega}{\partial s}\cdot\mathbf{u},$$

where $\mathbf{u}$ is the unique tangent vector field along $s \mapsto f(s,t)$ with $\mathbf{t},\mathbf{u}$ orthonormal and $(\mathbf{t},\mathbf{u})$ positively oriented. But $s \mapsto f(s,t)$ is an asymptotic curve, so the covariant derivative $D\mathbf{t}/ds$ in M is the same as the covariant derivative $D'\mathbf{t}/ds$ in S^3 (recall the equivalences on p.III.285). So we have

$$(1) \qquad \frac{D'\mathbf{t}}{ds} = -\frac{\partial\omega}{\partial s}\cdot\mathbf{u}.$$

This shows that

$$\mathbf{u} = \text{normal } \mathbf{n} \text{ to the curve } s \mapsto f(s,t)$$

$$\left|\frac{\partial\omega}{\partial s}(s,t)\right| = \text{curvature } \kappa(s) \text{ of the curve } s \mapsto f(s,t) .$$

On the other hand, Lemma 5-11 shows that ω satisfies

$$\frac{\partial^2\omega}{\partial s\partial t} = 0 ,$$

which implies that there are functions S and T with

$$\omega(s,t) = S(s) + T(t) ,$$

so that

$$\frac{\partial\omega}{\partial s}(s,t) = S'(s) \quad \text{and} \quad \frac{\partial\omega}{\partial t}(s,t) = T'(t) .$$

Thus the arclength parameterized curves $s \mapsto f(s,t)$ all have the <u>same</u> curvature functions $\kappa(s) = |S'(s)|$. Similarly, all curves $t \mapsto f(s,t)$ have the same curvature functions $|T'(t)|$.

But even more is true. For the Beltrami-Enneper Theorem (36) tells us that the torsion τ of the asymptotic curves $s \mapsto f(s,t)$ and $t \mapsto f(s,t)$ satisfies $\tau^2 = 1$ at points where $\kappa \neq 0$, and that the two asymptotic curves through a point have torsions of opposite signs if they both have $\kappa \neq 0$ at that point. We will first assume that for both sets of parameter curves κ is never 0. Then one set of parameter curves must have $\tau = 1$ everywhere, and the other set must have $\tau = -1$ everywhere. For definiteness, say that the curves $s \mapsto f(s,t)$ have $\tau = 1$. We now see that <u>all</u> curves $s \mapsto f(s,t)$ are congruent, and similarly <u>all</u> curves $t \mapsto f(s,t)$ are congruent.

Let A_s be the unique isometry of S^3 with $A_s(f(0,t)) = f(s,t)$ for all t. Under the family of isometries $\{A_s\}$, each point $f(0,t)$ moves

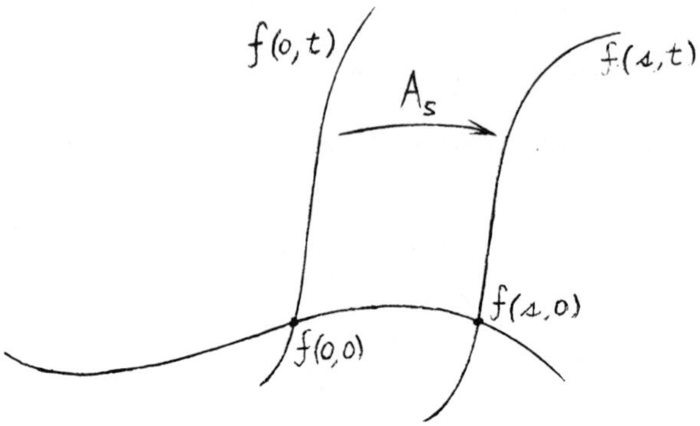

along the arclength parameterized curve $s \mapsto f(s,t)$. This strongly suggests that all the A_s are actually translations. In fact, we claim that all A_s are left-translations. To prove this, we consider the family of left translations $\{B_s\} = \{L_{f(s,0)f(0,0)^{-1}}\}$ which take $f(0,0)$ to $f(s,0)$. According to Theorem 44, B_{s*} takes the binormal $\mathbf{b}(0)$ of $s \mapsto f(s,0)$ at $s = 0$ into the binormal $\mathbf{b}(s)$ at s. Consequently, B_{s*} takes the osculating plane of this curve at 0 into the osculating plane at s. Hence we can write

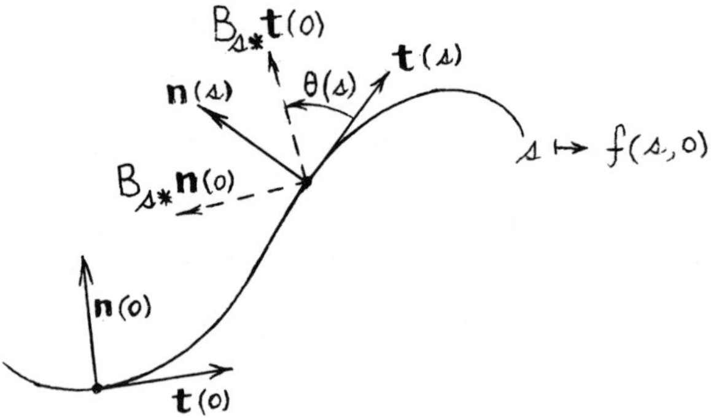

$$\mathbf{t}(s) = \cos\theta(s) \cdot B_{s*}\,\mathbf{t}(0) - \sin\theta(s) \cdot B_{s*}\,\mathbf{n}(0) \;,$$

where $\theta(s)$ is the oriented angle from $\mathbf{t}(s)$ to $B_{s*}\,\mathbf{t}(0)$. It is easy to compute $D'\mathbf{t}/ds$ in terms of θ: Without loss of generality assume that $f(0,0) = 1 \in S^3$, and that $\mathbf{t}(0)$ and $\mathbf{n}(0)$ are $X_1, X_2 \in \mathcal{L}(S^3)$. Then the functions f_i in equation (3) on p. 146 are just

$$f_1 = \cos\theta\,, \qquad f_2 = -\sin\theta\,, \qquad f_3 = 0\,,$$

so equation (6) on p. 147 gives

$$\begin{aligned}\frac{D'\mathbf{t}}{ds} &= -\theta'(s)[\sin\theta(s)\cdot B_{s*}\,\mathbf{t}(0) + \cos\theta\cdot B_{s*}\,\mathbf{n}(0)]\\ &= -\theta'(s)\cdot\mathbf{n}(s)\,.\end{aligned}$$

Comparing with equation (1) on p. 150, we see that $\theta' = S'$; since $\theta(0) = 0$, we find that

$$S(s) = \theta(s) + S(0)\,.$$

From this we easily see that

> B_{s*} takes the tangent vector to the curve $t \mapsto f(0,t)$ at $t = 0$
> to the tangent vector to the curve $t \mapsto f(s,t)$ at $t = 0$.

Moreover, these curves are asymptotic curves, so their osculating planes at $t = 0$ coincide with the osculating planes of the asymptotic curve $s \mapsto f(s,0)$ at 0 and s, respectively. Thus their binormals at $t = 0$ are the binormals $\mathbf{b}(0)$ and $\mathbf{b}(s)$ of the curve $s \mapsto f(s,t)$. Hence

> B_{s*} takes the binormal to the curve $t \mapsto f(0,t)$ at $t = 0$
> to the binormal to the curve $t \mapsto f(s,t)$ at $t = 0$.

These two facts show that B_s must be the isometry A_s. So A_s is indeed a left translation.

If we write $c(s) = f(s,0)$ and $\gamma(t) = f(0,t)$, we thus see that our surface M can be written as a collection of left translates of γ,

$$M = \{[c(s)\cdot c(0)^{-1}]\cdot\gamma(t)\} .$$

Notice that this can equally well be written as a collection of right translates of c,

$$\begin{aligned} M &= \{c(s)\cdot c(0)^{-1}\cdot\gamma(t)\} = \{c(s)\cdot\gamma(0)^{-1}\cdot\gamma(t)\} \\ &= \{c(s)\cdot[\gamma(0)^{-1}\cdot\gamma(t)]\} ; \end{aligned}$$

naturally we could have also deduced this description directly, by considering the isometries of the curves $s \longmapsto f(s,t)$, and applying the second part of Theorem 44.

Conversely, suppose we have any two curves c and γ with torsions $\tau = 1$ and $\tau = -1$, respectively. Suppose, moreover, that they are placed so that $c(0) = \gamma(0)$ and so that their osculating planes at 0 coincide. For simplicity, also assume that $c(0) = \gamma(0) = 1 \in S^3$. Then c and γ will not be tangent at 0, and we can consider the surface

$$M = \{c(s)\cdot\gamma(t)\} .$$

Applying Theorem 44 first to the curve c with $\tau = 1$, we find that the osculating plane of c at s coincides with the osculating plane of the curve $t \longmapsto c(s)\cdot\gamma(t)$ at $t = 0$; hence these osculating planes coincide with the tangent space of M at $c(s)$. Now applying Theorem 44 to the curves $t \longmapsto c(s)\cdot\gamma(t)$, all with torsions $\tau = -1$, we find that the tangent space

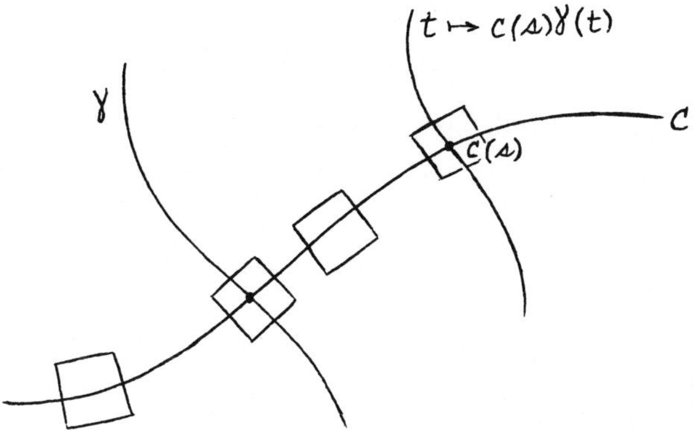

osculating planes coincide along c.

of M at any point $c(s_0)\cdot\gamma(t_0)$ coincides with the osculating planes of the parameter curves $s \longmapsto c(s)\cdot\gamma(t_0)$ at $s = s_0$ and $t \longmapsto c(s_0)\cdot\gamma(t)$ at $t = t_0$. Thus these parameter curves are asymptotic curves. So the Beltrami-Enneper Theorem shows that M has $K_{ext} = -1$.

We can also consider the case where c has torsion $\tau = 1$, but γ is a geodesic, and hence does not have a torsion defined anywhere. The first part of our argument still shows that the tangent space of M at points c(s) coincides with the osculating plane of c at s. In other words,

$$(1) \quad \frac{D'c'(s)}{ds} \text{ is a linear combination of } c'(s) \text{ and } L_{c(s)*}\gamma'(0) .$$

To show that the tangent space of M at $c(s_0)\cdot\gamma(t_0)$ coincides with the

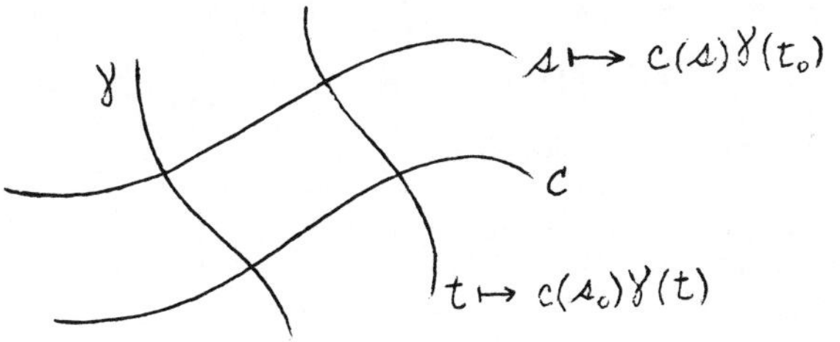

osculating plane of $s \longmapsto c(s)\cdot\gamma(t_0)$ at $s = s_0$, we must show that $\left.\frac{D'}{ds}\right|_{s=s_0} R_{\gamma(t_0)*}c'(s)$ is a linear combination of $R_{\gamma(t_0)*}c'(s_0)$ and $L_{c(s)*}\gamma'(t_0)$. Now we have

$$\left.\frac{D'}{ds}\right|_{s=s_0} R_{\gamma(t_0)*}c'(s) = R_{\gamma(t_0)*} \left.\frac{Dc'(s)}{ds}\right|_{s=s_0}$$

$$= \text{a linear combination of } R_{\gamma(t_0)*}c'(s_0) \text{ and } R_{\gamma(t_0)*}L_{c(s_0)*}\gamma'(0) , \quad \text{by (1)} .$$

So it suffices to observe that

$$R_{\gamma(t_0)*}L_{c(s_0)*}\gamma'(0) = L_{c(s_0)*}R_{\gamma(t_0)*}\gamma'(0)$$

$$= L_{c(s_0)*}\gamma'(t_0) ,$$

since the geodesic γ through $1 \in S^3$ is a 1-parameter subgroup, and hence the integral curve of a right invariant vector field (recall again the second proof Corollary I.10-8; although this proof deals with left invariant vector fields, it works just as well for right invariant vector fields). So our surface $M = \{c(s)\cdot\gamma(t)\}$ again has $K_{ext} = -1$.

Finally,* suppose that c and γ are both (distinct) geodesics through $1 \in S^3$. Then the surface $M = \{c(s)\cdot\gamma(t)\}$ still has $K_{ext} = -1$, or $K_{int} = 0$. To see this, we consider the parameter curves

---

*We will not consider the case where our asymptotic curves have curvature $\kappa(s) = 0$ for only certain s. The truly fanatical reader may wish to investigate this situation further.

$$\begin{array}{lcl} s \longmapsto c(s)\cdot\gamma(t_0) & \text{with tangent vectors} & R_{\gamma(t_0)*}c'(s_0) \\ t \longmapsto c(s_0)\cdot\gamma(t) & & L_{c(s_0)*}\gamma'(t_0)\ . \end{array}$$

We note that

$$\begin{aligned} \left\langle R_{\gamma(t_0)*}c'(s_0),\ L_{c(s_0)*}\gamma'(t_0)\right\rangle &= \left\langle R_{\gamma(t_0)*}L_{c(s_0)*}c'(0),\ L_{c(s_0)*}R_{\gamma(t_0)*}\gamma'(0)\right\rangle \\ &= \left\langle R_{\gamma(t_0)*}L_{c(s_0)*}c'(0),\ R_{\gamma(t_0)*}L_{c(s_0)*}\gamma'(0)\right\rangle \\ &= \langle c'(0),\ \gamma'(0)\rangle\ . \end{aligned}$$

Thus our surface has two families of geodesics intersecting at a constant angle, so it is flat by Proposition 4-6. In particular, the flat torus

$$\left\{x \in \mathbb{R}^4:\ x_1{}^2 + x_2{}^2 = \tfrac{1}{2} \text{ and } x_3{}^2 + x_4{}^2 = \tfrac{1}{2}\right\}$$

is of this form. It is generated by the two geodesics

$$\left\{\frac{1}{\sqrt{2}}(\cos\theta,\ \sin\theta,\ \cos\theta,\ \sin\theta)\right\}$$

$$\left\{\frac{1}{2}(\cos\phi+\sin\phi,\ \cos\phi-\sin\phi,\ -\cos\phi-\sin\phi,\ -\cos\phi+\sin\phi)\right\}\ ,$$

of which the second lies in the plane spanned by $(1,1,-1,-1)$, $(1,-1,-1,1)$ and the first in the orthogonal complement.

We now have a very general way of describing surfaces M in S^3 with $K_{ext} = -1$; we can take any "translation surface" $\{c(s)\cdot\gamma(t)\}$, where c and γ are curves of torsion 1 and -1 with $c(0) = \gamma(0) = 1 \in S^3$ and common osculating planes at 1. Since the curves c and γ are otherwise arbitrary, there are clearly a great number of such surfaces. We will describe some features of these surfaces in a little greater detail, and then indicate some open questions.

It will be very useful to introduce a famous creature of algebraic topology, the Hopf map $h\colon S^3 \to S^2$, which is defined as follows. We regard S^2 as the one-point compactification $\mathbb{C} \cup \{\infty\}$ of the complex numbers; the specific identification of S^2 and $\mathbb{C} \cup \{\infty\}$ will be given by means of stereographic projection, together with the identification of the north pole of S^2 with ∞. However, we will use a slightly different version of stereographic projection. We now regard S^2 as the standard unit sphere $\{p \in S^3\colon |p| = 1\}$, and map a point $p \in S^2 - \{(0,0,1)\}$ into the intersection $\sigma(p)$ of the (x,y)-plane with the straight line between $(0,0,1)$ and p. It is easy to check that

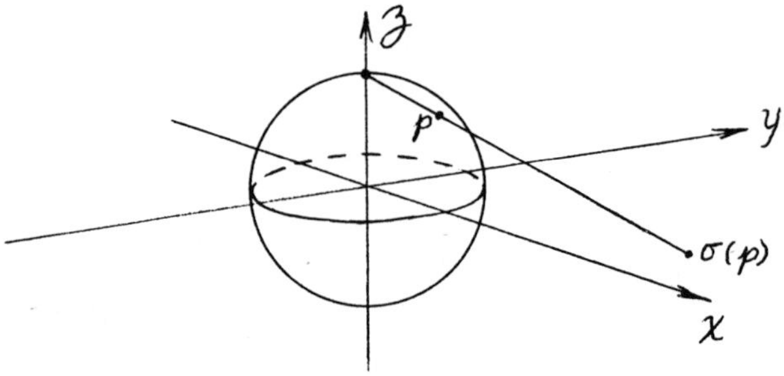

for our new σ we have

$$\sigma(a,b,c) = \left(\frac{a}{1-c}, \frac{b}{1-c}\right)$$

$$\sigma^{-1}(x,y) = \left(\frac{2x}{x^2+y^2+1}, \frac{2y}{x^2+y^2+1}, \frac{x^2+y^2-1}{x^2+y^2+1}\right) .$$

It is not hard to see that $S^2 = \mathbb{C} \cup \{\infty\}$ has a C^∞ atlas consisting of two maps

$$f_1 \colon \mathbb{C} \to \mathbb{C}$$
$$f_2 \colon \mathbb{C} - \{0\} \cup \{*\} \to \mathbb{C}$$

with f_1 = identity and

$$f_2(z) = \begin{cases} \frac{1}{z}, & z \neq \infty \\ 0, & z = \infty . \end{cases}$$

We consider S^3 as

$$\left\{(z_1,z_2) \in \mathbb{C} \times \mathbb{C} \colon |z_1|^2 + |z_2|^2 = 1\right\} .$$

Then $h\colon S^3 \to S^2$ is defined by

$$h(z_1,z_2) = \frac{z_1}{z_2} ,$$

where "z_1/z_2" $= \infty$ if $z_2 = 0$. This map is clearly C^∞ on the set where $z_2 \neq 0$, and also on the set where $z_1 \neq 0$, since we then have

$$f_2 \circ h(z_1,z_2) = \begin{cases} f_2\left(\frac{z_1}{z_2}\right), & z_2 \neq 0 \\ f_2(\infty), & z_2 = 0 \end{cases} = \begin{cases} \frac{z_2}{z_1}, & z_2 \neq 0 \\ 0, & z_2 = 0 \end{cases}$$
$$= \frac{z_2}{z_1} .$$

The inverse image $h^{-1}(z_0)$ of any point $z_0 \in \mathbb{C}$ is

$$h^{-1}(z_0) = \{(z_1,z_2) \in S^3 \colon z_1 = z_0 z_2\} .$$

If $z_j = x_j + iy_j$ for $j = 0,1,2$, this can be written as

$$h^{-1}(z_0) = \{(x_1,y_1,x_2,y_2) \in S^3\colon x_1 = x_0x_2 - y_0y_2 \text{ and } y_1 = x_0y_2 + x_2y_0\} ,$$

which is the intersection of S^3 with two hyperplanes through the origin. So $h^{-1}(z_0)$ is a great circle. Moreover,

$$h^{-1}(\infty) = \{(z_1,z_2) \in S^3\colon z_2 = 0\}$$

is also a great circle.

Now we need to know what the orthogonal maps $A\colon S^2 \to S^2$ look like when we consider them as maps $C \cup \{\infty\} \to C \cup \{\infty\}$. Elementary complex analysis tells that they must be maps of the form

$$f(z) = (az + b)/(cz + d) ,$$

for they must be one-one and have at most one pole, of order ≤ 1 (we can also use Problem 4-11 to reach the same conclusion). Some further calculations (Problem 21) shows that these maps, when normalized to have $ad\cdot bc = 1$, correspond to orthogonal maps if and only if

$$|a|^2 + |c|^2 = 1$$
$$|b|^2 + |d|^2 = 1$$
$$a\bar{b} = -c\bar{d} .$$

On the other hand, if these conditions are satisfied, then the map

$$g(z_1,z_2) = (az_1 + bz_2,\ cz_1 + dz_2)$$

is easily seen to be an isometry of $S^3 \subset \mathbb{C} \times \mathbb{C}$. Now for any set $X \subset S^2$ we have

$$\begin{aligned}
(z_1,z_2) \in h^{-1}(f^{-1}(X)) &\Leftrightarrow (z_1,z_2) \in S^3 \text{ and } \frac{z_1}{z_2} \in f^{-1}(X) \\
&\Leftrightarrow (z_1,z_2) \in S^3 \text{ and } \frac{a\frac{z_1}{z_2}+b}{c\frac{z_1}{z_2}+d} \in X \\
&\Leftrightarrow (z_1,z_2) \in S^3 \text{ and } \frac{az_1+bz_2}{cz_1+dz_2} \in X \\
&\Leftrightarrow (z_1,z_2) \in S^3 \text{ and } h(g(z_1,z_2)) \in X .
\end{aligned}$$

Thus

$$h^{-1}(f^{-1}(X)) = g^{-1}(h^{-1}(X)) .$$

In other words, if we want to know what $h^{-1}(X) \subset S^3$ looks like, up to an isometry of S^3, we can replace $X \subset S^2$ by any set related to X by an isometry of S^2. In particular, to find $h^{-1}(\Sigma)$ for $\Sigma \subset S^2$ a circle, we can assume that Σ is parallel to the (x,y)-plane, so that the stereographic projection of Σ in $\mathbb{C}$ is just a circle $\{z: |z| = R\}$. Then

$$h^{-1}(\{z: |z| = R\}) = \left\{(z_1,z_2): |z_1|^2 + |z_2|^2 = 1 \text{ and } \left|\frac{z_1}{z_2}\right| = R\right\}$$
$$= \left\{(z_1,z_2): |z_1| = \frac{R}{\sqrt{1+R^2}} \text{ and } |z_2| = \frac{1}{\sqrt{1+R^2}}\right\},$$

which is just a product torus. This shows that all product tori in S^3 are made up of a family of great circles, which are consequently asymptotic curves. When $R \neq 1$, the other asymptotic curves are not great circles. If they begin at one point of a great circle they will generally return to a different point of this great circle. This shows how a translation surface $\{c(s)\cdot\gamma(t)\}$ can

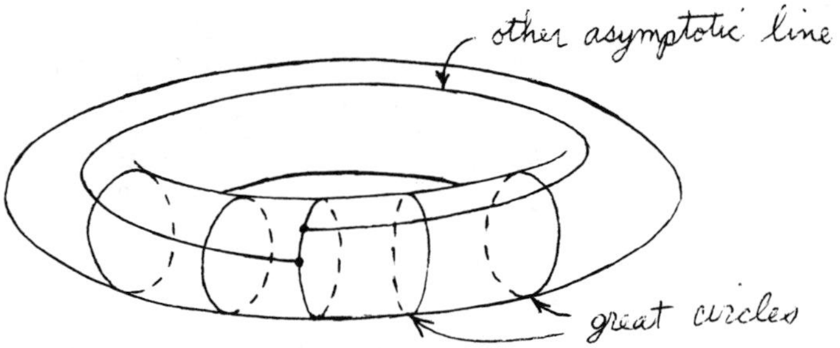

be compact even though the curve c or γ may not be closed.

Now let c be any immersed curve in S^3. We claim that the surface $h^{-1}(c)$ has $K_{ext} = -1$ everywhere. In fact, for any s_0, we can consider the osculating circle $\Sigma \subset S^2$ of c at s_0 (in other words, Σ is the circle in S^2 which is tangent to c at s_0 and whose curvature, as a curve in S^2, is the same as the curvature $\kappa(s_0)$ of c at s_0). Then Σ and c agree up to second order at $c(s_0)$, so $h^{-1}(\Sigma)$ and $h^{-1}(c)$ agree up to second order on the whole great circle $h^{-1}(c(s_0))$; since $h^{-1}(\Sigma)$ is a flat torus, with $K_{ext} = -1$ everywhere, $h^{-1}(c)$ must also have $K_{ext} = -1$ everywhere. Taking c to be an imbedded closed curve in S^2, we obtain an imbedded surface $h^{-1}(c)$ in S^3, with $K_{ext} = -1$, which is homeomorphic to a torus, but generally not a product torus. A non-geodesic asymptotic curve in $h^{-1}(c)$ will be a curve $\tilde{c}$ with $h \circ \tilde{c} = c$; it would be interesting (and probably very difficult) to determine for precisely which curves c this curve $\tilde{c}$ is closed. In this connection, we point out that there are certainly some closed curves in S^3 of constant torsion $\tau = 1$. In fact, just as cylindrical helices in $\mathbb{R}^3$ have constant torsion, the helices on product tori in S^3 are easily seen to have constant torsion, and in the latter case we can arrange for the helices to be closed. I do not know whether there are closed curves c and γ in S^3 of torsion $\tau = +1$ and $\tau = -1$ such that the translation surface $\{c(s)\cdot\gamma(t)\}$ is an imbedded torus (the helices on product tori give only immersed tori). Nor do I know the answer to the following problem, which seems quite hard; are there one-one curves c and γ in S^3 of torsion $\tau = +1$ and $\tau = -1$ such that the translation surface $\{c(s)\cdot\gamma(t)\}$ is a one-one map into S^3? Finally, one could try to analyze the non-orientable complete surfaces in S^3 with $K_{ext} = -1$.

Now we consider the case $N = H^3$, with constant curvature -1, so that (*) becomes

$$K_{int} = K_{ext} - 1 .$$

First we see that $K_{ext} < 0 \Longrightarrow K_{int} < 0$, so Theorem 41 shows that there are no complete surfaces immersed in H^3 with constant $K_{ext} < 0$. Since we also have $K_{ext} > 1 \Longrightarrow K_{int} > 0$, Theorem 39 implies that a complete surface immersed in H^3 with constant $K_{ext} > 1$ is all-umbilic; since $K_{int} > 0$, it must actually be a geodesic sphere.

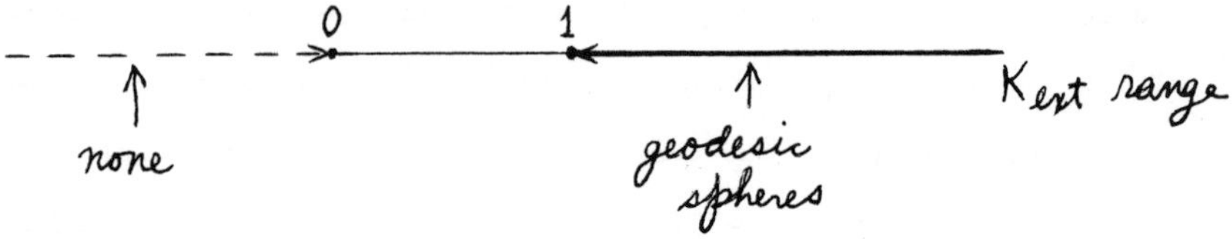

In the range $0 \le K_{ext} \le 1$ we have at least the totally geodesic spheres, the equidistant surfaces, and the horospheres, but we will find other examples also.

We consider first the upper range $K_{ext} = 1 \Rightarrow K_{int} = 0$. By considering the universal covering space of our immersed surface M with $K_{int} = 0$ we can assume that M is simply-connected. Thus M, with the induced metric, is isometric to $\mathbb{R}^2$ with its usual metric. Equivalently, we are considering isometric immersions $f\colon \mathbb{R}^2 \to H^3$, where $\mathbb{R}^2$ has its usual metric $dx \otimes dx + dy \otimes dy$, and H^3 has the metric $\langle\ ,\ \rangle$ of constant curvature -1. Let ℓ_{ij} be the coefficients of the second fundamental form II_f. In Gauss' equation,

$$\begin{aligned}\langle s(X,Z), s(Y,W)\rangle - \langle s(Y,Z), s(X,W)\rangle &= \langle R'(X,Y)Z,W\rangle - \langle R(X,Y)Z,W\rangle \\ &= -\,[\langle X,W\rangle\cdot\langle Y,Z\rangle - \langle X,Z\rangle\cdot\langle Y,W\rangle] - \langle R(X,Y)Z,W\rangle,\end{aligned}$$

we choose $X = Z = \partial/\partial x$ and $Y = W = \partial/\partial y$, to obtain

$$(1) \qquad \ell_{11}\ell_{22} - (\ell_{12})^2 = 1 .$$

In the Codazzi-Mainardi equations,

$$\begin{aligned} 0 &= (\nabla_X II)(Y,Z) - (\nabla_Y II)(X,Z) \\ &= X(II(Y,Z)) - Y(II(X,Z)) - \cdots + \cdots ,\end{aligned}$$

we take $X = \partial/\partial x$ and $Y = \partial/\partial y$, and then $Z = \partial/\partial x$ or $\partial/\partial y$ to obtain

(2) $$\frac{\partial \ell_{12}}{\partial x} = \frac{\partial \ell_{11}}{\partial y}, \qquad \frac{\partial \ell_{22}}{\partial x} = \frac{\partial \ell_{12}}{\partial y}.$$

These equations imply that there are functions $\alpha, \beta\colon \mathbb{R}^2 \to \mathbb{R}$ with

$$\text{(a)}\quad \frac{\partial \alpha}{\partial y} = \ell_{12} \qquad\qquad \text{(c)}\quad \frac{\partial \beta}{\partial y} = \ell_{22}$$

and

$$\text{(b)}\quad \frac{\partial \alpha}{\partial x} = \ell_{11} \qquad\qquad \text{(d)}\quad \frac{\partial \beta}{\partial x} = \ell_{12}.$$

Then (a) and (d) imply that there is a function $\phi\colon \mathbb{R}^2 \to \mathbb{R}$ with

$$\frac{\partial \phi}{\partial x} = \alpha \quad \text{and} \quad \frac{\partial \phi}{\partial y} = \beta.$$

Together with (b) and (c) we thus have

(3) $$\frac{\partial^2 \phi}{\partial x^2} = \ell_{11}, \qquad \frac{\partial^2 \phi}{\partial x \partial y} = \ell_{12}, \qquad \frac{\partial^2 \phi}{\partial y^2} = \ell_{22}.$$

Thus equation (1) yields

(*) $$\frac{\partial^2 \phi}{\partial x^2}\frac{\partial^2 \phi}{\partial y^2} - \left(\frac{\partial^2 \phi}{\partial x \partial y}\right)^2 = 1.$$

We now appeal to a strange result which is usually used in a completely different context (see Chapter 9):

<u>45. THEOREM (JÖRGENS)</u>. If $\phi\colon \mathbb{R}^2 \to \mathbb{R}$ is a function on the whole plane satisfying

(*) $$\frac{\partial^2\phi}{\partial x^2}\frac{\partial^2\phi}{\partial y^2} - \left(\frac{\partial^2\phi}{\partial x\partial y}\right)^2 = 1 ,$$

then ϕ is a quadratic polynomial in x and y.

<u>Proof</u>. We adopt the abbreviations

$$p = \frac{\partial\phi}{\partial x} \qquad q = \frac{\partial\phi}{\partial y}$$

$$r = \frac{\partial^2\phi}{\partial x^2} \qquad s = \frac{\partial^2\phi}{\partial x\partial y} \qquad t = \frac{\partial^2\phi}{\partial y^2} ,$$

so that our equation reads

(*) $$rt - s^2 = 1 .$$

This implies that $rt > 0$, so that r and t have the same sign. We can assume that $r, t > 0$ everywhere, by replacing ϕ by $-\phi$ if necessary.

For fixed (x_0,y_0) and (x_1,y_1), consider the function

$$h(\tau) = \phi(x_0 + \tau(x_1 - x_0),\ y_0 + \tau(y_1 - y_0)) .$$

We have

$$h'(\tau) = (x_1 - x_0)p + (y_1 - y_0)q ,$$
$$h''(\tau) = (x_1 - x_0)^2 r + 2(x_1 - x_0)(y_1 - y_0)s + (y_1 - y_0)^2 t ,$$

where p, q, r, s, t are evaluated at $(x_0 + \tau(x_1 - x_0),\ y_0 + \tau(y_1 - y_0))$. If $x_1 = x_0$, then $h''(\tau) = (y_1 - y_0)^2 t \geq 0$. If $x_1 \neq x_0$, then

$$h''(\tau) = (x_1 - x_0)^2\left[r - 2\left(\frac{y_1 - y_0}{x_1 - x_0}\right)s + \left(\frac{y_1 - y_0}{x_1 - x_0}\right)^2 t\right] .$$

The term in brackets is a quadratic polynomial in $(y_1 - y_0)/(x_1 - x_0)$ with discriminant $4s^2 - 4rt < 0$, by (*), so it is always positive. Thus we always have $h''(\tau) \geq 0$. This implies that

$$h'(1) \geq h'(0) ,$$

and thus

$$(1) \qquad (x_1 - x_0)(p_1 - p_0) + (y_1 - y_0)(q_1 - q_0) \geq 0 ,$$

where

$$p_i = p(x_i, y_i) , \qquad q_i = q(x_i, y_i) \qquad i = 0, 1 .$$

Consider the transformation of Lewy:

$$T(x,y) = (\xi(x,y), \eta(x,y)) = (x + p(x,y),\ y + q(x,y)) .$$

If we set

$$\xi_i = \xi(x_i, y_i) , \qquad \eta_i = \eta(x_i, y_i) \qquad i = 0, 1 ,$$

then equation (1) implies that

$$(\xi_1 - \xi_0)^2 + (\eta_1 - \eta_0)^2 \geq (x_1 - x_0)^2 + (y_1 - y_0)^2 .$$

Hence $T: \mathbb{R}^2 \to \mathbb{R}^2$ is distance-increasing, and, in particular, T is one-one. Moreover, the Jacobian of T is

$$\begin{pmatrix} \frac{\partial \xi}{\partial x} & \frac{\partial \xi}{\partial y} \\ \frac{\partial \eta}{\partial x} & \frac{\partial \eta}{\partial y} \end{pmatrix} = \begin{pmatrix} 1 + r & s \\ s & 1 + t \end{pmatrix} ,$$

with determinant

$$1 + r + t + rt - s^2 = 2 + r + t \qquad \text{by } (*)$$
$$\geq 2 \,,$$

so T is an immersion, and image T is open. But image T is also closed: For if $T(x_i, y_i) \to \alpha \in \mathbb{R}^2$, so that $\{T(x_i, y_i)\}$ is a Cauchy sequence, then $\{(x_i, y_i)\}$ is also a Cauchy sequence, since T is distance-increasing; thus $(x_i, y_i) \to \beta \in \mathbb{R}^2$, and $T(\beta) = \alpha$. So T is actually a diffeomorphism of $\mathbb{R}^2$ onto itself. It will be convenient to use classical ambiguous notation and denote the inverse map T^{-1} by $(\xi, \eta) \mapsto (x(\xi, \eta), y(\xi, \eta))$. Its Jacobian is

$$\begin{pmatrix} \frac{\partial x}{\partial \xi} & \frac{\partial x}{\partial \eta} \\ \frac{\partial y}{\partial \xi} & \frac{\partial y}{\partial \eta} \end{pmatrix} = \begin{pmatrix} 1+r & s \\ s & 1+t \end{pmatrix}^{-1}$$
$$= \frac{1}{2+r+t} \begin{pmatrix} 1+t & -s \\ -s & 1+r \end{pmatrix} ,$$

from which we can read off the partial derivatives of x and y.

Now define $F: \mathbb{R}^2 \to \mathbb{R}^2$ by

$$\begin{aligned} F(\xi, \eta) &= (U(\xi, \eta), V(\xi, \eta)) \\ &= (x - p, -y + q) \qquad \text{i.e.,} \\ &= (x(\xi, \eta) - p(x(\xi, \eta), y(\xi, \eta)), \; -y(\xi, \eta) + q(x(\xi, \eta), y(\xi, \eta))) \,. \end{aligned}$$

Then

$$\begin{aligned} \frac{\partial U}{\partial \xi} &= \frac{\partial x}{\partial \xi} - \frac{\partial p}{\partial x}\frac{\partial x}{\partial \xi} - \frac{\partial p}{\partial y}\frac{\partial y}{\partial \xi} \\ &= \frac{1}{2+r+t}[1 + t - r(1+t) - s(-s)] \\ &= \frac{t - r}{2+r+t} \,. \end{aligned}$$

Similarly, we find that

$$\frac{\partial V}{\partial \eta} = \frac{t - r}{2 + r + t} = \frac{\partial U}{\partial \xi}$$

and

$$\frac{\partial V}{\partial \xi} = \frac{2s}{2 + r + t} = - \frac{\partial U}{\partial \eta} .$$

Thus (U,V) satisfies the Cauchy-Riemann equations, so the map $F: \mathbb{C} \longrightarrow \mathbb{C}$ defined by

$$\begin{aligned} F(\xi + i\eta) &= U(\xi,\eta) + iV(\xi,\eta) \\ &= x - p + (-y + q)i \end{aligned}$$

is complex analytic, and for the complex derivative F' we have

$$\begin{aligned} F'(\xi + i\eta) &= \frac{\partial U}{\partial \xi} + i \frac{\partial V}{\partial \xi} \\ &= \frac{t - r + 2is}{2 + r + t} . \end{aligned} \tag{2}$$

Consequently,

$$\begin{aligned} |F'(\xi + i\eta)|^2 &= \frac{(t - r)^2 + 4s^2}{(2 + r + t)^2} \\ &= \frac{(t - r)^2 + 4rt - 4}{(2 + r + t)^2} \qquad \text{by } (*) \\ &= \frac{(t + r)^2 - 4}{(2 + r + t)^2} = \frac{-2 + r + t}{2 + r + t} , \end{aligned}$$

which gives

$$1 - |F'(\xi + i\eta)|^2 = \frac{4}{2 + r + t} > 0 . \tag{3}$$

Thus F' is bounded, and consequently constant, by Liouville's theorem. But equations (2) and (3) allow us to solve for r, s, t in terms of F':

$$s = \frac{2+r+t}{2} \cdot \operatorname{Im} F' = \frac{2 \cdot \operatorname{Im} F'}{1-|F'|^2}$$

$$\left.\begin{aligned} t - r &= \frac{4 \operatorname{Re} F'}{1-|F'|^2} \\ t + r &= \frac{4}{1-|F'|^2} - 2 \end{aligned}\right\} \Longrightarrow \begin{aligned} t &= \frac{1}{2}\left(\frac{4 \operatorname{Re} F'}{1-|F'|^2} + \frac{4}{1-|F'|^2} - 2\right) \\ r &= \frac{1}{2}\left(\frac{4}{1-|F'|^2} - 2 - \frac{4 \operatorname{Re} F'}{1-|F'|^2}\right) . \end{aligned}$$

Since F' is constant, so are r, s, t. ■

Applying Jörgens' Theorem to our ϕ, we find that the ℓ_{ij} are <u>constants</u>. We can assume, moreover, that $\ell_{12} = 0$, by means of an orthogonal transformation of $\mathbb{R}^2$. Then $\ell_{11} = k_1$ and $\ell_{22} = k_2$ are the principal curvatures of the immersed surface $f(\mathbb{R}^2)$, and $k_1 k_2 = 1$. By Theorem 21, the immersion f is determined, up to an isometry of H^3, by the pair $\{k_1, k_2\}$, with $k_1, k_2 > 0$. So in order to determine all such f, we just have to find one for each pair $\{k_1, k_2\}$ with $k_1 k_2 = 1$. For $k_1 = k_2 = 1$, all points are umbilics, and f must be a horosphere. For the others, consider the upper half space model $\mathbb{H}^3$. Our immersed surface $M \subset \mathbb{H}^3$ with constant k_1, k_2 must have isometries of $\mathbb{H}^3$ taking any point to any other. Now one simple case of isometries of $\mathbb{H}^3$ are the inversions with respect to a sphere around 0. These isometries take rays through 0 into themselves, and thus take cones through 0 into themselves. Moreover, if we consider only right circular cones, then there are clearly isometries of $\mathbb{H}^3$ taking any point on a circle parallel to the (x,y)-plane to any other point on this circle, and hence there are

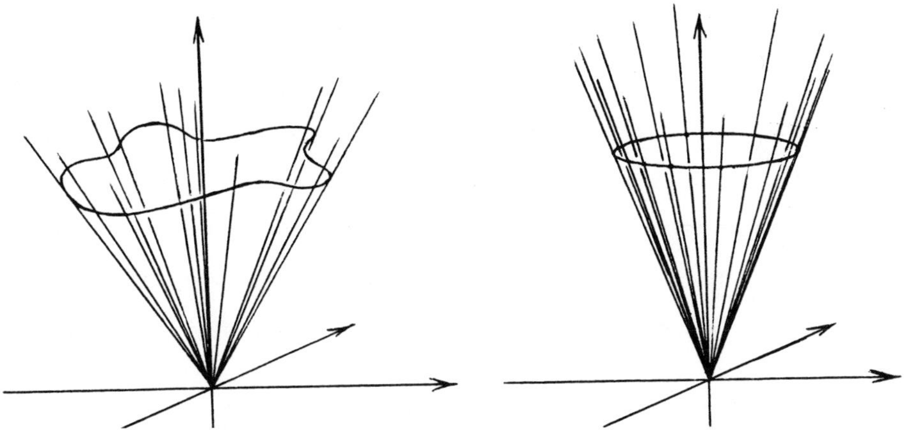

isometries of $\mathbb{H}^3$ taking any point on the cone to any other point. These cones thus have constant k_1, k_2. A simple calculation shows, in fact, that if the generators of the cone make an angle of θ with the z-axis, then the principal curvature k_1 for the principal vectors pointing along the generators is

$$k_1 = \sin\theta ,$$

while the principal curvature k_2 for the principal vectors pointing along the circles parallel to the (x,y)-plane is

$$k_2 = \frac{1}{\sin\theta} .$$

Thus $k_1 k_2 = 1$, and all pairs (k_1,k_2) are accounted for. We note, finally, that by the discussion on pp. 20-21, our cone is the set of points at a fixed distance from the z-axis. We thus have

46. THEOREM (VOLKOV AND VLADIMIROVA; SASAKI). A complete surface in H^3 with constant $K_{ext} = 1$ is either a horosphere or the set of points at a fixed distance from a geodesic.

Next we consider the lower range $K_{ext} = 0 \Longrightarrow K_{int} = -1$. We have already indicated that there are many complete surfaces $M \subset H^3$ with $K_{ext} = 0$, but now we will look more closely at their topological type. We know that if $B \subset \mathbb{R}^3$ is the projective model of H^3 (so B is the unit ball with a metric of constant curvature -1 whose geodesics are reparameterized straight lines of $\mathbb{R}^3$), then a surface $M \subset B$ has $K_{int} = -1$ if and only if M is flat, considered as a surface in $\mathbb{R}^3$ with the usual metric. Consider the intersection of a plane with B, and a portion P of this plane which is bounded by four non-intersecting geodesics $\gamma_1, \ldots, \gamma_4$. The geodesics γ_1 and γ_3 can be

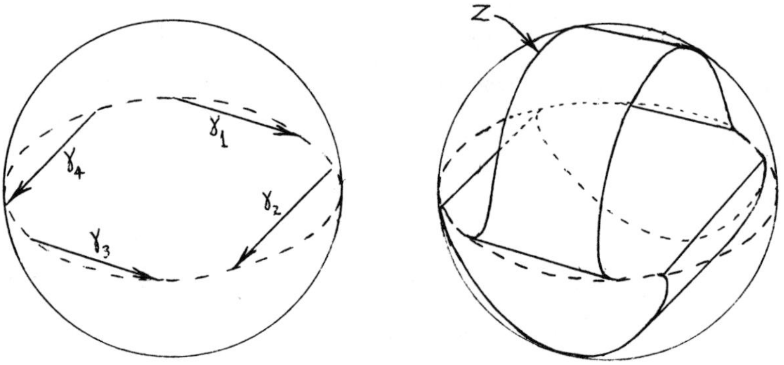

joined by a cylinder Z, and similarly γ_2 and γ_4 can be joined by a disjoint cylinder Z'. By choosing appropriate profile curves for these cylinders we can make a smooth surface $P \cup Z \cup Z'$, and it will have $K_{int} = -1$ everywhere. The resulting surface is topologically equivalent to a torus minus a disc (or a torus minus a point). By beginning with a portion P bounded by $2g$ non-intersecting geodesics, we can obtain surfaces homeomorphic to any compact surface with a point deleted. Notice that this construction produces only C^∞ surfaces, not analytic ones. It seems to me that all analytic flat surfaces

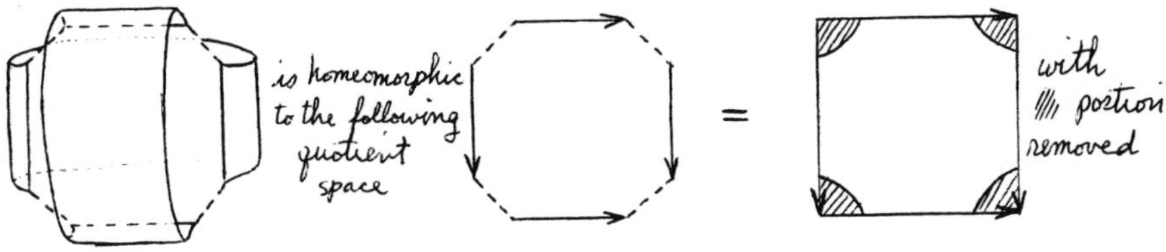

in $\mathbb{R}^3$, and a posteriori all complete analytic surfaces in B with $K_{int} = -1$, must be homeomorphic to a plane, cylinder, or Möbius strip; but I haven't tried to make a rigorous proof. If this does indeed turn out to be the case, it will be one of the rare instances where the requirements of smoothness and analyticity lead to different geometric conclusions.

We are still left with the complete surfaces in H^3 with $0 < K_{ext} < 1$. We can obtain infinitely many examples of such surfaces by looking at surfaces of revolution. Given a geodesic γ in the hyperbolic plane, we can describe a complete arc-length parameterized curve $s \mapsto c(s)$ in the hyperbolic plane in terms of the distance $r(s)$ from $c(s)$ to γ. A curve c can be found

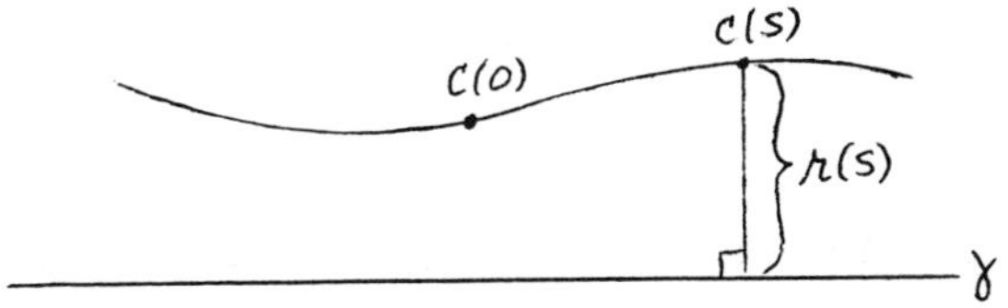

with a given function r provided that $|r'| \leq 1$, so that $|r(s_1) - r(s_0)| \leq s_1 - s_0$. If we rotate c around γ in hyperbolic 3-space, then the first fundamental form of our surface is (Problem 22)

$$I = \sinh^2 r(s)\, d\theta \otimes d\theta + ds \otimes ds ,$$

and we compute that its intrinsic curvature is

$$K_{int} = - \frac{1}{\sinh r(s)} \cdot \frac{d^2 \sinh r(s)}{ds^2} .$$

Setting $K_{int} = - c^2$, we obtain the strictly positive solution

$$\sinh r(s) = e^{cs} , \qquad \text{as well as} \qquad \sinh r(s) = a \cosh(cs) , \quad a > 0 .$$

Both solutions satisfy $|r'| < 1$ for $0 < c < 1$. Thus for each $K_{ext} = 1 - c^2$ with $0 < K_{ext} < 1$ we obtain a 1-parameter family of distinct surfaces, and one additional surface. In the model $(B^3, < , >)$ these surfaces look like

I do not know whether these are the only surfaces with $0 < K_{ext} < 1$, as in the case of surfaces with $K_{ext} = 1$, or if there are many others, as in the case $K_{ext} = 0$.

G. Hypersurfaces of Constant Curvature in Higher Dimensions

We now want to consider hypersurfaces $M^n \subset N^{n+1}$, where $(N, < , >)$ is a manifold of constant curvature K_0 and of dimension > 3. We are interested in the hypersurfaces M of constant curvature; since M is no longer a surface, there is no ambiguity of meaning here -- we are requiring that M, with the induced metric, have all sectional curvatures equal. After the exertions of the last section, it is a relief to find that everything is now much <u>easier</u>, and most of the results are essentially <u>local</u>. For example, we claim that there is no 3-dimensional manifold $M \subset \mathbb{R}^4$ with constant curvature -1, not even a non-complete one. In fact, if M has principal curvatures k_1, k_2, k_3 at p, then all products $k_i k_j$ must $= -1$, which is clearly impossible, since at least two of the k_i will have the same sign. More generally,

<u>47. THEOREM</u>. For $n > 2$, let N^{n+1} be a manifold of constant curvature K_0, and let $M^n \subset N^{n+1}$ be a hypersurface of constant curvature K. Then $K \geq K_0$. If $K > K_0$, then all points of M are umbilics, and if $K = K_0$, then at most one principal curvature is non-zero.

<u>Proof</u>. Let $k_1, \ldots, k_n$ be the principal curvatures at p. Gauss' equation shows that

$$K - K_0 = k_i k_j , \qquad i \neq j .$$

If $K = K_0$, then $k_i k_j = 0$ for all $i \neq j$, so if k_1, say, is $\neq 0$, then $k_2, \ldots, k_n = 0$. If $K - K_0 \neq 0$, then all $k_i \neq 0$, so the equation

$$k_1 k_i = k_1 k_j \qquad i, j \neq 1$$

implies that $k_2 = \cdots = k_n$. Similarly, $k_1 = \cdots = k_{n-1}$. Since $n > 2$, this implies that $k_1 = \cdots = k_n$. So all points are umbilics. Moreover, $K - K_0 = k_1 k_2 = (k_1)^2 > 0$. ■

We will examine the case $K = K_0$ in more detail later on. But first we indicate how more general results can be obtained by considering a certain covariant tensor of order 2, the Ricci tensor Ric of M. The map $\mathrm{Ric}(p)\colon M_p \times M_p \to \mathbb{R}$ is defined by

$$\mathrm{Ric}(p)\,(X_1,X_2) = \text{trace } Y \mapsto R(X_2,Y)X_1 \qquad Y \in M_p\,.$$

In terms of the components $R^i{}_{jk\ell}$ of R in a coordinate system $x^1,\ldots,x^n$, the tensor Ric is given by

$$\mathrm{Ric} = \sum_{j,k=1}^{n} \mathrm{Ric}_{jk}\, dx^j \otimes dx^k \qquad \text{for} \qquad \mathrm{Ric}_{jk} = \sum_{i=1}^{n} R^i{}_{jki}\ ;$$

thus Ric is obtained from R by contraction. If $X_1,\ldots,X_n$ is an orthonormal basis for M_p, then $\mathrm{Ric}(X_1,X_1)$ is the trace of the matrix $(\langle R(X_1,X_i)X_1,X_j\rangle)$. Therefore

$$\begin{aligned}\mathrm{Ric}(X_1,X_1) &= \sum_{i=1}^{n} \langle R(X_1,X_i)X_1,X_i\rangle \\ &= -\sum_{i=2}^{n} \langle R(X_i,X_1)X_1,X_i\rangle\ .\end{aligned}$$

So if $X \in M_p$ is any unit vector, then $-\mathrm{Ric}(X,X)$ is the sum of the sectional curvatures determined by X and any $n-1$ orthonormal vectors orthogonal to X. (We have defined Ric so that it agrees with the classical definition $\mathrm{Ric}_{jk} = \sum_i R^i{}_{jki}$; nowadays, the opposite sign is often used.) The following result is analogous to Schur's Theorem (II. 7-19).

48. THEOREM. If M is a connected Riemannian manifold of dimension $n \geq 3$ and

$$\mathrm{Ric}(X,Y) = \lambda\langle X,Y\rangle$$

for some function λ on M, then λ is constant.

Proof. Bianchi's second identity (II. 5-9), together with Ricci's Lemma, gives

(1) $$0 = R_{hijk;\ell} + R_{hik\ell;j} + R_{hi\ell j;k} = 0 .$$

Multiply by $\sum_{h,i,j,k} g^{hj}g^{ik}$. We have

$$\sum g^{hj}g^{ik}R_{hijk;\ell} = - \sum g^{hj}g^{ik}R_{ihjk;\ell} = - \sum_{h,j} g^{hj} \sum_{k} R^{k}{}_{hjk;\ell}$$

$$= - \sum_{h,j} (g^{hj}Ric_{hj})_{;\ell} = - \sum_{h,j} (g^{hj}g_{hj}\lambda)_{;\ell}$$

$$= - n \frac{\partial\lambda}{\partial x^{\ell}} ,$$

$$\sum g^{hj}g^{ik}R_{hik\ell;j} = \sum g^{hj}g^{ik}R_{ih\ell k;j} = \sum_{h,j} g^{hj} \sum_{k} R^{k}{}_{h\ell k;j}$$

$$= \sum_{h,j} (g^{hj}Ric_{h\ell})_{;j} = \sum_{h,j} (g^{hj}g_{h\ell}\lambda)_{;j}$$

$$= \frac{\partial\lambda}{\partial x^{\ell}} ,$$

$$\sum g^{ik}g^{hj}R_{hi\ell j;k} = \sum g^{ik}g^{hj}R_{j\ell ih;k} = \sum_{i,k} g^{ik} \sum_{h} R^{h}{}_{\ell ih;k}$$

$$= \sum_{i,k} (g^{ik}Ric_{\ell i})_{;k} = \sum_{i,k} (g^{ik}g_{\ell i}\lambda)_{;k}$$

$$= \frac{\partial\lambda}{\partial x^{\ell}} .$$

So (1) becomes

$$(n - 2) \frac{\partial\lambda}{\partial x^{\ell}} = 0 .$$

Since $n > 2$, we have $\partial\lambda/\partial x^{\ell} = 0$ for all ℓ. ■

A Riemannian manifold M with $\mathrm{Ric} = -\lambda\langle\ ,\ \rangle$ is called an <u>Einstein space</u>, and λ is sometimes called its <u>mean curvature</u> (not to be confused with the mean curvature H of a submanifold). If M has constant curvature K, then M is an Einstein space with mean curvature $\lambda = (n-1)K$. We note in passing that

<u>49. THEOREM</u>. A connected 3-dimensional Einstein space is a manifold of constant curvature.

<u>Proof</u>. Choose an orthonormal basis $X_1, X_2, X_3 \in M_p$, and let $K_{ij} = K_{ji}$ be the sectional curvature of the 2-dimensional subspace of M_p spanned by X_i and X_j. Then

$$-\mathrm{Ric}(X_1,X_1) = K_{12} + K_{13}$$
$$-\mathrm{Ric}(X_2,X_2) = K_{21} + K_{23}$$
$$-\mathrm{Ric}(X_3,X_3) = K_{31} + K_{32}\ .$$

Hence

$$-\mathrm{Ric}(X_1,X_1) - \mathrm{Ric}(X_2,X_2) + \mathrm{Ric}(X_3,X_3) = 2K_{12}\ .$$

Since all $\mathrm{Ric}(X_i,X_i) = -\lambda$, we have $K_{12} = \lambda/2$. ■

Now for a manifold $(N^{n+1}, \langle\ ,\ \rangle)$ of constant curvature K_0 we consider hypersurfaces $M \subset N$ which are Einstein spaces.

<u>50. THEOREM</u>. For $n > 2$, let N^{n+1} be a manifold of constant curvature K_0, and let $M^n \subset N^{n+1}$ be a hypersurface which is an Einstein space with

$Ric = -\lambda\langle\ ,\ \rangle$. If $\lambda > (n-1)K_0$, then all points of M are umbilics, and M is a manifold of constant curvature $K > K_0$. If $\lambda = (n-1)K_0$, then at most one principal curvature is non-zero, and M is a manifold of constant curvature K_0.

<u>Proof</u>. Let $X_1, \ldots, X_n \in M_p$ be principal directions with corresponding principal curvatures $k_1, \ldots, k_n$. Gauss' equation gives

$$(1) \qquad K_{ij} = k_i k_j + K_0 ,$$

where K_{ij} is the sectional curvature of the subspace of M_p spanned by X_i and X_j. Then

$$\lambda = \sum_{j\neq i} K_{ij} = \sum_{j\neq i} k_i k_j + (n-1)K_0$$
$$= (\sum_j k_j)k_i - (k_i)^2 + (n-1)K_0 .$$

Hence all principal curvatures k_i satisfy the equation

$$(*) \qquad x^2 - (\sum_j k_j)x + [\lambda - (n-1)K_0] = 0 .$$

If $\lambda = (n-1)K_0$, then every k_i is either 0 or the number $\sum_j k_j$. So there can clearly be only one $k_i \neq 0$. Then equation (1) shows that all K_{ij} equal K_0, so that M is a manifold of constant curvature K_0.

If $\lambda > (n-1)K_0$, then all k_i are one of the two roots α, β of (*), where

$$(2) \qquad \alpha\beta = \lambda - (n-1)K_0 > 0$$

$$(3) \qquad \alpha + \beta = \sum_j k_j .$$

If p of the k_i equal α, and the other $q = n - p$ of the k_i equal β, then equation (3) can be written

$$\alpha + \beta = p\alpha + q\beta \implies (p-1)\alpha + (q-1)\beta = 0 .$$

But α and β have the same sign, by (2), so either $p-1$ or $q-1$ is negative, which means that either p or q is zero. Thus all k_i are equal. Then equation (1) shows that all K_{ij} equal $K_0 + (k_1)^2 > K_0$, so that M is a manifold of constant curvature $K > K_0$. ■

Theorem 50 is not the best that can be obtained, for it is also known that if $\lambda < (n-1)K_0$, then K_0 must be > 0, and that in this case M must be one of a certain special class of hypersurfaces, with $\lambda = (n-2)K_0$. For the proof of this, the reader is referred to the original paper of Fialkow [1]; see also Ryan [1].

We now consider the critical case of an immersion $f: M^n \to N^{n+1}$, where M and N have the same constant curvature K_0, so that at most one principal curvature of $f(M)$ is non-zero at each point. If all principal curvatures are zero everywhere, so that the second fundamental form $s = 0$, then M is totally geodesic. Otherwise, we can consider the non-empty open set $U \subset M$ defined by

$$U = \{p \in M: s \neq 0 \text{ at } p\} .$$

Around any point $p \in U$, we choose an adapted orthonormal moving frame

$$X_1,\ldots,X_{n-1},X_n,X_{n+1}$$

such that $X_1,\ldots,X_{n-1}$ are principal vectors with principal curvatures 0, and X_n is a principal vector with non-zero principal curvature λ. [Our moving frame is really defined in a neighborhood of $f(p) \in N$, but for simplicity we will regard $M \subset N$ for all local arguments.] Let ϕ^α, ψ^α_β be the forms for this moving frame, and let θ^i, ω^i_j be the forms for $X_1,\ldots,X_n$. Then we have

$$(1) \qquad \psi^{n+1}_i = 0 \qquad i = 1,\ldots,n-1$$

$$(2) \qquad \psi^{n+1}_n = -\lambda\theta^n .$$

For $i = 1,\ldots,n-1$, the Codazzi-Mainardi equations give

$$0 = d\psi^{n+1}_i = \Psi^{n+1}_i - \sum_{j=1}^n \psi^{n+1}_j \wedge \omega^j_i = \Psi^{n+1}_i + \lambda\theta^n \wedge \omega^n_i .$$

Since $\Psi^{n+1}_i(X,Y) = 0$ for X, Y tangent to M, we find that $\theta^n \wedge \omega^n_i = 0$, or

$$(3) \qquad \omega^n_i \text{ is a multiple of } \theta^n \text{ (on } U) \qquad i = 1,\ldots,n-1 .$$

From this we derive a higher dimensional analogue of Proposition 5-4.

<u>51. PROPOSITION</u>. The distribution Δ on $U \subset M$ defined by

$$\begin{aligned}\Delta(p) &= \{X \in M_p\colon s(X,Y) = 0 \text{ for all } Y \in M_p\} \\ &= \{X \in M_p\colon X \text{ is a principal vector with principal curvature } 0\}\end{aligned}$$

is integrable. Every integral manifold of Δ is a totally geodesic submanifold of M, and is immersed as a totally geodesic submanifold in N.

Proof. Locally Δ is defined by $d\theta^1 = \cdots = d\theta^{n-1} = 0$. Now on U we have

$$d\theta^i = -\sum_{j=1}^{n} \omega_j^i \wedge \theta^j = -\sum_{j=1}^{n-1} \omega_j^i \wedge \theta^j \qquad \text{by (3)} .$$

So Proposition I.7-14 shows that Δ is integrable; the vector fields $X_1,\ldots,X_{n-1}$ are tangent along an integral manifold M_1 of Δ. Equation (3) says that

$$0 = \omega_i^n(X_j) = \langle \nabla'_{X_j} X_i, X_n \rangle \qquad i, j \le n-1 ,$$

i.e., that the second fundamental form of M_1 in M is zero. Since we also have

$$\langle \nabla'_{X_j} X_i, X_{n+1} \rangle = 0 \qquad i, j \le n-1$$

by the definition of Δ, it follows that M_1 is totally geodesic in N. ■

Now we want to study the function λ along a geodesic γ lying in an integral manifold M_1 of Δ.

52. LEMMA. Let γ be an arc-length parameterized geodesic in an integral manifold M_1 of Δ, and let $\lambda(s)$ be the value of the non-zero principal curvature at $\gamma(s)$. Then the function $\lambda(s)$ satisfies the differential equation

$$\left(\frac{1}{\lambda}\right)'' = -\frac{K_0}{\lambda} .$$

Proof. Choose the moving frame so that γ is an integral curve of X_1.

Equations (2) and (3) imply that there are g_i with

$$\omega_i^n = g_i \psi_n^{n+1} \, . \tag{3'}$$

The Codazzi-Mainardi equation for $i = n$ gives

$$\begin{aligned} d\psi_n^{n+1} &= \Psi_n^{n+1} - \sum_{j=1}^{n} \psi_j^{n+1} \wedge \omega_n^j \\ &= \Psi_n^{n+1} \, , \qquad \text{by (1)} \, . \end{aligned} \tag{4}$$

Thus

$$\begin{aligned} \Psi_n^{n+1} = d\psi_n^{n+1} &= - d\lambda \wedge \theta^n - \lambda \, d\theta^n \qquad \text{by (2)} \\ &= - d\lambda \wedge \theta^n + \lambda \sum_{i=1}^{n} \omega_i^n \wedge \theta^i \, , \end{aligned}$$

and therefore

$$d\lambda \wedge \theta^n = - \lambda \sum_{i=1}^{n} \theta^i \wedge \omega_i^n - \Psi_n^{n+1} \, .$$

Applying this to (X_1, X_n) gives

$$\begin{aligned} X_1(\lambda) &= - \lambda\omega_1^n(X_n) + \lambda\omega_1^n(X_1) \\ &= - \lambda\omega_1^n(X_n) && \text{by (3)} \\ &= - \lambda g_1 \psi_n^{n+1}(X_n) && \text{by (3')} \\ &= \lambda^2 g_1 && \text{by (2)} \, , \end{aligned}$$

which we can also write as

$$X_1\left(\frac{1}{\lambda}\right) = - g_1 \, . \tag{*}$$

Now on M we have the structural equation

$$d\omega_1^n = -\sum_{k=1}^{n-1} \omega_k^n \wedge \omega_1^k + \Omega_1^n ,$$

which by (3') becomes

$$d(g_1\psi_n^{n+1}) = -\sum_{k=1}^{n-1} g_k\psi_n^{n+1} \wedge \omega_1^k + \Omega_1^n ,$$

and thus by (2) and (4)

$$dg_1 \wedge \psi_n^{n+1} + g_1\Psi_n^{n+1} = -\lambda(\Sigma\ g_k\omega_1^k) \wedge \theta^n + \Omega_1^n .$$

Finally, we use (2) again to write our equation as

$$-\lambda\ dg_1 \wedge \theta^n + g_1\Psi_n^{n+1} = -\lambda(\Sigma\ g_k\omega_1^k) \wedge \theta^n + \Omega_1^n .$$

Applying this to (X_1, X_n) we get

$$-\lambda X_1(g_1) + 0 = 0 - K_0 ,$$

since all $\omega_1^k(X_1) = 0$. Thus (*) yields

$$X_1\left(X_1\left(\frac{1}{\lambda}\right)\right) = X_1(-g_1) = -\frac{K_0}{\lambda} . \quad \blacksquare$$

The solutions of the equation $(1/\lambda)'' = -K_0(1/\lambda)$ can be found explicitly -- $1/\lambda$ is linear if $K_0 = 0$, a linear combination of sin and cos if $K_0 > 0$, and a linear combination of sinh and cosh if $K_0 < 0$. In any case, $1/\lambda$ is bounded on any bounded interval.

53. COROLLARY. If M is complete, then the integral manifolds of Δ are complete.

Proof. We just have to show that a geodesic of an integral manifold M_1 cannot approach a boundary point of U. The argument is almost the same as that in the proof of Corollary 5-6. ∎

It is now a straightforward matter to generalize Theorem 5-9. We will make things easy for ourselves by choosing the simplest proof.

54. THEOREM. If M is a complete flat n-manifold and $f\colon M \to \mathbb{R}^{n+1}$ is an isometric immersion, then $f(M)$ is a generalized cylinder (it is congruent to a set of the form $\gamma \times \mathbb{R}^{n-1}$ for some curve $\gamma \subset \mathbb{R}^2$).

Proof. We can assume M is simply connected, and thus $\mathbb{R}^n$. If $f(M)$ is not totally geodesic, then the set $U \subset M$ is non-empty, so by Corollary 53 some hyperplane of M is mapped isometrically onto an $(n-1)$-dimensional plane of $\mathbb{R}^{n+1}$. Now apply the third proof of Theorem 5-9. ∎

We also obtain complete information for the case $K_0 > 0$.

55. THEOREM. If M^n is a complete manifold of constant curvature 1 and $f\colon M^n \to S^{n+1}$ is an isometric immersion, then $f(M)$ is a great n-sphere in S^{n+1}.

Proof. We can assume M is simply connected, and thus S^n. If $f(M)$ were not totally geodesic, then the set $U \subset M$ would be non-empty, so Corollary 53 would show that there are two disjoint complete totally geodesic $(n-1)$-dimensional submanifolds of $M = S^n$. This is impossible. ■

In the case $K_0 < 0$ we would not expect such good results, since even the case $n = 2$ is so complicated. Actually, the case $n = 2$ already contains essentially all the complexity there is, for one can show that if M^n is a complete manifold of constant curvature -1 in H^{n+1}, then the higher dimensional cohomology vanishes,

$$H^i(M) = 0 \qquad \text{for } i > 1 .$$

This is essentially a consequence of the analysis already provided, although technical details are required for a rigorous proof (see O'Neil [1]).

Addendum 1. The Laplacian

The material of these first 3 Addenda is essentially a part of intrinsic Riemannian geometry, and might thus seem out of place in this Chapter. But I felt that it was appropriate to put it here since this is the first time in a long while that we have seriously considered higher dimensional Riemannian manifolds. Moreover, the next Chapter will be devoted to material which is completely intrinsic in nature. Finally, some of the material covered here will be used when we return to the study of extrinsic geometry in Chapter 9.

In classical "vector analysis", there are three operators which play a crucial role. First of all, for every smooth function $f\colon \mathbb{R}^n \to \mathbb{R}$ we have a vector field, the gradient of f, defined by

$$\text{grad } f = \left(\frac{\partial f}{\partial x^1}, \ldots, \frac{\partial f}{\partial x^n}\right) = \sum_{i=1}^{n} \frac{\partial f}{\partial x^i} \cdot \frac{\partial}{\partial x^i} .$$

On the other hand, for every vector field X on $\mathbb{R}^n$, with

$$X = \sum_{i=1}^{n} a^i \frac{\partial}{\partial x^i} ,$$

we have a function, the divergence of X, defined by

$$\text{div } X = \sum_{i=1}^{n} \frac{\partial a_i}{\partial x^i} .$$

Finally, the Laplacian of f is the function*

*Classically, one introduced the operator $\nabla = \sum \frac{\partial}{\partial x^i} \cdot e_i$, where $e_i = \partial/\partial x^i$ is the ith basis vector of $\mathbb{R}^n$, and wrote (formally)

$$\begin{aligned} \text{grad } f &= \nabla f \\ \text{div } X &= \langle \nabla, X\rangle = \nabla \cdot X \\ \Delta f &= \langle \nabla, \nabla f\rangle = \nabla \cdot \nabla f . \end{aligned}$$

For this reason Δ was often denoted by ∇^2.

$$\Delta f = \operatorname{div}(\operatorname{grad} f) = \sum_{i=1}^{n} \frac{\partial^2 f}{\partial (x^i)^2} .$$

The operators grad, div, and Δ all have natural generalizations to an arbitrary Riemannian manifold $(M, \langle\ ,\ \rangle)$. Consider first the gradient of f. Notice that the components of grad f on $\mathbb{R}^n$ are just the coefficients of df in the expression $df = \Sigma(\partial f/\partial x^i)dx^i$. Consequently,

$$\left\langle \operatorname{grad} f, \sum_{i=1}^{n} b^i \frac{\partial}{\partial x^i} \right\rangle = \sum_{i=1}^{n} \frac{\partial f}{\partial x^i} b^i = df\left(\sum_{i=1}^{n} b^i \frac{\partial}{\partial x^i} \right) .$$

We can use this equation in any Riemannian manifold $(M, \langle\ ,\ \rangle)$ to define grad f as the unique vector field such that

$$\text{(I)} \qquad \langle \operatorname{grad} f, Y \rangle = df(Y) = Y(f) ,$$

for all vector fields Y on M. We easily see that

$$\text{(1)} \qquad \operatorname{grad}(fg) = f \cdot \operatorname{grad} g + g \cdot \operatorname{grad} f .$$

In terms of a coordinate system $x^1, \ldots, x^n$ on M we have

$$\operatorname{grad} f = \sum_{i=1}^{n} \left(\sum_{j=1}^{n} g^{ij} \frac{\partial f}{\partial x^j} \right) \cdot \frac{\partial}{\partial x^i} .$$

The divergence of a vector field X on M may be defined as

$$\text{(II)} \qquad (\operatorname{div} X)(p) = \operatorname{trace} Y \longmapsto \nabla_Y X \qquad Y \in M_p$$

$$= \sum_{i=1}^{n} \langle \nabla_{Y_i} X, Y_i \rangle \qquad Y_1, \ldots, Y_n \in M_p \text{ orthonormal} .$$

This clearly coincides with the original definition in Euclidean space. It is easy to check that

(2) $$\operatorname{div}(fX) = X(f) + f\cdot\operatorname{div} X = df(X) + f\cdot\operatorname{div} X\ .$$

In terms of a coordinate system $x^1,\ldots,x^n$ we have

$$X = \sum_{i=1}^{n} a^i \frac{\partial}{\partial x^i} \implies \operatorname{div} X = \sum_{i=1}^{n} a^i{}_{;i} = \sum_{i=1}^{n}\left(\frac{\partial a^i}{\partial x^i} + \sum_{j=1}^{n} a^j \Gamma^i_{ji}\right).$$

We can also define $\operatorname{div}\omega$ when ω is a 1-form, for the connection ∇ on vector fields gives rise to a connection ∇ on 1-forms (Chapter II.6), and we can set

(III) $$(\operatorname{div}\omega)(p) = \sum_{i=1}^{n} (\nabla_{X_i}\omega)(X_i) \qquad X_1,\ldots,X_n \in M_p \text{ orthonormal}\ .$$

It is easily checked that this definition does not depend on the choice of the orthonormal basis $X_1,\ldots,X_n$, but we can also give a completely invariant definition. We note that every bilinear map $\alpha\colon V\times V \to \mathbb{R}$ gives rise to a map $\alpha'\colon V \to V^*$ by $\alpha'(v)(w) = \alpha(v,w)$. If we also have an inner product on V, then we have an isomorphism $V^* \to V$, and thus we obtain a linear map

$$V \xrightarrow{\alpha'} V^* \to V\ .$$

It is easily seen that the trace of this composition is the same as

$$\sum_{i=1}^{n} \alpha(X_i,X_i) \qquad X_1,\ldots,X_n \text{ an orthonormal basis for } V\ .$$

To apply this to the case at hand, we consider the tensor $\nabla\omega$, with

$$(\nabla\omega)(X,Y) = (\nabla_X\omega)(Y) .$$

Then

$$(\text{div } \omega)(p) = \text{trace of the composition } M_p \xrightarrow{(\nabla\omega)'} M_p^* \to M_p ,$$

where the isomorphism $M_p^* \to M_p$ comes from the metric. The analogue of equation (2) is

$$(3) \qquad \text{div}(f\omega) = \langle df,\omega\rangle + f \text{ div } \omega ,$$

where the inner product $\langle\ ,\ \rangle$ on M_p^* comes from the inner product $\langle\ ,\ \rangle$ on M_p in the standard way.

More generally, consider a tensor A which is covariant of order k. We define $\text{div } A$ to be a covariant tensor of order $k-1$ by the (admittedly asymmetric) formula

$$(\text{III}') \quad \text{div } A(p)(Y_2,\ldots,Y_k) = \sum_{i=1}^{n} (\nabla_{X_i} A)(X_i,Y_2,\ldots,Y_k) \qquad X_1,\ldots,X_n \text{ orthonormal in } M_p .$$

The reader may easily work out a completely invariant definition.

In Problem I.9-13 we introduced the Divergence Theorem for n-dimensional submanifolds-with-boundary of $\mathbb{R}^n$. Now that we have generalized the definition of div, we would like to generalize this theorem also. An examination of the proof hinted at in that Problem leads us to hope that the following alternative definition of div is valid (the symbol $\lrcorner$ is defined in Problem I.7-4).

<u>56. LEMMA</u>. Let M be an oriented n-dimensional Riemannian manifold, with volume element dV (which can be considered to be an n-form, since M is oriented). Then for every vector field X on M we have

$$(*) \qquad d(X \lrcorner\, dV) = (\text{div } X)\cdot dV .$$

Proof. If (*) holds for X_1 and X_2, then it clearly holds for $X_1 + X_2$. Moreover, if (*) holds for X, then

$$\begin{aligned} d(fX \lrcorner dV) &= d(f \cdot (X \lrcorner dV)) \\ &= df \wedge (X \lrcorner dV) + f \cdot d(X \lrcorner dV) \\ &= df \wedge (X \lrcorner dV) + f \cdot \operatorname{div} X \cdot dV . \end{aligned}$$

Now Problem I.7-4(f) gives

$$\begin{aligned} 0 = X \lrcorner (df \wedge dV) &= (X \lrcorner df) \wedge dV - df \wedge (X \lrcorner dV) \\ &= X(f) \cdot dV - df \wedge (X \lrcorner dV) , \end{aligned}$$

so our formula becomes

$$\begin{aligned} d(fX \lrcorner dV) &= X(f) \cdot dV + f \cdot \operatorname{div} X \cdot dV \\ &= (\operatorname{div} fX) \cdot dV \qquad \text{by } (2) . \end{aligned}$$

Thus (*) is also true for fX.

Now let $X_1, \ldots, X_n$ be a positively oriented orthonormal moving frame, with dual forms $\theta^1, \ldots, \theta^n$, so that $dV = \theta^1 \wedge \ldots \wedge \theta^n$. By the considerations of the previous paragraph, it suffices to prove (*) when X is some X_i, and we might as well take $X = X_1$. We easily see that

$$X_1 \lrcorner dV = X_1 \lrcorner (\theta^1 \wedge \ldots \wedge \theta^n) = \theta^2 \wedge \ldots \wedge \theta^n .$$

So

$$\begin{aligned} d(X_1 \lrcorner dV) = d(\theta^2 \wedge \ldots \wedge \theta^n) &= \sum_{j=2}^{n} (-1)^j \, \theta^2 \wedge \ldots \wedge d\theta^j \wedge \ldots \wedge \theta^n \\ &= - \sum_{j=2}^{n} (-1)^j \, \theta^2 \wedge \ldots \wedge \left(\sum_{i=1}^{n} \omega_i^j \wedge \theta^i \right) \wedge \ldots \wedge \theta^n \end{aligned}$$

$$= - \sum_{j=2}^{n} (-1)^j \theta^2 \wedge \dots \wedge (\omega_1^j \wedge \theta^1) \wedge \dots \wedge \theta^n$$

$$= - \sum_{j=2}^{n} (-1)^j \omega_1^j \wedge \theta^1 \wedge \dots \wedge \hat{\theta^j} \wedge \dots \wedge \theta^n .$$

But

$$\omega_1^j = \sum_{k=1}^{n} \omega_1^j(X_k)\cdot\theta^k = \sum_{k=1}^{n} <\nabla_{X_k} X_1, X_j>\theta^k ,$$

so we obtain

$$d(X_1 \lrcorner \, dV) = \sum_{j=2}^{n} <\nabla_{X_j} X_1, X_j> \theta^1 \wedge \dots \wedge \theta^n$$

$$= (\text{div } X_1) dV . \blacksquare$$

As an easy corollary we now obtain

57. THEOREM (THE DIVERGENCE THEOREM). Let M be a compact oriented n-dimensional Riemannian manifold-with-boundary, with outward pointing unit normal ν on ∂M. Denote the volume element of M by dV_n, and that of ∂M by dV_{n-1}. Let X be a vector field on M. Then

$$\int_M \text{div } X \, dV_n = \int_{\partial M} <X,\nu> dV_{n-1} .$$

Proof. This follows from Stokes' Theorem, and the easily verified fact that $X \lrcorner \, dV$ equals $<X,\nu> dV_{n-1}$ on ∂M. $\blacksquare$

58. COROLLARY (GREEN'S THEOREM). If M is a compact oriented n-dimensional Riemannian manifold without boundary, and X is any vector field on M, then

$$\int_M \operatorname{div} X \, dV_n = 0 .$$

Notice that even when M is not orientable, equation (*) in Lemma 56 can be used to define $\operatorname{div} X$, for both sides of the equation change sign when the orientation is reversed, so locally the formula defines $\operatorname{div} X$ unambiguously.

We now define the Laplacian Δf of f on M by

$$\text{(IV)} \qquad \Delta f = \operatorname{div}(\operatorname{grad} f) .$$

For a coordinate system $x^1,\dots,x^n$ on M we have

$$\Delta f = \sum_{i=1}^{n}\left(\sum_{j=1}^{n} g^{ij}\frac{\partial f}{\partial x^j}\right)_{;i} = \sum_{i=1}^{n}\left(\sum_{j=1}^{n} g^{ij} f_{;j}\right)_{;i}$$

with the notation of Chapter II.5

$$= \sum_{i,j=1}^{n} g^{ij} f_{;ji} = \sum_{i,j=1}^{n} g^{ij} f_{;ij}$$

$$= \sum_{i,j=1}^{n} g^{ij}\left(\frac{\partial^2 f}{\partial x^i \partial x^j} - \sum_{k=1}^{n}\frac{\partial f}{\partial x^k}\Gamma_{ij}^k\right) .$$

If $x^1,\dots,x^n$ is a normal coordinate system at p, so that $\Gamma_{ij}^k(p) = 0$, and $g_{ij}(p) = \delta_{ij}$, then

$$\Delta f(p) = \sum_{i=1}^{n} \frac{\partial^2 f}{\partial (x^i)^2}(p) .$$

We can also state a more precise result along these lines. Suppose that

$X_1,\dots,X_n$ are vector fields which are orthonormal at p. Then

$$\begin{aligned}
\text{(IV}'\text{)}\quad \Delta f(p) &= \operatorname{div}(\operatorname{grad} f)(p) \\
&= \operatorname{trace} X \longmapsto \nabla_X \operatorname{grad} f \qquad X \in M_p \\
&= \sum_{i=1}^{n} \langle \nabla_{X_i(p)} \operatorname{grad} f, X_i(p)\rangle \\
&= \sum_{i=1}^{n} X_i(p)(\langle \operatorname{grad} f, X_i\rangle) - \sum_{i=1}^{n} \langle (\operatorname{grad} f)(p), \nabla_{X_i(p)} X_i\rangle \\
&= \sum_{i=1}^{n} X_i(p)(df(X_i)) - \sum_{i=1}^{n} \langle (\operatorname{grad} f)(p), \nabla_{X_i(p)} X_i\rangle \\
&= \sum_{i=1}^{n} (X_i X_i f)(p) - \sum_{i=1}^{n} \langle (\operatorname{grad} f)(p), \nabla_{X_i(p)} X_i\rangle .
\end{aligned}$$

So we have

$$\text{(IV}''\text{)}\quad \Delta f(p) = \sum_{i=1}^{n} (X_i X_i f)(p) \qquad \text{when } \begin{cases} X_1,\dots,X_n & \text{are orthonormal at } p \\ \nabla_{X_i} X_i = 0 & \text{at } p . \end{cases}$$

We ought to mention that the Laplacian Δf on a surface was first introduced by Beltrami, so Δ is often called the Laplace-Beltrami operator. For reasons that will appear in the next Addendum, the Laplacian is often defined as the negative of the Laplacian as defined here. There is no general agreement on the proper sign, so whenever a lecturer states that his Laplacian is the usual one (or the negative of the usual one), one half of the audience (or the other half) raises their eyebrows and murmurs disgruntedly "hmmph, so he calls _that_ the usual Laplacian!"

A simple calculation (using normal coordinates, or equation (IV"), to make things even easier) shows that

(4) $$\Delta(fg) = f\cdot\Delta g + g\cdot\Delta f + 2\langle \text{grad } f, \text{grad } g\rangle .$$

We will use this formula to derive a result of importance later on.

59. PROPOSITION. Let M be a compact oriented n-dimensional Riemannian manifold-with-boundary, with outward pointing unit normal ν on ∂M. Then

$$\int_M [f\Delta f + \langle \text{grad } f, \text{grad } f\rangle]\, dV_n = \int_{\partial M} \langle f \text{ grad } f, \nu\rangle\, dV_{n-1} .$$

In particular, if $f = 0$ on ∂M [and, a posteriori if $\partial M = \emptyset$], then

$$\int_M f\Delta f\, dV_n = -\int_M \langle \text{grad } f, \text{grad } f\rangle\, dV_n .$$

Proof. The Divergence Theorem (57) gives

$$\int_M \Delta(f^2)\, dV_n = \int_M \text{div}(\text{grad } f^2)\, dV_n = \int_{\partial M} \langle \text{grad } f^2, \nu\rangle\, dV_{n-1} .$$

Then equations (1) and (4) give the result. ■

As a corollary we have

60. LEMMA (BOCHNER'S LEMMA). Let M be a compact connected Riemannian manifold (without boundary). If $f\colon M \to \mathbb{R}$ has $\Delta f \geq 0$ everywhere, then f is a constant function (and $\Delta f = 0$).

Proof. We can assume that M is orientable, by taking the orientable two-fold covering space of M if necessary. First of all, Corollary 58 gives

$$\int_M \Delta f \; dV = \int_M \operatorname{div}(\operatorname{grad} f) \; dV = 0 \; .$$

Since $\Delta f \geq 0$ on M, this already implies that $\Delta f = 0$ on M. Now (the second part of) Proposition 59 gives

$$0 = \int_M f\Delta f \; dv = - \int_M \langle \operatorname{grad} f, \operatorname{grad} f \rangle \; dV \; .$$

So we must have $\operatorname{grad} = 0 \implies df = 0 \implies f$ is constant. ∎

An alternative proof of Lemma 60 is given in Addendum 2 to Chapter 10.

A couple of explicit calculations of the Laplacian will be used at various times. Our first calculation is most easily carried out in a coordinate system. Consider a 1-form

$$\omega = \sum_i a_i dx^i \; ,$$

and the vector field

$$X = \sum_i a^i \frac{\partial}{\partial x^i} \; , \qquad a^i = \sum_j g^{ij} a_j \; .$$

This vector field X is described intrinsically by the equation

$$\langle X,Y \rangle = \omega(Y) \qquad \text{for all vector fields } Y \; ,$$

so that, in particular,

$$\langle X,X \rangle = \omega(X) = \sum_i a^i a_i \; .$$

Now

$$\Delta\Big(\sum_i a^i a_i\Big) = \sum_{j,k} g^{jk}\Big(\sum_i a^i a_i\Big)_{;jk}$$
$$= \sum_{j,k} g^{jk} \sum_i \Big(a^i{}_{;j}a_i + a^i a_{i;j}\Big)_{;k}$$
$$= \sum_{j,k} g^{jk} \sum_i \Big(a^i{}_{;jk}a_i + a^i{}_{;j}a_{i;k} + a^i{}_{;k}a_{i;j} + a^i a_{i;jk}\Big) .$$

Since

$$\sum_{i,j,k} g^{jk}a^i{}_{;jk}a_i = \sum_{i,j,k} \sum_{\ell,m} g^{jk}g^{i\ell}a_{\ell;jk}g_{im}a^m$$
$$= \sum_{j,k,m} g^{jk}a_{m;jk}a^m = \sum_{i,j,k} g^{jk}a_{i;jk}a^i ,$$

and

$$\sum_{i,j,k} g^{jk}a^i{}_{;j}a_{i;k} = \sum_{i,j,k} \sum_{\ell} g^{jk}g^{i\ell}a_{\ell;j}a_{i;k} ,$$

we have, finally,

$$(5) \qquad \Delta\Big(\sum_i a^i a_i\Big) = 2\Big[\sum_{i,j,k} g^{jk}a^i a_{i;jk} + \sum_{i,j,k,\ell} g^{jk}g^{i\ell}a_{i;k}a_{\ell;j}\Big] .$$

Our second calculation is easily carried out in a coordinate-free way. Consider an immersion $f\colon M^n \to \mathbb{R}^m$, so that M has a Riemannian metric $I_f = f^*\langle\ ,\ \rangle$. We will compute Δf with respect to this metric (the fact that f is $\mathbb{R}^m$-valued causes no difficulty, for we can compute the Laplacian of each of its component functions -- for simplicity we suppress the various components and simply use formula (IV") for $\mathbb{R}^m$-valued f). It will make things conceptually easier to think of M^n as a subset of $\mathbb{R}^m$, so that f is the inclusion map. Let $X_1,\ldots,X_n$ be vector fields on M satisfying the conditions of (IV"). We first want to figure out what the $\mathbb{R}^m$-valued function $X_i(f)$ is. Now

$$X_i(f) = df(X_i) = \text{"the vector part" of } f_*(X_i) \qquad \text{by Problem I.4-3}$$
$$= X_i \quad \text{(considered as a point of } \mathbb{R}^m) .$$

Therefore,

$$X_i(X_i f)(p) = \nabla'_{X_i} X_i(p)$$
$$= \nabla_{X_i} X_i(p) + s(X_i(p), X_i(p))$$
$$= s(X_i(p), X_i(p)) \qquad \text{by our conditions on } X_1, \ldots, X_n .$$

Thus we see that

$$(6) \qquad \Delta f(p) = n \cdot \eta(p) ,$$

where η is the mean curvature normal. Notice, in particular, that if M has $\eta = 0$, then $\Delta f = 0$. Lemma 60 then implies that M cannot be compact (for $n \geq 1$), which reproves Corollary 31. In the particular case of a hypersurface, we have

$$(7) \qquad \Delta f = nH \cdot \nu ,$$

where ν is the unit normal field.

Notice the correspondence between equation (7) and equation (II') on p. III.162, which can be written in the simple form

$$\Delta_{\langle,\rangle} f = \mathfrak{n} ,$$

where $\Delta_{\langle,\rangle}$ indicates the Laplacian with respect to the metric $\langle,\rangle$ on M. Since the Laplacian is such a natural operator on a Riemannian manifold, it is not surprising to find $\Delta_{\langle,\rangle} f$ related to $\mathfrak{n}$. (Note also that $\Delta_{\langle,\rangle}$

involves g_{ij} and Christoffel symbols Γ^k_{ij}, and thus third derivatives of f, just like $\mathfrak{n}$). As a matter of fact, this equation was originally used as the <u>definition</u> of $\mathfrak{n}$ (it is clearly a special linear affine invariant!).

The Laplacian can be generalized in two very important ways. One such generalization is treated in the next Addendum. A different generalization, important in Chapter 9, is suggested by the next to last line of equation (IV'), which can be written

$$\Delta f(p) = \sum_{i=1}^{n} X_i(p)(df(X_i)) - \sum_{i=1}^{n} df(\nabla_{X_i(p)} X_i) \qquad X_1,\ldots,X_n \text{ orthonormal at } p .$$

Now Corollary II.6-5 says that the covariant derivative ∇df is given by

$$(\nabla_{X_p} df)(Y_p) = X_p(df(Y)) - df(\nabla_{X_p} Y)$$

for any vector fields X, Y extending X_p, Y_p. Thus we can write

$$\Delta f(p) = \sum_{i=1}^{n} (\nabla_{X_i} df)(X_i) \qquad X_1,\ldots,X_n \in M_p \text{ orthonormal} .$$

Thus, using (III) we can just as well define Δf by

$$\Delta f = \operatorname{div}(df) .$$

[Naturally, one could, with some work, demonstrate the equation $\operatorname{div}(\operatorname{grad} f) = \operatorname{div}(df)$ directly from the completely invariant definitions.]

The nice thing about this new definition of Δf is that it can be generalized immediately. Consider a vector bundle $\varpi\colon E \to M$, where M has a metric $\langle\ ,\ \rangle$, and E has some connection D. If ξ is any section of E, then $D_{X_p}\xi \in \varpi^{-1}(p)$ for $X_p \in M_p$. We can therefore think of $D\xi$ as a section of the bundle Hom(TM,E) whose fibre at p is $\mathrm{Hom}(M_p, \varpi^{-1}(p))$. Now the connection ∇ on M determined by $\langle\ ,\ \rangle$, together with the connection D on E, determines a connection $\tilde{\nabla}$ on Hom(TM,E). This is defined as on p. 54, except that the situation is even simpler. As in that case, we easily see that for any vector fields X, Y and any section ψ on Hom(TM,E), we have

$$(8) \qquad (\tilde{\nabla}_{X_p} \psi)(Y_p) = D_{X_p}(\psi(Y)) - \psi(\nabla_{X_p} Y) .$$

[If $E = M \times \mathbb{R}$, so that the sections of E are functions $f\colon M \to \mathbb{R}$, and we define $D_X f$ to be $df(X)$, then $\tilde{\nabla}$ will just be the connection ∇ on 1-forms.] Naturally $\tilde{\nabla}\psi$ will denote the section of $\mathrm{Hom}(TM \times TM, E)$ with

$$(\tilde{\nabla}\psi)(X,Y) = (\tilde{\nabla}_X \psi)(Y) .$$

For a section ξ of E we can now define

$$(\mathrm{V}) \qquad \Delta\xi(p) = \sum_{i=1}^{n} (\tilde{\nabla}_{X_i} D\xi)(X_i) \qquad X_1,\ldots,X_n \in M_p \text{ orthonormal} .$$

(A completely invariant definition is easily formulated, as before.) If we let $X_1,\ldots,X_n$ be vector fields which are orthonormal at p, then

$$
\begin{aligned}
(\mathrm{V}') \qquad \Delta\xi(p) &= \sum_{i=1}^{n} (\tilde{\nabla}_{X_i(p)} D\xi)(X_i(p)) \\
&= \sum_{i=1}^{n} D_{X_i(p)}(D\xi(X_i)) - \sum_{i=1}^{n} D\xi(\nabla_{X_i(p)} X_i) \qquad \text{by (8)} \\
&= \sum_{i=1}^{n} D_{X_i(p)} D_{X_i} \xi - \sum_{i=1}^{n} D\xi(\nabla_{X_i(p)} X_i) .
\end{aligned}
$$

So we have, in complete analogy with equation (IV"),

$$\text{(V")} \quad \Delta\xi(p) = \sum_{i=1}^{n} (D_{X_i} D_{X_i} \xi)(p) \qquad \text{when} \begin{cases} X_1,\ldots,X_n & \text{are orthonormal at } p \\ \nabla_{X_i} X_i = 0 & \text{at } p\,. \end{cases}$$

Addendum 2. The * Operator and the Laplacian on Forms; Hodge's Theorem

Let V be an oriented n-dimensional vector space with an inner product $\langle\ ,\ \rangle$. The $*$ operator, from alternating k-linear functions $\Omega^k(V)$ to $\Omega^{n-k}(V)$, is usually defined as follows. Let $v_1,\ldots,v_n$ be a positively oriented orthonormal basis of V, and let $\phi_1,\ldots,\phi_n$ be the dual basis. Then

$$*(\phi_{i_1}\wedge\ldots\wedge\phi_{i_k}) = \pm\,\phi_{j_1}\wedge\ldots\wedge\phi_{j_{n-k}},$$

where $i_1,\ldots,i_k$ are k distinct numbers from $1,\ldots,n$, and $j_1,\ldots,j_{n-k}$ are the other $n-k$ numbers of this set, arranged in some order; we use the $+$ sign if $v_{i_1},\ldots,v_{i_k},v_{j_1},\ldots,v_{j_{n-k}}$ is positively oriented, and the $-$ sign otherwise. We also set $*\,1 = \pm\,\phi_1\wedge\ldots\wedge\phi_n$, where $1 \in \Omega^0(V) = \mathbb{R}$, and $*(\phi_1\wedge\ldots\wedge\phi_n) = \pm\,1$. It is easy to see, first of all, that this definition is consistent, for a fixed basis $v_1,\ldots,v_n$, and then that the definition is also independent of the orthonormal basis. An invariant definition can be given as follows. We always have a map

$$\Omega^k(V)\times\Omega^{n-k}(V)\xrightarrow{\ \wedge\ }\Omega^n(V)\ .$$

An orientation and inner product on V gives us an isomorphism $\Omega^n(V)\xrightarrow{\approx}\mathbb{R}$, so we have a bilinear map

$$\{\ ,\ \}\colon \Omega^k(V)\times\Omega^{n-k}(V)\to\mathbb{R}\ .$$

Then we can define

$$A\colon \Omega^k(V)\to\left(\Omega^{n-k}(V)\right)^*$$

by

$$A(\omega)(\eta) = \{\omega,\eta\} \qquad \omega \in \Omega^k(V)\ , \quad \eta \in \Omega^{n-k}(V)\ .$$

Now the inner product on V also gives us an isomorphism $V \to V^*$ from which we derive an isomorphism $\left(\Omega^{n-k}(V)\right)^* \to \Omega^{n-k}(V)$. One easily checks that the composition

$$\Omega^k(V) \xrightarrow{A} \left(\Omega^{n-k}(V)\right)^* \to \Omega^{n-k}(V)$$

is precisely $*$. Straightforward calculations give

(1) $** = *\circ*\colon \Omega^k(V) \to \Omega^k(V)$ is $(-1)^{k(n-k)}$ times the identity .

In Chapter I.9 we mentioned that the inner product on V gives inner products on all vector spaces $\Omega^k(V)$, although we did not describe most of these inner products explicitly. The inner product on $\Omega^1(V) = V^*$ can be described by the condition that the dual basis $\phi^1,\dots,\phi^n$ is orthonormal if and only if $v_1,\dots,v_n$ is orthonormal in V. Using the inner product $\langle\ ,\ \rangle$ thus defined on $\Omega^1(V)$, we can describe the inner product on $\Omega^k(V)$ as the unique one with

(2) $$\langle\phi_1\wedge\dots\wedge\phi_k,\ \psi_1\wedge\dots\wedge\psi_k\rangle = \det(\langle\phi_i,\psi_j\rangle)$$

for $\phi_i,\ \psi_j \in V^*$. In particular, if $\phi_1,\dots,\phi_n$ is orthonormal in V^*, then

$$\langle\phi_{i_1}\wedge\dots\wedge\phi_{i_k},\ \phi_1\wedge\dots\wedge\phi_k\rangle$$

$$= \det\begin{pmatrix} \delta_{i_1 1} & \cdots & \delta_{i_1 k} \\ \vdots & & \vdots \\ \delta_{i_k 1} & \cdots & \delta_{i_k k} \end{pmatrix}$$

$$= \begin{cases} 0 & \text{if } \{i_1,\ldots,i_k\} \neq \{1,\ldots,k\} \\ \operatorname{sgn}\pi & \text{if } i_\alpha = \pi(\alpha) \text{ for some permutation } \pi \text{ of } \{1,\ldots,k\} . \end{cases}$$

Since the naming of the indices was purely arbitrary, we have, just as well,

$$(3) \quad \langle \phi_{i_1} \wedge \cdots \wedge \phi_{i_k}, \phi_{j_1} \wedge \cdots \wedge \phi_{j_k} \rangle = \begin{cases} 0 & \text{if } \{i_1,\ldots,i_k\} \neq \{j_1,\ldots,j_k\} \\ \operatorname{sgn}\pi & \text{if } j_\alpha = \pi(i_\alpha) . \end{cases}$$

So we can also describe the inner product on $\Omega^k(V)$ as the one which makes the $\phi_{i_1} \wedge \cdots \wedge \phi_{i_k}$ $(i_1 < \cdots < i_k)$ an orthonormal basis, for any orthonormal basis $\phi_1,\ldots,\phi_n$ of V^*.

Now note that

$$\phi_{i_1} \wedge \cdots \wedge \phi_{i_k} \wedge *(\phi_1 \wedge \cdots \wedge \phi_k) = \phi_{i_1} \wedge \cdots \wedge \phi_{i_k} \wedge \pm\, \phi_{k+1} \wedge \cdots \wedge \phi_n$$

$$= \begin{cases} 0 & \text{if } \{i_1,\ldots,i_k\} \neq \{1,\ldots,k\} \\ (\operatorname{sgn}\pi)\cdot *1 & \text{if } i_\alpha = \pi(\alpha) . \end{cases}$$

Again, since the naming of the indices was arbitrary, we have, just as well,

$$(4) \quad \phi_{i_1} \wedge \cdots \wedge \phi_{i_k} \wedge *(\phi_{j_1} \wedge \cdots \wedge \phi_{j_k}) = \begin{cases} 0 & \text{if } \{i_1,\ldots,i_k\} \neq \{j_1,\ldots,j_k\} \\ (\operatorname{sgn}\pi)\cdot *1 & \text{if } j_\alpha = \pi(i_\alpha) . \end{cases}$$

Comparing (3) and (4), we see that for $\omega, \eta \in \Omega^k(V)$ we have

$$(5) \qquad \langle\omega,\eta\rangle = *(\omega \wedge *\eta) = *(\eta \wedge *\omega) .$$

Now everything which we have done can be extended to k-forms on an oriented Riemannian n-manifold $(M, \langle\ ,\ \rangle)$. We have an operator $*$ taking k forms to $(n-k)$-forms, and $** = (-1)^{k(n-k)}$ on k-forms. It is easy to check (using the dual forms to an orthonormal moving frame, for example) that $*$ takes C^∞ forms to C^∞ forms. Note that the volume element dV on M is just $*1$ for the constant function (0-form) 1.

We also have, for two k-forms, ω and η, a function $\langle\omega,\eta\rangle$ on M. We would like a formula for $\langle\omega,\eta\rangle$ when we have coordinate expressions

$$\text{(a)} \qquad \omega = \sum_{i_1<\cdots<i_k} a_{i_1\cdots i_k}\, dx^{i_1}\wedge\cdots\wedge dx^{i_k}$$

$$\text{(b)} \qquad \eta = \sum_{j_1<\cdots<j_k} b_{j_1\cdots j_k}\, dx^{j_1}\wedge\cdots\wedge dx^{j_k} .$$

For this, and later, purposes, it will be convenient to express a form in terms of tensor products of the dx^i, instead of wedge products. Recall (Theorem I.7-2(3)) that

$$\begin{aligned} dx^{i_1}\wedge\cdots\wedge dx^{i_k} &= \frac{(1+\cdots+1)!}{1!\cdots1!}\,\mathrm{Alt}(dx^{i_1}\otimes\cdots\otimes dx^{i_k}) \\ &= \sum_{\sigma\in S_k} \operatorname{sgn}\sigma\; dx^{\sigma(i_1)}\otimes\cdots\otimes dx^{\sigma(i_k)} . \end{aligned}$$

This shows that the expression (a) can also be written

$$\omega = \sum_{i_1,\ldots,i_k} a_{i_1\cdots i_k}\, dx^{i_1}\otimes\cdots\otimes dx^{i_k} ,$$

where the new $a_{i_1\cdots i_k}$ are skew-symmetric in the indices $i_1,\ldots,i_k$ and agree with the old $a_{i_1\cdots i_k}$ when $i_1<\cdots<i_k$. Now let g_{ij} be the components of $\langle\ ,\ \rangle$ in our coordinate system, so that g^{ij} are the components

of $\langle\ ,\ \rangle$ on the dual space. With any tensor, covariant of order k,

$$A = \sum_{i_1,\ldots,i_k} a_{i_1\ldots i_k}\, dx^{i_1} \otimes \cdots \otimes dx^{i_k} ,$$

we can associate the tensor, contravariant of order k,

$$\tilde{A} = \sum_{j_1,\ldots,j_k} a^{j_1\ldots j_k} \frac{\partial}{\partial x^{j_1}} \otimes \cdots \otimes \frac{\partial}{\partial x^{j_k}} ,$$

where

$$(6) \qquad a^{j_1\ldots j_k} = \sum_{i_1,\ldots,i_k} g^{i_1 j_1} \cdots g^{i_k j_k}\, a_{j_1\ldots j_k} .$$

In the special case where

$$A = \sum_i a_i dx^i ,$$

it is clear how $\tilde{A}$ is described invariantly: if we think of $\tilde{A}(p)$ as a linear function on $M_p{}^*$, then

$$\tilde{A}(p)(\phi) = A(p)(S(\phi)) ,$$

where $S\colon M_p{}^* \longrightarrow M_p$ is the isomorphism given by the metric. In general,

$$\tilde{A}(p)(\phi_1,\ldots,\phi_k) = A(p)(S(\phi_1),\ldots,S(\phi_k)) .$$

Notice that if the $a_{i_1\ldots i_k}$ are skew-symmetric in the indices, then so are the $a^{j_1\ldots j_k}$. So if ω is given by (a), then $\tilde{\omega}$ is also given by

$$\tilde{\omega} = \sum_{j_1<\cdots<j_k} a^{j_1\cdots j_k} \frac{\partial}{\partial x^{j_1}} \wedge \cdots \wedge \frac{\partial}{\partial x^{j_k}}$$

[note, however, that the $a^{j_1\cdots j_k}$ are computed from (6), in which $a_{i_1\cdots i_k}$ is defined, by skew-symmetry, for all $i_1,\ldots,i_k$]. We now claim that for ω, η given by (a) and (b), we have

$$(7)\quad \langle\omega,\eta\rangle = \sum_{i_1<\cdots<i_k} a_{i_1\cdots i_k} b^{i_1\cdots i_k} = \sum_{i_1<\cdots<i_k} a^{i_1\cdots i_k} b_{i_1\cdots i_k}$$
$$= \frac{1}{k!}\sum_{i_1\cdots i_k} a_{i_1\cdots i_k} b^{i_1\cdots i_k} = \frac{1}{k!}\sum_{i_1\cdots i_k} a^{i_1\cdots i_k} b_{i_1\cdots i_k} .$$

To prove this, we note that the last two expressions can be defined invariantly as contractions (traces) of $\omega\otimes\tilde{\eta}$ or $\tilde{\omega}\otimes\eta$. So it suffices to check that (7) holds at a point p where $dx^1,\ldots,dx^n$ are orthonormal. In this case $g^{ij} = \delta^{ij}$ at p, so $a^{i_1\cdots i_k} = a_{i_1\cdots i_k}$ at p. The desired result then follows immediately from equation (3).

On the oriented Riemannian n-manifold M we can also do something else. Since we have the map d, which raises the degree of a form, we can define a map δ, which lowers the degree of a form, by

$$\delta = (-1)^{n(k+1)+1} {*d*} \qquad \text{from } k\text{-forms to } (k-1)\text{-forms} .$$

We clearly have $\delta^2 = 0$, and $\delta = 0$ on functions (0-forms). Note that on k-forms we have

$$(8) \qquad *\delta = (-1)^{n(k+1)+1}(**)d*$$
$$= (-1)^{n(k+1)+1}\cdot(-1)^{(n-k+1)(k-1)}d*$$

by (1) [since $d*$ of a k-form is an $n-k+1$ form]

$$= (-1)^k d* ,$$

and similarly

$$(9) \qquad \delta* = (-1)^{k+1}*d .$$

Notice that δ can really be defined even when M is not orientable, for its definition is local, and changing the orientation of M reverses the sign of $*$, so leaves δ unchanged. We now define an operator Δ from k-forms to k-forms by

$$\Delta = \delta d + d\delta .$$

The reader may check that on 0-forms this Δ is the negative of the one in the previous Addendum. [N.B. The connection ∇ on M gives rise in a natural way to a connection ∇ on the bundle of k-forms on M, so the final definition of the previous Addendum also gives us a Laplacian on k-forms. But that Laplacian is <u>not</u> related to the one defined here.] Simple computations, using (8) and (9) for the last equation, give

$$(10) \qquad d\Delta = \Delta d , \qquad \delta\Delta = \Delta\delta , \qquad *\Delta = \Delta* .$$

On a compact oriented manifold M we can define the inner product (ω,η) of two k-forms ω, η by

$$(\omega,\eta) = \int_M \langle\omega,\eta\rangle dV = \int_M \omega \wedge *\eta \qquad \text{by (5)} .$$

This inner product $(\ ,\ )$ is clearly symmetric and positive definite. Now if ω is a $(k-1)$-form and η is a k-form, then

$$\begin{aligned} d(\omega \wedge *\eta) &= d\omega \wedge *\eta + (-1)^{k-1}\omega \wedge d{*\eta} \\ &= d\omega \wedge *\eta - \omega \wedge *\delta\eta \qquad \text{by (8)} . \end{aligned}$$

So Stoke's Theorem gives

$$0 = \int_M d(\omega \wedge *\eta) = \int_M d\omega \wedge *\eta - \int_M \omega \wedge *\delta\eta ,$$

or

$$(11) \qquad (d\omega,\eta) = (\omega,\delta\eta) .$$

Thus δ is the "adjoint" of d for the inner product $(\ ,\ )$, and this property characterizes $\delta\eta$, since $(\ ,\ )$ is positive definite. From this we easily see that Δ is self-adjoint with respect to the inner product $(\ ,\ )$ on k-forms,

$$(12) \qquad (\Delta\omega,\eta) = (\omega,\Delta\eta) .$$

In Euclidean space, a function f with $\Delta f = 0$ is called harmonic. In an oriented Riemannian manifold $(M, \langle\ ,\ \rangle)$ we call a k-form ω <u>harmonic</u> if $\Delta\omega = 0$. When M is compact, we can write

$$(\Delta\omega,\omega) = ([d\delta + \delta d]\omega, \omega) = (\delta\omega,\delta\omega) + (d\omega,d\omega) ,$$

which shows that

$$(13) \qquad \Delta\omega = 0 \implies d\omega = 0 \text{ and } \delta\omega = 0 , \qquad M \text{ compact}$$

(the converse is trivial). If ω, η are k-forms, and $\Delta\omega = 0$, then equation (12) gives

$$(\Delta\eta,\omega) = (\eta,\Delta\omega) = 0 \ .$$

So the vector space of all harmonic k-forms (the kernel of Δ) is orthogonal to the image of Δ. The fundamental result on harmonic forms states that these two orthogonal subspaces of the k-forms span the whole vector space of k-forms:

<u>THE HODGE DECOMPOSITION THEOREM</u>. If M is a compact oriented Riemannian n-manifold, then for each k with $0 \leq k \leq n$, the vector space H^k of harmonic k-forms is finite dimensional, and the vector space $E^k(M)$ of all k-forms on M can be written as an orthogonal direct sum decomposition

$$E^k(M) = \Delta(E^k(M)) \oplus H^k(M) \ .$$

For a proof of this result, which is completely analytic in nature, the reader is referred to Warner {1}; the proof given there is elementary and completely self-contained. We will merely indicate the consequences of the theorem for the de Rham cohomology. The orthogonal decomposition of $E^k(M)$ gives two projection maps

$$E^k(M) \xrightarrow{H^k} H^k(M), \qquad E^k(M) \xrightarrow{h^k} \Delta(E^k(M)) \ .$$

For any $\alpha \in E^k(M)$, the form $h^k(\alpha) = \alpha - H^k(\alpha)$ is uniquely $\Delta\omega$ for some ω. Set

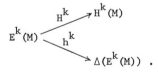

$$G(\alpha) = \text{the unique } \omega \text{ with } \Delta\omega = \alpha - H^k(\alpha) ,$$

so that

$$G = [\Delta|\ \Delta(E^k(M))]^{-1} \circ h^k .$$

Now consider any linear map $T\colon E^k(M) \longrightarrow E^\ell(M)$ with $T\Delta = \Delta T$ [e.g., $T = d, \delta, \Delta$]. We easily see that

$$T(H^k) \subset H^\ell , \qquad T(\Delta(E^k(M)) \subset \Delta(E^\ell(M)) .$$

So

$$T \circ h^k = h^\ell \circ T , \qquad T \circ [\Delta|\ \Delta(E^k(M))] = [\Delta|\ \Delta(E^\ell(M))] \circ T .$$

From this we see that $GT = TG$. In particular, G commutes with d.

Now let ω be any k-form. Then we have

$$\begin{aligned} \alpha &= \Delta G\alpha + H^k(\alpha) \\ &= d\delta G\alpha + \delta dG\alpha + H^k(\alpha) \\ &= d\delta G\alpha + \delta G d\alpha + H^k(\alpha) . \end{aligned}$$

So if $d\alpha = 0$, then

$$\alpha = d\delta G\alpha + H^k(\alpha) .$$

Thus $H^k(\alpha)$ is a harmonic k-form in the same de Rham cohomology class as α. On the other hand, suppose α_1 and α_2 are two harmonic k-forms in the same de Rham cohomology class, so that

$$\alpha_1 - \alpha_2 = d\beta$$

for some β. Then

$$(d\beta,d\beta) = (d\beta,\alpha_1-\alpha_2) = (\beta,\delta\alpha_1-\delta\alpha_2) \quad \text{by (11)}$$
$$= 0 \quad \text{by (13)} .$$

So $d\beta = 0$, or $\alpha_1 = \alpha_2$. Thus there is a unique harmonic form in each de Rham cohomology class. In other words, the k-dimensional de Rham cohomology vector space is isomorphic to the vector space $H^k(M)$ of harmonic k-forms.

We will give a simple application of this result in a moment. First we would like to observe that both d and δ can be defined in terms of the connection ∇ on M. For d this is easy.

61. PROPOSITION. If ω is a k-form on a Riemannian manifold, then

$$d\omega = (-1)^k(k+1)\cdot \text{Alt}\ \nabla\omega .$$

Proof. Let $x^1,\dots,x^n$ be a normal coordinate system at p, and let

$$\omega = \sum_{i_1<\dots<i_k} a_{i_1\dots i_k}\, dx^{i_1}\wedge\dots\wedge dx^{i_k}$$
$$= \sum_{i_1,\dots,i_k} a_{i_1\dots i_k}\, dx^{i_1}\otimes\dots\otimes dx^{i_k}, \quad \text{as on p. 205.}$$

Then (p. II.246)

$$\nabla\omega = \sum_{i_1,\dots,i_k}\sum_h a_{i_1\dots i_k;h}\, dx^{i_1}\otimes\dots\otimes dx^{i_k}\otimes dx^h .$$

So

$$(k+1)!\ \mathrm{Alt}\ \nabla\omega = \sum_{i_1,\dots,i_k} \sum_h a_{i_1\cdots i_k;h}(k+1)!\ \mathrm{Alt}(dx^{i_1}\otimes\cdots\otimes dx^{i_k}\otimes dx^h)$$

$$= \sum_{i_1,\dots,i_k} \sum_h a_{i_1\cdots i_k;h}\ dx^{i_1}\wedge\dots\wedge dx^{i_k}\wedge dx^h$$

$$= k! \sum_{i_1<\cdots<i_k} \sum_h a_{i_1\cdots i_k;h}\ dx^{i_1}\wedge\dots\wedge dx^{i_k}\wedge dx^h ,$$

by skew-symmetry of the $a_{i_1\cdots i_k}$. So

$$(*)\qquad (-1)^k(k+1)\ \mathrm{Alt}\ \nabla\omega = \sum_{i_1<\cdots<i_k} \sum_h a_{i_1\cdots i_k;h}\ dx^h\wedge dx^{i_1}\wedge\dots\wedge dx^{i_k} .$$

But at p we have (Proposition II.5-1)

$$a_{i_1\cdots i_k;h}(p) = \frac{\partial}{\partial x^h} a_{i_1\cdots i_k}(p) .$$

So

$$(-1)^k(k+1)\ \mathrm{Alt}\ \nabla\omega(p) = \sum_{i_1<\cdots<i_k} \sum_h \frac{\partial}{\partial x^h} a_{i_1\cdots i_k}(p)\ dx^h\wedge dx^{i_1}\wedge\dots\wedge dx^{i_k}(p)$$

$$= d\omega(p) .\ \blacksquare$$

Naturally, the use of a normal coordinate system at p was merely a simplifying device; in an arbitrary coordinate system we would obtain the same result with a little more calculation -- the Christoffel symbols in (*) all cancel out after we write all $dx^h\wedge dx^{i_1}\wedge\dots\wedge dx^{i_k}$ in terms of increasing sequences of indices. The formula

$$d\omega = \sum_{i_1<\cdots<i_k} \sum_h a_{i_1,\ldots,i_k;h}\, dx^h \wedge dx^{i_1} \wedge \ldots \wedge dx^{i_k},$$

which follows from (*) and the final result of the theorem, can be rewritten as follows:

$$d\omega = \sum_{j_1<\cdots<j_{k+1}} b_{j_1\cdots j_{k+1}}\, dx^{j_1} \wedge \ldots \wedge dx^{j_{k+1}},$$

for
$$b_{j_1\cdots j_{k+1}} = \sum_{\mu=1}^{k+1} (-1)^{\mu+1}\, a_{j_1\cdots\hat{j}_\mu\cdots j_{k+1};j_\mu}.$$

Notice that if the a's are skew-symmetric, then the b's <u>will be also</u>.

Now suppose that we have a $(k+1)$-form

$$\eta = \sum_{j_1<\cdots<j_{k+1}} c_{j_1\cdots j_{k+1}}\, dx^{j_1} \wedge \ldots \wedge dx^{j_{k+1}}.$$

Then equation (7) gives

$$\begin{aligned}
(14)\quad \langle d\omega,\eta\rangle &= \frac{1}{(k+1)!} \sum_{j_1,\ldots,j_{k+1}} b_{j_1\cdots j_{k+1}}\, c^{j_1\cdots j_{k+1}} \\
&= \frac{1}{(k+1)!} \sum_{j_1,\ldots,j_{k+1}} \sum_{\mu=1}^{k+1} (-1)^{\mu+1}\, a_{j_1\cdots\hat{j}_\mu\cdots j_{k+1};j_\mu}\, c^{j_1\cdots j_{k+1}} \\
&= \frac{1}{(k+1)!} \sum_{j_1,\ldots,j_{k+1}} \sum_{\mu=1}^{k+1} (-1)^{\mu+1} \cdot -\, a_{j_1\cdots\hat{j}_\mu\cdots j_{k+1}}\, c^{j_1\cdots j_{k+1}}{}_{;j_\mu} \\
&\quad + \frac{1}{(k+1)!} \sum_{j_1,\ldots,j_{k+1}} \sum_{\mu=1}^{k+1} (-1)^{\mu+1} \left(a_{j_1\cdots\hat{j}_\mu\cdots j_{k+1}}\, c^{j_1\cdots j_{k+1}}\right)_{;j_\mu} \\
&= \Sigma_1 + \Sigma_2, \quad \text{say.}
\end{aligned}$$

Now it is easy to see that

$$\sum_{j_1,\dots,j_{k+1}} a_{j_1\cdots\hat{j}_\mu\cdots j_{k+1}} c^{j_1\cdots j_{k+1}}{}_{;j_\mu}$$

$$= \sum_{j_1,\dots,j_{k+1}} a^{j_1\cdots\hat{j}_\mu\cdots j_{k+1}} \sum_\rho g^{j_\mu\rho} c_{j_1\cdots j_{k+1};\rho}$$

$$= \sum_{j_1,\dots,j_{k+1}} (-1)^{\mu-1} a^{j_1\cdots\hat{j}_\mu\cdots j_{k+1}} \sum_\rho g^{j_\mu\rho} c_{j_\mu j_1\cdots\hat{j}_\mu\cdots j_{k+1};\rho} .$$

Hence

$$\Sigma_1 = \frac{-1}{(k+1)!} \sum_{j_1,\dots,j_{k+1}} \sum_{\mu=1}^{k+1} a^{j_1\cdots\hat{j}_\mu\cdots j_{k+1}} \sum_\rho g^{j_\mu\rho} c_{j_\mu j_1\cdots\hat{j}_\mu\cdots j_{k+1};\rho}$$

$$= -\frac{1}{(k+1)!} \sum_{j_1,\dots,j_{k+1}} \sum_{\mu=1}^{k+1} a^{j_1\cdots\hat{j}_\mu\cdots j_{k+1}} \gamma_{j_1\cdots\hat{j}_\mu\cdots j_{k+1}}, \quad \text{say}$$

$$= -\frac{k+1}{(k+1)!} \sum_{\ell_1,\dots,\ell_k} a^{\ell_1\cdots\ell_k} \gamma_{\ell_1\cdots\ell_k}$$

$$= -\frac{1}{k!} \sum_{\ell_1,\dots,\ell_k} a^{\ell_1\cdots\ell_k} \gamma_{\ell_1\cdots\ell_k} .$$

Now the γ's are simply the components of the tensor $\operatorname{div}\eta$, defined by (III') on p. 190. So

(15) $$\Sigma_1 = -\langle\omega, \operatorname{div}\eta\rangle .$$

On the other hand, we obtain $k+1$ well-defined vector fields X_μ with

$$X_\mu = \sum_{\rho=1}^{n}\left(\sum_{j_1,\dots,j_{k+1}} a_{j_1\cdots\hat{j}_\mu\cdots j_{k+1}} c^{j_1\cdots j_{\mu-1}\rho j_{\mu+1}\cdots j_{k+1}}\right)\frac{\partial}{\partial x^\rho} .$$

Then

$$(16) \qquad \Sigma_2 = \operatorname{div}\left(\frac{1}{(k+1)!}\sum_{\mu=1}^{k+1} X_\mu\right) = \operatorname{div} Y \ , \quad \text{say.}$$

Combining (14), (15), (16), we have

$$(*) \qquad \langle d\omega,\eta\rangle = -\langle\omega, \operatorname{div}\eta\rangle + \operatorname{div} Y \ .$$

From this we conclude

62. PROPOSITION. If η is a $(k+1)$-form on an oriented Riemannian manifold M, then

$$\delta\eta = -\operatorname{div}\eta \ .$$

Proof. First suppose M is compact. Equation (*) gives, for any k-form ω,

$$\begin{aligned}(d\omega,\eta) &= \int_M \langle d\omega,\eta\rangle \, dV = \int_M \langle\omega, -\operatorname{div}\eta\rangle \, dV + \int_M \operatorname{div} Y \, dV \\ &= \int_M \langle\omega, -\operatorname{div}\eta\rangle \, dV + 0 \qquad \text{by Corollary 58} \\ &= (\omega, -\operatorname{div}\eta) \ .\end{aligned}$$

Since $\delta\eta$ is the unique form with $(d\omega,\eta) = (\omega,\delta\eta)$ for all k-forms ω, it follows that $\delta\eta = -\operatorname{div}\eta$.

If M is not compact, we can still conclude, from Theorem 57, that $(d\omega,\eta) = (\omega, -\operatorname{div}\eta)$ for all k-forms ω with compact support. This is still sufficient to imply that $\delta\eta = -\operatorname{div}\eta$. ■

Now consider a 1-form $\omega = \Sigma\, a_i dx^i$. Propositions 61 and 62 say that

$$d\omega = 0 \iff a_{i;j} = a_{j;i}$$

$$\delta\omega = 0 \iff 0 = \sum_{i,j} g^{ij} a_{i;j} = \sum_i a^i{}_{;i} \,.$$

Suppose that $d\omega = \delta\omega = 0$, and consider the expression (5) on p. 197 for $\Delta(\sum_i a^i a_i)$. For the first term in parentheses we have

$$\begin{aligned}
\sum_{i,j,k} g^{jk} a^i a_{i;jk} &= \sum_{i,j,k} g^{jk} a^i a_{j;ik} && \text{since } d\omega = 0 \\
&= \sum_{i,j,k} g^{jk} a^i (a_{j;ki} + \sum_\ell a_\ell R^\ell{}_{jik}) && \text{by Ricci's identity} \\
&= \sum_{i,j,k} a^i (g^{jk} a_{j;k})_{;i} + \sum_{i,j,k,\ell} g^{jk} a^i a_\ell R^\ell{}_{jik} \\
&= 0 + \sum_{i,j,k,\ell} g^{jk} a^i a_\ell R^\ell{}_{jik} && \text{since } \delta\omega = 0 \\
&= \sum_{i,j,k,\ell,\mu} g^{jk} a^i g_{\ell\mu} a^\mu R^\ell{}_{jik} \\
&= \sum_{i,j,k,\mu} a^i a^\mu g^{jk} R_{\mu jik} \\
&= -\sum_{i,j,k,\mu} a^i a^\mu g^{jk} R_{j\mu ik} \\
&= -\sum_{i,k,\mu} a^i a^\mu R^k{}_{\mu ik} \\
&= -\sum_{i,\mu} \mathrm{Ric}_{\mu i} a^i a^\mu \,.
\end{aligned}$$

Thus we obtain

$$(*) \qquad \Delta(\sum_i a^i a_i) = 2\left[-\sum_{i,j} \mathrm{Ric}_{ij} a^i a^j + \sum_{i,j,k,\ell} g^{jk} g^{i\ell} a_{i;k} a_{\ell;j}\right].$$

63. THEOREM (BOCHNER). Let M be a compact oriented Riemannian manifold with $-\operatorname{Ric}(X,X) > 0$ for all $X \neq 0$. (This holds, in particular, if all sectional curvatures of M are > 0.) Then the 1-dimensional de Rham cohomology of M is zero.

Proof. Let ω be any 1-form on M with $\Delta\omega = 0$. Then also $d\omega = \delta\omega = 0$, by (13). Then (*) shows that $\Delta(\sum_i a^i a_i) \geq 0$, since the second sum on the right is clearly ≥ 0. Recall (p. 196) that $\sum_i a^i a_i$ is a well-defined function f on M. So Lemma 60 implies that $\Delta(\sum_i a^i a_i) = 0$. By the hypothesis on Ric, this implies that $\omega = 0$. In other words, 0 is the only harmonic 1-form. Since the vector space of all harmonic 1-forms is isomorphic to the 1-dimensional de Rham cohomology of M, the theorem follows. ■

In the next Chapter we will prove that a compact Riemannian manifold M with $-\operatorname{Ric}(X,X) > 0$ for all $X \neq 0$ actually has a finite fundamental group $\pi_1(M)$. Then the result of Theorem 63 follows by algebraic topology [the Hurewicz theorem shows that the first homology group $H_1(M;\mathbb{Z})$ is finite, and the universal coefficient theorem then implies that $H^1(M;\mathbb{R}) = \operatorname{Hom}((H_1(M);\mathbb{Z}),\mathbb{R}) = 0$]. However, Theorem 63 has many generalizations, proved using similar techniques, that have never been strengthened in the same way.

Addendum 3. When are two Riemannian Manifolds Isometric?

Suppose we are given two Riemmanian manifolds M, $\bar{M}$ of the same dimension n. We would like a way of finding out whether they are locally isometric. In other words, we ask if there is a point $p \in M$, a point $\bar{p} \in \bar{M}$, and an isometry $\alpha\colon U \to \bar{U}$ of a neighborhood U of p onto a neighborhood $\bar{U}$ of $\bar{p}$. We have a slightly different problem if we are already given p and $\bar{p}$, and merely seek U and $\bar{U}$. Admittedly, both of these problems are a little strange, for we are not very likely to be given two explicit Riemannian metrics just out of the clear blue sky; specific metrics which actually come up in practice are so special, and the requirements of isometry so stringent, that there is usually no difficulty seeing whether they are isometric. As a matter of fact, I know of no instance where the (complicated) general methods which we will develop are actually used. But it is nevertheless quite significant that we can now settle the question of isometry in the category of Riemannian manifolds, for this shows that any intrinsic invariant of a Riemannian manifold can be defined in terms of the various invariants (like the curvature tensor) which we have already discovered.

The theory is rather special, and quite pleasant, in the 2-dimensional case. First some preliminaries. For functions f, g on a Riemannian manifold $(M, \langle\ ,\ \rangle)$, we introduce the classical notation*

$$\Delta_1(f,g) = \langle \operatorname{grad} f, \operatorname{grad} g\rangle\ , \qquad \Delta_1(f) = \Delta_1(f,f)\ .$$

It is clear that if $\alpha\colon M \to \bar{M}$ is an isometry, and $f, g\colon \bar{M} \to \mathbb{R}$, then

$$(1) \qquad \bar{\Delta}_1(f\circ\alpha,\ g\circ\alpha) = \Delta_1(f,g)\ ,$$

---

*In classical differential geometry books, the Laplacian Δf was denoted by $\Delta_2 f$ (and, worst of all, $\Delta_1 f$ was sometimes written as Δf), but we will stick with Δf for the Laplacian.

where $\bar{\Delta}_1$ is formed with respect to the metric on $\bar{M}$. For a metric

$$\langle\ ,\ \rangle = E\, du\otimes du + F[du\otimes dv + dv\otimes du] + G\, dv\otimes dv$$

on a 2-dimensional manifold, we easily compute that

$$\Delta_1(f,g) = \frac{1}{EG-F^2}\left[E\frac{\partial f}{\partial v}\frac{\partial g}{\partial v} - F\left(\frac{\partial f}{\partial v}\frac{\partial g}{\partial u} + \frac{\partial f}{\partial u}\frac{\partial g}{\partial v}\right) + G\frac{\partial f}{\partial u}\frac{\partial g}{\partial u}\right] .$$

In particular, we have

$$\Delta_1 u = \frac{G}{EG-F^2}, \qquad \Delta_1(u,v) = \frac{-F}{EG-F^2}, \qquad \Delta_1 v = \frac{E}{EG-F^2} .$$

This gives

$$\frac{1}{EG-F^2} = \Delta_1 u\cdot\Delta_1 v - \Delta_1(u,v)^2 = \Theta^2(u,v), \quad \text{say,}$$

and thus

$$(2) \qquad E = \frac{\Delta_1(v)}{\Theta^2(u,v)}, \qquad F = \frac{-\Delta_1(u,v)}{\Theta^2(u,v)}, \qquad G = \frac{\Delta_1(u)}{\Theta^2(u,v)} .$$

This equation shows that the metric $\langle\ ,\ \rangle$ is determined once we know $\Delta_1(u)$, $\Delta_1(u,v)$, and $\Delta_1(v)$ for any coordinate system (u,v). We can formalize the contents of this equation as follows.

64. <u>LEMMA</u>. Let $\alpha\colon M \to \bar{M}$ be a diffeomorphism of 2-dimensional Riemannian manifolds, and for each coordinate system $(\bar{u},\bar{v})$ on $\bar{M}$, define $(u,v) = (\bar{u},\bar{v})\circ\alpha$ on M. If α is an isometry, then

$$(*) \quad \Delta_1 u = (\bar{\Delta}_1\bar{u})\circ\alpha, \qquad \Delta_1(u,v) = \bar{\Delta}_1(\bar{u},\bar{v})\circ\alpha, \qquad \Delta_1(v) = (\bar{\Delta}_1\bar{v})\circ\alpha .$$

Conversely, if these equations holds for some collection of coordinate systems $(\bar{u},\bar{v})$ whose domains cover $\bar{M}$, then α is an isometry.

Proof. Since $u = \bar{u}\circ\alpha$ and $v = \bar{v}\circ\alpha$, the first part of the theorem follows immediately from equation (1). To prove the converse, let the metrics on M and $\bar{M}$ be

$$\langle\ ,\ \rangle = E\, du\otimes du + \cdots \qquad \text{and} \qquad \langle\ ,\ \rangle^{-} = \bar{E}\, d\bar{u}\otimes d\bar{u} + \cdots .$$

Since

$$du = d(\bar{u}\circ\alpha) = \alpha^*(d\bar{u}) , \qquad dv = \alpha^*(d\bar{v}) ,$$

we have

$$\alpha^*(\bar{E}\, d\bar{u}\otimes d\bar{u} + \cdots) = (\bar{E}\circ\alpha)\, du\otimes du + \cdots .$$

But the hypothesis (*), together with equation (2), gives $\bar{E}\circ\alpha = E$, etc. ∎

Now consider two metrics $\langle\ ,\ \rangle$ and $\langle\ ,\ \rangle^{-}$ on two 2-dimensional Riemannian manifolds M and $\bar{M}$. We want to know if there is locally an isometry $\alpha\colon M \to \bar{M}$. It might happen that the curvature K of $\langle\ ,\ \rangle$ is constant. Then α exists if and only if the curvature $\bar{K}$ of $\langle\ ,\ \rangle^{-}$ is the same constant; if this is the case, then there is a 2-parameter family of isometries α. Suppose instead that the curvature K of $\langle\ ,\ \rangle$ is not constant. We consider a region where the sets K = constant give a foliation,

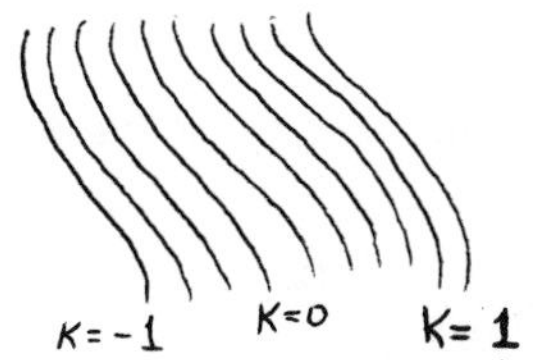

and we will try to decide whether the isometry exists in this region. To be sure, the sets $K = \text{constant}$ might look much worse; for example $K = 0$ might be a single point, or a whole set with interior, etc., etc. But in general, the more complicated the decomposition we obtain, the easier it will be to handle the problem, for then the sets $\bar{K} = \text{constant}$ must look just as complicated. At any rate, we will, over and over again, restrict our attention to the "general" case, and not worry about the exceptional situations. When the sets $K = \text{constant}$ give a foliation, then the sets $\bar{K} = \text{constant}$ must also, if the required isometry α is to exist. Moreover, the isometry α must take the set $K = c$ onto the set $\bar{K} = c$. However this still leaves a lot of leeway, and does not yet determine α. We now consider the function $\Delta_1 K$. This function might not give us any new information at all, for $\Delta_1 K$ might be a constant on each of the sets $K = \text{constant}$. We will first consider the case where $\Delta_1 K$ is not constant on these sets. In fact, we want to assume that $\Delta_1 K$ varies monotonically on each set $K = \text{constant}$. Then it will "generally" be the case that $(K, \Delta_1 K)\colon M \to \mathbb{R}^2$ is a local coordinate system for M. This is the situation which we will actually consider. If the isometry α is to exist, then $(\bar{K}, \bar{\Delta}_1 \bar{K})\colon \bar{M} \to \mathbb{R}^2$ must also be a local coordinate system for $\bar{M}$. Suppose this also occurs. Then clearly the isometry α must, in fact, be the composition

$$\alpha = (\bar{K}, \bar{\Delta}_1 \bar{K})^{-1} \circ (K, \Delta_1 K) ,$$

defined in some open set $U \subset M$. Now the question arises: how do we know whether this α is actually an isometry? There is an easy answer to this question: Lemma 64 tells us that α is an isometry if and only if

$$\Delta_1 K = \bar{\Delta}_1 \bar{K} \circ \alpha , \qquad \Delta_1 (K, \Delta_1 K) = \bar{\Delta}_1 (\bar{K}, \bar{\Delta}_1 \bar{K}) \circ \alpha , \qquad \Delta_1 (\Delta_1 K) = \bar{\Delta}_1 (\bar{\Delta}_1 \bar{K}) \circ \alpha .$$

Moreover, the first of these equations is automatic, by the definition of α.

Now consider the opposite extreme, where $\Delta_1 K$ is a function of K. If α exists, then $\bar{\Delta}_1\bar{K}$ must be the same function of $\bar{K}$,

$$\text{(a)} \qquad \Delta_1 K = f\circ K \ , \qquad \bar{\Delta}_1\bar{K} = f\circ\bar{K} \ .$$

We look at the Laplacians ΔK and $\bar{\Delta}\bar{K}$. If $(K,\Delta K)$ is a local coordinate system, then $(\bar{K},\bar{\Delta}\bar{K})$ must be also, and α must be

$$\alpha = (\bar{K},\bar{\Delta}\bar{K})^{-1} \circ (K,\Delta K) \ .$$

This α is an isometry if and only if

$$\Delta_1(K,\Delta K) = \bar{\Delta}_1(\bar{K},\bar{\Delta}\bar{K})\circ\alpha \ , \qquad \Delta_1(\Delta K) = \bar{\Delta}_1(\bar{\Delta}\bar{K})\circ\alpha \ ;$$

the extra condition $\Delta_1 K = \bar{\Delta}_1\bar{K}\circ\alpha$ follows from (a) and the definition of α. This still leaves us with the case where ΔK also fails to be independent of K in the worst possible way, so that in addition to (a) we have

$$\text{(b)} \qquad \Delta K = g\circ K \ , \qquad \bar{\Delta}\bar{K} = g\circ\bar{K} \ .$$

Then it turns out (Problem 24) that the surfaces are isometric, and there is a 1-parameter family of isometries between them.

For higher dimensional manifolds the treatment will be more systematic, but correspondingly less concrete. We already know (Corollary II.7-13) that the metric in a normal coordinate system determined by an orthonormal frame $X_{1p},\dots,X_{np}$ is completely determined by knowing $\langle R(X_i,X_j)X_k,X_\ell\rangle$, where $X_1,\dots,X_n$ is the moving frame adapted to $X_{1p},\dots,X_{np}$. This result gives us a criterion for determining when a neighborhood of $p \in M$ is isometric to a neighborhood of $\bar{p} \in \bar{M}$, but it cannot be regarded as a reasonable solution of our problem, for we may not be able to compute the geodesics, or the parallel translations along these geodesics. All we can compute is the equations for the geodesics and for parallel translation -- usually we will not be able to solve these equations explicitly. What we want is a criterion involving only quantities directly computable in a coordinate system -- like curvature, covariant derivatives of tensors which have already been computed, etc.

Recall the map $\Phi\colon \mathbb{R} \times M_p \longrightarrow M$ (p. II.300) defined by

$$\Phi(t,X_p) = \exp(tX_p) \ .$$

From the discussion on pp. II.300-310 [c.f. especially Corollary 9 and Theorem 12] we see that the metric in the normal coordinate system determined by $X_{1p},\ldots,X_{np}$ is completely determined once the functions $\mathbf{R}^i{}_{jk\ell}\circ\Phi$ are known. Now suppose that the metric is analytic. Then its form in normal coordinates is known once we know

$$\frac{\partial(\mathbf{R}^i{}_{jk\ell}\circ\Phi)}{\partial t}(0,X_p) \ , \qquad \frac{\partial^2(\mathbf{R}^i{}_{jk\ell}\circ\Phi)}{\partial t^2}(0,X_p) \ ,\ldots \qquad \text{all } X_p \in M_p \ .$$

Now

$$(1) \quad \frac{\partial(\mathbf{R}^i{}_{jk\ell}\circ\Phi)}{\partial t}(t,X_p) = \lim_{h\to 0} \frac{\mathbf{R}^i{}_{jk\ell}(\Phi(t+h,X_p)) - \mathbf{R}^i{}_{jk\ell}(\Phi(t,X_p))}{h} \ .$$

Let $\mathcal{R}$ be the tensor

$$\mathcal{R}(X,Y,Z,W) = \langle R(X,Y)Z,W\rangle \ ,$$

so that (c.f. p. II.309)

$$\mathbf{R}^i{}_{jk\ell} = \mathcal{R}(X_k, X_\ell, X_j, X_i) \ .$$

Since $\Phi(t,X_p) = \exp(tX_p)$, and since the X_i are defined by parallel translating the X_{ip} along geodesics, equation (1) can be written

$$\begin{aligned}
&\frac{\partial(\mathbf{R}^i{}_{jk\ell}\circ\Phi)}{\partial t}(t,X_p) \\
&= (\nabla_{X(\Phi(t,X_p))}\mathcal{R})\Big(X_k(\Phi(t,X_p)),\ X_\ell(\Phi(t,X_p)),\ X_j(\Phi(t,X_p)),\ X_i(\Phi(t,X_p))\Big) \\
&= (\nabla\mathcal{R})\Big(X_k(\Phi(t,X_p)),\ X_\ell(\Phi(t,X_p)),\ X_j(\Phi(t,X_p)),\ X_i(\Phi(t,X_p)),\ X(\Phi(t,X_p))\Big),
\end{aligned}$$

where X is also defined by parallel translating X_p along geodesics. In general, we have

$$\begin{aligned}
&\frac{\partial^d(\mathbf{R}^i{}_{jk\ell}\circ\Phi)}{\partial t^d}(t,X_p) \\
&= (\overbrace{\nabla\cdots\nabla}^{d}\mathcal{R})\Big(X_k(\Phi(t,X_p)),\ldots,X_i(\Phi(t,X_p)),\ \overbrace{X(\Phi(t,X_p)),\ldots,X(\Phi(t,X_p))}^{d}\Big) \ .
\end{aligned}$$

In particular,

$$\frac{\partial^d(\mathbf{R}^i{}_{jk\ell}\circ\Phi)}{\partial t^d}(0,X_p) = (\nabla^d\mathcal{R})(X_{pk},X_{p\ell},X_{pj},X_{pi},\ \overbrace{X_p,\ldots,X_p}^{d}) \ .$$

Thus, in the analytic case, the metric in normal coordinates around p is determined completely by knowing all $\nabla^d\mathcal{R}$ at p.

Thus we have a criterion for deciding when some neighborhood of a given point $p \in M$ can be taken isometrically onto a neighborhood of a point $\bar{p} \in \bar{M}$.

This criterion works only for analytic metrics, but its real defect is the fact that we have to compute infinitely many quantities $(\nabla^d \mathcal{R})(p)$. Now we will explain how one can decide whether some open set in M is isometric to some open set of $\bar{M}$, without being given points p, $\bar{p}$ in advance, without assuming the metric is analytic, and by computing only finitely many covariant derivatives $\nabla^d \mathcal{R}$. True, we will have to compute the $\nabla^d \mathcal{R}$ on all of M, not just at one point p, but in practice the only way to compute the $(\nabla^d \mathcal{R})(p)$ is to compute the $\nabla^d \mathcal{R}$ in a whole coordinate system, anyway.

First we consider a general problem having nothing to do with metrics. Suppose we are given two manifolds M^N and $\bar{M}^N$ of the same dimension N. Let $\omega^1,\dots,\omega^N$ be N everywhere linearly independent 1-forms on (a subset of) M, and let $\bar{\omega}^1,\dots,\bar{\omega}^N$ be similar 1-forms on $\bar{M}$. We will find a way of deciding when there is locally a diffeomorphism $\alpha\colon M \to \bar{M}$ such that $\omega^i = \alpha^*\bar{\omega}^i$ for $i = 1,\dots,N$. First of all, let us write

$$d\omega^i = \sum_{j<k} C^i_{jk}\,\omega^j \wedge \omega^k$$

$$d\bar{\omega}^i = \sum_{j<k} \bar{C}^i_{jk}\,\bar{\omega}^j \wedge \bar{\omega}^k$$

for certain functions C^i_{jk} and $\bar{C}^i_{jk}$. If α exists, then we will have $C^i_{jk} = \bar{C}^i_{jk}\circ\alpha$. Now suppose that among the functions C^i_{jk} there are N which form a coordinate system $(u_1,\dots,u_N)$ on M. Then if α exists, the corresponding N functions $\bar{C}^i_{jk}$ must form a coordinate system $(\bar{u}_1,\dots,\bar{u}_N)$ on $\bar{M}$. In this case, the diffeomorphism α must be

$$(1) \qquad \alpha = (\bar{u}_1,\dots,\bar{u}_N)^{-1} \circ (u_1,\dots,u_N)\ .$$

For this α we certainly have

$$(2) \qquad C^i_{jk} = \bar{C}^i_{jk} \circ \alpha$$

when C^i_{jk} is one of the u's, and we may add the extra condition that equation (2) hold in all cases. [In the "general" case, the other C^i_{jk} will be functions of the u's,

$$C^i_{jk} = f^i_{jk} \circ (u_1, \ldots, u_N) ,$$

so we are demanding that the other $\bar{C}^i_{jk}$ be the same functions of the $\bar{u}$'s.] We still have to decide when α given by (1) is the required diffeomorphism. For this we write

$$dC^i_{jk} = \sum_\ell C^i_{jk,\ell}\,\omega^\ell$$

$$d\bar{C}^i_{jk} = \sum_\ell \bar{C}^i_{jk,\ell}\,\bar{\omega}^\ell$$

for certain functions $C^i_{jk,\ell}$ and $\bar{C}^i_{jk,\ell}$. If α has the desired properties, then we must also have

$$(3) \qquad C^i_{jk,\ell} = \bar{C}^i_{jk,\ell} \circ \alpha .$$

Conversely, suppose equation (3) holds. Then

$$(4) \qquad \begin{aligned} \sum_\ell C^i_{jk,\ell} \cdot (\omega^\ell - \alpha^*\bar{\omega}^\ell) &= \sum_\ell C^i_{jk,\ell}\,\omega^\ell - \sum_\ell (\bar{C}^i_{jk,\ell} \circ \alpha) \cdot \alpha^*\bar{\omega}^\ell \\ &= \sum_\ell C^i_{jk,\ell}\,\omega^\ell - \alpha^*\Big(\sum_\ell \bar{C}^i_{jk,\ell} \cdot \bar{\omega}^\ell\Big) \end{aligned}$$

$$= dC^i_{jk} - \alpha^*(d\bar{C}^i_{jk})$$

$$= dC^i_{jk} - d(\bar{C}^i_{jk}\circ\alpha)$$

$$= 0 \qquad\qquad \text{by (2).}$$

This is a set of N^3 equations in N unknowns. It can be written in terms of the $N^3 \times N$ matrix $(C^i_{jk,\ell})$ in which ℓ denotes the column, and $(^{\,i}_{jk})$ denotes the row. This matrix contains the $N \times N$ submatrix $(u_{i,\ell})$, which is non-singular, since $(u_1,\ldots,u_N)$ is a coordinate system. So the matrix $(C^i_{jk,\ell})$ has rank N. This means that the only solution of our equations is the zero solution. Thus, $\omega^\ell = \alpha^*\bar{\omega}^\ell$ for $\ell = 1,\ldots,N$.

Suppose, on the contrary, that we can choose only $N_1 < N$ functions C^i_{jk} which are independent (meaning that for any coordinate system $x^1,\ldots,x^N$, the $N_1 \times N$ matrix $(\partial C^i_{jk}/\partial x^\ell)$ has rank N_1; or equivalently, that the $N_1 \times N$ matrix $(C^i_{jk,\ell})$ has rank N_1). We now look at the functions $C^i_{jk,\ell}$. Among these we may be able to choose N_2 functions $C^i_{jk,\ell}$ which together with the N_1 functions C^i_{jk} are independent. If $N_1 + N_2 < N$, then we look at the functions $C^i_{jk,\ell m}$ defined by

$$dC^i_{jk,\ell} = \sum_m C^i_{jk,\ell m}\omega^m .$$

Among these we may be able to pick N_3 which can be added to the $N_1 + N_2$ functions already obtained. Suppose that, after continuing in this way, we eventually obtain $N_1 + \cdots + N_\mu = N$ independent functions. Then we can determine what α must be; moreover, we can decide whether this α really works by seeing if α satisfies

$$C^i_{jk,\ell_1,\ldots,\ell_{\mu+1}} = \bar{C}^i_{jk,\ell_1,\ldots,\ell_{\mu+1}} \circ \alpha \ .$$

On the other hand, it may happen that we never obtain N independent functions. In the general case this will happen because at some stage, the functions

$$C^i_{jk,\ell_1,\ldots,\ell_{\mu+1}}$$

are all functions of the previously chosen functions. [Notice that once this happens at stage μ, it will happen at all later stages. So in general, the integers $N_1, N_2, \ldots$ which we picked in the previous case are all ≥ 1. Thus we either obtain N independent functions in $\leq N$ stages, or we arrive at the present situation in $\leq N$ stages.] We now have $N' = N_1 + \cdots + N_\mu < N$ independent functions. If α exists, then it must satisfy the N' equations

$$(*) \qquad \begin{cases} C^i_{jk} = \bar{C}^i_{jk} \circ \alpha \\ \vdots \\ C^i_{jk,\ell_1,\ldots,\ell_\mu} = \bar{C}^i_{jk,\ell_1,\ldots,\ell_\mu} \circ \alpha \ . \end{cases}$$

In the same way that we obtained equations (4), we can use equations (*) to deduce N' linear equations for $\omega^1 - \bar{\omega}^1, \ldots, \omega^N - \bar{\omega}^N$. Moreover, the rank of the matrix for these equations is N', so we can solve for $N - N'$ of the unknowns in terms of the other N'. Without loss of generality, we can assume that these equations can be solved for the last $N - N'$ of the $\omega^i - \bar{\omega}^i$ in terms of the first N' of the $\omega^i - \bar{\omega}^i$. Then clearly the diffeomorphism α has the desired properties if it satisfies (*) as well as

$$(**) \qquad \omega^i = \alpha^* \bar{\omega}^i \qquad i = 1, \ldots, N' \ .$$

We claim that there are always such diffeomorphisms α, in fact, an $(N-N')$-parameter family of them. To prove this, we look for the graph of α, as a subset of $M \times \bar{M}$. Let $\pi\colon M \times \bar{M} \longrightarrow M$ and $\bar{\pi}\colon M \times \bar{M} \longrightarrow \bar{M}$ be the projections. Since the $C^i_{jk},\ldots,$ and $\bar{C}^i_{jk},\ldots$ in (*) are independent functions, the set

$$\mathcal{M} = \{x \in M \times \bar{M}\colon C_{ij} \circ \pi(x) = \bar{C}^i_{jk} \circ \bar{\pi}(x), \ldots\}$$

is a submanifold, of dimension $2N - N'$. We will denote the restrictions of $\pi^*\omega^i$ and $\bar{\pi}^*\bar{\omega}^i$ to $\mathcal{M}$ simply by $\pi^*\omega^i$ and $\bar{\pi}^*\bar{\omega}^i$. Consider the ideal $\mathcal{I}$ on $\mathcal{M}$ generated by the 1-forms

$$\pi^*\omega^i - \bar{\pi}^*\bar{\omega}^i , \qquad i = 1,\ldots,N' .$$

We have

$$\begin{aligned} d(\pi^*\omega^i - \bar{\pi}^*\bar{\omega}^i) &= \Sigma(C^i_{jk} \circ \pi)\pi^*\omega^i \wedge \pi^*\omega^j \\ &\quad - \Sigma(\bar{C}^i_{jk} \circ \bar{\pi})\bar{\pi}^*\bar{\omega}^i \wedge \bar{\pi}^*\bar{\omega}^j \\ &= \Sigma(C^i_{jk} \circ \pi)[\pi^*\omega^i \wedge \pi^*\omega^j - \bar{\pi}^*\bar{\omega}^i \wedge \bar{\pi}^*\bar{\omega}^j] \qquad (\text{on } \mathcal{M}) \\ &= \Sigma(C^i_{jk} \circ \pi)[(\pi^*\omega^i \wedge \bar{\pi}^*\bar{\omega}^i) \wedge \pi^*\omega^j \\ &\qquad - \bar{\pi}^*\bar{\omega}^i \wedge (\pi^*\omega^k - \bar{\pi}^*\bar{\omega}^k)] , \end{aligned}$$

which is in $\mathcal{I}$. Thus there is a submanifold $\mathcal{M}'$ of $\mathcal{M}$ on which the forms $\pi^*\omega^i - \bar{\pi}^*\bar{\omega}^i$ all vanish. This submanifold $\mathcal{M}'$ has dimension $(2N - N') - N' = 2(N - N')$. There is an $(N - N')$-parameter family of N-dimensional submanifolds $\mathcal{M}''$ of $\mathcal{M}'$ which project one-one onto M. Each of these is the graph of an appropriate α.

Finally, let us return to the case of two Riemannian manifolds M^n and

$\bar{M}^n$. We immediately pass to the principal bundles $O(TM)$ and $O(T\bar{M})$ of orthonormal frames. On these bundles we have forms $\boldsymbol{\theta} = (\boldsymbol{\theta}^i)$, $\boldsymbol{\omega} = (\boldsymbol{\omega}^i_j)$ and $\bar{\boldsymbol{\theta}} = (\bar{\boldsymbol{\theta}}^i)$, $\bar{\boldsymbol{\omega}} = (\bar{\boldsymbol{\omega}}^i_j)$. Recall that for $u = (u_1,\dots,u_n) \in O(TM)$ and a tangent vector $Y \in O(TM)_u$, we have

$$\pi_* Y_u = \sum_{i=1}^{n} \boldsymbol{\theta}^i(Y_u)\cdot u_i \ ,$$

where $\pi\colon O(TM) \longrightarrow M$ is the projection map. In particular, $\boldsymbol{\theta}(Y_u) = 0$ if and only if $\pi_* Y_u = 0$. Now any isometry $\alpha\colon M \longrightarrow \bar{M}$ gives rise to a diffeomorphism $\tilde{\alpha}\colon O(TM) \longrightarrow O(T\bar{M})$, and $\tilde{\alpha}^*\bar{\boldsymbol{\theta}} = \boldsymbol{\theta}$, $\tilde{\alpha}^*\bar{\boldsymbol{\omega}} = \boldsymbol{\omega}$. Conversely, suppose we have a diffeomorphism $\beta\colon O(TM) \longrightarrow O(T\bar{M})$ with $\beta^*\bar{\boldsymbol{\theta}} = \boldsymbol{\theta}$. If c is any curve in the fibre of $O(TM)$ at p, then for all t we have $\pi_* c'(t) = 0$, and thus

$$\begin{aligned} 0 = \boldsymbol{\theta}(c'(t)) &= \beta^*\bar{\boldsymbol{\theta}}(c'(t)) \\ &= \bar{\boldsymbol{\theta}}(\beta_* c'(t)) \\ \Longrightarrow\ 0 = \bar{\pi}_*\beta_* c'(t) &= (\bar{\pi}\circ\beta\circ c)'(t) \ . \end{aligned}$$

Thus $\bar{\pi}\circ\beta\circ c$ is constant. This shows that β takes fibres to fibres, so there is a map $\alpha\colon M \longrightarrow \bar{M}$ with $\bar{\pi}\circ\beta = \alpha\circ\pi$. Moreover, if $u = (u_1,\dots,u_n) \in O(TM)$ and $Y_u \in O(TM)_u$ satisfies $\pi_* Y_u = u_j$, then $\boldsymbol{\theta}^i(Y_u) = \delta^i_j$, so

$$\begin{aligned} \delta^i_j &= \beta^*\bar{\boldsymbol{\theta}}^i(Y_u) \\ &= \bar{\boldsymbol{\theta}}^i(\beta_* Y_u) \\ &= i^{\text{th}} \text{ component of } \bar{\pi}_*\beta_* Y_u \text{ with respect to } \beta(u) \\ &= \quad '' \quad '' \quad '' \quad \alpha_*\pi_* Y_u \quad '' \quad '' \quad '' \quad '' \\ &= \quad '' \quad '' \quad '' \quad \alpha_*(u_j) \quad '' \quad '' \quad '' \quad '' \ . \end{aligned}$$

Thus β must be

$$\beta(u) = (\alpha_* u_1, \ldots, \alpha_* u_n) ,$$

i.e., $\beta = \tilde{\alpha}$. In particular, α is an isometry. Thus we see that the existence of an isometry $\alpha\colon M \longrightarrow \bar{M}$ is equivalent to the existence of a diffeomorphism $\beta\colon O(TM) \longrightarrow O(T\bar{M})$ such that $\beta^* \bar{\boldsymbol{\theta}}^i = \boldsymbol{\theta}^i$. Hence it is also equivalent to the existence of a diffeomorphism $\beta\colon O(TM) \longrightarrow O(T\bar{M})$ such that $\beta^* \bar{\boldsymbol{\theta}}^i = \boldsymbol{\theta}^i$ and $\beta^* \bar{\boldsymbol{\omega}}^i_j = \boldsymbol{\omega}^i_j$. We have just seen how to decide whether such a β exists, since the $\boldsymbol{\theta}^i$ and $\boldsymbol{\omega}^i_j$ are everywhere linearly independent and span the 1-forms, and similarly for the $\bar{\boldsymbol{\theta}}^i$ and $\bar{\boldsymbol{\omega}}^i_j$. The first step is to compute the $d\boldsymbol{\theta}^i$ and $d\boldsymbol{\omega}^i_j$ in terms of the $\boldsymbol{\theta}^i$ and $\boldsymbol{\omega}^i_j$. We already have the structural equations (p. II.378),

$$(1) \qquad d\boldsymbol{\theta}^i = -\sum_j \boldsymbol{\omega}^i_j \wedge \boldsymbol{\theta}^j$$

$$(2) \qquad d\boldsymbol{\omega}^i_j = -\sum_k \boldsymbol{\omega}^i_k \wedge \boldsymbol{\omega}^k_j + \boldsymbol{\Omega}^i_j$$

$$= -\sum_k \boldsymbol{\omega}^i_k \wedge \boldsymbol{\omega}^k_j - \sum_{k<\ell} A_{ijk\ell}\, \boldsymbol{\theta}^k \wedge \boldsymbol{\theta}^\ell , \qquad \text{say.}$$

These functions $A_{ijk\ell}$ are the first set which we have to examine. Now if $s = (X_1, \ldots, X_n)$ is an orthonormal moving frame, then its dual forms and connection forms are $\theta^i = s^* \boldsymbol{\theta}^i$ and $\omega^i_j = s^* \boldsymbol{\omega}^i_j$. So s^* of equation (2) gives

$$d\omega^i_j = -\sum_k \omega^i_k \wedge \omega^k_j - \sum_{k<\ell} (A_{ijk\ell} \circ s)\theta^k \wedge \theta^\ell .$$

On the other hand, we have

$$d\omega_j^i = -\sum_k \omega_k^i \wedge \omega_j^k + \Omega_j^i$$

$$= -\sum_k \omega_k^i \wedge \omega_j^k + \sum_{k<\ell} \langle R(X_k,X_\ell)X_j,X_i\rangle \theta^k \wedge \theta^\ell .$$

Thus we see that

$$A_{ijk\ell}(u) = -\langle R(u_k,u_\ell)u_j,u_i\rangle = \mathcal{R}(u_i,u_j,u_k,u_\ell) .$$

The next set of functions which we need to look at are those appearing in the expansion

$$dA_{ijk\ell} = \sum_{\mu,\nu} (\qquad) \boldsymbol{\omega}_\nu^\mu + \sum_\mu A_{ijk\ell,\mu} \boldsymbol{\theta}^\mu .$$

Taking s^* of this equation, and evaluating at a tangent vector X of M, we get

$$X(\mathcal{R}(X_i,X_j,X_k,X_\ell)) = X(A_{ijk\ell}\circ s) = d(A_{ijk\ell}\circ s)(X) = s^*(dA_{ijk\ell})(X)$$

$$= \sum_{\mu,\nu} [(\qquad)\circ s]\omega_\nu^\mu(X) + \Sigma[A_{ijk\ell,\mu}\circ s]\theta^\mu(X) .$$

Since

$$X(\mathcal{R}(X_i,X_j,X_k,X_\ell)) = (\nabla\mathcal{R})(X_i,X_j,X_k,X_\ell,X) + \mathcal{R}(\nabla_X X_i,\ldots) + \mathcal{R}(X_i,\nabla_X X_j,\ldots) + \cdots ,$$

we see that we must have

$$dA_{ijk\ell} = \sum_\mu A_{\mu jk\ell} \boldsymbol{\omega}_i^\mu + \cdots + \sum_\mu A_{ijk\mu} \boldsymbol{\omega}_\ell^\mu + \sum_\mu A_{ijk\ell,\mu} \boldsymbol{\theta}^\mu ,$$

where

$$A_{ijk\ell,\mu}(u) = (\nabla\mathcal{R})(u_i,u_j,u_k,u_\ell,u_\mu) .$$

Similarly, we get

$$dA_{ijk\ell,\mu} = \sum_\nu A_{\nu jk\ell,\mu}\,\boldsymbol{\omega}_i^\nu + \cdots + \sum_\nu A_{ijk\nu,\mu}\,\boldsymbol{\omega}_\ell^\nu + \sum_\nu A_{ijk\ell,\nu}\,\boldsymbol{\omega}_\mu^\nu + \sum_\nu A_{ijk\ell,\mu\nu}\,\boldsymbol{\theta}^\nu ,$$

where

$$A_{ijk\ell,\mu\nu}(u) = (\nabla\nabla\mathcal{R})(u_i,u_j,u_k,u_\ell,u_\mu,u_\nu) ,$$

and so on. So we see that after computing a finite number of the functions $\mathcal{R}$, $\nabla\mathcal{R}$, $\nabla\nabla\mathcal{R}$,..., we can finally decide if the desired isometry α exists (provided we can keep track of what in the world we are doing, which doesn't seem very likely).

Addendum 4. Better Imbedding Invariants

There is a theory, due to Burstin, Mayer, and Allendoerfer, which shows that certain tensors are a complete set of invariants for submanifolds $M^n \subset N^m$ of a manifold N of constant curvature. (As in the theory of curves, we have to impose certain conditions on M, which say, roughly speaking, that at each point M bends in the same number of directions.) One almost never sees any applications of this theory nowadays, but perhaps that is partly because the classical expositions make it so inaccessible. In our presentation, we will first consider a special case of the general problem, so as not to be overwhelmed with details at the beginning.

For a Riemannian manifold $(M, \langle\ ,\ \rangle)$ we know, by the "Fundamental Lemma of Riemannian geometry," that there is a unique connection ∇ on TM which is compatible with the metric and also symmetric. The following Lemma gives, under certain conditions, an analogous characterization of the normal connection D on the normal bundle Nor M of M in N.

65. LEMMA. Let $M \subset N$ have normal bundle Nor M and second fundamental form s. Suppose that $s\colon M_p \times M_p \to M_p{}^{\perp}$ is onto for all p. Then the normal connection D in Nor M is the unique connection δ such that

(1) δ is compatible with the metric in Nor M:

$$X(\langle \xi, \eta \rangle) = \langle \delta_X \xi, \eta \rangle + \langle \xi, \delta_X \eta \rangle \qquad \text{for sections } \xi, \eta \text{ of Nor } M\,,$$

(2) δ satisfies the Codazzi-Mainardi equations:

$$\begin{aligned}\perp(R'(X,Y)Z) = {} & [\delta_X(s(Y,Z)) - s(\nabla_X Y, Z) - s(Y, \nabla_X Z)] \\ & - [\delta_Y(s(X,Z)) - s(\nabla_Y X, Z) - s(X, \nabla_Y Z)]\,.\end{aligned}$$

Proof. Consider the expression

$$\langle \delta_X s(Y_1,Y_2),\ s(Z_1,Z_2)\rangle - \langle \delta_{Y_1} s(X,Y_2),\ s(Z_1,Z_2)\rangle$$
$$- \langle \delta_{Y_1} s(Z_1,Z_2),\ s(X,Y_2)\rangle + \langle \delta_{Z_1} s(Y_1,Z_2),\ s(X,Y_2)\rangle$$
$$+ \langle \delta_{Z_1} s(X,Y_2),\ s(Y_1,Z_2)\rangle - \langle \delta_X s(Z_1,Y_2),\ s(Y_1,Z_2)\rangle \ .$$

Equation (2) shows that each row can be expressed in terms of the vector fields X, Y_i, Z_i. But equation (1) shows that the sum of the two terms involving δ_{Y_1} or δ_{Z_1} can also be expressed in this way. Thus we can write

(3) $$\langle \delta_X s(Y_1,Y_2),\ s(Z_1,Z_2)\rangle - \langle \delta_X s(Z_1,Y_2),\ s(Y_1,Z_2)\rangle = \cdots ,$$

where $\cdots$ can be expressed in terms of the vector fields. So we have as well

(3') $$\langle \delta_X s(Y_2,Z_1),\ s(Z_2,Y_1)\rangle - \langle \delta_X s(Z_2,Z_1),\ s(Y_2,Y_1)\rangle = \cdots .$$

Adding (3) and (3'), we obtain

(4) $$\langle \delta_X s(Y_1,Y_2),\ s(Z_1,Z_2)\rangle - \langle \delta_X s(Z_1,Z_2),\ s(Y_1,Y_2)\rangle = \cdots .$$

But by (1) we also have

(5) $$\langle \delta_X s(Y_1,Y_2),\ s(Z_1,Z_2)\rangle + \langle \delta_X s(Z_1,Z_2),\ s(Y_1,Y_2)\rangle = \cdots .$$

So by adding (4) and (5) we obtain

(*) $$<\delta_X s(Y_1,Y_2),\ s(Z_1,Z_2)> = \cdots .$$

Since $s\colon M_p \times M_p \to M_p{}^{\perp}$ is onto, this shows that $\delta_{X_p} s(Y_1,Y_2)$ is uniquely determined by X_p, Y_1, Y_2. Since every section of Nor M is a linear combination, over the C^∞ functions, of sections of the form $s(Y_1,Y_2)$, this shows that δ is uniquely determined. ∎

<u>Remark</u>. Naturally we are mainly interested in the case where N has constant curvature, in which case the left side of (2) is 0. Now given any bundle $\varpi\colon E \to M$ with a metric $<\ ,\ >$, and a symmetric section s of $\mathrm{Hom}(TM \times TM, E)$, we can consider the "Codazzi-Mainardi equations"

$$[\delta_X(s(Y,Z)) - s(\nabla_X Y,Z) - s(Y,\nabla_X Z)]$$
$$= [\delta_Y(s(X,Z)) - s(\nabla_Y X,Z) - s(X,\nabla_Y Z)] .$$

The proof of Lemma 65 shows that if $s\colon M_p \times M_p \to \varpi^{-1}(p)$ is always onto, then there is at most one δ compatible with $<\ ,\ >$ which satisfies this equation. However, there may not be any such δ (unless s is always one-one).

Now for a submanifold $M \subset N$ we will denote the induced metric $<\ ,\ >$ on M by $\mathcal{F}_0$, and define a tensor $\mathcal{F}_1$ by

$$\mathcal{F}_1(X_1,X_2,Y_1,Y_2) = <s(X_1,X_2),\ s(Y_1,Y_2)> .$$

66. PROPOSITION. Let $M, \bar{M} \subset N$ be connected submanifolds of a complete simply connected manifold N of constant curvature. Suppose that the second fundamental forms $s\colon M_p \times M_p \to M_p{}^{\perp}$ and $\bar{s}\colon \bar{M}_q \times \bar{M}_q \to \bar{M}_q{}^{\perp}$ are onto at all points. Let $\phi\colon M \to \bar{M}$ be a diffeomorphism such that

$$\phi^*\bar{\mathscr{F}}_0 = \mathscr{F}_0 \qquad \text{and} \qquad \phi^*\bar{\mathscr{F}}_1 = \mathscr{F}_1 .$$

Then ϕ is the restriction of an isometry of N.

Proof. First of all, since $\phi^*\bar{\mathscr{F}}_0 = \mathscr{F}_0$, we have

$$(1) \qquad \phi_*(\nabla_X Y) = \bar{\nabla}_{\phi_* X}\phi_* Y .$$

Now let $\{X_\alpha\}$ be any set of vectors in M_p which span M_p, and let $\bar{X}_\alpha = \phi_*(X_\alpha) \in \bar{M}_{f(p)}$. Consider the vectors $s(X_\alpha,X_\beta) \in M_p{}^{\perp}$, and the corresponding vectors $\bar{s}(\bar{X}_\alpha,\bar{X}_\beta) \in \bar{M}_{f(p)}{}^{\perp}$. By hypothesis, we have

$$\langle s(X_\alpha,X_\beta),\ s(X_\gamma,X_\delta)\rangle = \langle \bar{s}(\bar{X}_\alpha,\bar{X}_\beta),\ \bar{s}(\bar{X}_\gamma,\bar{X}_\delta)\rangle .$$

Since the second fundamental forms are onto $M_p{}^{\perp}$ and $\bar{M}_{f(p)}{}^{\perp}$, this implies (Problem 25) that there is a unique inner product preserving isomorphism $M_p{}^{\perp} \to \bar{M}_{f(p)}{}^{\perp}$ which takes $s(X_\alpha,X_\beta) \mapsto \bar{s}(\bar{X}_\alpha,\bar{X}_\beta)$. This isomorphism cannot depend on the $\{X_\alpha\}$, for if we also have spanning vectors $\{Y_\alpha\}$, we can consider the set $\{X_\alpha\} \cup \{Y_\alpha\}$.

By applying this construction for all $p \in M$, we obtain a bundle isomorphism $\tilde{\phi}$: Nor $M \longrightarrow$ Nor $\bar{M}$ covering ϕ such that $\tilde{\phi}$ preserves inner products and second fundamental forms:

$$(2) \qquad \langle\tilde{\phi}(\xi), \tilde{\phi}(\eta)\rangle = \langle\xi, \eta\rangle \quad \text{for sections } \xi, \eta \text{ of Nor } M$$

$$(3) \qquad \tilde{\phi}(s(X,Y)) = \bar{s}(\phi_*X, \phi_*Y) .$$

We claim that $\tilde{\phi}$ also preserves the normal connections:

$$(4) \qquad \tilde{\phi}(D_X\xi) = \bar{D}_{\phi_*X}(\tilde{\phi}(\xi)) .$$

To prove this we note that since every section of Nor $\bar{M}$ is uniquely of the form $\tilde{\phi}(\xi)$, and every tangent vector of $\bar{M}$ is uniquely of the form ϕ_*X, we can define a connection δ on Nor $\bar{M}$ with

$$\delta_{\phi_*X}(\tilde{\phi}(\xi)) = \tilde{\phi}(D_X\xi) .$$

Now the connection D is compatible with the metric and satisfies the Codazzi-Mainardi equations; applying the equations (1)-(3), we find that δ is compatible with the metric and satisfies the Codazzi-Mainardi equations (for $\bar{M}$). Hence $\delta = \bar{D}$, by Lemma 65. This proves (4).

The desired result now follows from Theorem 20. ∎

When we have a manifold $M \subset N$ whose second fundamental form does not fill up the normal bundle, we will have to differentiate more times, precisely as in the case of curves. Notice that the subspace of N_p spanned by M_p and $s_p(M_p \times M_p) \subset M_p^{\perp}$ can also be described as the space spanned by all X_p and $\nabla'_{X_p} Y$ for vector fields X, Y on M. But we can also consider $\nabla'_{X_p}(\nabla'_Y Z)$, etc., and thereby obtain more vectors in N_p. To simplify the notation, we will write

$$\nabla'(X,Y) = \nabla'_X Y$$

$$\nabla'(X,Y,Z) = \nabla'_X(\nabla'_Y Z) \ , \qquad \text{etc.}$$

We define the kth <u>osculating space</u> $\overset{k}{\mathrm{Osc}}\, M_p \subset N_p$ <u>of</u> M <u>at</u> p to be the subspace of N_p which is spanned by all

$$X_1(p),\ \nabla'(X_1,X_2)(p),\ \ldots,\ \nabla'(X_1,\ldots,X_k)(p) \ ,$$

for vector fields $X_1,\ldots,X_k$ on M. Thus the 1st osculating space $\overset{1}{\mathrm{Osc}}\, M_p$ is just M_p. It will also be convenient to define $\overset{0}{\mathrm{Osc}}\, M_p$ to be the $\{0\}$ subspace of M_p.

A submanifold $M \subset N$ will be called <u>nicely curved</u> if the dimension of each osculating space $\overset{k}{\mathrm{Osc}}\, M_p$ is the same for all $p \in M$. (A curve c in N is nicely curved if and only if it has the property that if some curvature function κ_k is non-zero at one point, then κ_k is non-zero everywhere.) Henceforth we will consider only nicely curved submanifolds $M \subset N$. It is easy to see that for each k we then have a vector bundle $\overset{k}{\mathrm{Osc}}\, M$ over M, whose fibre over p is $\overset{k}{\mathrm{Osc}}\, M_p$. If ξ is a smooth section of $\overset{k}{\mathrm{Osc}}\, M$, then ξ is locally a sum of terms $f\cdot\nabla'(X_1,\ldots,X_r)$ for smooth f and X_i, and $r \leq k$. Since

$$\nabla'_X(f\nabla'(X_1,\ldots,X_r)) = X(f)\cdot\nabla'(X_1,\ldots,X_r) + f\cdot\nabla'(X,X_1,\ldots,X_r) ,$$

we see that

$$(*) \qquad \xi \text{ a section of } \overset{k}{\mathrm{Osc}}\, M \implies \nabla'_{X_p}\xi \in \overset{k+1}{\mathrm{Osc}}\, M_p .$$

It is easy to see that if $\overset{k}{\mathrm{Osc}}\, M = \overset{k+1}{\mathrm{Osc}}\, M$, then also $\overset{k+1}{\mathrm{Osc}}\, M = \overset{k+2}{\mathrm{Osc}}\, M = \cdots$. So there is some $\ell \geq 1$ with

$$\overset{0}{\mathrm{Osc}}\, M \subsetneq \overset{1}{\mathrm{Osc}}\, M \subsetneq \cdots \subsetneq \overset{\ell}{\mathrm{Osc}}\, M = \overset{\ell+1}{\mathrm{Osc}}\, M = \cdots .$$

The letter ℓ will always have this significance. Notice that $\overset{\ell}{\mathrm{Osc}}\, M_p$ need not be all of N_p; the dimension of $\overset{\ell}{\mathrm{Osc}}\, M_p$ (for any $p \in M$) will be called the <u>formal imbedding number</u> $\#(M)$ <u>of</u> M.

<u>67. PROPOSITION</u>. If $M \subset N$ is nicely curved, then the distribution $p \mapsto \overset{\ell}{\mathrm{Osc}}\, M_p$ on M is parallel along every curve in M (as defined on p. 40).

Consequently, if N is a manifold of constant curvature, and M is connected, then M is contained in some $\#(M)$-dimensional totally geodesic submanifold of N (but not in any lower dimensional one).

<u>Proof</u>. To prove the first part, it obviously suffices to work locally. In a neighborhood U of any point $p \in M$ we can choose smooth linearly independent sections $\xi_1,\ldots,\xi_{\#(M)}$ of $\overset{\ell}{\mathrm{Osc}}\, M$. For a curve c in U, let V_μ be the vector field along c given by $V_\mu(t) = \xi_\mu(c(t))$. Then $(*)$ says that

$$\frac{D'V_\mu}{dt} \in \overset{\ell+1}{\mathrm{Osc}}\, M_{c(t)} = \overset{\ell}{\mathrm{Osc}}\, M_{c(t)} ;$$

thus there are smooth functions $f_{\mu\nu}$ such that

$$\frac{D'V_\mu}{dt} = \sum_\nu f_{\mu\nu} V_\nu \ .$$

Now let W be any vector field along c with $D'W/dt = 0$. Then

$$\frac{d}{dt}\langle W, V_\mu\rangle = \left\langle \frac{D'W}{dt}, V_\mu \right\rangle + \left\langle W, \frac{D'V_\mu}{dt}\right\rangle$$
$$= \langle 0, V_\mu\rangle + \langle W, \Sigma f_{\mu\nu} V_\nu\rangle$$
$$= \Sigma\, f_{\mu\nu}\langle W, V_\nu\rangle \ .$$

This is a system of differential equations for the functions $\langle W, V_\mu\rangle$. One solution is $\langle W, V_\mu\rangle = 0$ for all μ. So by uniqueness of solutions we see that if $W(0)$ is perpendicular to $\overset{\ell}{\mathrm{Osc}}\, M_{c(0)}$, then $W(t)$ is perpendicular to $\overset{\ell}{\mathrm{Osc}}\, M_{c(t)}$ for all t. This proves that $\overset{\ell}{\mathrm{Osc}}\, M$ is parallel along c.

The second part follows from Corollary 11. ■

We now define the kth ***normal space*** $\overset{k}{\mathrm{Nor}}\, M_p$ ***of*** M ***at*** p to be the orthogonal complement of $\overset{k}{\mathrm{Osc}}\, M_p$ in $\overset{k+1}{\mathrm{Osc}}\, M_p$. Thus we have

$$\overset{k+1}{\mathrm{Osc}}\, M_p = \overset{k}{\mathrm{Osc}}\, M_p \oplus \overset{k}{\mathrm{Nor}}\, M_p \ .$$

Notice that $\overset{0}{\mathrm{Nor}}\, M_p$ is just M_p, while $\overset{k}{\mathrm{Nor}}\, M_p$ has dimension 0 for $k \geq \ell - 1$. It will also be convenient to let $\overset{-1}{\mathrm{Nor}}\, M_p$ be the $\{0\}$ subspace of M_p. Each $\overset{k}{\mathrm{Nor}}\, M_p \subset N_p$ has an orthogonal complement in N_p, and thus we have two projections

$$\mathbf{T}^k\colon N_p \longrightarrow \overset{k}{\text{Nor}}\, M_p$$

$$\perp^k\colon N_p \longrightarrow \text{orthogonal complement of } \overset{k}{\text{Nor}}\, M_p \text{ in } N_p\,.$$

Notice that $\mathbf{T}^0 = \mathbf{T}\colon N_p \longrightarrow M_p$ and $\perp^0 = \perp\colon N_p \longrightarrow M_p{}^{\perp}$ (but note that $\mathbf{T}^k$ goes into a subspace of $M_p{}^{\perp}$ for $k > 0$). We shall actually use only the projections $\mathbf{T}^k$.

For nicely curved $M \subset N$ we clearly have, for each k, a vector bundle $\overset{k}{\text{Nor}}\, M$ whose fibre at p is $\overset{k}{\text{Nor}}\, M_p$. The bundle $\overset{0}{\text{Nor}}\, M$ is just the tangent bundle TM, while the bundles $\overset{k}{\text{Nor}}\, M$ for $k > 0$ are all subbundles of the normal bundle Nor M. There are natural Riemannian metrics $\langle\ ,\ \rangle$ on all bundles $\overset{k}{\text{Nor}}\, M$, since they are all subbundles of $(TN)|M$.

The 1st normal space $\overset{1}{\text{Nor}}\, M_p$ is the subspace of N_p spanned by all $s(X_p, Y_p)$ for $X_p, Y_p \in M_p$. In general, given vector fields $X_1, \ldots, X_{k+1}$ on M, consider

$$\mathbf{T}^k(\nabla'(X_1, \ldots, X_{k+1}))\ .$$

It is easily checked that this expression is linear in each X_i <u>over the</u> C^∞ <u>functions</u> (compare p.III.5). So its value at p depends only on the values of the X_i at p and we can define

$$s^k(X_{1_p}, \ldots, X_{k+1_p}) = \mathbf{T}^k(\nabla'(X_1, \ldots, X_{k+1})(p)) \in \overset{k}{\text{Nor}}\, M_p$$

for any vector fields X_i extending X_{i_p}. Clearly $\overset{k}{\text{Nor}}\, M_p$ is spanned by the image of s^k. It seems reasonable to let s^0 denote the identity map of M_p into $\overset{0}{\text{Nor}}\, M_p = M_p$.

68. LEMMA. If $M \subset N$, where N has constant curvature, then s^k is symmetric.

Proof. First we have

$$\begin{aligned}\nabla'(X_1,\dots,X_k,X_{k+1})(p) &- \nabla'(X_1,\dots,X_{k+1},X_k)(p)\\ &= \nabla'(X_1,\dots,X_{k-1},[X_k,X_{k+1}])(p) \in \overset{k}{\mathrm{Osc}}\, M_p ,\end{aligned}$$

so $\mathbf{T}^k$ of the left side is 0, which proves that s^k is symmetric in X_k and X_{k+1}. We also have, for example,

$$\begin{aligned}\nabla'(X_1,\dots,X_{k-1},X_k,X_{k+1})(p) &- \nabla'(X_1,\dots,X_k,X_{k-1},X_{k+1})(p)\\ &= \nabla'\Big(X_1,\dots,X_{k-2},\nabla'_{X_{k-1}}(\nabla'_{X_k}X_{k+1}) - \nabla'_{X_k}(\nabla'_{X_{k-1}}X_k)\Big)(p)\\ &= \nabla'\Big(X_1,\dots,X_{k-2},\nabla'_{[X_{k-1},X_k]}X_{k+1} + R'(X_{k-1},X_k)X_{k+1}\Big)(p) .\end{aligned}$$

Since $R'(X_{k-1},X_k)X_{k+1}$ is tangent to M, this is in $\overset{k-1}{\mathrm{Osc}}\, M_p$, so again $\mathbf{T}^k$ of the left side is 0. Similarly, s^k is symmetric under interchange of any two adjacent arguments. ∎

For vector fields $X_1,\dots,X_{k+1}$ on a nicely curved submanifold $M \subset N$ we can write

$$s^k(X_1,\dots,X_{k+1}) = \nabla'(X_1,\dots,X_{k+1}) + \xi \qquad \xi \text{ a section of } \overset{k}{\mathrm{Osc}}\, M .$$

Then

$$\mathbf{T}^{k+1}\nabla'_{X_p} s^k(X_1,\dots,X_{k+1}) = \mathbf{T}^{k+1}\nabla'_{X_p}\nabla'(X_1,\dots,X_{k+1}) ,$$

since $\nabla'_{X_p}\xi \in \overset{k+1}{\mathrm{Osc}}\, M_p$ by (*), on p. 241. Thus we have

$$(**)\qquad \mathbf{T}^{k+1}\nabla'_{X_p} s^k(X_1,\ldots,X_{k+1}) = s^{k+1}(X_p, X_1(p),\ldots,X_{k+1}(p)) .$$

Now suppose we have vector fields $\{Y_\alpha\}$ which span the tangent space of M in a neighborhood of p. Every element of $\overset{1}{\mathrm{Nor}}\, M_p$, for example, can be written as

$$\Sigma\; c_{\alpha\beta}s(Y_\alpha(p),Y_\beta(p)) .$$

This expression is usually not unique (even if the $Y_\alpha(p)$ are linearly independent). But suppose that we have constants $c_{\alpha\beta}$ with

$$\Sigma\; c_{\alpha\beta}s(Y_\alpha(p),Y_\beta(p)) = 0 .$$

Let $\mathscr{S}$ be a collection of pairs (α,β) such that

$$\{s(X_\alpha(p),X_\beta(p)) \colon (\alpha,\beta) \in \mathscr{S}\}$$

is a basis of $\overset{1}{\mathrm{Nor}}\, M_p$. Since M is nicely curved, it follows that $\{s(X_\alpha(q),X_\beta(q)) \colon (\alpha,\beta) \in \mathscr{S}\}$ is a basis of $\overset{1}{\mathrm{Nor}}\, M_q$ for all points q in a neighborhood of p. Now consider the section

$$\Sigma\; c_{\alpha\beta}s(Y_\alpha,Y_\beta)$$

of $\overset{1}{\mathrm{Nor}}\, M$, where the $c_{\alpha\beta}$ denote constant functions. In a neighborhood of p we can write

$$\Sigma\; c_{\alpha\beta}s(Y_\alpha,Y_\beta) = \sum_{(\alpha,\beta)\in\mathscr{S}} f_{\alpha\beta}\cdot s(Y_\alpha,Y_\beta)$$

for unique smooth functions $f_{\alpha\beta}$. Clearly $f_{\alpha\beta}(p) = 0$. Applying ∇'_{X_p} to the above equation we thus obtain

$$\Sigma\, c_{\alpha\beta}\nabla'_{X_p} s(Y_\alpha, Y_\beta) = \sum_{(\alpha,\beta)\in\mathscr{S}} X_p(f_{\alpha\beta})\cdot s(Y_\alpha(p), Y_\beta(p)) + 0$$

$$\in \overset{1}{\text{Nor}}\, M_p \ .$$

Consequently,

$$0 = \Sigma\, c_{\alpha\beta}\mathbf{T}^2\nabla'_{X_p} s(Y_\alpha, Y_\beta) = \Sigma\, c_{\alpha\beta} s(X_p, Y_\alpha(p), Y_\beta(p)) \qquad \text{by } (**) \ .$$

Thus we see that

$$\Sigma\, c_{\alpha\beta} s(Y_\alpha(p), Y_\beta(p)) = 0 \implies \Sigma\, c_{\alpha\beta} s(X_p, Y_\alpha(p), Y_\beta(p)) = 0 \ .$$

It follows that there is a well-defined map from $\overset{1}{\text{Nor}}\, M_p$ to $\overset{2}{\text{Nor}}\, M_p$ which takes

$$\Sigma\, c_{\alpha\beta} s(Y_\alpha(p), Y_\beta(p)) \longmapsto \Sigma\, c_{\alpha\beta} s(X_p, Y_\alpha(p), Y_\beta(p)) \ .$$

This map doesn't depend on the choice of $\{Y_\alpha\}$, for if we also have spanning vector fields $\{Z_\alpha\}$, we can apply the above argument to the collection $\{Y_\alpha\}\cup\{Z_\alpha\}$. The argument clearly works for all k, so we see that there is a well-defined bi-linear map

$$\mathbf{s}^k \colon M_p \times \overset{k}{\text{Nor}}\, M_p \longrightarrow \overset{k+1}{\text{Nor}}\, M_p$$

such that

$$\mathbf{s}^k(X_p, s^k(X_{1_p}, \ldots, X_{k+1_p})) = s^{k+1}(X_p, X_{1_p}, \ldots, X_{k+1_p}) .$$

Now suppose we have any section ξ of $\overset{k}{\text{Nor}}\, M$. Locally ξ can be written as a sum of terms $f \cdot s^k(X_1, \ldots, X_{k+1})$. Since

$$\nabla'_{X_p}(f \cdot s^k(X_1, \ldots, X_{k+1})) = X_p(f) \cdot s^k(X_{1_p}, \ldots, X_{k+1_p}) + f(p) \cdot \nabla'_{X_p} s^k(X_1, \ldots, X_{k+1}) ,$$

we see that

$$\mathbf{T}^{k+1}\nabla'_{X_p}\left(f \cdot s^k(X_1, \ldots, X_{k+1})\right) = f(p) \cdot \mathbf{T}^{k+1}\nabla'_{X_p} s^k(X_1, \ldots, X_{k+1}) .$$

Then (**) shows that

$$(***) \qquad \mathbf{T}^{k+1}\nabla'_{X_p}\xi = \mathbf{s}^k(X_p, \xi_p) \qquad \xi \text{ a section of } \overset{k}{\text{Nor}}\, M .$$

Now we will consider all the other components of $\nabla'_{X_p}\xi$ for ξ a section of $\overset{k}{\text{Nor}}\, M$. First we note

---

<u>69. LEMMA</u>. If ξ is a section of $\overset{k}{\text{Nor}}\, M$ and $X_p \in M_p$, then

$$\nabla'_{X_p}\xi \in \overset{k-1}{\text{Nor}}\, M_p \oplus \overset{k}{\text{Nor}}\, M_p \oplus \overset{k+1}{\text{Nor}}\, M_p .$$

<u>Proof</u>. Since ξ is a section of $\overset{k+1}{\text{Osc}}\, M$, we have $\nabla'_{X_p}\xi \in \overset{k+2}{\text{Osc}}\, M_p$. Now if η is any section of $\overset{j}{\text{Osc}}\, M$ for $j < k$, then $\langle\xi, \eta\rangle = 0$, so

$$0 = X_p(\langle\xi, \eta\rangle) = \langle\nabla'_{X_p}\xi, \eta(p)\rangle + \langle\xi(p), \nabla'_{X_p}\eta\rangle .$$

If we also have $j < k-1$, then $\nabla'_{X_p}\eta \in \overset{k}{\mathrm{Osc}}\, M_p$, so $\langle \xi(p), \nabla'_{X_p}\eta\rangle = 0$, and hence $\langle \nabla'_{X_p}\xi, \eta(p)\rangle = 0$. ■

Thus we see that for a section ξ of $\overset{k}{\mathrm{Nor}}\, M$ we can write

$$\nabla'_{X_p}\xi = \mathbf{T}^{k-1}(\nabla'_{X_p}\xi) + \mathbf{T}^{k}(\nabla'_{X_p}\xi) + \mathbf{T}^{k+1}(\nabla'_{X_p}\xi) .$$

The third term of this decomposition is already given by (***). For the first term we have a result which generalizes Proposition 12.

<u>70. PROPOSITION</u>. If ξ is a section of $\overset{k}{\mathrm{Nor}}\, M$ and $X_p \in M_p$, then the vector $\mathbf{T}^{k-1}(\nabla'_{X_p}\xi) \in \overset{k-1}{\mathrm{Nor}}\, M_p$ satisfies

$$\langle \mathbf{T}^{k-1}(\nabla'_{X_p}\xi), \eta_p\rangle = \langle \nabla'_{X_p}\xi, \eta_p\rangle = -\langle \xi(p), \mathbf{s}^{k-1}(X_p,\eta_p)\rangle$$

$$\text{for all } \eta_p \in \overset{k-1}{\mathrm{Nor}}\, M_p .$$

Consequently, $\mathbf{T}^{k-1}(\nabla'_{X_p}\xi)$ depends only on X_p and ξ_p.

<u>Proof</u>. If η is a section of $\overset{k-1}{\mathrm{Nor}}\, M$ extending η_p, then $\langle \xi, \eta\rangle = 0$, so

$$\begin{aligned} 0 = X_p(\langle \xi, \eta\rangle) &= \langle \nabla'_{X_p}\xi, \eta_p\rangle + \langle \xi(p), \nabla'_{X_p}\eta\rangle \\ &= \langle \nabla'_{X_p}\xi, \eta_p\rangle + \langle \xi(p), \mathbf{T}^{k}\nabla'_{X_p}\eta\rangle , \end{aligned}$$

since $\xi(p) \in \overset{k}{\mathrm{Nor}}\, M_p$, by assumption. Now apply (***). ■

For any vector $\xi_p \in \overset{k}{\text{Nor}}\, M_p$ we can now let

$$A^k_{\xi_p}(X_p) = -\,\mathbf{T}^{k-1}(\nabla'_{X_p}\xi) \,,$$

for any section ξ of $\overset{k}{\text{Nor}}\, M$ extending ξ_p, so that we have a map

$$A^k_{\xi_p} : M_p \longrightarrow \overset{k-1}{\text{Nor}}\, M_p$$

satisfying

$$\langle A^k_{\xi_p}(X_p), \eta_p\rangle = \langle \xi_p, \mathbf{s}^{k-1}(X_p,\eta_p)\rangle \,.$$

(Note that for $k = 0$ we are dealing with the 0 map.) For convenience, we will sometimes write

$$A^k(\xi_p;X_p) \quad \text{for} \quad A^k_{\xi_p}(X_p) \,.$$

Finally, for the expression $\mathbf{T}^k(\nabla'_{X_p}\xi)$ we introduce a new symbol,

$$D^k{}_{X_p}\xi = \mathbf{T}^k(\nabla'_{X_p}\xi) \qquad \xi \text{ a section of } \overset{k}{\text{Nor}}\, M \,.$$

It is easy to check that D^k _is a connection on_ $\overset{k}{\text{Nor}}\, M$ _which is compatible with the metric_ $\langle\ ,\ \rangle$ _on_ $\overset{k}{\text{Nor}}\, M$. We can now write our decomposition of $\nabla'_{X_p}\xi$ as

The Frenet Equations

$$\nabla'_{X_p}\xi = -A^k_{\xi_p}(X_p) + D^k_{X_p}\xi + \mathbf{s}^k(X_p,\xi_p),$$

for ξ a section of $\overset{k}{\text{Nor}}\, M$ and $X_p \in M_p$.

The terms $A^k_{\xi_p}(X_p)$ and $\mathbf{s}^k(X_p,\xi_p)$ are completely determined by the maps s^{k-1}, s^k and s^{k+1}. These Frenet equations essentially contain the Frenet equations for a curve when M is 1-dimensional; in general, they contain the Gauss equations (for $k = 0$) and (part of) the Weingarten equations for $k = 1$.

<u>71. THEOREM</u>. Let $M^n, \bar{M}^n \subset N^m$ be connected nicely curved submanifolds of a complete simply connected manifold N of constant curvature. Let $\phi\colon M \to \bar{M}$ be an isometry. Suppose that for all $k \geq 1$ there are bundle isomorphisms $\tilde{\phi}^k\colon \overset{k}{\text{Nor}}\, M \to \overset{k}{\text{Nor}}\, \bar{M}$ covering ϕ which preserve inner products, second fundamental forms s^k, and connections D^k. Then there is an isometry A of N such that $\phi = A|M$ and $\tilde{\phi}^k = A_*|\overset{k}{\text{Nor}}\, M$.

Proof. We obviously want to reduce this to Theorem 20. Notice first that since $\overset{\ell}{\text{Osc}}\, M_p = \overset{0}{\text{Nor}}\, M_p \oplus \cdots \oplus \overset{\ell-1}{\text{Nor}}\, M_p$, and similarly for $\bar{M}$, the formal imbedding dimension $\#(M)$ must equal $\#(\bar{M})$. Taking into account Proposition 67, we see that there is no loss of generality in assuming that $\#(M) = \#(\bar{M}) = m$. Then the bundle isomorphisms $\tilde{\phi}^1,\ldots,\tilde{\phi}^{\ell-1}$ combine to give a bundle isomorphism $\tilde{\phi}\colon \text{Nor}\, M \to \text{Nor}\, \bar{M}$. Clearly $\tilde{\phi}$ preserves inner products, second fundamental forms s^k, and connections D^k. In particular $\tilde{\phi}$ takes the second form s $(= s^1)$ to $\bar{s}$ $(= \bar{s}^1)$. To prove that

$$(*) \qquad \tilde{\phi}(D_X\xi) = \bar{D}_{\phi_*X}(\tilde{\phi}(\xi))$$

for all sections ξ of E, it suffices to consider separately sections ξ of $\overset{k}{\text{Nor}}\, M$. Then the Frenet equations give

$$D_X\xi = \perp(\nabla'_X\xi) = \begin{cases} D^1{}_X\xi + \mathbf{s}^1(X,\xi) & k = 1 \\ -A^k_\xi(X) + D^k{}_X\xi + \mathbf{s}^k(X,\xi) & k > 1\,, \end{cases}$$

with corresponding formulas for $\bar{D}_{\phi_*X}\tilde{\phi}(\xi)$. Since $\tilde{\phi}$ preserves D^k, as well as A^k and $\mathbf{s}^k$ (for they are determined by s^{k-1}, s^k and s^k), we see that equation (*) does indeed hold. ■

Now we need certain equations satisfied by the connections D^k. We will state these in terms of vector fields on M and sections of $\overset{k}{\text{Nor}}\, M$. After the proof we will give another formulation, in terms of tangent vectors in M_p, and vectors in $\overset{k}{\text{Nor}}\, M_p$, which will make the result appear as a genuine generalization of the Codazzi-Mainardi equations for D.

<u>72. THEOREM</u>. Let $M \subset N$ be nicely curved. Then for all vector fields X, Y on M and sections ξ of $\overset{k}{\text{Nor}}\, M$ $(k \geq 0)$ we have

<u>The Generalized Codazzi-Mainardi Equations</u>

$$\mathbf{T}^{k+1}R'(X,Y)\xi = [D^{k+1}{}_X(\mathbf{s}^k(Y,\xi)) - \mathbf{s}^k(\nabla_XY,\xi) - \mathbf{s}^k(Y,D^k{}_X\xi)]$$
$$- [D^{k+1}{}_Y(\mathbf{s}^k(X,\xi)) - \mathbf{s}^k(\nabla_YX,\xi) - \mathbf{s}^k(X,D^k{}_Y\xi)]\,.$$

When N has constant curvature, the left side is zero.

Proof. By the Frenet equations we have

$$\nabla'_Y\xi = -A^k_\xi(Y) + D^k{}_Y\xi + \mathbf{s}^k(Y,\xi) .$$

Since $A^k_\xi(Y)$ is a section of $\mathrm{Nor}^{k-1} M$, Lemma 69 implies that

$$\mathbf{T}^{k+1}\nabla'_X\nabla'_Y\xi = \mathbf{T}^{k+1}\nabla'_X D^k{}_Y\xi + \mathbf{T}^{k+1}\nabla'_X\, \mathbf{s}^k(Y,\xi) .$$

Using (***) on p. 247, and the definition of D^{k+1}, we thus have

(1) $$\mathbf{T}^{k+1}\nabla'_X\nabla'_Y\xi = \mathbf{s}^k(X,D^k{}_Y\xi) + D^{k+1}{}_X(\mathbf{s}^k(Y,\xi)) ,$$

(1') $$\mathbf{T}^{k+1}\nabla'_Y\nabla'_X\xi = \mathbf{s}^k(Y,D^k{}_X\xi) + D^{k+1}{}_Y(\mathbf{s}^k(X,\xi)) .$$

We also have, by (***),

$$\mathbf{T}^{k+1}\nabla'_{[X,Y]}\xi = \mathbf{s}^k([X,Y],\xi) ,$$

and thus

(2) $$\mathbf{T}^{k+1}\nabla'_{[X,Y]}\xi = \mathbf{s}^k(\nabla_X Y,\xi) - \mathbf{s}^k(\nabla_Y X,\xi) .$$

Substituting (1), (1'), (2) into the formula $R'(X,Y)\xi = \nabla'_X\nabla'_Y\xi - \nabla'_Y\nabla'_X\xi - \nabla'_{[X,Y]}\xi$, we obtain the desired result.

When N has constant curvature K_0 we have

$$R'(X,Y)\xi = K_0[\langle Y,\xi\rangle X - \langle X,\xi\rangle Y] ,$$

which is tangent to M. So $\mathbf{T}^{k+1}R'(X,Y)\xi = 0$. ■

It is easily checked that in these generalized Codazzi-Mainardi equations, each of the expressions in brackets is linear in X, Y, and ξ <u>over the</u> C^∞ <u>functions</u>, and thus its value at p depends only on X_p, Y_p, ξ_p. To give this value explicitly, we note that we can consider $\mathbf{s}^k$ as a section of the bundle $\mathrm{Hom}(TM \times \overset{k}{\mathrm{Nor}}\, M, \overset{k+1}{\mathrm{Nor}}\, M)$. Using the connections ∇, D^k and D^{k+1} on TM, $\overset{k}{\mathrm{Nor}}\, M$, and $\overset{k+1}{\mathrm{Nor}}\, M$, we can define a natural connection $\tilde{\nabla}$ on this bundle (compare p. 54 for the case $k = 0$). It is easily seen that Theorem 72 can be written

$$\mathbf{T}^{k+1}R'(X_p,Y_p)\xi_p = (\tilde{\nabla}_{X_p}\, \mathbf{s}^k)(Y_p,\xi_p) - (\tilde{\nabla}_{Y_p}\, \mathbf{s}^k)(X_p,\xi_p)$$

$$X_p,\ Y_p \in M_p, \quad \text{and} \quad \xi_p \in \overset{k}{\mathrm{Nor}}\, M_p .$$

Now we can state the proper form of Lemma 65.

<u>73. LEMMA (FUNDAMENTAL LEMMA OF RIEMANNIAN SUBMANIFOLD THEORY)</u>. Let $M \subset N$ be nicely curved. Then the set of normal connections D^k in $\overset{k}{\mathrm{Nor}}\, M$ is the unique set of connections δ^k on $\overset{k}{\mathrm{Nor}}\, M$ such that

(0) $\delta^0 = \nabla$,

(1) δ^k is compatible with the metric in $\overset{k}{\mathrm{Nor}}\, M$:

$$X(\langle\xi, \eta\rangle) = \langle\delta^k{}_X\xi, \eta\rangle + \langle\xi, \delta^k{}_X\eta\rangle \qquad \text{for sections } \xi, \eta \text{ of } \overset{k}{\mathrm{Nor}}\, M ,$$

(2) The δ^k satisfy the Codazzi-Mainardi equations:

$$\mathbf{T}^{k+1}R'(X,Y)\xi = [\delta^{k+1}{}_X(\mathbf{s}^k(Y,\xi)) - \mathbf{s}^k(\nabla_X Y,\xi) - \mathbf{s}^k(Y,\delta^k{}_X\xi)]$$
$$- [\delta^{k+1}{}_Y(\mathbf{s}^k(X,\xi)) - \mathbf{s}^k(\nabla_Y X,\xi) - \mathbf{s}^k(X,\delta^k{}_Y\xi)] .$$

Proof. We will show that if $\delta^k = D^k$, then $\delta^{k+1} = D^{k+1}$. Since $\delta^0 = \nabla = D^0$, this will prove the result. We begin by considering the expression

$$\begin{aligned}&\langle\delta^{k+1}{}_X\, \mathbf{s}^k(Y_1,\xi),\ \mathbf{s}^k(Z_1,\eta)\rangle - \langle\delta^{k+1}{}_{Y_1}\, \mathbf{s}^k(X,\xi),\ \mathbf{s}^k(Z_1,\eta)\rangle\\ &- \langle\delta^{k+1}{}_{Y_1}\, \mathbf{s}^k(Z_1,\eta),\ \mathbf{s}^k(X,\xi)\rangle + \langle\delta^{k+1}{}_{Z_1}\, \mathbf{s}^k(Y_1,\eta),\ \mathbf{s}^k(X,\xi)\rangle\\ &+ \langle\delta^{k+1}{}_{Z_1}\, \mathbf{s}^k(X,\xi),\ \mathbf{s}^k(Y_1,\eta)\rangle - \langle\delta^{k+1}{}_X\, \mathbf{s}^k(Z_1,\xi),\ \mathbf{s}^k(Y_1,\eta)\rangle ,\end{aligned}$$

where ξ, η are sections of $\overset{k}{\text{Nor}}\,M$. As in the proof of Lemma 65, we are led to the conclusion that we can write

$$\langle\delta^{k+1}{}_X\, \mathbf{s}^k(Y_1,\xi),\ \mathbf{s}^k(Z_1,\eta)\rangle - \langle\delta^{k+1}{}_X\, \mathbf{s}^k(Z_1,\xi),\ \mathbf{s}^k(Y_1,\eta)\rangle = \cdots ,$$

where $\cdots$ can be expressed in terms of X, Y_1, Z_1, ξ, η and $\delta^k = D^k$. In particular, if we choose $\xi = s^k(Y_2,\ldots,Y_{k+2})$ and $\eta = s^k(Z_2,\ldots,Z_{k+2})$, then we obtain

$$(3)\quad \langle\delta^{k+1}{}_X s^{k+1}(Y_1,\ldots,Y_{k+2}),\ s^{k+1}(Z_1,\ldots,Z_{k+2})\rangle$$
$$- \langle\delta^{k+1}{}_X s^{k+1}(Z_1,Y_2,\ldots,Y_{k+2}),\ s^{k+1}(Y_1,Z_2,\ldots,Z_{k+2})\rangle = \cdots .$$

We will abbreviate the left side of this equation by

$$\{Y_1,\ldots,Y_{k+2};\ Z_1,\ldots,Z_{k+2}\} - \{Z_1,Y_2,\ldots,Y_{k+2};\ Y_1,Z_2,\ldots,Z_{k+2}\} .$$

Now consider the following expressions (the pattern becomes apparent by looking at the terms in the second column):

$$\{Y_1,Y_2,Y_3,\ldots,Y_{k+2};\ Z_1,Z_2,Z_3,\ldots,Z_{k+2}\} - \{Z_1,Y_2,Y_3,\ldots,Y_{k+2};\ Y_1,Z_2,Z_3,\ldots,Z_{k+2}\}$$

$$\{Y_2,Z_1,Y_3,\ldots,Y_{k+2};\ Z_2,Y_1,Z_3,\ldots,Z_{k+2}\} - \{Z_2,Z_1,Y_3,\ldots,Y_{k+2};\ Y_2,Y_1,Z_3,\ldots,Z_{k+2}\}$$

$$\{Y_3,Z_2,Z_1,\ldots,Y_{k+2};\ Z_3,Y_2,Y_1,\ldots,Z_{k+2}\} - \{Z_3,Z_2,Z_1,\ldots,Y_{k+2};\ Y_3,Y_2,Y_1,\ldots,Z_{k+2}\}$$

$$\vdots$$

$$\{Y_{k+1},Z_k,\ldots,Z_1,Y_{k+2};\ Z_{k+1},Y_k,\ldots,Y_1,Z_{k+2}\} - \{Z_{k+1},\ldots,Z_1,Y_{k+2};\ Y_{k+1},\ldots,Y_1,Z_{k+2}\}$$

$$\{Y_{k+2},Z_{k+1},\ldots,Z_1;\ Z_{k+2},Y_{k+1},\ldots,Y_1\} - \{Z_{k+2},\ldots,Z_1;\ Y_{k+2},\ldots,Y_1\} .$$

Notice that the second term in each line is the same as the first term in the next line, since s^{k+1} is symmetric. So adding all the equations (3) having the above expressions on the left we obtain

$$(4)\quad \langle \delta^{k+1}{}_X s^{k+1}(Y_1,\ldots,Y_{k+2}),\ s^{k+1}(Z_1,\ldots,Z_{k+2})\rangle$$
$$- \langle \delta^{k+1}{}_X s^{k+1}(Z_1,\ldots,Z_{k+2}),\ s^{k+1}(Y_1,\ldots,Y_{k+2})\rangle = \cdots .$$

But by (1) we also have

$$(5)\quad \langle \delta^{k+1}{}_X s^{k+1}(Y_1,\ldots,Y_{k+2}),\ s^{k+1}(Z_1,\ldots,Z_{k+2})\rangle$$
$$+ \langle \delta^{k+1}{}_X s^{k+1}(Z_1,\ldots,Z_{k+2}),\ s^{k+1}(Y_1,\ldots,Y_{k+2})\rangle = \cdots .$$

So by adding (4) and (5) we obtain

$$(*)\qquad \langle \delta^{k+1}{}_X s^{k+1}(Y_1,\ldots,Y_{k+2}),\ s^{k+1}(Z_1,\ldots,Z_{k+2})\rangle = \cdots .$$

Since $\overset{k+1}{\text{Nor}}\ M_p$ is spanned by image s^{k+1}, this proves, as in Lemma 65, that δ^{k+1} is uniquely determined. ■

Now for a manifold $M \subset N$ we define tensors $\mathcal{F}_k$ by

$$\mathcal{F}_k(X_1,\dots,X_{k+1}, Y_1,\dots,Y_{k+1}) = \langle s^k(X_1,\dots,X_{k+1}), s^k(Y_1,\dots,Y_{k+1})\rangle .$$

If $X_1,\dots,X_n \in M_p$ is a basis, then we can form the $n^{k+1} \times n^{k+1}$ matrix

$$(\mathcal{F}_k(X_{i_1},\dots,X_{i_{k+1}}, X_{j_1},\dots,X_{j_{k+1}})) .$$

It is easy to see that this matrix is positive semi-definite and that its rank is just the dimension of $\overset{k}{\text{Nor}}\ M_p$.*

<u>74. THEOREM</u>. Let $M, \bar{M} \subset N$ be connected nicely curved submanifolds of a complete simply connected manifold N of constant curvature. Let $\phi\colon M \to \bar{M}$ be an isometry such that

$$\phi^*\bar{\mathcal{F}}_k = \mathcal{F}_k \qquad \text{for all } k .$$

Then ϕ is the restriction of an isometry of N.

<u>Proof</u>. The preceeding remarks show that the dimension of $\overset{k}{\text{Nor}}\ M$ must equal the dimension of $\overset{k}{\text{Nor}}\ \bar{M}$. Since s^k is onto $\overset{k}{\text{Nor}}\ M$, the procedure used in

*For those who know about tensor products of vector spaces this can be expressed more simply. We can regard s^k as a linear map $s^k\colon M_p \otimes \cdots \otimes M_p \to \overset{k}{\text{Nor}}\ M_p$, so $\mathcal{F}_k$ is a bilinear map $\mathcal{F}_k\colon (M_p \otimes \cdots \otimes M_p) \times (M_p \otimes \cdots \otimes M_p) \to \mathbb{R}$. The matrix considered above is the matrix of this bilinear map with respect to the basis $\{X_{i_1} \otimes \cdots \otimes X_{i_{k+1}}\}$ of $M_p \otimes \cdots \otimes M_p$.

the proof of Proposition 66 allows us to construct bundle isomorphisms $\tilde{\phi}^k\colon \overset{k}{\mathrm{Nor}}\, M \longrightarrow \overset{k}{\mathrm{Nor}}\, \bar{M}$ which preserve inner products and second fundamental forms s^k. Again arguing as in Proposition 66, but using Lemma 73 in place of Lemma 65, we see that the $\tilde{\phi}^k$ also preserves the connections D^k. So we can apply Theorem 71. ∎

We would also like to discuss when a given set of tensors $\{\mathcal{F}_k\}$ on a manifold M come from an imbedding of M in a complete manifold N of constant curvature. The Codazzi-Mainardi equations represent only one set of integrability conditions, and we still have to consider the other components of $\nabla'_X\nabla'_Y\xi - \nabla'_Y\nabla'_X\xi - \nabla'_{[X,Y]}\xi$. If ξ is a section of $\overset{k}{\mathrm{Nor}}\, M$, then the only components we have to consider are $\mathbf{T}^{k-2},\ldots,\mathbf{T}^{k+2}$, where $\mathbf{T}^{k+1}$ is already taken care of by the Codazzi-Mainardi equations.

First consider $\mathbf{T}^{k+2}$. From the Frenet equations

$$\nabla'_Y\xi = -A^k_\xi(Y) + D^k{}_Y\xi + \mathbf{s}^k(Y,\xi)$$

we obtain

$$\mathbf{T}^{k+2}\nabla'_X\nabla'_Y\xi = \mathbf{T}^{k+2}\nabla'_X\, \mathbf{s}^k(Y,\xi) = \mathbf{s}^{k+1}(X, \mathbf{s}^k(Y,\xi)) \qquad \text{by } (***).$$

Also

$$\mathbf{T}^{k+2}\nabla'_Y\nabla'_X\xi = \mathbf{s}^{k+1}(Y, \mathbf{s}^k(X,\xi))$$

$$\mathbf{T}^{k+2}\nabla'_{[X,Y]}\xi = 0\ .$$

So we have

$$\mathbf{T}^{k+2}R'(X,Y)\xi = \mathbf{s}^{k+1}(X, \mathbf{s}^k(Y,\xi)) - \mathbf{s}^{k+1}(Y, \mathbf{s}^k(X,\xi)) .$$

In a space of constant curvature, the left side is 0. On the other hand, the right hand side is clearly always 0, since s^{k+2} is symmetric. Thus we do not obtain any new condition for imbedding in a manifold of constant curvature by looking at $\mathbf{T}^{k+2}$.

Next consider $\mathbf{T}^{k-2}$. The Frenet equations give us [recall the alternative notation $A^k(\xi;X)$ for $A^k_\xi(X)$]

$$\mathbf{T}^{k-2}\nabla'_X\nabla'_Y\xi = -\mathbf{T}^{k-2}\nabla'_X A^k_\xi(Y) = A^{k-1}(A^k_\xi(Y);X)$$

$$\mathbf{T}^{k-2}\nabla'_Y\nabla'_X\xi = -\mathbf{T}^{k-2}\nabla'_Y A^k_\xi(X) = A^{k-1}(A^k_\xi(X);Y)$$

$$\mathbf{T}^{k-2}\nabla'_{[X,Y]}\xi = 0 .$$

So we obtain

$$\mathbf{T}^{k-2}R'(X,Y)\xi = A^{k-1}(A^k_\xi(Y);X) - A^{k-1}(A^k_\xi(X);Y) .$$

In a space of constant curvature the left side is 0 (this is clear for $k > 2$, since $R'(X,Y)\xi$ is tangent to M; it is true even for $k = 2$, since $R'(X,Y)\xi = K_0[\langle Y,\xi\rangle X - \langle X,\xi\rangle Y]$, and $\langle X,\xi\rangle = \langle Y,\xi\rangle = 0$). On the other hand, for any section η of Nor^{k-2} we have

$$\begin{aligned}\langle A^{k-1}(A^k_\xi(Y);X), \eta\rangle &= \langle A^k_\xi(Y), \mathbf{s}^{k-1}(X,\eta)\rangle \\ &= \langle \xi, \mathbf{s}^k(Y, \mathbf{s}^{k-1}(X,\eta))\rangle ,\end{aligned}$$

so we see that the right side of our equation is always 0. So, once again, we obtain no new conditions for imbedding M in a manifold of constant curvature.

Now consider $\mathbf{T}^{k-1}$. We have

$$\mathbf{T}^{k-1}\nabla'_X\nabla'_Y\xi = -\mathbf{T}^{k-1}\nabla'_X A^k_\xi(Y) + \mathbf{T}^{k-1}\nabla'_X D^k{}_Y\xi$$

$$= -D^{k-1}{}_X A^k_\xi(Y) - A^k(D^k{}_Y\xi;X)$$

$$\mathbf{T}^{k-1}\nabla'_Y\nabla'_X\xi = -D^{k-1}{}_Y A^k_\xi(X) - A^k(D^k{}_X\xi;Y)$$

$$\mathbf{T}^{k-1}\nabla'_{[X,Y]}\xi = -A^k_\xi([X,Y]) = -A^k_\xi(\nabla_X Y) + A^k_\xi(\nabla_Y X) .$$

Thus we obtain

$$-\mathbf{T}^{k-1}R'(X,Y)\xi = [D^{k-1}{}_X A^k_\xi(Y) - A^k(D^k{}_X\xi;Y) - A^k_\xi(\nabla_X Y)]$$

$$- [D^{k-1}{}_Y A^k_\xi(X) - A^k(D^k{}_Y\xi;X) - A^k_\xi(\nabla_Y X)] .$$

Taking the inner product with a section η of $\operatorname{Nor}^{k-1} M$, we obtain the equivalent equation

$$\text{(a)} \quad -\langle R'(X,Y)\xi, \eta\rangle = [\langle D^{k-1}{}_X A^k_\xi(Y), \eta\rangle - \langle D^k{}_X\xi, \mathbf{s}^{k-1}(Y,\eta)\rangle - \langle\xi, \mathbf{s}^{k-1}(\nabla_X Y,\eta)\rangle]$$

$$- [\langle D^{k-1}{}_Y A^k_\xi(X), \eta\rangle - \langle D^k{}_Y\xi, \mathbf{s}^{k-1}(X,\eta)\rangle$$

$$- \langle\xi, \mathbf{s}^{k-1}(\nabla_Y X,\eta)\rangle] .$$

But we also have

$$\langle A^k_\xi(Y), \eta\rangle = \langle\xi, \mathbf{s}^{k-1}(Y,\eta)\rangle$$

$$\Downarrow$$

$$X(\langle A^k_\xi(Y), \eta\rangle) = X(\langle\xi, \mathbf{s}^{k-1}(Y,\eta)\rangle$$

$$\Downarrow$$

$$\Downarrow$$

$$\langle D^{k-1}{}_X A^k_\xi(Y),\ \eta\rangle + \langle A^k_\xi(Y),\ D^{k-1}{}_X\eta\rangle = \langle D^k{}_X\xi,\ \mathbf{s}^{k-1}(Y,\eta)\rangle + \langle \xi,\ D^k{}_X\,\mathbf{s}^{k-1}(Y,\eta)\rangle$$

$$\Downarrow$$

$$\langle D^{k-1}{}_X A^k_\xi(Y),\ \eta\rangle - \langle D^k{}_X\xi,\ \mathbf{s}^{k-1}(Y,\eta)\rangle = \langle \xi,\ D^k{}_X\,\mathbf{s}^{k-1}(Y,\eta)\rangle - \langle \xi,\ \mathbf{s}^{k-1}(Y,D^{k-1}{}_X\eta)\rangle\ .$$

Therefore the right side of (a) can be written

$$[\langle \xi,\ D^k{}_X\,\mathbf{s}^{k-1}(Y,\eta)\rangle - \langle \xi,\ \mathbf{s}^{k-1}(Y,D^{k-1}{}_X\eta)\rangle - \langle \xi,\ \mathbf{s}^{k-1}(\nabla_X Y,\eta)\rangle]$$

$$- [\langle \xi,\ D^{k-1}{}_Y\,\mathbf{s}^{k-1}(X,\eta)\rangle - \langle \xi,\ \mathbf{s}^{k-1}(X,D^{k-1}{}_Y\eta)\rangle - \langle \xi,\ \mathbf{s}^{k-1}(\nabla_Y X,\eta)\rangle]$$

$$= \langle R'(X,Y)\eta,\ \xi\rangle\ ,\quad \text{by the Codazzi-Mainardi equations.}$$

So equation (a) follows from the Codazzi-Mainardi equations; we obtain no new conditions by looking at $\mathbf{T}^{k-1}$.

Finally, we have to look at $\mathbf{T}^k$. We have

$$\mathbf{T}^k\nabla'_X\nabla'_Y\xi = -\ \mathbf{T}^k\nabla'_X A^k_\xi(Y) + \mathbf{T}^k\nabla'_X D^k{}_Y\xi + \mathbf{T}^k\nabla'_X\,\mathbf{s}^k(Y,\xi)$$

$$= -\ \mathbf{s}^{k-1}(X,A^k_\xi(Y)) + D^k{}_X D^k{}_Y\xi - A^{k+1}(\,\mathbf{s}^k(Y,\xi);X)$$

$$\mathbf{T}^k\nabla'_Y\nabla'_X\xi = -\ \mathbf{s}^{k-1}(Y,A^k_\xi(X)) + D^k{}_Y D^k{}_X\xi - A^{k+1}(\,\mathbf{s}^k(X,\xi);Y)$$

$$\mathbf{T}^k\nabla'_{[X,Y]}\xi = D^k{}_{[X,Y]}\xi\ .$$

So we obtain

$$\text{(b)}\quad \mathbf{T}^k R'(X,Y)\xi = D^k{}_X D^k{}_Y \xi - D^k{}_Y D^k{}_X \xi - D^k{}_{[X,Y]}\xi$$
$$+\ \mathbf{s}^{k-1}(X, A^k_\xi(X)) - \mathbf{s}^{k-1}(X, A^k_\xi(Y))$$
$$+\ A^{k+1}(\mathbf{s}^k(X,\xi);Y) - A^{k+1}(\mathbf{s}^k(Y,\xi);X)\ .$$

When $k = 0$, the terms involving $\mathbf{s}^{k-1}$ do not appear. In this case, if we take $\xi = Z$ to be a section of $\overset{0}{\mathrm{Nor}}\, M = TM$ we obtain

$$R'(X,Y)Z = R(X,Y)Z + A^1(s(X,Z);Y) - A^1(s(Y,Z);X)$$
$$\Downarrow$$
$$\langle R'(X,Y)Z, W\rangle = \langle R(X,Y)Z, W\rangle + \langle s(X,Z), s(Y,W)\rangle - \langle s(X,W), s(Y,Z)\rangle$$

i.e. Gauss' equation. But for $k > 0$ we obtain an unsavory hybrid between Gauss' equation and the Ricci equations. We can obtain a nicer looking set of equations by considering the bundles $\overset{k}{\mathrm{Osc}}\, M$. There is a projection $\mathbf{T}^{[k]}\colon N_p \to \overset{k}{\mathrm{Osc}}\, M_p$, defined by means of the orthogonal complement of $\overset{k}{\mathrm{Osc}}\, M_p$ in N_p, and we can thus define a connection $D^{[k]}$ on $\overset{k}{\mathrm{Osc}}\, M$ by

$$D^{[k]}{}_X\xi = \mathbf{T}^{[k]}\nabla'_X\xi \qquad \xi \text{ a section of } \overset{k}{\mathrm{Osc}}\, M\ .$$

This connection has a curvature tensor $R^{[k]}$ defined by

$$R^{[k]}(X,Y)\xi = D^{[k]}{}_X D^{[k]}{}_Y\xi - D^{[k]}{}_Y D^{[k]}{}_X\xi - D^{[k]}{}_{[X,Y]}\xi\ .$$

<u>75. PROPOSITION</u>. Let $M \subset N$ be nicely curved. Then for all vectors $X, Y \in M_p$ and $\xi \in \overset{k}{\mathrm{Osc}}\, M_p$ we have the

Generalized Gauss Equation

$$\mathbf{T}^{[k]}R'(X,Y)\xi = R^{[k]}(X,Y)\xi + A^k(\mathbf{s}^{k-1}(X,\mathbf{T}^{k-1}\xi);Y) - A^k(\mathbf{s}^{k-1}(Y,\mathbf{T}^{k-1}\xi);X) .$$

So for $\xi, \eta \in \overset{k}{\mathrm{Osc}}\, M_p$ we have

$$\langle R'(X,Y)\xi, \eta\rangle = \langle R^{[k]}(X,Y)\xi, \eta\rangle + \langle \mathbf{s}^{k-1}(X,\mathbf{T}^{k-1}\xi), \mathbf{s}^{k-1}(Y,\mathbf{T}^{k-1}\eta)\rangle$$
$$- \langle \mathbf{s}^{k-1}(Y,\mathbf{T}^{k-1}\xi), \mathbf{s}^{k-1}(X,\mathbf{T}^{k-1}\eta)\rangle .$$

Proof. We have

$$\begin{aligned}\nabla'_Y\xi &= D^{[k]}{}_Y\xi + \mathbf{T}^k\nabla'_Y\xi \\ &= D^{[k]}{}_Y\xi + \mathbf{T}^k(\nabla'_Y\mathbf{T}^{k-1}\xi) \\ &= D^{[k]}{}_Y\xi + \mathbf{s}^{k-1}(Y,\mathbf{T}^{k-1}\xi) .\end{aligned}$$

Therefore

$$\begin{aligned}\nabla'_X\nabla'_Y\xi &= D^{[k]}{}_X D^{[k]}{}_Y\xi + \mathbf{s}^{k-1}(X,\mathbf{T}^{k-1}\cdot D^{[k]}{}_Y\xi) \\ &\quad + D^{[k]}{}_X\, \mathbf{s}^{k-1}(Y,\mathbf{T}^{k-1}\xi) + 0 .\end{aligned}$$

So

$$\begin{aligned}(1)\qquad \mathbf{T}^{[k]}\nabla'_X\nabla'_Y\xi &= D^{[k]}{}_X D^{[k]}{}_Y\xi + D^{[k]}{}_X\, \mathbf{s}^{k-1}(Y,\mathbf{T}^{k-1}\xi) \\ &= D^{[k]}{}_X D^{[k]}{}_Y\xi + \mathbf{T}^{k-1}\nabla'_X\, \mathbf{s}^{k-1}(Y,\mathbf{T}^{k-1}\xi) \\ &= D^{[k]}{}_X D^{[k]}{}_Y\xi - A^k(\mathbf{s}^{k-1}(Y, \mathbf{s}^{k-1}\xi);X) .\end{aligned}$$

Also

$$(2) \qquad \mathbf{T}^{[k]}\nabla'_Y\nabla'_X\xi = D^{[k]}{}_Y D^{[k]}{}_X\xi - A^k(\mathbf{s}^{k-1}(X, \mathbf{T}^{k-1}\xi);Y)$$

$$(3) \qquad \mathbf{T}^{[k]}\nabla'_{[X,Y]}\xi = D^{[k]}{}_{[X,Y]}\xi \ .$$

Equations (1)-(3) give the result. ■

Although we derived Gauss' equation from scratch, it is important to note that it is formally equivalent to equation (b) on p.261 , in the following sense. For a section ξ of $\overset{k}{\mathrm{Osc}}\, M$ we could define $D^{[k]}{}_X\xi$ as

$$\begin{aligned} D^{[k]}{}_X\xi = {} & [D^0{}_X\mathbf{T}^0\xi + \mathbf{s}^0(X, \mathbf{T}^0\xi)] \\ & + [-A^1(\mathbf{T}^1\xi;X) + D^1{}_X\mathbf{T}^1\xi + \mathbf{s}^1(X,\xi)] \\ & \vdots \\ & + [-A^{k-2}(\mathbf{T}^{k-2}\xi;X) + D^{k-2}{}_X\mathbf{T}^{k-2}\xi + \mathbf{s}^{k-2}(X,\xi)] \\ & + [-A^{k-1}(\mathbf{T}^{k-1}\xi;X) + D^{k-1}{}_X\mathbf{T}^{k-1}\xi] \ . \end{aligned}$$

Then the equations of Proposition 75, together with the Codazzi-Mainardi equations, imply equations (b) on p. 261; the verification of this claim is left to the reader. So the Codazzi-Mainardi equations and Gauss' equation are the full set of integrability conditions for the Frenet equations. But we still have a lot of work to do before we can decide when a set of tensors $\{\mathcal{F}_k\}$ on M come from an imbedding of M in a space of constant curvature.

First we claim that if ℓ has its usual significance, then

$$R^{[\ell]}(X,Y)\xi = \mathbf{T}^{[\ell]}R'(X,Y)\xi$$
$$= 0 , \quad \text{when } N \text{ has constant curvature.}$$

This follows immediately from Proposition 67, which shows that $D^{[\ell]} = \nabla'$ on $\overset{\ell}{\text{Osc}}$. We could also note that $R^{[\ell]} = R^{[\ell+1]}$, and that the terms $A^{\ell+1}$ which then arise in Gauss' equation are 0, since they lie in $\overset{\ell}{\text{Nor}}\, M_p$.

Now we have to establish certain important identities for the curvature tensors $R^{[k]}$, analogous to those for $R = R^{[1]}$. Recall that we have

(1) $$R(X,Y)Z + R(Y,X)Z = 0 ,$$

(2) $$\langle R(X,Y)Z, W\rangle + \langle R(X,Y)W, Z\rangle = 0 ,$$

(3) $$\mathfrak{S}\{R(X,Y)Z\} = R(X,Y)Z + R(Y,Z)X + R(Z,X)Y = 0 ,$$

(4) $$\langle R(X,Y)Z, W\rangle + \langle R(Z,W)X, Y\rangle = 0 .$$

When we are dealing with a submanifold M of another Riemannian manifold N, these identities follow immediately from Gauss' equation

$$\langle R'(X,Y)Z, W\rangle = \langle R(X,Y)Z, W\rangle + \langle s(X,Z), s(Y,W)\rangle - \langle s(Y,Z), s(X,W)\rangle ,$$

and the corresponding identities for R'. Similarly, we have

<u>76. PROPOSITION</u>. Let $M \subset N$ be nicely curved. Then

(1) $$R^{[k]}(X,Y)\xi + R^{[k]}(Y,X)\xi = 0 ,$$

(2) $$\langle R^{[k]}(X,Y)\xi, \eta\rangle + \langle R^{[k]}(X,Y)\eta, \xi\rangle = 0 ,$$

(3) $$\mathfrak{S}\{R^{[k]}(X,Y)\cdot \mathbf{s}^{k-2}(Z,\zeta)\} = 0 \qquad \zeta \in \overset{k-2}{\text{Nor}}\, M_p ,$$

(3') $$\mathfrak{S}'\langle R^{[k]}(X,Y)\cdot s^{k-1}(Z_1,\dots,Z_k),\ s^{k-1}(W_1,\dots,W_k)\rangle = 0$$

where $\mathfrak{S}'$ indicates a cyclic sum over $(Y,Z_1,\dots,Z_k,\ W_1,\dots,W_k)$,

(4) $$\begin{aligned} 0 = {} & \langle R^{[k]}(X_1,Y_1)\cdot s^{k-1}(X_2,\dots,X_{k+1}),\ s^{k-1}(Y_2,\dots,Y_{k+1})\rangle \\ & + \langle R^{[k]}(X_2,Y_2)\cdot s^{k-1}(Y_1,X_3,\dots,X_{k+1}),\ s^{k-1}(X_1,Y_3,\dots,Y_{k+1})\rangle \\ & + \langle R^{[k]}(X_3,Y_3)\cdot s^{k-1}(Y_1,Y_2,X_4,\dots,X_{k+1}),\ s^{k-1}(X_1,X_2,Y_4,\dots,Y_{k+1})\rangle \\ & \ \vdots \\ & + \langle R^{[k]}(X_{k+1},Y_{k+1})\cdot s^{k-1}(Y_1,\dots,Y_k),\ s^{k-1}(X_1,\dots,X_k)\rangle . \end{aligned}$$

Moreover, these identities follow formally from Gauss' equation for $R^{[k]}$ (and the properties of the curvature tensor R' for the ambient manifold).

Proof. An easy computation. ■

More important for us will be the (second) Bianchi identity

$$\mathfrak{S}\{(\nabla_Z R)(X,Y,W)\} = 0$$

[where we write R as $(X,Y,W) \mapsto R(X,Y,W)$] .

Although we have derived this identity for the curvature tensor of a (symmetric) connection on the tangent bundle, it is actually more general:

77. PROPOSITION. Let ∇ be a connection on TM, with torsion tensor T, and let D be a connection on a bundle $\varpi\colon E \to M$ with curvature tensor $R = R_D$. Let $\tilde{\nabla}$ be the natural connection on the bundle $\mathrm{Hom}(TM \times TM \times E,\ E)$ determined

by the connections ∇ on TM and D on E. Then

$$\mathfrak{S}\{(\tilde{\nabla}_Z R)(X,Y,\xi)\} + \mathfrak{S}\{R(T(X,Y),Z)\xi\} = 0 .$$

In particular, if $T = 0$, then

$$\mathfrak{S}\{(\tilde{\nabla}_Z R)(X,Y,\xi)\} .$$

<u>Proof</u>. Exactly the same as the proof on p. II.266-267, replacing W by ξ throughout. ■

Note that when E is TM, the connection $\tilde{\nabla}$ is just denoted by ∇, in conformity with previous usage.

<u>78. COROLLARY</u>. Let $M \subset N$ be nicely curved, and let $\tilde{\nabla}$ be the natural connection on $\text{Hom}(TM \times TM \times \overset{k}{\text{Osc}}\, M, \overset{k}{\text{Osc}}\, M)$ determined by the connections ∇ on TM and $D^{[k]}$ on $\overset{k}{\text{Osc}}\, M$. Then

$$(1) \qquad \mathfrak{S}\{(\tilde{\nabla}_Z R^{[k]})(X,Y,\xi)\} = 0 .$$

In addition,

$$(2) \quad \mathfrak{S}''\{<(\tilde{\nabla}_{X_1} R^{[k]})(X,Y_1,s^{k-1}(X_2,\dots,X_{k+1}),\ s^{k-1}(Y_2,\dots,Y_{k+1})>\} = 0$$

where $\mathfrak{S}''$ indicates a cyclic sum over $(X_1,\dots,X_{k+1},\ Y_1,\dots,Y_{k+1})$.

Moreover, equation (2) follows formally from (1), Gauss' equation for $R^{[k]}$, and the fact that the connection $D^{[k]}$ on $\overset{k}{\text{Osc}}\, M$ is compatible with the metric (and the properties of the curvature tensor R' for the ambient manifold).

Proof. To obtain equation (2), we apply X to both sides of equation (4) in Proposition 76. We have, for example,

$$X(\langle R^{[k]}(X_1,Y_1,s^{k-1}(X_2,\dots,X_{k+1})),\ s^{k-1}(Y_2,\dots,Y_{k+1})\rangle$$

$$= \langle D^{[k]}{}_X\big(R^{[k]}(X_1,Y_1,s^{k-1}(X_2,\dots,X_{k+1}))\big),\ s^{k-1}(Y_2,\dots,Y_{k+1})\rangle$$

$$+ \langle R^{[k]}(X_1,Y_1,s^{k-1}(X_2,\dots,X_{k+1})),\ D^{[k]}{}_X s^{k-1}(Y_2,\dots,Y_{k+1})\rangle\ ,$$

which by Corollary II.6-5 is

$$= \langle(\tilde{\nabla}_X R^{[k]})(X_1,Y_1,s^{k-1}(X_2,\dots,X_{k+1})),\ s^{k-1}(Y_2,\dots,Y_{k+1})\rangle$$

$$+ \langle R^{[k]}(\nabla_X X_1,Y_1,s^{k-1}(X_2,\dots,X_{k+1})),\ s^{k-1}(Y_2,\dots,Y_{k+1})\rangle$$

$$+ \langle R^{[k]}(X_1,\nabla_X Y_1,s^{k-1}(X_2,\dots,X_{k+1})),\ s^{k-1}(Y_2,\dots,Y_{k+1})\rangle$$

$$+ \langle R^{[k]}(X_1,Y_1,D^{[k]}{}_X s^{k-1}(X_2,\dots,X_{k+1})),\ s^{k-1}(Y_2,\dots,Y_{k+1})\rangle$$

$$+ \langle R^{[k]}(X_1,Y_1,s^{k-1}(X_2,\dots,X_{k+1})),\ D^{[k]}{}_X s^{k-1}(Y_2,\dots,Y_{k+1})\rangle\ .$$

Using (1) we can replace the term involving $(\tilde{\nabla}_X R^{[k]})$ by two terms, involving $\tilde{\nabla}_{X_1} R^{[k]}$ and $\tilde{\nabla}_{Y_1} R^{[k]}$. After performing this substitution, and summing all the terms thus arising from equation (4) of Proposition 76, everything cancels out except for the terms which constitute equation (2). ■

Corollary 78 will play an especially important role in our theory. To begin with, consider the case $R = R^{[1]}$, which depends only on the connection ∇ on TM. In the Remark after Lemma 65, we pointed out that for any bundle $\varpi\colon E \to M$ with a metric $\langle\ ,\ \rangle$ and a symmetric section s of $\mathrm{Hom}(TM\times TM, E)$, we can consider the "Codazzi-Mainardi equations" for a connection δ on E. The proof of Lemma 65 shows that if δ is to be compatible with the metric $\langle\ ,\ \rangle$ and also satisfy this equation, then $\langle \delta_X s(Y_1,Y_2), s(Z_1,Z_2)\rangle$ is completely determined, by equation (*) in the proof. However, if we are given a δ which does satisfy (*), it is by no means clear that δ is compatible with the metric and satisfies the Codazzi-Mainardi equations. To see what is happening here, we need to examine the formulas much more closely. If the reader returns to the proof of Lemma 65 he will see that when explicitly written out, equation (3) in the proof reads

$$\begin{aligned}
&\langle \delta_X s(Y_1,Y_2), s(Z_1,Z_2)\rangle - \langle \delta_X s(Z_1,Y_2), s(Y_1,Z_2)\rangle \\
&\quad = \langle s(\nabla_{Y_1}X,Y_2) - s(\nabla_X Y_1,Y_2) + s(X,\nabla_{Y_1}Y_2) - s(Y_1,\nabla_X Y_2), s(Z_1,Z_2)\rangle \\
&\qquad - \langle s(\nabla_{Z_1}Y_1,Y_2) - s(\nabla_{Y_1}Z_1,Y_2) + s(Y_1,\nabla_{Z_1}Z_2) - s(Z_1,\nabla_{Y_1}Z_2), s(X,Y_2)\rangle \\
&\qquad + \langle s(\nabla_X Z_1,Y_2) - s(\nabla_{Z_1}X,Y_2) + s(Z_1,\nabla_X Y_2) - s(X,\nabla_{Z_1}Y_2), s(Y_1,Z_2)\rangle \\
&\qquad + Y_1(\langle s(X,Y_2), s(Z_1,Z_2)\rangle) - Z_1(\langle s(X,Y_2), s(Y_1,Z_2)\rangle) \\
&\quad = \mathcal{E}(X,Y_1,Y_2,Z_1,Z_2)\ , \quad \text{say.}
\end{aligned}$$

Following the proof a little further along, we arrive at the explicit formula

$$\begin{aligned}
2\langle \delta_X s(Y_1,Y_2), s(Z_1,Z_2)\rangle = {}& \mathcal{E}(X,Y_1,Y_2,Z_1,Z_2) + \mathcal{E}(X,Y_2,Z_1,Z_2,Y_1) \\
& + X(\langle s(Y_1,Y_2), s(Z_1,Z_2)\rangle)\ .
\end{aligned}$$

Now we can form

$$\text{(A)}\qquad 2\langle \delta_U s(V,X),\ s(Y,Z)\rangle - 2\langle \delta_V s(U,X),\ s(Y,Z)\rangle$$

$$= \mathcal{E}(U,V,X,Y,Z)$$

$$+ \mathcal{E}(U,X,Y,Z,V)$$

$$- \mathcal{E}(V,U,X,Y,Z)$$

$$- \mathcal{E}(V,X,Y,Z,U)$$

$$+ U(\langle s(V,X),\ s(Y,Z)\rangle) - V(\langle s(U,X),\ s(Y,Z)\rangle)$$

$$= V(\langle s(U,X),\ s(Y,Z)\rangle) - Y(\langle s(U,X),\ s(V,Z)\rangle) + \cdots$$

$$+ X(\langle s(U,Y),\ s(V,Z)\rangle) - Z(\langle s(U,Y),\ s(X,V)\rangle) + \cdots$$

$$- U(\langle s(V,X),\ s(Y,Z)\rangle) - Y(\langle s(V,X),\ s(U,Z)\rangle) + \cdots$$

$$- X(\langle s(V,Y),\ s(U,Z)\rangle) + Z(\langle s(V,Y),\ s(X,U)\rangle) + \cdots$$

$$+ U(\langle s(V,X),\ s(Y,Z)\rangle) - V(\langle s(U,X),\ s(Y,Z)\rangle)$$

$$= \mathfrak{S}\{Z(\langle s(X,U),\ s(V,Y)\rangle - \langle s(X,V),\ s(Y,U)\rangle)\} + \cdots$$

where $\mathfrak{S}$ indicates a cyclic sum over (X,Y,Z)

$$= \mathfrak{S}\{Z(\langle R(X,Y)V,\ U\rangle)\} + \cdots$$

$$= \mathfrak{S}\{\langle \nabla_Z(R(X,Y)V),\ U\rangle\} + \cdots$$

$$= \mathfrak{S}\{\langle (\nabla_Z R)(X,Y,V),\ U\rangle\} + \cdots .$$

We have not troubled ourselves to write down all the $\cdots$ terms, but, as you may suspect, when we apply Corollary 76(2) [for $k = 1$] we find that equation (A) comes down to precisely the Codazzi-Mainardi equations! In deriving this, we use only Gauss' equation for R, and the fact that ∇ is compatible with the metric (and properties of R' for the ambient manifold).

Similarly, we may form

(B) $$2\langle\delta_X s(X_1,X_2),\ s(Y_1,Y_2)\rangle + 2\langle s(X_1,X_2),\ \delta_X s(Y_1,Y_2)\rangle$$

$$= \mathcal{E}(X,X_1,X_2,Y_1,Y_2)$$

$$+ \mathcal{E}(X,X_2,Y_1,Y_2,X_1)$$

$$+ \mathcal{E}(X,Y_1,Y_2,X_1,X_2)$$

$$+ \mathcal{E}(X,Y_2,X_1,X_2,Y_1)$$

$$+ 2X(\langle s(X_1,X_2),\ s(Y_1,Y_2)\rangle)$$

$$= X_1(\langle s(X,X_2),\ s(Y_1,Y_2)\rangle) - Y_1(\langle s(X,X_2),\ s(X_1,Y_2)\rangle) + \cdots$$

$$+ X_2(\langle s(X,Y_1),\ s(Y_2,X_1)\rangle) - Y_2(\langle s(X,Y_1),\ s(X_2,X_1)\rangle) + \cdots$$

$$+ Y_1(\langle s(X,Y_2),\ s(X_1,X_2)\rangle) - X_1(\langle s(X,Y_2),\ s(Y_1,X_2)\rangle) + \cdots$$

$$+ Y_2(\langle s(X,X_1),\ s(X_2,Y_1)\rangle) - X_2(\langle s(X,X_1),\ s(Y_2,Y_1)\rangle) + \cdots$$

$$+ 2X(\langle s(X_1,X_2),\ s(Y_1,Y_2)\rangle)$$

$$= \mathfrak{S}''\{X_1(\langle s(X,X_2),\ s(Y_1,Y_2)\rangle - \langle s(X,Y_2),\ s(Y_1,X_2)\rangle)\} + \cdots$$

where $\mathfrak{S}''$ indicates a cyclic sum over (X_1,X_2,Y_1,Y_2)

$$= -\ \mathfrak{S}''\{X_1(\langle R(X,Y_1)X_2,\ Y_2\rangle)\} + \cdots$$

$$= -\ \mathfrak{S}''\{\langle\nabla_{X_1}(R(X,Y_1)X_2),\ Y_2\rangle\} + \cdots$$

$$= -\ \mathfrak{S}''\{\langle(\nabla_{X_1}R)(X,Y_1,X_2),\ Y_2\rangle\} + \cdots .$$

When we apply Corollary 78(2), it turns out that everything on the right side of (B) cancels, except the term $2X\big(\langle s(X_1,X_2),\ s(Y_1,Y_2)\rangle\big)$. So we see that δ is compatible with the metric!

More generally, we have

79. PROPOSITION. The fact that D^{k+1} satisfies the Codazzi-Mainardi equations and is compatible with the metric follows formally from equation (*) in the proof of Lemma 73, Gauss' equation for $R^{[k+1]}$, and the fact that D^k is compatible with the metric (and the properties of R' for the ambient manifold).

Proof. An abominable calculation. ■

We are finally ready to consider the general imbedding question. The situation is rather complicated, and we will merely outline the results, without going into details. We are given a simply connected manifold M^n and tensors $\mathcal{F}_0, \ldots, \mathcal{F}_{\ell-1}$ on M, the tensor $\mathcal{F}_k$ being covariant of order $2(k+1)$ and symmetric in the first $k+1$ arguments, in the last $k+1$ arguments, and under interchange of the first $k+1$ arguments with the last $k+1$ arguments. We assume that $\mathcal{F}_0$ is positive definite, and thus a Riemannian metric on M; we will also denote $\mathcal{F}_0$ by $\langle\ ,\ \rangle$. For $k \geq 1$ we assume that $\mathcal{F}_k$ is positive semi-definite of constant rank $r_k > 0$. Set $m = n + r_1 + \cdots + r_{\ell-1}$. We want to know when these tensors come from an immersion of M into a given complete m-dimensional manifold N of constant curvature K_0. As usual, we can reduce this to a local problem, so we assume that M is diffeomorphic to $\mathbb{R}^n$, and we choose a basis $X_1, \ldots, X_n$ for the vector fields on M. For $1 \leq k \leq \ell - 1$ we take as our "k^{th} normal bundle" $E^k = M \times \mathbb{R}^{r_k}$. Similarly, for our "$k^{th}$ osculating bundle" 0^k we take the trivial bundle whose fibre over p is $0^k{}_p = M_p \oplus E^1{}_p \oplus \cdots \oplus E^{k-1}{}_p$. Each E^k has r_k natural sections $p \longmapsto (p, (0, \ldots, 0, 1, 0, \ldots, 0))$, and we give E^k the Riemannian metric which makes these orthonormal; these metrics will all be denoted by $<\ ,\ >$. We now define $s^k\colon TM \times \cdots \times TM \to E^k$ rather arbitrarily. By hypothesis, the $n^{k+1} \times n^{k+1}$ matrix

$$(\mathcal{F}_k(X_{i_1},\ldots,X_{i_{k+1}},X_{j_1},\ldots,X_{j_{k+1}}))$$

has rank r_k at each point. Making M smaller if necessary, we can assume that there is a set $\mathcal{S}$ of exactly r_k $(k+1)$-tuples $(\alpha_1,\ldots,\alpha_{k+1})$ such that the corresponding r_k rows of this matrix are everywhere linearly independent. Then for $(\alpha_1,\ldots,\alpha_{k+1}) \in \mathcal{S}$ we define $s^k(X_{\alpha_1},\ldots,X_{\alpha_{k+1}})$ to be one of the r_k natural sections of E^k (choosing an arbitrary correspondence between the elements of $\mathcal{S}$ and the r_k natural sections of E^k). There is now a unique way to define $s^k(X_{i_1},\ldots,X_{i_{k+1}})$ in general so that

$$<s^k(X_{i_1},\ldots,X_{i_{k+1}}),\ s^k(X_{j_1},\ldots,X_{j_{k+1}})> = \mathcal{F}_k(X_{i_1},\ldots,X_{i_{k+1}},X_{j_1},\ldots,X_{j_{k+1}}) .$$

Now we would like to define maps

$$s^k\colon TM \times E^k \to E^{k+1}$$

such that

$$s^k(X_i,\ s^k(X_{i_1},\ldots,X_{i_{k+1}})) = s^{k+1}(X_i,X_{i_1},\ldots,X_{i_{k+1}}) .$$

But in this abstract set-up there is no way to prove that this map is well-defined. Instead we have to assume

(I) For each i and j, the $n^{k+1} \times 2n^{k+1}$ matrix

$$\left(\mathcal{F}_k(X_{i_1},\ldots,X_{i_{k+1}},X_{j_1},\ldots,X_{j_{k+1}}),\ \mathcal{F}_{k+1}(X_i,X_{i_1},\ldots,X_{i_{k+1}},X_j,X_{j_1},\ldots,X_{j_{k+1}})\right)$$

is of rank r_k.

[The $(k+1)$-tuple $(i_1,\ldots,i_{k+1})$ determines a row of this matrix, and the $(k+1)$-tuple $(j_1,\ldots,j_{k+1})$ determines 2 different columns.]

With this assumption we can define $\mathbf{s}^k$. We can thus also define the maps A^k_ξ for ξ an element of E^k.

Now we want to define connections D^k on the E^k. Consider first D^1. The proof of Lemma 73 tells us that we have to define D^1 so that

$$(a_1)\quad \langle {D^1}_{X_i} s^1(X_{i_1},X_{i_2}),\ s^1(X_{j_1},X_{j_2})\rangle = E_1(X_i,X_{i_1},X_{i_2},X_{j_1},X_{j_2})\ ,$$

where E_1 is some explicit expression we could work out. In order to know that we can define D^1 so that this formula holds, we must assume

(II_1) For each i, i_1, i_2, the $n^2 \times 2n^2$ matrix

$$\left(\mathcal{F}_1(X_{h_1},X_{h_2},X_{j_1},X_{j_2}),\ E_1(X_i,X_{i_1},X_{i_2},X_{j_1},X_{j_2})\right)$$

is of rank r_2.

[(h_1,h_2) determines a row, and (j_1,j_2) determines 2 columns.]

With this assumption we can define D^1 so that equation (a) holds.

Of course, we already have the connection $D^0 = \nabla$ on TM determined by the metric $\mathcal{F}_0 = \langle\ ,\ \rangle$, and we want to assume that its curvature tensor $R = R^{[1]}$ satisfies

$$\langle R'(X,Y)Z,\ W\rangle = \langle R(X,Y)Z,\ W\rangle + \langle s^1(X,Z),\ s^1(Y,W)\rangle - \langle s^1(X,W),\ s^1(Y,Z)\rangle\ ,$$

i.e.,

$$(\mathrm{III}_1)\quad K_0\cdot[\langle X,W\rangle\cdot\langle Y,Z\rangle - \langle X,Z\rangle\cdot\langle Y,W\rangle]$$
$$= \langle R(X,Y)Z,\ W\rangle + \mathcal{F}_1(X,Z,Y,W) - \mathcal{F}_1(X,W,Y,Z)\ .$$

Proposition 79 then shows that D^1 satisfies the Codazzi-Mainardi equations and is compatible with the metric in E^1. We can now define $D^{[2]}$ on O^2 by the formula on p. 263, and it therefore makes sense to assume the generalized Gauss equation for $R^{[2]}$. Actually, it suffices to assume the special case

$$0 = \langle R^{[2]}(X,Y)s^1(X_1,X_2),\ s^1(Y_1,Y_2)\rangle$$
$$+ \langle s^2(X,X_1,X_2),\ s^2(Y,Y_1,Y_2)\rangle - \langle s^2(Y,X_1,X_2),\ s^2(Y,Y_1,Y_2)\rangle\ ,$$

i.e.

$$(\mathrm{III}_2)\quad 0 = \langle R^{[2]}(X,Y)s^1(X_1,X_2),\ s^1(Y_1,Y_2)\rangle$$
$$+ \mathcal{F}_2(X,X_1,X_2,Y,Y_1,Y_2) - \mathcal{F}_2(Y,X_1,X_2,Y,Y_1,Y_2)\ .$$

Now we want to define D^2 so that

$$(a_2)\quad \langle D^2{}_{X_i}s^2(X_{i_1},X_{i_2},X_{i_3}),\ s^2(X_{j_1},X_{j_2},X_{j_3})\rangle = E_2(X_i,X_{i_1},X_{i_2},X_{i_3},X_{j_1},X_{j_2},X_{j_3})\ ,$$

where E_2 is an explicit expression we could work out (it involves D^1, but we already have an expression for D^1). In order to know that we can define D^2 so that this formula holds, we must assume

(II_2) For each i, i_1, i_2, i_3, the $n^3 \times 2n^3$ matrix

$$(\mathcal{F}_2(X_{h_1},X_{h_2},X_{h_3},X_{j_1},X_{j_2},X_{j_3}),\ E_2(X_i,X_{i_1},X_{i_2},X_{i_3},X_{j_1},X_{j_2},X_{j_3}))$$

is of rank r_3.

With this assumption we can define D^2 so that (a_2) holds. Then Proposition 79 shows that D^2 satisfies the Codazzi-Mainardi equations and is compatible with the metric in E^2. We can now define $D^{[3]}$ on 0^3 and it makes sense to assume the Gauss equation for $R^{[3]}$. Continuing in this way, we can formulate conditions

(II_k) $\qquad 1 \leq k \leq \ell - 1$

(III_k) $\qquad 1 \leq k \leq \ell - 1$.

Finally, we can formulate

(IV) $\qquad R^{[\ell]} = 0$.

Standard arguments about integrability conditions show that if the conditions (I), $\{(II_k)\}$, $\{(III_k)\}$, and (IV) hold, then the tensors $\mathcal{F}_0, \ldots, \mathcal{F}_{\ell-1}$ on M come from an immersion of M into N.

PROBLEMS

1. Let $f\colon \mathbb{R}^n \longrightarrow \mathbb{R}^n$ be a map with $\langle f(v), f(w)\rangle = \langle v,w\rangle$, where $\langle\ ,\ \rangle$ is a non-degenerate inner product on $\mathbb{R}^n$. Show that

$$\langle f(\Sigma\, a_i e_i),\ f(e_j)\rangle = \langle \Sigma\, a_i f(e_i),\ f(e_j)\rangle$$

for all j, and conclude that f is linear.

2. Consider $\mathbb{R}^{n+1}$ with the metric

$$-\,dx^0 \otimes dx^0 + dx^1 \otimes dx^1 + \dots + dx^n \otimes dx^n .$$

(a) For the Levi-Civita connection (compare p. II.394), the geodesics are the ordinary straight lines.

(b) If $g\colon \mathbb{R}^{n+1} \longrightarrow \mathbb{R}^{n+1}$ is an isometry (with respect to this metric) with $g(0) = 0$ and g_{*0} = identity, then g = identity. [This can also be derived, as in Problem 1-5, from an appropriate generalization of Corollary II.7-13.]

(c) If $f\colon \mathbb{R}^{n+1} \longrightarrow \mathbb{R}^{n+1}$ is an isometry with $f(0) = 0$, then $f = f_{*0}$.

(d) Every isometry of $\mathbb{R}^{n+1}$ is of the form $p \longmapsto A(p) + q$ for $A \in O^1(n+1)$, and $q \in \mathbb{R}^{n+1}$.

3. Determine the geodesics of H^n by the same method used for S^n in Chapter I.9 (reflection through a 2-dimensional plane $P \subset \mathbb{R}^{n+1}$ is an isometry).

4. A <u>linear fractional transformation</u> is a map

$$z \longmapsto \frac{az + b}{az + d} \qquad a, b, c, d \in \mathbb{C}, \qquad ad - bc \neq 0,$$

of the extended complex plane $\mathbb{C} \cup \{\infty\}$ to itself.

(a) The set of linear fractional transformations is a group under composition.

(b) For distinct z_1, z_2, z_3, the transformation

$$z \longmapsto \frac{z - z_2}{z - z_3} \Big/ \frac{z_1 - z_2}{z_1 - z_3}$$

takes z_1 to 1, and z_2 to 0, and z_3 to ∞.

(c) There is a linear fractional transformation taking any three distinct points $z_1, z_2, z_3 \in \mathbb{C} \cup \{\infty\}$ to any other three distinct points w_1, w_2, w_3.

(d) If a linear fractional transformation keeps 1, 0, and ∞ fixed, then it is the identity.

(e) There is a unique linear fractional transformation taking z_1, z_2, z_3 to 1, 0, ∞.

(f) The transformation of part (c) is unique.

(g) The linear fractional transformations which take the real axis to itself are precisely those with $a, b, c, d \in \mathbb{R}$.

(h) The linear fractional transformations which take the upper half plane onto itself are

$$f(z) = \frac{az + b}{cz + d} ,$$

a, b, c, d $\in \mathbb{R}$ and $ad - bc > 0$. We can then clearly assume that $ad - bc = 1$.

5. For distinct z_1, z_2, z_3, the <u>cross ratio</u> (z,z_1,z_2,z_3) is defined as

$$(z,z_1,z_2,z_3) = \frac{z - z_2}{z - z_3} \bigg/ \frac{z_1 - z_2}{z_1 - z_3} ;$$

thus (z,z_1,z_2,z_3) is $f(z)$ where f is the linear fractional transformation taking z_1, z_2, z_3 to $1, 0, \infty$.

(a) If g is a linear fractional transformation, then

$$(g(z),g(z_1),g(z_2),g(z_3)) = (z,z_1,z_2,z_3).$$

(b) If $\theta = \arg w$ denotes an angle between the positive x-axis and the ray from 0 to w, so that $w = |w|e^{i\theta}$, then

$$\arg(z,z_1,z_2,z_3) = \arg \frac{z - z_2}{z - z_3} - \arg \frac{z_1 - z_2}{z_1 - z_3}$$

$$= \theta_1 - \theta_2 \quad \text{in the picture below.}$$

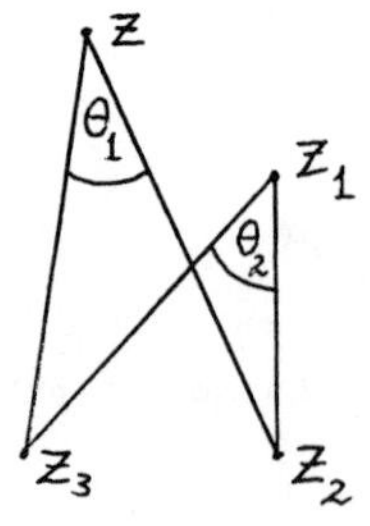

Conclude that (z,z_1,z_2,z_3) is real if and only if z,z_1,z_2,z_3 lie on a circle or straight line.

(c) A linear fractional transformation takes circles and straight lines into circles and straight lines.

6. In this problem we will use the notation on pp. 462-463.

(a) The metric on the upper half plane can be written

$$< \, , > \;=\; \frac{dz \otimes d\bar{z}}{(\operatorname{Im} z)^2} \, .$$

(b) For the linear fraction transformation f of Problem 4(h), we have

$$\operatorname{Im} f(z) = \frac{\operatorname{Im} z}{|cz + d|^2} \, ,$$

$$f_z = \frac{\partial}{\partial z}\left(\frac{az + b}{cz + d}\right) = \frac{1}{(cz + d)^2} \, ,$$

$$f^*(dz) = df = f_z dz + f_{\bar{z}} d\bar{z} = \frac{dz}{(cz + d)^2} \, ,$$

$$f^*(d\bar{z}) = d\bar{f} = \frac{d\bar{z}}{(c\bar{z} + d)^2} \, .$$

(c) Conclude that f is an isometry of the upper half plane.

(d) There is such an isometry taking any given point z to any other.

Hint: Consider the linear fraction transformation taking z_1, z_2, z in the figure below to w_1, w_2, w.

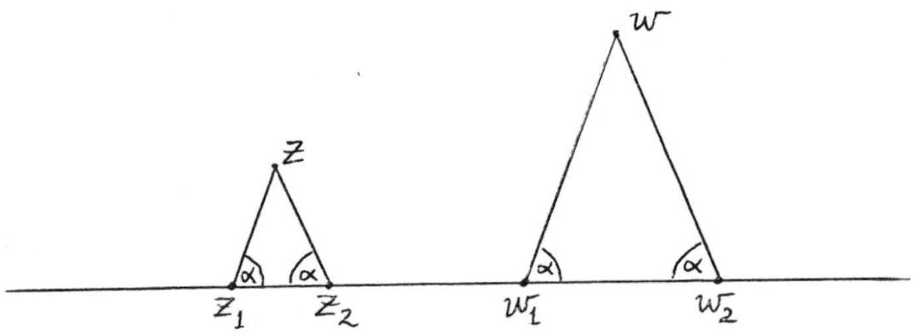

(e) In the B^2 model, the linear fractional transformations keeping S = boundary B^2 fixed are isometries, and there are such isometries taking any point to any other. Conclude that these isometries are all the orientation preserving isometries of B^2, by noting that rotations about the origin are linear fractional transformations.

(f) The geodesic circles around 0 in B^2 are clearly ordinary circles. Conclude that all geodesic circles are ordinary circles, and that the same result holds in the upper half plane. (The converse can be proved exactly as in the higher dimensional case.)

7. (a) In the upper half plane, the distance between $z_1 = x + iy_1$ and $z_2 = x + iy_2$ is

$$d(z_1,z_2) = \left|\int_{y_1}^{y_2} \frac{dy}{y}\right| = \left|\log \frac{y_2}{y_1}\right| = |\log(z_0,z_1,z_2,z_3)|$$

where $z_0 = x$ and $z_3 = \infty$.

(b) Let z_1,z_2 be any two points of the upper half plane and let the semi-circle through z_1,z_2 perpendicular to the x-axis meet the x-axis at z_0 and z_3. Then

$$d(z_1,z_2) = |\log\,(z_0,z_1,z_2,z_3)|.$$

(c) Similarly, find the formula for $d(z_1,z_2)$ in B^2.

8. (a) The only geodesic maps $f\colon \mathbb{R}^n \longrightarrow \mathbb{R}^n$ defined on all of $\mathbb{R}^n$ are the affine maps. <u>Hint</u>: Assume $f(0) = 0$, and recall the parallelogram law for addition, as on p. III.312.

(b) Every geodesic map from S^{n+} to S^{n+} is of the form $\phi^{-1}\circ A\circ\phi$, where $\phi\colon S^{n+} \longrightarrow \mathbb{R}^n$ is the standard geodesic map, and $A\colon \mathbb{R}^n \longrightarrow \mathbb{R}^n$ is affine.

9. In this Problem we will determine all geodesic maps $f\colon U \longrightarrow V$ where U and V are open subsets of $\mathbb{R}^n$. We will use material from projective geometry - the reader is referred to Hartshorne {1} for all terms and theorems*.

*For an analytic derivation see Scheffers {1; V.2, 429-432}.

We need the fact that every $A = (a_{ij}) \in GL(n+1,\mathbb{R})$ determines a geodesic map $\bar{A}\colon \mathbb{P}^n \longrightarrow \mathbb{P}^n$, and that every such map comes from some $A \in GL(n+1,\mathbb{R})$, unique up to multiplication by a real number. If we regard $\mathbb{R}^n \subset \mathbb{P}^n$, then the action of $\bar{A}$ on $\mathbb{R}^n$ is easily seen to be $\bar{A}(x^1,\ldots,x^n) = (y^1,\ldots,y^n)$, where

$$y^i = \frac{\sum_{j=1}^{n} a_{ij}x^j + a_{i,n+1}}{\sum_{j=1}^{n} a_{n+1,j}x^j + a_{n+1,n+1}}$$

(points where the denominator vanish go into the line at infinity). We will also use Desargue's Theorem and its converse (= its dual).

(a) Given any point $0 \in \mathbb{R}^n$, and three lines ℓ_1, ℓ_2, ℓ_3 through 0 which intersect U, show that there is a Desargue configuration with all other points in U. Hint: In the figure below, points A, A', and P are fixed, while B and B', and C and C', are chosen close together.

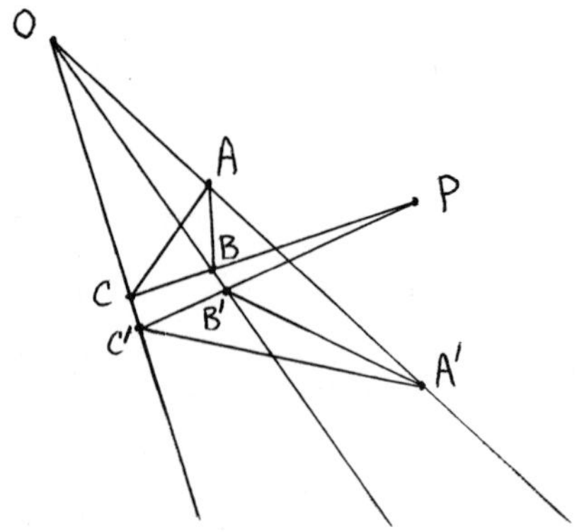

Conclude that the lines containing the $f(\ell_i \cap U)$ are concurrent. Thus show that there is a well-defined extension $\tilde{f}: \mathbb{P}^n \longrightarrow \mathbb{P}^n$ with the property that if $P \varepsilon \ell_1 \cap \ell_2$ where ℓ_1 and ℓ_2 intersect U, then $\tilde{f}(P)$ is the intersection of the lines containing $f(\ell_1 \cap U)$ and $f(\ell_2 \cap U)$. Show also that $\tilde{f}$ is one-one and onto.

(b) Let P, Q, R be three collinear points of $\mathbb{P}^n$. In the figure below, we first choose A,A' ε U, and then B, B' ε U, so that the lines AA' and BB' intersect at a point 0 ε U. Show that we can also arrange for QA and RB

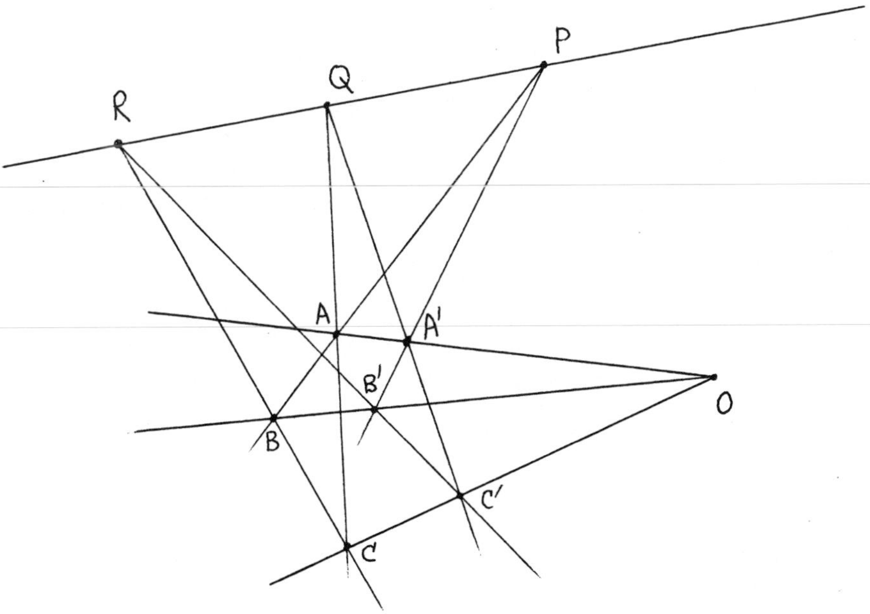

to intersect at a point $C \in U$ and for RB' and QA' to intersect at a point $C' \in U$. Then show that AA' and BB' and CC' intersect at $0 \in U$, so that we have a Desargue configuration with all points except P, Q, R in U. Conclude that $\tilde{f}(P)$, $\tilde{f}(Q)$, and $\tilde{f}(R)$ are collinear.

(c) Every geodesic map $f\colon U \to V$, where $U, V \in \mathbb{R}^n$ are open connected sets, is the restriction of some map $\bar{A}$ for $A \in GL(n+1,\mathbb{R})$.

(d) Every geodesic map from H^n into H^n is of the form $\phi^{-1}\circ\bar{A}\circ\phi$, where $\phi\colon H^n \to B^n(1)$ is the standard geodesic map, and $\bar{A}\colon B^n(1) \to B^n(1)$ is a geodesic map which takes $B^n(1)$ into $B^n(1)$.

10. (a) Let $f\colon \mathbb{P}^2 \to \mathbb{P}^2$ be a geodesic map which takes a circle $\Sigma \subset \mathbb{R}^2 \subset \mathbb{P}^2$ into itself. Show that f is determined by knowing $f(P)$, $f(Q)$, $f(R)$ for distinct points, $P, Q, R \in \Sigma$. Hint: Consider the tangent lines at P and Q, which intersect at some point S.

(b) Show that there is such an f for any given values of $f(P)$, $f(Q)$, $f(R)$. (You will need to use the fact that a conic is determined by three points and 2 tangents — see a book on projective geometry which treates conics.)

(c) Parts (a) and (b) show that the group of all geodesic maps $f\colon \mathbb{P}^2 \to \mathbb{P}^2$ with $f(\Sigma) = \Sigma$ has dimension 3. Using Problem 9, conclude that every geodesic map of H^2 onto itself is an isometry.

(d) Generalize to higher dimensions. Also consider the geodesic maps of S^n onto itself.

(e) Use the geodesic maps $H^n \longrightarrow B^n(1)$ and $B^n(2) \longrightarrow B^n(1)$ to describe an isometry between H^n and $B^n(2)$.

11. For vectors $v_1,\dots,v_{m-1}$ in $\mathbb{R}^m$, we define $v_1 \times \dots \times v_{m-1}$ to be the unique vector with

$$\langle v_1 \times \dots \times v_{m-1}, w\rangle = \det \begin{pmatrix} v_1 \\ \vdots \\ v_{m-1} \\ w \end{pmatrix}$$

for all $w \in \mathbb{R}^m$.

(a) If $T\colon \mathbb{R}^m \longrightarrow \mathbb{R}^m$ is an orientation preserving isometry, then

$$T(v_1 \times \dots \times v_{m-1}) = T(v_1) \times \dots \times T(v_{m-1}).$$

(b) Show how to define $v_1 \times \dots \times v_{m-1}$ for $v_1,\dots,v_{m-1}$ in an oriented m-dimensional vector space V with an inner product $\langle\ ,\ \rangle$.

12. (a) Let c be an arc-length parameterized curve in $(N, \langle\ ,\ \rangle)$, with $\kappa_1,\dots,\kappa_{m-1} \neq 0$, and Frenet frame $\mathbf{v}_1,\dots,\mathbf{v}_{m-1}$. Using $\nu_r = \mathbf{v}_r$ as a trivialization of the normal bundle of image c, show that

$$II^r(\mathbf{v}_1, \mathbf{v}_1) = \kappa_1\delta_2^r$$

$$\beta_r^s(\mathbf{v}_1) = -\kappa_{r-1}\delta_{r-1}^s + \kappa_r\delta_{r+1}^s .$$

Hence II^r and β^s_r are expressible in terms of $\kappa_1,\dots,\kappa_{m-1}$, and conversely, $\kappa_1,\dots,\kappa_{m-1}$ are expressible in terms of the II^r and β^s_r.

(b) Derive Corollary 4 from Theorem 20.

(c) Prove the assertion at the top of p. 75 by showing that $\phi\circ c = c$ for every curve $c\colon [0,1]\to M$ with $c(0) = p$.

13. Let $M^n, \bar{M}^n \subset S^m \subset \mathbb{R}^{m+1}$, with corresponding covariant differentiations $\nabla, \bar{\nabla}, \nabla', \boldsymbol{\nabla}'$ (as in the proof of Theorem 27). Let $\phi\colon M\to\bar{M}$ be an isometry, and $\tilde{\phi}\colon \text{Nor } M\to\text{Nor } \bar{M}$ a bundle isomorphism covering ϕ between the normal bundles in S^m which preserves $\langle\ ,\ \rangle$, s, and D. Let ν be the unit normal on S^m.

(a) The normal bundle **Nor** M of M in $\mathbb{R}^{m+1}$ has fibre $M_p^{\boldsymbol{\perp}} = M_p^{\perp} \oplus R\cdot\nu(p)$, and similarly for **Nor** $\bar{M}$. Define $\tilde{\boldsymbol{\phi}}\colon$ **Nor** $M \longrightarrow$ **Nor** $\bar{M}$ extending $\tilde{\phi}$ by $\tilde{\boldsymbol{\phi}}(\nu(p)) = \nu(\phi(p))$. Then $\tilde{\boldsymbol{\phi}}$ is inner product preserving.

(b) The second fundamental form **s** of M in $\mathbb{R}^{m+1}$ is given by

$$\mathbf{s}(X,Y) = s(X,Y) + \langle X,Y\rangle\nu .$$

(c) The normal connection **D** of **Nor** M is given by

$$\mathbf{D}_X\xi = D_X\xi \qquad \xi \text{ is a section of Nor } M$$

$$\mathbf{D}_X\nu = 0.$$

(d) The bundle isomorphism $\tilde{\phi}$ preserves $\mathbf{s}$ and $\mathbf{D}$, so there is a Euclidean motion $A: \mathbb{R}^{m+1} \longrightarrow \mathbb{R}^{m+1}$ with $\phi = A|M$ and $\tilde{\phi} = A_*|$ **Nor** M.

(e) From the action of $\tilde{\phi}$ on $\nu(p)$ conclude that A keeps 0 fixed, so that it also represents an isometry of S^m.

(f) Treat the case of two manifolds $M^n, \bar{M}^n \subset H^m$ similarly.

14. Let $(M, \langle\!\langle\ ,\ \rangle\!\rangle)$ be as in part (2) of Theorem 19, except with Gauss' Equation as on p.80, with $K_0 = 1$. Let $\boldsymbol{\varpi}: \mathbf{E} \longrightarrow M$ be the bundle whose fibre at p is $\varpi^{-1}(p) \oplus \mathbb{R}$, and extend $\{\ ,\ \}$ to a metric $\boldsymbol{\{\ ,\ \}}$ by

$$\boldsymbol{\{}(v,a),\ (w,b)\boldsymbol{\}} = \{v,w\} + ab\ .$$

Define a symmetric section $\boldsymbol{\sigma}$ of $\mathrm{Hom}(TM \times TM, \mathbf{E})$ by

$$\boldsymbol{\sigma}(X,Y) = (\sigma(X,Y), \langle\!\langle X,Y\rangle\!\rangle),$$

and define a connection $\boldsymbol{\delta}$ on E compatible with $\boldsymbol{\{\ ,\ \}}$ by

$$\boldsymbol{\delta}_X\xi = \delta_X\xi \qquad \xi \text{ is a section of } E$$

$$\boldsymbol{\delta}_X\zeta = 0 \qquad \text{where } \zeta \text{ is the section}$$

$$\zeta(p) = (0,1) \in \varpi^{-1}(p) \oplus \mathbb{R}.$$

(a) Gauss' equation, in the form with $K_0 = 0$, holds for $\boldsymbol{\sigma}$.

(b) The Codazzi-Mainardi equations hold for $\tilde{\nabla}\boldsymbol{\sigma}$.

(c) The Ricci equations hold for $R_{\boldsymbol{\delta}}$, $\boldsymbol{\sigma}$, $\mathbf{A}_\xi$.

(d) Let $f\colon M \longrightarrow \mathbb{R}^{m+1}$ be the isometric immersion given by Theorem 19, for M, $\mathbf{E}$, $\{,\}$, σ, δ. Regard f as an imbedding (by working locally), and let ν be the vector field $\tilde{f}(\zeta)$ on $f(M)$. Then for all tangent vectors X, Y of $f(M)$ we have

$$\langle s(X,Y), \nu\rangle = \langle X,Y\rangle \quad \Rightarrow \quad \nabla'_X \nu = -X.$$

(e) Let $p \in f(M)$ be a fixed point. Changing f by a translation, we can assume that $\nu(p) = -p$ (identifying tangent vectors of $\mathbb{R}^{m+1}$ with elements of $\mathbb{R}^{m+1}$, as usual). Let $c\colon [0,1] \longrightarrow f(M)$ be a curve with $c(0) = p$. Then

$$\frac{d\nu(c(t))}{dt} = -c'(t),$$

and consequently $\nu(c(t)) = -c(t)$ for all t. Conclude that $f(M) \subset S^m$.

(f) Treat the case $K_0 = -1$ similarly.

15. The Lie algebra $\mathfrak{gl}(m,\mathbb{R})$ has as a basis the metrices E_α^β which have zeros everywhere except for a 1 in the ith <u>row</u> and j^{th} <u>column</u>, so that

$$(E_\alpha^\beta)_\sigma^\rho = \delta_\alpha^\rho \delta_\sigma^\beta .$$

Let $\{\psi_\alpha^\beta\}$ be the dual basis, and let $\tilde{\psi}_\alpha^\beta$ be the left-invariant 1-forms on $GL(m,\mathbb{R})$ which extend the ψ_α^β.

(a) Show that

$$d\tilde{\psi}^{\beta}_{\alpha} = - \sum_{\gamma=1}^{m} \tilde{\psi}^{\beta}_{\gamma} \wedge \tilde{\psi}^{\gamma}_{\alpha} ,$$

either by computing the brackets of the E^{β}_{α} and using the first equation on p. I.538, or, more easily, by using the last equation on p. I.549.

(b) The Lie algebra $\mathfrak{o}(m)$ has as a basis the matrices $E^{\beta}_{\alpha} - E^{\alpha}_{\beta}$ $(\alpha < \beta)$. The dual basis is

$$(1) \qquad \phi^{\beta}_{\alpha} = \frac{\psi^{\beta}_{\alpha} - \psi^{\alpha}_{\beta}}{2} \qquad \alpha < \beta .$$

Define $\phi^{\beta}_{\alpha} = - \phi^{\alpha}_{\beta}$ for $\alpha > \beta$ and $\phi^{\alpha}_{\alpha} = 0$. Note that equation (1) still holds. Verify that we now have

$$d\tilde{\phi}^{\beta}_{\alpha} = - \sum_{\gamma=1}^{m} \tilde{\phi}^{\beta}_{\gamma} \wedge \tilde{\phi}^{\gamma}_{\alpha}$$

(c) Derive Theorem 19 as a consequence of Theorems I.10-17 and I.10-18.

16. Use Problem I.7-14(a) to show that the even powers of λ in the characteristic polynomial $\chi(\lambda)$ of A can be expressed in terms of the determinants of the 2×2 submatrices of A.

17. For a hypersurface $M \subset \mathbb{R}^{n+1}$, generalize Proposition 2-6 so as to express the $(n+1)$st fundamental form in terms of the first n fundamental forms and the elementary symmetric curvatures.

18. For an immersion $f:M^n \longrightarrow \mathbb{R}^{n+1}$ with normal map $n_f = \nu\circ f$, show that we still have

$$\mathrm{III}_f = \mathrm{I}_{n_f} = -\,\mathrm{II}_{n_f} .$$

19. Let c be a curve in a hypersurface $M \subset N$ of a manifold of constant curvature K_0, and let X be a vector field of N along M. Then $\nabla'_{c'(s)}X$ is always a multiple of $c'(s)$ if and only if the ruled surface $\{\exp_{c(s)} tX(c(s))\}$ has constant intrinsic curvature K_0.

20. Let $\sigma\colon S^n - \{\text{north pole}\} \to \mathbb{R}^n$ be the version of stereographic projection in which S^n denotes the standard unit sphere around 0.

(a) For this σ we have

$$\sigma(p) = \left(\frac{p^1}{1-p^{n+1}}, \ldots, \frac{p^n}{1-p^{n+1}}\right)$$

$$\sigma^{-1}(y) = \left(\frac{2y^1}{1+\Sigma(y^i)^2}, \ldots, \frac{2y^n}{1+\Sigma(y^i)^2}, \frac{\Sigma(y^i)^2-1}{1+\Sigma(y^i)^2}\right) .$$

(b) Let $c\colon [0,2\pi] \to \mathbb{R}^n$ be a curve, parameterized proportionally to arc-length, which goes once around a circle centered at 0 and passing through y, so that c' has squared length $|y|^2$. Then $(\sigma^{-1}\circ c)'$ has squared length

$$\frac{4|y|^2}{[1 + |y|^2]^2} .$$

Thus $\sigma^{-1}{}_*$ multiplies lengths of tangent vectors at y by $2/(1 + |y|^2)$.

21. Let $\sigma: S^2 \to \mathbb{C} \cup \{\infty\}$ be stereographic projection, where S^2 is the standard unit sphere around $(0,0,0)$.

(a) If R_θ is rotation of S^2 through an angle of θ around the z-axis, then $\sigma \circ R_\theta \circ \sigma^{-1}: \mathbb{C} \cup \{\infty\} \to \mathbb{C} \cup \{\infty\}$ is just $z \mapsto e^{i\theta} z$.

(b) If R'_θ is rotation through an angle of θ around the y-axis, calculate that $\sigma \circ R'_\theta \circ \sigma^{-1}$ is

$$z \mapsto \frac{(1 + \cos\theta)z - \sin\theta}{(\sin\theta)z + (1 + \cos\theta)} .$$

(c) The group $SO(3)$ is generated by all R_θ and R'_θ. (A direct proof can be given, or one can note that $SO(3)$ is 3-dimensional, and the R_θ and R'_ϕ do not commute.) The group of all 4×4 complex matrices $\begin{pmatrix} a & b \\ c & d \end{pmatrix}$ satisfying the conditions on p. 161 is also 3-dimensional. Conclude that this group is precisely the group of all $\sigma \circ A \circ \sigma^{-1}$ for $A \in O(3)$.

22. Consider $B^2(2)$, with the matric on p. 10. From p. II.7-54 we see that the geodesic circle of radius r is given by

$$c(\theta) = 2 \tanh \frac{r}{2} (\cos\theta, \sin\theta) \qquad 0 \leq \theta \leq 2\pi .$$

(a) Calculate that

$$|c'(\theta)| = \sinh r .$$

(b) Then verify the formula for I given on p. 173.

23. (a) For a coordinate system u,v on a 2-dimensional Riemannian manifold, show that the formula on p. 193 can be written

$$\Delta f = \frac{1}{W}\left\{\frac{\partial}{\partial u}\left(\frac{G\frac{\partial f}{\partial u} - F\frac{\partial f}{\partial v}}{W}\right) + \frac{\partial}{\partial v}\left(\frac{E\frac{\partial f}{\partial v} - F\frac{\partial f}{\partial u}}{W}\right)\right\},$$

where $W = \sqrt{EG - F^2}$.

(b) If (u,v) is isothermal (this means that $E = G$ and $F = 0$; compare p. II.334), then

$$\Delta f = \frac{1}{E}\left(\frac{\partial^2 f}{\partial u^2} + \frac{\partial^2 f}{\partial v^2}\right) .$$

(c) A coordinate system (x,y) on a 2-dimensional Riemannian manifold is isothermal if and only if $\Delta_1 x = \Delta_1 y$ and $\Delta_1(x,y) = 0$.

(d) If (x,y) is an isothermal coordinate system then $\Delta x = \Delta y = 0$.

(e) If $\Delta x = 0$, then there is locally a function y with

$$dy = \frac{F\frac{\partial x}{\partial u} - E\frac{\partial x}{\partial v}}{W}\,du + \frac{G\frac{\partial x}{\partial u} - F\frac{\partial x}{\partial v}}{W}\,dv$$

(here E, F, G are the components of $\langle\ ,\ \rangle$ in the (u,v) coordinate system). The functions x and y satisfy $\Delta_1 x = \Delta_1 y$ and $\Delta_1(x,y) = 0$, so (x,y) is an isothermal coordinate system.

24. Let h be a function on a 2-dimensional Riemannian manifold such that the sets h = constant give a foliation of the manifold.

(a) Suppose that there is an isothermal coordinate system (x,y) such that one family of parameter curves lie along the curves h = constant; thus $x = f\circ h$ for some function f. Use Problem 23 to show that

$$\Delta h\cdot(f'\circ h) + \Delta_1 h\cdot(f''\circ h) = 0.$$

Hence $\Delta h/\Delta_1 h$ is some function composed with h.

(b) Conversely, if $\Delta h/\Delta_1 h = F\circ h$ for some function F, and we set $x = f\circ h$ for

$$f' = e^{-\int f},$$

then $\Delta x = 0$, and the function y of Problem 23(e) satisfies

$$\Delta_1 y = e^{-2\int F}\Delta_1 h\ .$$

(c) So

$$< \, , \, > = \frac{1}{\Delta_1 h}\left(dh \otimes dh + e^{2\int F} dy \otimes dy \right).$$

(d) If we have equations (a) and (b) on p. 223, then the corresponding metrics are

$$\frac{1}{f\circ K}(dK \otimes dK + e^{2\int g/f} dy \otimes dy)$$

$$\frac{1}{f\circ \bar{K}}(d\bar{K} \otimes d\bar{K} + e^{2\int g/f} d\bar{y} \otimes d\bar{y})$$

So there is a one-parameter family of isometries between the surfaces.

(e) There is a function x with

$$dx \otimes dx = \frac{1}{f\circ K} dK \otimes dK ,$$

are similarly for $\bar{x}$. Hence, each surface is isometric to a surface of revolution (see formula (4) on p. III.231).

25. Let V and W be two inner product spaces of the same dimension. Let $\{v_\rho\}$ be an indexed set of (not necessarily distinct) vectors which span V, and let $\{\bar{v}_\rho\}$ span W. Suppose that

$$\langle v_\rho, v_\sigma \rangle = \langle \bar{v}_\rho, \bar{v}_\sigma \rangle \qquad \text{for all } \rho, \sigma .$$

Show that $\Sigma c_\rho v_\rho = 0 \Rightarrow \Sigma c_\rho \bar{v}_\rho = 0$, and conclude that there is a unique inner product preserving isomorphism $V \longrightarrow W$ which takes v_ρ to $\bar{v}_\rho$.

Chapter 8. The Second Variation

In this chapter we return to the study of the calculus of variations, and introduce an important (essentially classical) construction, which has surprizingly significant consequences for differential geometry. Recall that the calculus of variations was first invoked in order to find paths which locally minimize the length function L for a Riemannian manifold M. In the course of our investigations we found that the energy function was more convenient to work with, and that the critical paths for the length function are precisely the same as those for the energy function, except that the latter are necessarily parameterized proportionally to arc length. These critical points for E are, of course, the geodesics on M, and at present we know only that sufficiently small pieces of geodesics are paths of minimal length. We now want to develop conditions which determine when a given geodesic is, in its entirety, a path of smaller length than nearby paths. We recall one fact from Problem I.9-31: For a piecewise C^∞ curve $\gamma\colon [a,b] \to M$ we always have

$$[L_a^b(\gamma)]^2 \le (b-a)E_a^b(\gamma) \ ,$$

with equality precisely when γ is parameterized proportionally to arclength. From this it is easy to see that a geodesic γ has minimal length among all nearby paths between $\gamma(a)$ and $\gamma(b)$ precisely when it has minimal energy among all such paths. Thus we lose no information by restricting all our considerations to the energy function E.

We begin with a brief summary of the results which we already have. Consider a piecewise C^∞ path $\gamma\colon [a,b] \to M$ and a piecewise C^∞ variation $\alpha\colon (-\varepsilon,\varepsilon) \times [a,b] \to M$ of γ. We define

$$W(t) = \frac{\partial\alpha}{\partial u}(0,t) \qquad \text{the "variation vector field"}$$

$$V(t) = \frac{d\gamma}{dt} \qquad \text{the "velocity vector field of } \gamma\text{"}$$

$$A(t) = \frac{D}{dt}V(t) \qquad \text{the "acceleration vector field of } \gamma\text{"} ,$$

and if $a = t_0 < \cdots < t_N = b$ includes all discontinuity points of V, we set

$$\Delta_{t_i} V = V(t_i+) - V(t_i-) \qquad i = 1,\ldots,N-1$$

$$\Delta_{t_0} V = V(t_0+)$$

$$\Delta_{t_N} V = -V(t_N-) .$$

We have the following formula (Theorem II.6-14) for the "first variation" of E:

$$\left.\frac{dE(\bar{\alpha}(u))}{du}\right|_{u=0} = -\int_a^b \langle W(t), A(t)\rangle\, dt - \sum_{i=0}^{N} \langle W(t_i), \Delta_{t_i} V\rangle ;$$

for variations keeping the endpoints fixed, the sum can be written from 1 to $N-1$. From this formula we found that γ is a geodesic $(A(t) = 0)$ if and only if γ is a critical point for E.

Recall that if $f\colon M \to \mathbb{R}$ is a real-valued function, then $f_*\colon M_p \to \mathbb{R}_{f(p)}$ may be determined as follows. Given $X_p \in M_p$, we choose a path $c\colon (-\varepsilon,\varepsilon) \to M$ with $c'(0) = X_p$; then

$$f_*(X_p) = \text{tangent vector of } f\circ c \text{ at } 0 = \left.\frac{df(c(u))}{du}\right|_{u=0} \cdot \left.\frac{d}{dt}\right|_{f(p)} .$$

This suggests some notation which is exactly analogous, except that we will be sloppy and throw away the uninteresting d/dt term. For any piecewise C^∞ vector field W along γ, we define

$$E_*(W) = \left.\frac{dE(\bar{\alpha}(u))}{du}\right|_{u=0},$$

where α is some piecewise C^∞ variation of γ with W as its variation vector field. The first variation formula shows that the right side depends only on W, so that $E_*(W)$ is really well-defined; the formula also shows that E_* is linear. Perhaps we should explicitly make the observation that any piecewise C^∞ vector field W is the variation vector field of some α; for example, we can take

$$\alpha(u,t) = \exp u\cdot W(t) .$$

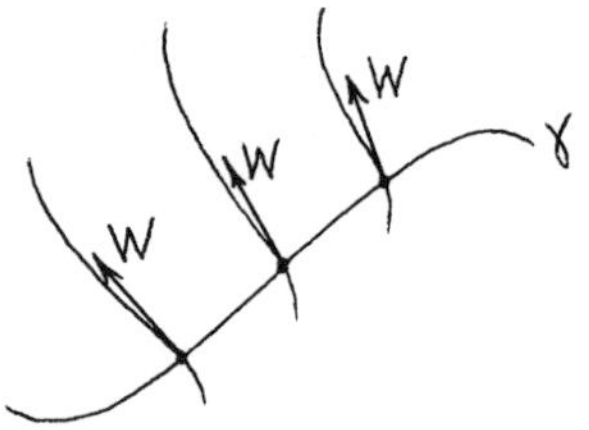

As this example shows, we can even arrange for α to be a variation keeping endpoints fixed if $W(a) = W(b) = 0$. The notation $E_*(W)$ suggests that piecewise C^∞ vector fields W along γ may be thought of as "tangent vectors" to the curve γ. Actually, it will be convenient to restrict this terminology to those W which vanish at a and b. So if Ω denotes the set of all piecewise C^∞ paths $\gamma\colon [a,b] \to M$ between two fixed points p and q, we will define Ω_γ, the "tangent space of Ω at γ", to be the vector space

$$\Omega_\gamma = \{W\colon W \text{ is a piecewise } C^\infty \text{ vector field along } \gamma \text{ with } W(a) = W(b) = 0\} .$$

We know that if $E: \Omega \to \mathbb{R}$ has a minimum, or even a local minimum, at γ, then γ must be a geodesic, so $E_*: \Omega_\gamma \to \mathbb{R}$ must be 0. This is a necessary condition, analogous to the necessary condition $D_i f(x) = 0$ for a function $f: \mathbb{R}^n \to \mathbb{R}$ to have a local maximum or minimum at $x \in \mathbb{R}^n$. We also want to find sufficient conditions for a geodesic γ to be a minimum for E; as a guide, we will first recall what is known in the case of functions $f: \mathbb{R}^n \to \mathbb{R}$.

In the one variable case, there are very easy sufficient conditions for a function $f: \mathbb{R} \to \mathbb{R}$ to have a local maximum or minimum:

(1) If $f'(x) = 0$ and $f''(x) > 0$, then f has a (strict) local minimum at x.

(2) If $f'(x) = 0$ and $f''(x) < 0$, then f has a (strict) local maximum at x.

To prove (1), for example, we simply note that if $f'(x) = 0$ and $f''(x) > 0$, then we must have $f'(x+h) > 0$ for small $h > 0$, and $f'(x+h) < 0$ for small $h < 0$. So f is strictly decreasing in some interval $(x-\varepsilon, x]$, and strictly increasing on some inerval $[x, x+\varepsilon)$. We also obtain, automatically, the following partial converses:

(1') If f has a local minimum at x, and $f''(x)$ exists, then $f''(x) \geq 0$.

(2') If f has a local maximum at x, and $f''(x)$ exists, then $f''(x) \leq 0$.

[Proof of (1'): If we had $f''(x) < 0$, then f would have a strict local maximum at x, by (2), contradicting the hypothesis that it has a local minimum at x.]

As soon as we consider functions $f: \mathbb{R}^2 \to \mathbb{R}$, the situation becomes more complicated. We certainly cannot expect to conclude that a critical point x of f is a local minimum simply because

$$D_{1,1}f(x) > 0 \quad \text{and} \quad D_{2,2}f(x) > 0 \ ;$$

this condition merely implies that x is a local minimum for f along the lines through x which are parallel to one of the axes. Even the condition that every second order directional derivative is positive,

[see p. III.448, §4]

$$\left.\frac{d^2}{dt^2}\right|_{t=0} f(x+tv) > 0 \qquad \text{all} \ v \in \mathbb{R}^2 ,$$

does not guarantee that f will have a local minimum at x. For example, consider the region $A \subset \mathbb{R}^2$ shown below. We can find a C^∞ function

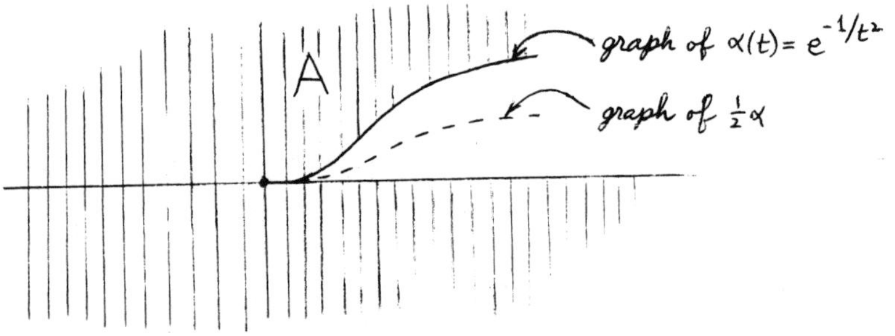

$f\colon \mathbb{R}^2 \to \mathbb{R}$ such that $f(x,y) = x^2 + y^2$ on A, but such that f is negative on the dashed curve. Every straight line through 0 contains some open segment I with $0 \in I \subset A$, so all second order directional derivatives will be positive; nevertheless, f does not have a local minimum at 0. (Of course, this function f is rather weird, but it is not a priori clear that the energy function E is not equally "weird" on the infinite dimensional space Ω of piecewise C^∞ paths.)

If we use mixed partial derivatives, then we do have a sufficient condition that a critical point x be either a (strict) local maximum or a (strict) local minimum, namely

$$\text{(I)} \qquad \det\left(\frac{\partial^2 f}{\partial x_i \partial x_j}(x)\right) > 0 .$$

We have essentially already proved this in Chapter 2, for this inequality is exactly the condition that x be an elliptic point of the surface $\{(x_1,x_2,f(x_1,x_2))\}$, and therefore lie on one side of its tangent plane at x; this tangent plane is just the (x_1,x_2)-plane, since x is a critical point. If condition (I) is satisfied, we can then distinguish between a local maximum and a local minimum merely by examining the sign of $\partial^2 f/\partial(x_1)^2$ at x. If, instead of condition (I), we have

$$\text{(II)} \qquad \det\left(\frac{\partial^2 f}{\partial x_i \partial x_j}(x)\right) < 0 ,$$

then x definitely is not either a local maximum or minimum for f. This also follows from the considerations of Chapter 2, for in this case the surface $\{(x_1,x_2,f(x_1,x_2))\}$ lies on both sides of its tangent plane. When the determinant is 0, we are in the borderline case where no conclusions can be drawn. Essentially the same considerations hold for functions $f\colon \mathbb{R}^n \to \mathbb{R}$, except that it is no longer so easy to find out if the eigenvalues of

$$\left(\frac{\partial^2 f}{\partial x_i \partial x_j}(x)\right)$$

all have the same sign, which is precisely the condition that f have either a local maximum or a local minimum at x.

Notice that the analogues of (1') and (2') require no modification: If $f\colon \mathbb{R}^n \to \mathbb{R}$ has a local minimum at x, then surely

$$\left.\frac{d^2}{dt^2}\right|_{t=0} f(x+tv) \geq 0$$

for all $v \in \mathbb{R}^n$ for which this limit exists. In fact, if the opposite equality held for some $v \in \mathbb{R}^n$, then f would have a strict local maximum at x along the line $\{x+tv\colon t \in \mathbb{R}\}$.

Our aim now is to see what information we can get when we generalize these considerations of elementary calculus, and examine the second derivative $d^2E(\bar{\alpha}(u))/du^2(0)$, for all variations α of a geodesic $\gamma\colon [a,b] \to M$; classically, this second derivative was called the "second variation" of E. Our remarks about n-dimensional calculus might suggest that it would be even more useful to consider "mixed partial derivatives," and even if they don't suggest it, mixed partial derivatives do turn out to be the thing to look at. We first define a <u>2-parameter variation</u> α of γ to be a function

$$\alpha\colon U \times [a,b] \to M ,$$

for some neighborhood U of $0 \in \mathbb{R}^2$, such that

1) $\alpha(0,t) = \gamma(t)$

2) there is a partition $a = t_0 < \cdots < t_N = b$ of $[a,b]$ so that α is C^∞ on each $U \times [t_{i-1},t_i]$.

We say that α is a variation keeping endpoints fixed if

3) $\alpha(u,a) = \gamma(a)$
$\alpha(u,b) = \gamma(b)$ for all $u \in U$.

As before, we let $\bar{\alpha}(u)$ be the path $t \mapsto \alpha(u,t)$. A 2-parameter variation α of γ gives rise to two "variation vector fields" W_1 and W_2 along γ, defined by

$$W_i(t) = \frac{\partial\alpha}{\partial u_i}(0,0,t) .$$

Notice that W_i may be only piecewise C^∞ vector fields along γ even if γ itself is everywhere C^∞.

1. THEOREM (SECOND VARIATION FORMULA). Let $\gamma: [a,b] \to M$ be a geodesic, with velocity vector field $V(t) = d\gamma/dt$, and let $\alpha: U \times [a,b] \to M$ be a 2-parameter variation of γ, with variation vector fields

$$W_i(t) = \frac{\partial\alpha}{\partial u_i}(0,0,t) .$$

Choose $a = t_0 < \cdots < t_N = b$ to include all discontinuity points of DW_1/dt, and let

$$\Delta_{t_i}\frac{DW_1}{dt} = \frac{DW_1}{dt}(t_i+) - \frac{DW_1}{dt}(t_i-) \qquad i = 1,\dots,N-1$$

$$\Delta_{t_0}\frac{DW_1}{dt} = \frac{DW_1}{dt}(t_0+)$$

$$\Delta_{t_N}\frac{DW_1}{dt} = -\frac{DW_1}{dt}(t_N-) .$$

Then

$$\left.\frac{\partial^2 E(\bar{\alpha}(u))}{\partial u_1 \partial u_2}\right|_{(u_1,u_2)=(0,0)} = -\int_a^b \left\langle W_2(t), \frac{D^2 W_1}{dt^2} + R(W_1(t), V(t))V(t)\right\rangle dt - \sum_{i=0}^{N} \left\langle W_2(t_i), \Delta_{t_i} \frac{DW_1}{dt}\right\rangle .$$

(When α is a variation keeping endpoints fixed, the sum can be written from 1 to $N-1$.)

<u>Proof</u>. By the first variation formula (Theorem II.6-14), we have

$$\left.\frac{\partial E(\bar{\alpha}(u))}{\partial u_2}\right|_{u_2=0} = -\int_a^b \left\langle \frac{\partial \alpha}{\partial u_2}, \frac{D}{\partial t}\frac{\partial \alpha}{\partial t}\right\rangle dt - \sum_{i=0}^{N} \left\langle \frac{\partial \alpha}{\partial u_2}, \Delta_{t_i} \frac{\partial \alpha}{\partial t}\right\rangle ,$$

where all terms on the right side are to be evaluated at $(t, u_1, 0)$. So

$$(1) \quad \left.\frac{\partial^2 E(\bar{\alpha}(u))}{\partial u_1 \partial u_2}\right|_{u_2=0} = -\int_a^b \left\langle \frac{D}{\partial u_1}\frac{\partial \alpha}{\partial u_2}, \frac{D}{\partial t}\frac{\partial \alpha}{\partial t}\right\rangle dt - \int_a^b \left\langle \frac{\partial \alpha}{\partial u_2}, \frac{D}{\partial u_1}\frac{D}{\partial t}\frac{\partial \alpha}{\partial t}\right\rangle dt - \sum_{i=0}^{N} \left\langle \frac{D}{\partial u_1}\frac{\partial \alpha}{\partial u_2}, \Delta_{t_i}\frac{\partial \alpha}{\partial t}\right\rangle - \sum_{i=0}^{N} \left\langle \frac{\partial \alpha}{\partial u_2}, \frac{D}{\partial u_1}\Delta_{t_i}\frac{\partial \alpha}{\partial t}\right\rangle .$$

Now when $u_1 = 0$ we have

$$\frac{D}{\partial t}\frac{\partial \alpha}{\partial t}(t,0,0) = 0 \quad \text{and} \quad \Delta_{t_i}\frac{\partial \alpha}{\partial t}(t,0,0) = 0 ,$$

since $t \mapsto \alpha(t,0,0) = \gamma(t)$ is a geodesic. So the first and third terms of equation (1) are zero for $u_1 = 0$. After a simple manipulation with the fourth term we then have

$$(2)\qquad \left.\frac{\partial^2 E(\bar{\alpha}(u))}{\partial u_1 \partial u_2}\right|_{(u_1,u_2)=(0,0)} = -\int_a^b \left\langle W_2(t),\ \frac{D}{\partial u_1}\frac{D}{\partial t}V\right\rangle dt - \sum_{i=0}^{N}\left\langle W_2(t_i),\ \Delta_{t_i}\frac{DW_1}{dt}\right\rangle,$$

where all terms on the right are now evaluated at $(t,0,0)$. Now we can use Proposition II.6-10 to write

$$\frac{D}{\partial u_1}\frac{D}{\partial t}V - \frac{D}{\partial t}\frac{D}{\partial u_1}V = R\left(\frac{\partial\alpha}{\partial u_1}, \frac{\partial\alpha}{\partial t}\right)V = R(W_1,V)V .$$

Moreover, Proposition II.6-9 gives us

$$\frac{D}{\partial u_1}V = \frac{D}{\partial u_1}\frac{\partial\alpha}{\partial t} = \frac{D}{\partial t}\frac{\partial\alpha}{\partial u_1} = \frac{D}{dt}W_1 ,$$

so we have

$$\frac{D}{\partial u_1}\frac{D}{\partial t}V = \frac{D^2 W_1}{dt^2} + R(W_1,V)V .$$

Substituting into (2), we obtain the desired result. ■

Suppose we are given two piecewise C^∞ vector fields W_1 and W_2 along a geodesic $\gamma\colon [a,b] \longrightarrow M$. We can always find at least one variation α with these as variation vector fields, namely

$$\alpha(u_1,u_2,t) = \exp[u_1W_1(t) + u_2W_2(t)] .$$

Extending the notation introduced previously, we define

$$E_{**}(W_1,W_2) = \left.\frac{\partial^2 E(\bar{\alpha}(u))}{\partial u_1\ \partial u_2}\right|_{(u_1,u_2)=(0,0)} ,$$

for any variation α with variation vector fields W_1 and W_2; the second variation formula shows that $E_{**}(W_1,W_2)$ does not depend on the choice of α. The notation $E_{**}(W_1,W_2)$ is used only when W_1 and W_2 are vector fields along a geodesic; otherwise the second derivative will depend on the choice of α (compare p. I.223 and Problem I.5-17). It is clear from the second variation formula that E_{**} is bilinear. It is also true that E_{**} is symmetric, $E_{**}(W_1,W_2) = E_{**}(W_2,W_1)$; this is not at all clear from the second variation formula, but it follows immediately from the fact that $E(\bar{\alpha}(u))$ is a C^∞ function of u, and consequently

$$\frac{\partial^2 E(\bar{\alpha}(u))}{\partial u_1 \partial u_2} = \frac{\partial^2 E(\bar{\alpha}(u))}{\partial u_2 \partial u_1} .$$

The second variation formula reveals a hitherto unsuspected significance of curvature, and turns out to be responsible for many of the deeper consequences which we will be able to draw from assumptions about the curvature of M. We begin the program which will uncover these results by formulating questions about local minima for E in terms of E_{**}. Notice that if $\alpha\colon (-\varepsilon,\varepsilon)\times[a,b] \longrightarrow M$ is a one-parameter variation of γ, and we define a 2-parameter variation β by

$$\beta(u_1,u_2,t) = \alpha(u_1+u_2,\ t) ,$$

then

$$\left.\frac{\partial^2 E(\bar{\alpha}(u))}{\partial u^2}\right|_{u=0} = \left.\frac{\partial^2 E(\bar{\beta}(u))}{\partial u_1 \partial u_2}\right|_{(u_1,u_2)=(0,0)} .$$

If γ has variation vector field W, then β clearly has variation vector fields $W_1 = W_2 = W$. Consequently,

$$\left.\frac{\partial^2 E(\bar{\alpha}(u))}{\partial u^2}\right|_{u=0} = E_{**}(W,W) .$$

Thus, if γ is going to be a local minimum for energy, then we must have $E_{**}(W,W) \geq 0$ for all $W \in \Omega_\gamma$. Briefly expressed:

If γ is a local minimum, then E_{**} is positive semi-definite.

We also <u>hope</u> that γ actually will be a local minimum whenever we have the strict inequality $E_{**}(W,W) > 0$ for all non-zero $W \in \Omega_\gamma$. Briefly expressed:

If E_{**} is positive definite, then we <u>hope</u> that γ is a local minimum.

Our approach to this problem will be somewhat roundabout; we first investigate the vector fields $W \in \Omega_\gamma$ which satisfy $E_*(W,W_2) = 0$ for all $W_2 \in \Omega_\gamma$, and hence represent something of a borderline between positive definiteness and positive semi-definiteness.

A piecewise C^∞ vector field W along γ is called a <u>Jacobi field</u> if it satisfies the "Jacobi equation"

$$\frac{D^2 W}{dt^2} + R(W,V)V = 0 , \qquad V = d\gamma/dt .$$

In a coordinate system this equation becomes a <u>linear</u> second order differential equation. Or, if we choose parallel vector fields $Y_1,\ldots,Y_n$ along γ which are orthonormal at 0, and hence orthonormal everywhere along γ, and set $W(t) = \Sigma f^i(t)Y_i(t)$, then our equation becomes

$$0 = \frac{d^2f^i}{dt^2} + \sum_{j=1}^{n} a^i_j(t)f^j(t) \qquad i = 1,\ldots,n ,$$

where $a^i_j = \langle R(Y_j,V)V,\ Y_i\rangle$. The solutions of this equation are everywhere C^∞ and, since the equation is linear, every solution can be defined on all of γ. It is also clear from the linearity of the equation that the set of all Jacobi fields W along γ forms a vector space. The dimension of this vector space is 2n, since each Jacobi field W is determined by its initial conditions

$$W(0),\ \frac{DW}{dt}(0) \in M_{\gamma(0)} .$$

2. PROPOSITION. Let $\gamma\colon [a,b] \to M$ be a geodesic and let $W \in \Omega_\gamma$. Then W is a Jacobi field if and only if

$$E_{**}(W,W_2) = 0$$

for all $W_2 \in \Omega_\gamma$.

Proof. If $W \in \Omega_\gamma$ is a Jacobi field, then the second variation formula shows immediately that

$$E_{**}(W,W_2) = -\int_a^b \langle W_2, 0\rangle dt - \sum_{i=1}^{N-1} \langle W_2(t_i),\ 0\rangle = 0 .$$

Conversely, suppose that $W \varepsilon \Omega_\gamma$ and that $E_{**}(W,W_2) = 0$ for all $W_2 \varepsilon \Omega_\gamma$. Choose $a = t_0 < \cdots < t_N = b$ so that each $W|[t_{i-1},t_i]$ is smooth, and let $f: [a,b] \longrightarrow [0,1]$ be a C^∞ function with $f(t_i) = 0$ and $f > 0$ otherwise. If we define

$$W_2 = f \cdot \left(\frac{D^2W}{dt^2} + R(W,V)V\right) ,$$

then

$$0 = E_{**}(W,W_2) = - \int_a^b f \cdot \left\| \frac{D^2W}{dt^2} + R(W,V)V \right\|^2 dt \quad - \quad \sum_{i=0}^{N} \langle 0, \Delta_{t_i} \frac{DW}{dt} \rangle .$$

This implies that

$$(1) \qquad \frac{D^2W}{dt^2} + R(W,V)V = 0 \qquad \text{on each} \quad (t_{i-1},t_i) ,$$

so each $W|[t_{i-1},t_i]$ is a Jacobi field.

Next choose W_2 to be any vector field along γ with $W_2(a) = W_2(b) = 0$ and $W_2(t_i) = \Delta_{t_i} DW/dt$ for $i = 1,\ldots,N-1$. Then by (1) we have

$$0 = E_{**}(W,W_2) = - \int_a^b \langle W, 0 \rangle \, dt \quad - \quad \sum_{i=1}^{N} \left\| \Delta_{t_i} \frac{DW}{dt} \right\|^2 ,$$

so each $\Delta_{t_i} DW/dt = 0$. This means that the Jacobi fields $W|[t_{i-1},t_i]$ for two consecutive intervals have the same values of DW/dt on the intersection of the intervals. Since a Jacobi field is determined by its initial values, this shows that W is actually a Jacobi field on all of γ. ∎

Notice that there may not exist any non-trivial Jacobi fields W along γ which vanish at both a and b (indeed we hope to find conditions under which E_{**} is positive definite, which certainly excludes the possibility of non-zero Jacobi fields). When there *is* a non-zero Jacobi field W along γ with $W(a) = W(b) = 0$, we say that a and b are <u>conjugate values</u> along γ, and we define the <u>multiplicity</u> of a and b as conjugate values to be the dimension of the vector space consisting of all such Jacobi fields. We also say that $\gamma(a)$ and $\gamma(b)$ are <u>conjugate points</u> of γ, but this terminology is ambiguous when γ has self-intersections.

Since a Jacobi field W is determined by its initial values $W(a)$, $DW/dt(a)$ at any point a, the multiplicity of two conjugate values a and b is clearly $\leq n$. Actually, it is always $\leq n-1$. To prove this, we just have to produce a Jacobi field along γ which is 0 at a but nowhere else. The vector field $W(t) = (t-a)V(t)$ has this property; it is a Jacobi field because

$$\frac{DW}{dt} = V(t) + (t-a)\frac{DV}{dt} = V(t) \ ,$$

$$\frac{D^2W}{dt^2} + R(W,V)V = \frac{DV}{dt} + (t-a)R(V,V)V = 0 \ .$$

More generally, we have

3. PROPOSITION. Let γ be a geodesic, with velocity vector field $V = d\gamma/dt$.
(1) The vector field fV along γ is a Jacobi field if and only if f is linear.
(2) Every Jacobi field W along γ can be written uniquely as $fV + W^{\perp}$, where f is linear and $W^{\perp}$ is a Jacobi field perpendicular to γ.

(3) If a Jacobi field W along γ is perpendicular to γ at two points a and b, then W is perpendicular to γ everywhere. In particular, if $W(a) = W(b) = 0$, then W is perpendicular to γ everywhere.

Proof. (1) If $W = fV$, then $D^2W/dt^2 = f''V$, so the Jacobi equation for W is

$$0 = \frac{D^2W}{dt^2} + R(W,V)V = f''V + fR(V,V)V = f''V .$$

(2) Given a Jacobi field W along γ, we can write $W = fV + W^{\perp}$ for some f and some vector field $W^{\perp}$ perpendicular to γ. The Jacobi equation for W gives

$$\text{(a)} \qquad 0 = \frac{D^2W}{dt^2} + R(W,V)V = f''V + \frac{D^2W^{\perp}}{dt^2} + R(W^{\perp},V)V .$$

Now

$$0 = \langle W^{\perp},V\rangle \implies 0 = \left\langle \frac{DW^{\perp}}{dt} , V\right\rangle \implies 0 = \left\langle \frac{D^2W^{\perp}}{dt^2}, V\right\rangle$$

and we also have

$$0 = \langle R(W^{\perp},V)V, V\rangle .$$

So (a) implies that $f'' = 0$, and therefore that $W^{\perp}$ is a Jacobi field. Uniqueness is obvious.

(3) Write $W = fV + W^{\perp}$ as in (2). Then the linear function f must satisfy $f(a) = f(b) = 0$. So $f = 0$. ∎

Proposition 3 shows that for the purposes of investigating conjugate values, we need consider only perpendicular Jacobi fields. In particular, when M is a surface, and Y is a unit normal vector field along the geodesic $\gamma: [a,b] \to M$, any normal vector field W can be written uniquely as $W = gY$. We have $DY/dt = 0$, since γ is a geodesic and Y makes a constant angle with the parallel vector field $d\gamma/dt$. So the Jacobi equation for W becomes

$$g''(t)Y(t) + g(t)R(Y(t), V(t))V(t) = 0 ,$$

which is equivalent to

$$g''(t) + g(t)\langle R(Y(t),V(t))V(t), Y(t)\rangle = 0 ,$$

since we obtain $0 = 0$ when we take the inner product of the original equation with V. When the tangent vector $V = d\gamma/dt$ has length 1, we can write our equation as

$$g''(t) + K(\gamma(t))\cdot g(t) = 0 ,$$

where K is the Gaussian curvature; this is the classical "Jacobi equation" for M.

The next theorem, basically due to Jacobi, gives a geometric way of obtaining Jacobi fields.

4. PROPOSITION. Let $\gamma: [a,b] \to M$ be a geodesic and let $\alpha: (-\varepsilon,\varepsilon) \times [a,b] \to M$ be a variation of γ through geodesics, so that each $\bar{\alpha}(u): [a,b] \to M$ is also a geodesic. Then the variation vector field $W(t) = \partial\alpha/\partial u(0,t)$ is a Jacobi field along γ.

Proof. Since α is a variation of γ through geodesics, we have

$$\frac{D}{\partial t}\frac{\partial\alpha}{\partial t} = 0 \ .$$

Therefore

$$0 = \frac{D}{\partial u}\frac{D}{\partial t}\frac{\partial\alpha}{\partial t} = \frac{D}{\partial t}\frac{D}{\partial u}\frac{\partial\alpha}{\partial t} + R\left(\frac{\partial\alpha}{\partial u}, \frac{\partial\alpha}{\partial t}\right)\frac{\partial\alpha}{\partial t} \quad \text{by Proposition II.6-10}$$

$$= \frac{D^2}{\partial t^2}\frac{\partial\alpha}{\partial u} + R\left(\frac{\partial\alpha}{\partial u}, \frac{\partial\alpha}{\partial t}\right)\frac{\partial\alpha}{\partial t} \quad \text{by Proposition II.6-9,}$$

which shows that $\partial\alpha/\partial u$ is a Jacobi field. ■

Thus one way of obtaining Jacobi fields is to move geodesics around. In

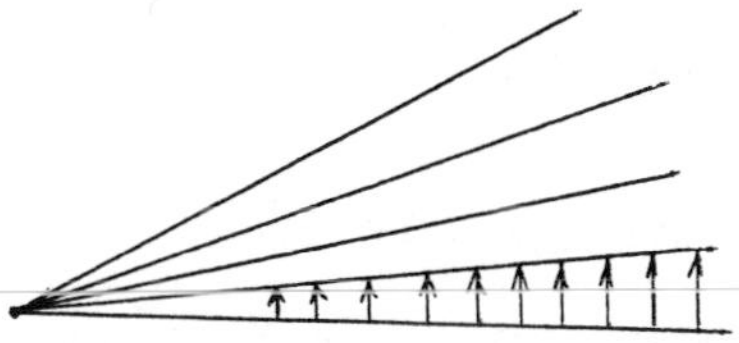

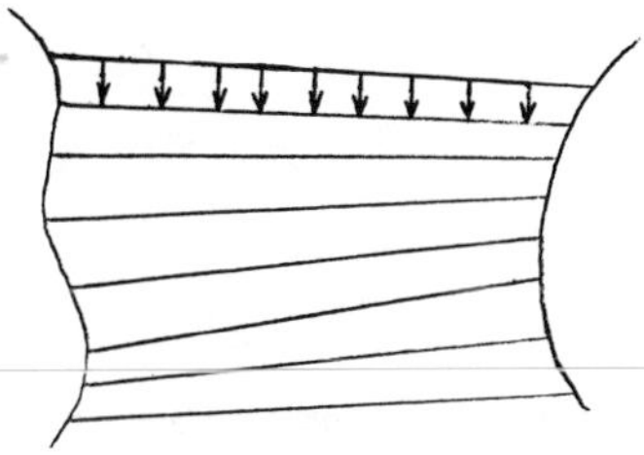

particular, if γ is a semi great circle on S^n, joining two antipodal points p and q, then a rotation of S^n keeping p and q fixed yields a variation vector field along γ which is a Jacobi field vanishing at p and q. Since

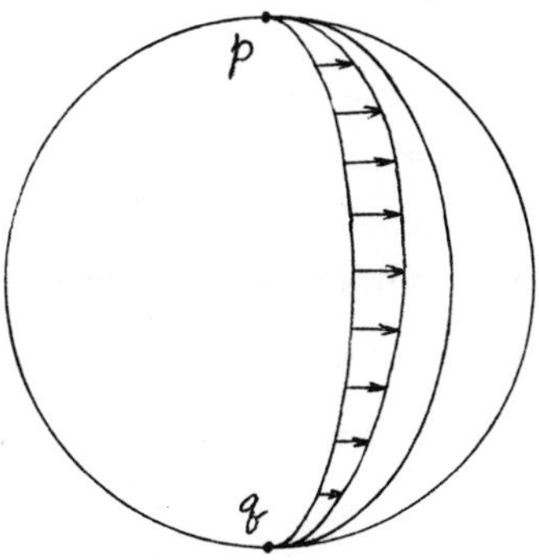

we can rotate in $n-1$ different directions, the points p and q have multiplicity $n-1$, the theoretical maximum.

5. PROPOSITION. Every Jacobi field along a geodesic $\gamma\colon [a,b] \to M$ is the variation vector field of a variation of γ through geodesics.

Proof. First suppose that γ lies completely inside an open set $U \subset M$ such that any two points $p, q \in U$ are joined by a unique geodesic in U, depending smoothly on p and q, of length $d(p,q)$. Given two vectors $W(a) \in M_{\gamma(a)}$ and $W(b) \in M_{\gamma(b)}$, choose curves $c_a, c_b\colon (-\varepsilon,\varepsilon) \to U$ such that

$$c_a(0) = \gamma(a)\,, \qquad c_b(0) = \gamma(b)$$
$$c_a{}'(0) = W_a\,, \qquad c_b{}'(0) = W_b\,.$$

Define $\alpha\colon (-\varepsilon,\varepsilon) \times [a,b] \to M$ by letting $\bar{\alpha}(u)\colon [a,b] \to M$ be the unique geodesic in U from $c_a(u)$ to $c_b(u)$ of length $d(c_a(u), c_b(u))$. Then $W(t) = \partial\alpha/\partial u\,(0,t)$ is a Jacobi field along γ, by Proposition 4. To show that all Jacobi fields arise in this way, simply consider the map

$$\Phi\colon \{\text{Jacobi fields along } \gamma\} \to M_{\gamma(a)} \times M_{\gamma(b)}$$

given by

$$W \mapsto (W(a), W(b))\,.$$

We have just shown that Φ is onto. Since the domain and range of Φ both have dimension $2n$, the linear map Φ must also be one-one. Thus W is determined by $W(a), W(b)$; this shows that when the above construction is applied to $W(a)$ and $W(b)$, the resulting Jacobi field $\partial\alpha/\partial u\ (0,t)$, obtained by a variation through geodesics, is precisely the given Jacobi field W.

For a general geodesic γ, we note that for sufficiently small δ, the restricted geodesic $\gamma|[a, a+\delta]$ will lie in an appropriate set U, by Theorem I.9-14. This gives us a variation through geodesics $\alpha\colon (-\varepsilon,\varepsilon)\times[a, a+\delta]$ with $\partial\alpha/\partial u\ (0,t)$ equal to the given Jacobi field $W(t)$ for $t \in [a, a+\delta]$. Using compactness of $[a,b]$, it is easy to see that if ε is made sufficiently small, then each geodesic $\bar{\alpha}(u)$ can be extended to a geodesic $\bar{\bar{\alpha}}(u)\colon [a,b] \longrightarrow M$. Then $(u,t) \longmapsto \bar{\bar{\alpha}}(u)(t)$ is the required variation through geodesics. ■

An examination of the proof of Proposition 5 shows that if W is a Jacobi field along a geodesic $\gamma\colon [a,b] \to M$ with $W(a) = 0$, then we can even find a variation $\alpha\colon (-\varepsilon,\varepsilon)\times[a,b] \to M$ of γ through geodesics such that

$$\frac{\partial\alpha}{\partial u}(0,t) = W(t)$$

$$\alpha(u,a) = \gamma(a) \qquad \text{for all } u \in (-\varepsilon,\varepsilon)\ .$$

However, if $W(b) = 0$ for some other point b, we may not be able to choose α so that we also have $\alpha(u,b) = \gamma(b)$ for all u; we will merely have this condition "up to first order", that is,

$$\frac{\partial\alpha}{\partial u}(0,b) = 0\ .$$

Thus a conjugate point of $p = \gamma(a)$ is a place where some 1-parameter family of geodesics starting from p "nearly" intersect. This description of conjugate points shows why they play such an important role in the study of local minima for length, for it is easy to give an intuitive argument to prove that a geodesic

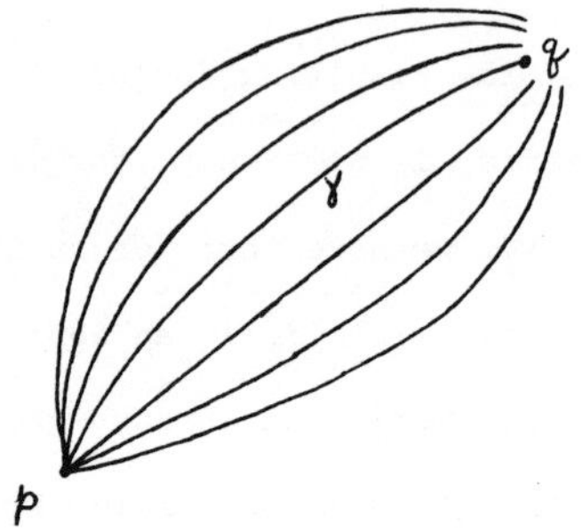

$\gamma: [a,b] \to M$ cannot locally minimize length if there is some $\tau \in (a,b)$ conjugate to a. In fact, suppose we have another geodesic η from $\gamma(a)$ to $\gamma(\tau)$ with nearly the same length as $\gamma|[a,\tau]$. Then γ has nearly the same length as η followed by $\gamma|[\tau,b]$. But this compound curve has a corner,

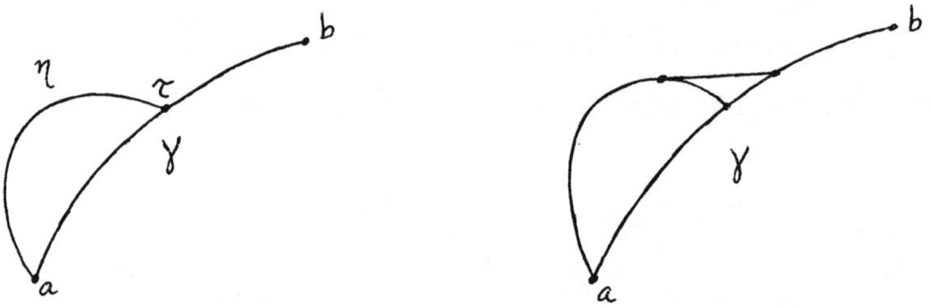

and can clearly be made shorter by replacing the corner with a minimal geodesic. Therefore, γ is not a curve of minimal length. This reasoning turns out to be perfectly valid, provided that one works infinitesimally:

6. THEOREM. Let $\gamma: [a,b] \to M$ be a geodesic, and suppose that there is a number $\tau \in (a,b)$ which is conjugate to a along γ. Then there is some $W \in \Omega_\gamma$ with $E_{**}(W,W) < 0$. Consequently, γ is not a local minimum for E.

Proof. Since τ is conjugate to a along γ, there is a non-zero Jacobi field J along γ such that $J(a) = J(\tau) = 0$. Let $\tilde{J}$ be the vector field along γ with

$$\tilde{J}(t) = J(t) \qquad a \leq t \leq \tau$$
$$\tilde{J}(t) = 0 \qquad \tau \leq t \leq b \ .$$

Notice that the discontinuity of $D\tilde{J}/dt$ at τ is

$$\Delta_\tau \frac{D\tilde{J}}{dt} = \frac{DJ}{dt}(\tau) \neq 0 \ ,$$

the inequality following from the fact that $J(\tau) = 0$, but J is non-zero. Choose a vector field X along γ which vanishes at a and b and which satisfies

(1) $$\langle X(\tau),\ \Delta_\tau D\tilde{J}/dt\rangle = 1 \ .$$

Now let c be a small number and consider the vector field

$$W = \frac{1}{c}\tilde{J} - cX \ .$$

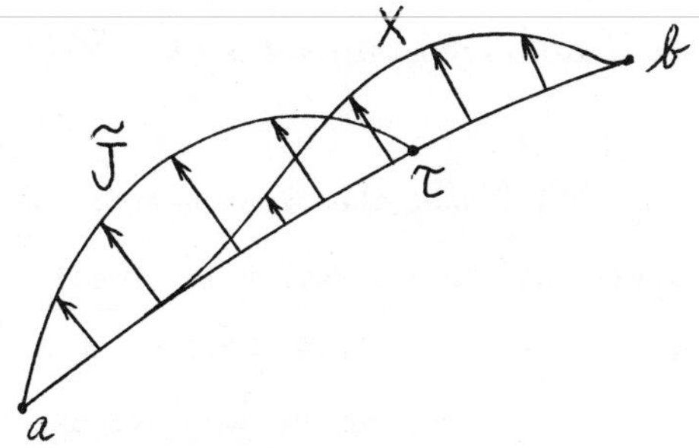

We have

$$E_{**}(W,W) = \frac{1}{c^2}E_{**}(\tilde{J},\tilde{J}) - 2E_{**}(\tilde{J},X) + c^2E_{**}(X,X) \ .$$

Using the second variation formula, this becomes

$$E_{**}(W,W) = 0 - 2\langle X(\tau),\ \Delta_\tau D\tilde{J}/dt\rangle + c^2E_{**}(X,X)$$
$$= -\ 2 + c^2E_{**}(X,X) \qquad \text{by (1)} \ .$$

For sufficiently small c this is negative, which proves the first part of the theorem.

We have really already observed that the first part of the theorem implies the second, but we repeat the reasoning here. Suppose we have $W \in \Omega_\gamma$ with $E_{**}(W,W) < 0$. Consider the variations

$$\alpha(u,t) = \exp uW(t)$$
$$\beta(u_1,u_2,t) = \alpha(u_1 + u_2,\ t) = \exp(u_1 + u_2)W(t) .$$

Then

$$\left.\frac{\partial^2 E(\bar{\alpha}(u))}{\partial u^2}\right|_{u=0} = \left.\frac{\partial^2 E(\bar{\beta}(u))}{\partial u_1 \partial u_2}\right|_{(u_1,u_2)=(0,0)}$$
$$= E_{**}(W,W) < 0 .$$

So $u \mapsto E(\bar{\alpha}(u))$ has a strict relative maximum at $u = 0$. Therefore γ is not a relative minimum for E. ■

Notice that the first part of this proof makes crucial use of the discontinuity of DW/dt, which is closely related to the kink in the "intuitive proof". (Once we have obtained this W, however, we can always smooth it out to obtain an everywhere C^∞ vector field W with $E_{**}(W,W) < 0$.)

Our next hope is that a geodesic <u>does</u> minimize length among nearby paths if there are no conjugate points. In order to consider this case, we first need a result which contains essentially the same information as Propositions 4 and 5, but in a form that is much easier to use; for simplicity we state it for a geodesic defined on $[0,1]$.

7. THEOREM. Let $\gamma\colon [0,1] \to M$ be a geodesic with $\gamma(0) = p \in M$ and $\gamma'(0) = v \in M_p$, so that γ can be described as $t \mapsto \exp tv$ for the map

$$\exp = \exp_p\colon M_p \to M .$$

Then 0 and 1 are conjugate values for γ if and only if v is a critical point of exp.

Proof. Suppose that v is a critical point for exp. Then $\exp_*(X) = 0$ for some non-zero $X \in (M_p)_v$ = the tangent space of M_p at v. Let c be a path in M_p with $c(0) = v$ and $c'(0) = X$, and define

$$\alpha(u,t) = \exp tc(u) \qquad 0 \le t \le 1 .$$

Then α is a variation of γ through geodesics, so the vector field

$$W(t) = \left.\frac{\partial}{\partial u}\right|_{u=0} \exp tc(u)$$

is a Jacobi field along γ. We clearly have $W(0) = 0$, and also

$$\begin{aligned} W(1) &= \left.\frac{\partial}{\partial u}\right|_{u=0} \exp c(u) = \exp_* c'(0) \\ &= \exp_* X = 0 . \end{aligned}$$

Moreover,

$$\begin{aligned} \frac{DW}{dt}(0) &= \left.\frac{D}{\partial t}\right|_{t=0} \left.\frac{\partial}{\partial u}\right|_{u=0} \exp tc(u) \\ &= \left.\frac{D}{\partial u}\right|_{u=0} \left.\frac{\partial}{\partial t}\right|_{t=0} \exp tc(u) \qquad \text{by Proposition II.6-9} \\ &= \left.\frac{D}{\partial u}\right|_{u=0} c(u) ; \end{aligned}$$

this last expression is the covariant derivative of the vector field $u \mapsto c(u)$ along the constant curve $u \mapsto p$. Hence

$$\frac{DW}{dt}(0) = c'(0) = X \neq 0 .$$

In particular, W is not identically 0, which shows that 0 and 1 are conjugate values for γ.

Now suppose v is not a critical point for $\exp$. If $X_1, \ldots, X_n \in (M_p)_v$ are n linearly independent vectors, then $\exp_*(X_1), \ldots, \exp_*(X_n) \in M_{\gamma(1)}$ are also linearly independent. Choose paths $c_1, \ldots, c_n$ in M_p with $c_i(0) = v$ and $c_i'(0) = X_i$, and consider the variations

$$\alpha_i(u,t) = \exp tc_i(u) ,$$

with variation vector fields W_i. Then the W_i are Jacobi fields along γ which vanish at 0. Moreover, the $W_i(1) = \exp_*(X_i)$ are independent, so no non-trivial linear combination of the W_i can vanish at 1. Since the vector space of Jacobi fields along γ which vanish at 0 has dimension exactly n, it follows that no non-zero Jacobi field along γ vanishes at 0 and also at 1. ■

Since the points in M_p where $\exp_*$ is zero form a closed set, Theorem 7 shows that the numbers τ conjugate to 0 along a geodesic $\gamma\colon [0,\infty) \to M$ also form a closed set. In particular, if there is any such τ, then there is a <u>first</u> τ conjugate to 0. Actually, much more is true, for the set of τ conjugate to 0 consists only of isolated points, so there are only finitely many τ conjugate to 0 in any interval $[0,b]$. We will not prove this here, but it is included in another result which we will state later on.

It is now a simple matter to prove the local length-minimizing property of a geodesic $\gamma\colon [a,b] \to M$ satisfying the condition that no number $\tau \in (a,b]$ is a conjugate value of a along γ. For simplicity, we will call such a γ a geodesic "without conjugate points".

8. THEOREM. Let $\gamma\colon [a,b] \to M$ be a one-one geodesic with no conjugate points. Then γ has strictly smaller length than all sufficiently nearby paths between $p = \gamma(a)$ and $q = \gamma(b)$ (except for those which are merely reparameterizations of γ).

Proof. By reparameterizing, we may assume that $[a,b] = [0,1]$. If $v = \gamma'(0)$, then by Theorem 7 the map $\exp = \exp_p\colon M_p \to M$ is regular on the set $\{tv\colon 0 \le t \le 1\} \subset M_p$. By Lemma I.9-19 there is an open set $U \supset L$ on which $\exp$ is a diffeomorphism. The result then follows from Problem I.9-29. ∎

Remark. Theorem 8 clearly remains true even for geodesics γ with self-intersections, provided that "nearby" paths refer to paths c with $c(t)$ close to $\gamma(t)$ for all t.

Let us test out Theorems 6 and 8 on the 2-sphere $S^2(r)$ of radius r, with $\gamma\colon [0,L] \to S^2(r)$ a portion of a great circle starting from a point p. We take γ to be parameterized by arclength, so that $V = d\gamma/dt$ has length 1.

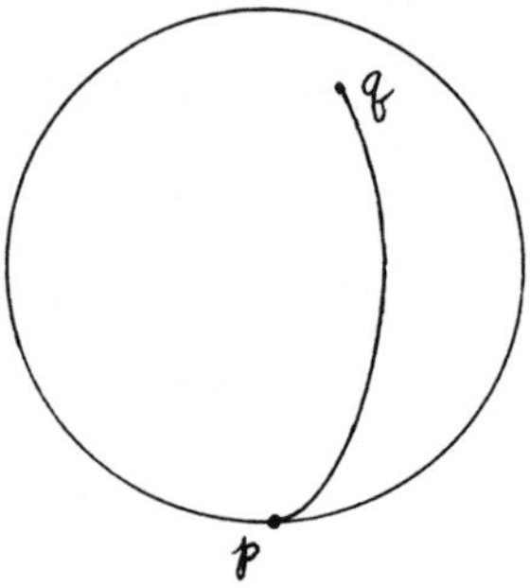

Proposition 3(3) shows that in order to investigate conjugate points along γ, it suffices to consider Jacobi fields which are perpendicular to γ. If Y is a unit normal vector field along γ, then the Jacobi equation for $W = gY$ is (compare p. 312)

$$g''(t) + \frac{1}{r^2} g(t) = 0 .$$

The solutions vanishing at $t = 0$ are all multiples of

$$g(t) = \sin\left(\frac{t}{r}\right) ,$$

which has its first positive 0 at πr. So if $L > \pi r$, then γ contains a conjugate point, and Theorem 6 shows that γ does not locally minimize length. This is easy to see from the picture; in fact, in this case the intuitive proof of Theorem 6 works exactly. If $L < \pi r$, then Theorem 8 shows that γ does locally minimize length.

We have exactly the same situation for any compact surface of revolution M, when we take p to be one of the points where M intersects the z-axis I_z.

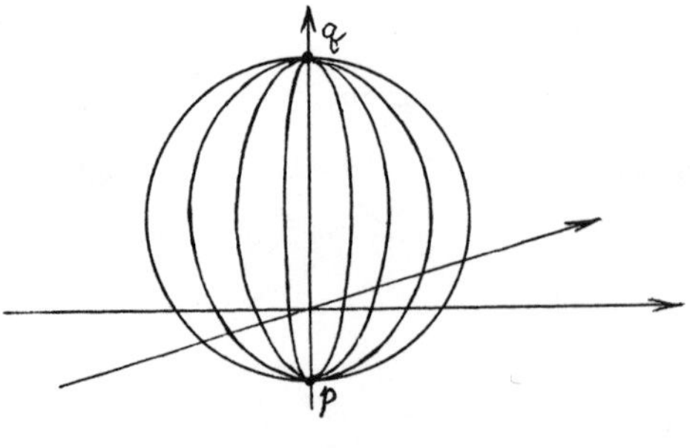

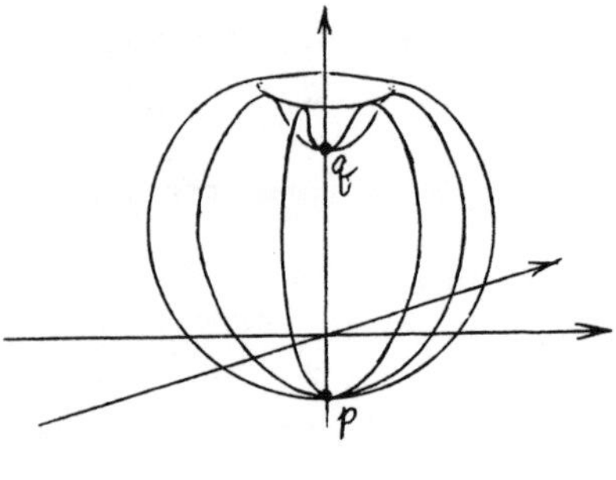

The geodesics through p are the meridians, and it is clear, just by looking at the picture, that the only point conjugate to p along any geodesic is the other point q of $M \cap I_z$. Geodesics which do not reach q are local strict minima for length, and geodesics which extend past q are not local minima.

In this example it is clear that a geodesic γ which does not reach q is actually a minimum among all paths. [Proof: A minimum path between p and the other end of γ exists, since M is complete, and this path must be a geodesic; we know what all geodesics through p are, and γ is clearly the shortest.] However, it is easy to concoct examples where the non-existence of conjugate points implies only that γ is a local minimum for E. For example, we can round off the edges of the surface shown below (the boundary of part of a spherical wedge). Since the surface is a sphere in a neighborhood of γ,

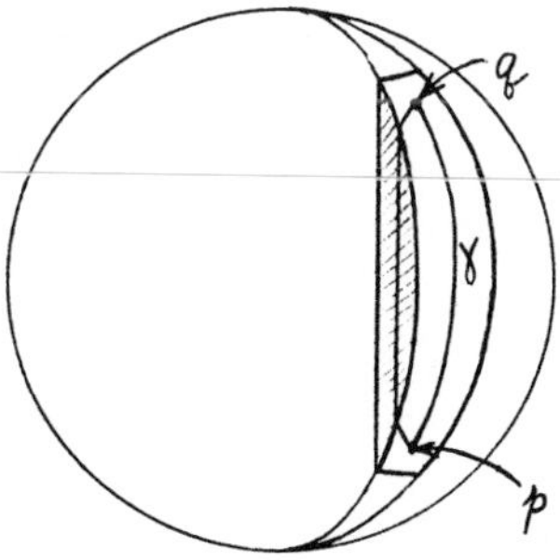

it is still the case that no two points of γ are conjugate, and Theorem 8 still applies. On the other hand, there is clearly a shorter path between p and q if the wedge is thin enough.

A little more interesting situation arises for an ellipsoid. For the geodesic γ shown below, the first point q conjugate to p along γ occurs past the point p' on the opposite end of the axis. (To establish this fact one has to examine the Jacobi equations for γ with some care).

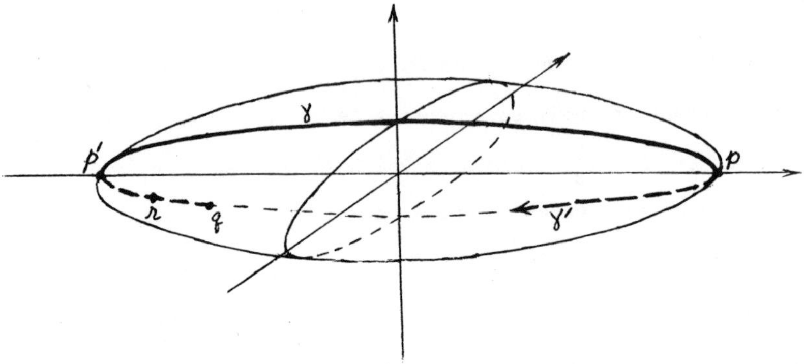

If r is a point between p' and q, then the portion of γ between p and r is a local minimum for E, but clearly not a global minimum, since the extension γ' of γ in the other direction past p has shorter length from p to r (on the other hand, γ' is the only other geodesic from p to r which is shorter than γ).

Notice that Theorems 6 and 8 do not cover the case where b is the only point in $(a,b]$ which is conjugate to a. This is the borderline case for which no conclusions can be drawn. It may happen, first of all, that γ is a local minimum for length, but not a strict local minimum. This is illustrated, of course, by taking γ to be half of a great circle on the sphere S^2. Now consider an ellipsoid, with three unequal axis $a > b > c$, and let p be a point at one end of the largest axis. The figure below shows the conjugate points of the geodesics starting from p (it is the envelope of these geodesics -- compare the Addendum to Chapter 3); this set is a curve with four cusps. The geodesic from p to A is a strict global minimum, while the geodesic going from p to p' and then on to B is a strict local minimum.

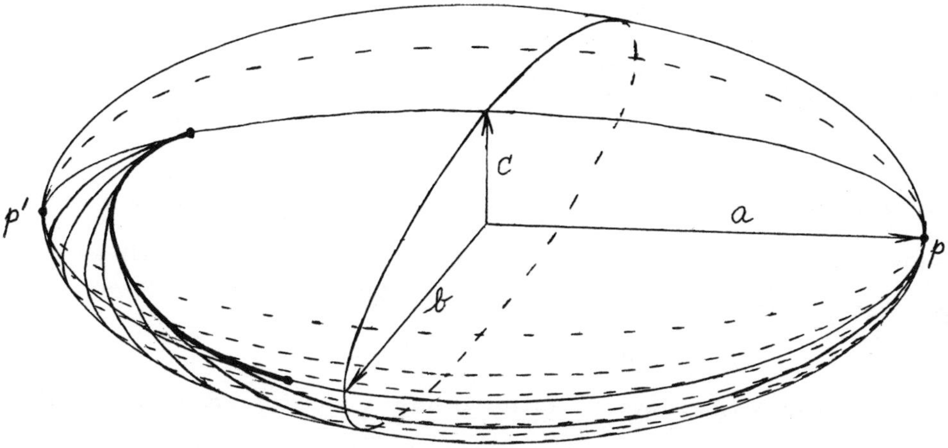

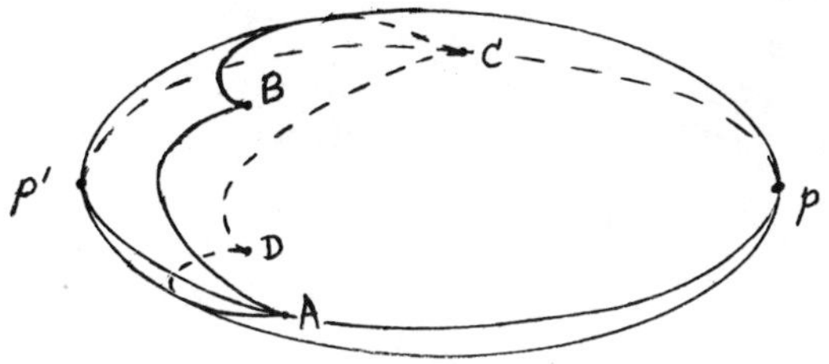

The next result complements Theorem 8 so that it appears to parallel Theorem 6 more closely.

<u>9. PROPOSITION</u>. Let $\gamma: [a,b] \longrightarrow M$ be a geodesic without conjugate points. Then $E_{**}(W,W) > 0$ for every non-zero $W \in \Omega_\gamma$.

<u>Proof</u>. Theorem 8 (and the Remark following it) implies that $E_{**}(W,W) \geq 0$. For if $E_{**}(W,W) < 0$, then γ would not be a local minimum for E (by the argument in the proof of Theorem 6).

Now suppose we had $E_{**}(W_1,W_1) = 0$ for some non-zero $W_1 \in \Omega_\gamma$. Then for any $W_2 \in \Omega_\gamma$ we would have

$$0 \leq E_{**}(W_1 + cW_2, W_1 + cW_2)$$
$$= 0 + 2c\, E_{**}(W_1, W_2) + c^2 E_{**}(W_2, W_2) .$$

Since this is supposed to be true for all c, it is clear that we would have to have $E_{**}(W_1, W_2) = 0$. Thus W_1 would be a Jacobi field, contradicting the fact that b is not a conjugate value of a. ∎

More generally, we have the following result, which plays an important role later on.

10. COROLLARY. Let $\gamma\colon [a,b] \longrightarrow M$ be a geodesic without conjugate points, W a Jacobi field along γ, and X a piecewise C^∞ vector field along γ with

$$X(a) = W(a) , \qquad X(b) = W(b) .$$

Then

$$E_{**}(X,X) \geq E_{**}(W,W) ,$$

and equality holds only when $X = W$.

Proof. The second variation formula shows that for all piecewise C^∞ vector fields W_2 along γ we have

$$(1) \qquad E_{**}(W,W_2) = \left\langle W_2, \frac{DW}{dt} \right\rangle \Big|_a^b = \left\langle W_2(b), \frac{DW}{dt}(b) \right\rangle - \left\langle W_2(a), \frac{DW}{dt}(a) \right\rangle .$$

Moreover, since $X - W \in \Omega_\gamma$, Proposition 9 shows that

$$0 \le E_{**}(X-W,\ X-W)$$
$$= E_{**}(X,X) + E_{**}(W,W) - 2E_{**}(W,X)$$
$$= E_{**}(X,X) + \left\langle W, \frac{DW}{dt}\right\rangle\Big|_a^b - 2\left\langle X, \frac{DW}{dt}\right\rangle\Big|_a^b \qquad \text{by (1)}$$
$$= E_{**}(X,X) - \left\langle W, \frac{DW}{dt}\right\rangle\Big|_a^b \qquad \text{since } W = X \text{ at } a \text{ and } b$$
$$= E_{**}(X,X) - E_{**}(W,W) \qquad \text{by (1) again .}$$

Moreover, it is clear that equality holds only if $X-W = 0$. ■

Theorem 6 and Proposition 9 show that for a geodesic $\gamma\colon [a,b] \to M$, the existence of conjugate points is practically equivalent to the existence of vector fields $W \in \Omega_\gamma$ with $E_{**}(W,W) < 0$:

(A) If there is some $\tau \in (a,b)$ conjugate to a, then there is some $W \in \Omega_\gamma$ with $E_{**}(W,W) < 0$ (Theorem 6);

(B) If there is some $W \in \Omega_\gamma$ with $E_{**}(W,W) < 0$, then there is some $\tau \in (a,b]$ conjugate to a (Proposition 9).

We will see later that it can be very convenient to replace questions about conjugate points by questions about vector fields $W \in \Omega_\gamma$ with $E_{**}(W,W) < 0$. Actually, the situation is even better than we have indicated, because statement (B) can be strengthened: if $E_{**}(W,W) < 0$ for some $W \in \Omega_\gamma$, then there is $\tau \in (a,b)$ conjugate to a. In fact, there is a far-reaching generalization of these results. We say that E_{**} is <u>negative-definite</u> on a subspace $\mathcal{V} \subset \Omega_\gamma$ if $E_{**}(W,W) < 0$ for all non-zero $W \in \mathcal{V}$, and we define the <u>index</u> of E_{**} to be the largest dimension of any subspace $\mathcal{V} \subset \Omega_\gamma$ on which E_{**} is negative definite (compare p. 4). Then we have the celebrated

Morse Index Theorem. The index of E_{**} for a geodesic γ: $[a,b] \to M$ is the number of $\tau \in (a,b)$ which are conjugate to a, each conjugate value being counted with its multiplicity. This index is always finite.

In terms of the index of E_{**}, our Theorem 6 can be reformulated as follows: if the number of conjugate values is ≥ 1, then the index is ≥ 1. Fro the Morse Index Theorem we need the more general assertion, that the index of $(E_a^t)_{**}$ increases by at least ν as t passes a conjugate value τ with multiplicity ν. This is the only point in the proof that does not involve simple general considerations, and it may be handled by essentially the same trick which was used in the present proof of Theorem 6. I hope that by clearing this path right up to the proof of the Index Theorem, I may have enticed you into reading the proof in Milnor { 2 }, which also describes some of the beautiful applications of these differential geometric ideas to topology.

In order to obtain interesting differential geometric consequences of our foundational results, we need to find hypotheses which imply something about the solutions of Jacobi equations. These hypotheses usually involve the sectional curvature $K(P)$ of 2-dimensional subspaces $P \subset M_p$; recall that $K(P) = \langle R(X,Y)Y, X\rangle$ for orthonormal $X, Y \in P$. Clearly all sectional curvatures of M are ≤ 0 if and only if $\langle R(X,Y)Y, X\rangle \leq 0$ for all pairs X, Y of vectors at the same point of M.

11. PROPOSITION. If all sectional curvatures of M are ≤ 0, then no two points of M are conjugate along any geodesic.

Proof. If γ is a geodesic with velocity vector field $V = d\gamma/dt$, and W is a Jacobi field along γ, then

$$\frac{D^2W}{dt^2} + R(W,V)V = 0 ,$$

so

$$\left\langle \frac{D^2W}{dt^2}, W \right\rangle = - \langle R(W,V)V, W\rangle \geq 0 .$$

Therefore

$$\frac{d}{dt}\left\langle \frac{DW}{dt}, W \right\rangle = \left\langle \frac{D^2W}{dt^2}, W \right\rangle + \left\langle \frac{DW}{dt}, \frac{DW}{dt} \right\rangle \geq 0 ,$$

which means that $\langle DW/dt, W\rangle$ is increasing.

Now if W vanishes at two points, t_0 and t_1, then $\langle DW/dt, W\rangle$ vanishes at t_0 and t_1, so $\langle DW/dt, W\rangle$ must be 0 on the interval $[t_0,t_1]$. This clearly implies that DW/dt also vanishes at t_0. Hence $W = 0$. ■

Although Proposition 11 shows that all geodesic segments on M are local minima for length, this does not mean that they are necessarily global minima. In fact, if we consider a compact surface M with everywhere negative curvature (Chapter 6, Addendum 1), it is clear that no geodesic $\gamma\colon \mathbb{R} \to M$ can be a global minimum for length on arbitrarily large segments.

The most interesting consequence of Proposition 11 is obtained by combining it with the following general result.

12. THEOREM. Let M be a complete, connected, n-dimensional Riemannian manifold, and let p be a point of M such that no point of M is conjugate to p along any geodesic. Then $\exp = \exp_p\colon M_p \to M$ is a covering map. In particular, if M is simply connected, then M is diffeomorphic to $\mathbb{R}^n$.

13. COROLLARY (HADAMARD-CARTAN). A complete, simply connected, n-dimensional Riemannian manifold with all sectional curvatures ≤ 0 is diffeomorphic to $\mathbb{R}^n$.

Proof. The Corollary follows immediately from the Theorem and Proposition 11. To prove the Theorem, let $\langle\ ,\ \rangle$ be the Riemannian metric on M, and consider the tensor $\exp^*\langle\ ,\ \rangle$ on M_p. Since there are no points conjugate to p, the map $\exp_*$ is always one-one, so $\exp^*\langle\ ,\ \rangle$ is a Riemannian metric on M_p. We claim that M_p is complete in the metric $\exp^*\langle\ ,\ \rangle$. To prove this, we just note that the straight lines through 0 in M_p are clearly geodesics for the metric $\exp^*\langle\ ,\ \rangle$, since their images under the local isometry $\exp\colon M_p \to M$ are geodesics in M. Since all geodesics through $0 \in M_p$ can be defined for all t, it follows from Problem I.9-43 that M_p is complete. The Theorem then follows from

14. LEMMA. Let M and N be connected Riemannian manifolds with M complete, and let $\phi\colon M \to N$ be a local isometry. Then N is complete and ϕ is a covering map onto N.

Proof. Let $p_0 \in M$. Given a geodesic $\gamma\colon (-\varepsilon,\varepsilon) \to N$ with $\gamma(0) = \phi(p_0)$, let c be the geodesic in M with $c(0) = p_0$ and $\phi_* c'(0) = \gamma'(0)$. Then $\gamma = \phi \circ c$, since ϕ is a local isometry. Since c can be defined on all of $\mathbb{R}$, we can extend γ to all of $\mathbb{R}$ as $\phi \circ c$. Thus N is complete, by Problem I.9-43.

To prove that ϕ is onto N it suffices to prove that $\phi(M)$ is closed (for $\phi(M)$ is open, since ϕ is everywhere regular). Let $q \in \overline{\phi(M)}$, and let V be a convex neighborhood of 0 in N_q on which $\exp_q$ is a diffeomorphism. There is a point $q' \in \exp_q(V)$ of the form $q' = \phi(p')$ for $p' \in M$. Let γ be the geodesic in $\exp_q(V)$ with $\gamma(0) = q'$ and $\gamma(1) = q$. Consider the geodesic c in M with $c(0) = p'$ and $\phi_* c'(0) = \gamma'(0)$. Then $\gamma = \phi \circ c$, as before. The point $p = c(1)$ is defined and $\phi(p) = \phi(c(1)) = \gamma(1) = q$. Thus $\overline{\phi(M)} \subset \phi(M)$, so $\phi(M)$ is closed. Hence ϕ is onto N.

The proof that ϕ is a covering map will be similar to the proof on pp.III.383-385. For fixed $q \in N$, let

$$V = \{Y \in N_q\colon \|Y\| < 2\varepsilon\} \subset N_q$$

be a neighborhood of 0 in N_q on which $\exp_q$ is a diffeomorphism. Suppose that $p \in \phi^{-1}(q)$. Consider the map

$$f = \exp_p \circ {\phi_{p*}}^{-1} \circ (\exp_q(V))^{-1} ,$$

$$f\colon \exp_q(V) \longrightarrow \exp_p\big(\{X \in M_p\colon \|X\| < 2\varepsilon\}\big) \subset M ;$$

this map is defined since M is complete. It is easy to see that

$$\phi\colon \exp_p\big(\{X \in M_p\colon \|X\| < 2\varepsilon\}\big) \longrightarrow \exp_q(V)$$

is a diffeomorphism with inverse f. Now let

$$W = \exp_q\big(\{Y \in N_q\colon \|Y\| < \varepsilon\}\big) \subset N ,$$

and for each $p \in M$, let

$$W_p = \exp_p\big(\{X \in M_p\colon \|X\| < \varepsilon\}\big) \subset M .$$

We claim that

$$\phi^{-1}(W) = \bigcup_{p \in \phi^{-1}(q)} W_p .$$

In fact, given $p' \in \phi^{-1}(W)$, let γ be the geodesic in W of length $d(\phi(p'),q)$ with $\gamma(0) = \phi(p')$ and $\gamma(1) = q$. Let c be the geodesic with $c(0) = p'$ and $\phi_*c'(0) = \gamma'(0)$. Then c is defined on $[0,1]$, since M is complete, and $\phi \circ c = \gamma$ on $[0,1]$. In particular, $p = c(1) \in \phi^{-1}(q)$, and it is easy to see that all points of $c([0,1])$ are in W_p. Thus $p' = c(0) \in W_p$.

To complete the proof we just have to show that the W_p are disjoint. Now if $W_{p_1} \cap W_{p_2} \neq \emptyset$, then we clearly have

$$p_2 \in \exp_{p_1}\left(\{X \in M_{p_1} : \|X\| < 2\varepsilon\}\right) .$$

But we know that ϕ is a diffeomorphism on this set. Since $\phi(p_1) = \phi(p_2)$, it follows that $p_1 = p_2$. ■

Proposition 11 is but a special case of more general results involving manifolds whose sectional curvatures satisfy certain inequalities. These results all follow from one theorem, but the mere statement of this theorem tends to overwhelm one with its complexity. So we will approach it rather gingerly by first proving special cases, all of which represent important steps in the historical evolution of the final result.

The first theorem of this type depends on a surprisingly simple proposition about second order differential equations. Remember that a solution ϕ of such an equation is determined by $\phi(a)$ and $\phi'(a)$. Consequently, a non-zero solution ϕ must have isolated zeros.

15. THEOREM (THE STURM COMPARISON THEOREM). Let f and h be two continuous functions satisfying $f(t) \leq h(t)$ for all t in an interval I, and let ϕ and η be two functions satisfying the differential equations

(1) $$\phi'' + f\phi = 0$$

(2) $$\eta'' + h\eta = 0$$

on I. Assume that ϕ is not the zero function, and let $a, b \in I$ be two consecutive zeros of ϕ.

(1) The function η must have a zero in (a,b), unless $f = h$ everywhere on $[a,b]$ and η is a constant multiple of ϕ on $[a,b]$.

(2). Suppose that we have $\eta(a) = 0$, and also $\eta'(a) = \phi'(a) > 0$ [which can be achieved by choosing a suitable multiple of η, and changing ϕ to $-\phi$, if necessary]. If τ is the smallest zero of η in $(a,b]$, then

$$\phi(t) \geq \eta(t) \quad \text{for } a \leq t \leq \tau ,$$

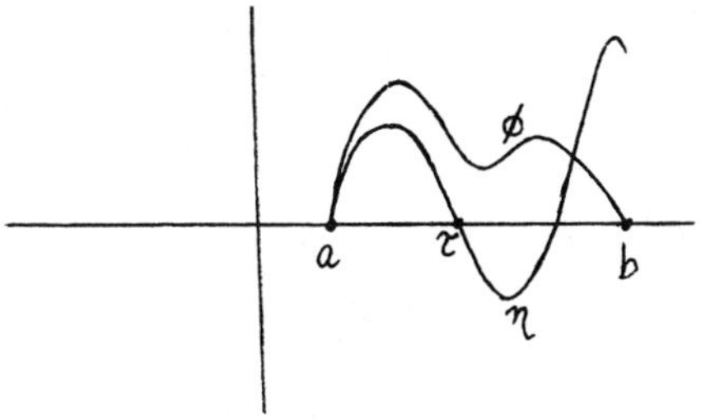

and equality holds for some t only if $f = h$ on $[a,t]$.

Proof. Equations (1) and (2) give

(3) $$\phi''\eta - \eta''\phi = (h - f)\phi\eta .$$

Suppose that η were nowhere zero on (a,b). It is easy to see that there is no loss of generality in assuming that

$$(4) \qquad \eta,\ \phi > 0 \qquad \text{on} \quad (a,b) .$$

Then (3) gives

$$0 \leq \phi''\eta - \eta''\phi ,$$

so

$$(5) \quad 0 \leq \int_a^b \phi''\eta - \eta''\phi = \int_a^b (\phi'\eta - \eta'\phi)'$$

$$= \phi'(b)\eta(b) - \phi'(a)\eta(a) , \qquad \text{since} \quad \phi(a) = \phi(b) = 0 .$$

On the other hand, (4) clearly implies that

$$(6) \left\{ \begin{matrix} \phi'(a) > 0, & \phi'(b) < 0 \\ \eta(a), \ \eta(b) \geq 0 & \end{matrix} \right\} \Longrightarrow \phi'(b)\eta(b) - \phi'(a)\eta(a) \leq 0 .$$

If $f \neq h$, then we actually have strict inequality in (5), which contradicts the second part of (6). This contradiction shows that η must have a zero on (a,b).

If $f = h$ on $[a,b]$, then equality holds in (5), and the first part of (6) implies that we must have $\eta(a) = \eta(b) = 0$. Since ϕ and η then satisfy the same second order equation on $[a,b]$ and $\phi(a) = \eta(a)$, the solution η must be a constant multiple of ϕ on $[a,b]$.

Now suppose that $\eta(a) = 0$ and $\eta'(a) = \phi'(a) > 0$. If τ is the smallest zero of η in $(a,b]$, then $\phi,\ \eta > 0$ on (a,τ), so (3) gives

$$0 \leq \phi''\eta - \eta''\phi = (\phi'\eta - \eta'\phi)' \qquad \text{on} \quad (a,\tau) .$$

This implies that

$$0 \leq \phi'\eta - \eta'\phi \qquad \text{on} \quad (a,\tau) ,$$

since $[\phi'\eta - \eta'\phi](a) = 0$. Using positivity of η on (a,τ) again, this gives

$$0 \leq \left(\frac{\phi}{\eta}\right)' \quad \text{on} \quad (a,\tau) . \tag{7}$$

But

$$\lim_{t\to a} \frac{\phi(t)}{\eta(t)} = \lim_{t\to a} \frac{\phi'(t)}{\eta'(t)} \qquad \text{by L'Hospital's Rule}$$

$$= 1 , \qquad \text{by assumption.}$$

Therefore

$$\frac{\phi}{\eta} \geq 1 \quad \text{on} \quad (a,\tau) ,$$

which is the desired inequality. The proof of the final statement is left to the reader. ■

Remark 1. Since $\phi'(a)$ and $\eta'(a)$ exist, and $\eta'(a) \neq 0$, we really used only a trivial case of L'Hospital's Rule; we could have simply written

$$\lim_{t\to a} \frac{\phi(t)}{\eta(t)} = \lim_{t\to a} \frac{\phi(t) - \phi(a)}{\eta(t) - \eta(a)} = \lim_{t\to a} \frac{\dfrac{\phi(t) - \phi(a)}{t - a}}{\dfrac{\eta(t) - \eta(a)}{t - a}}$$

$$= \frac{\phi'(a)}{\eta'(a)} = 1 .$$

Remark 2. In our applications, we will be interested only in the case where $\eta(a) = 0$. The reasoning for part 1 is then unnecessary, because part 2 shows that $\phi \geq \eta$ on any interval (a,τ) on which $\phi, \eta > 0$; this clearly implies that η vanishes somewhere on $(a,b]$. Moreover, if b were the first zero of η, then we would have $\phi(b) = \eta(b) = 0$, so we would have $f = h$ on [a,b], by the final statement in part 2. Nevertheless, part 1 is still of interest; here is one consequence:

16. COROLLARY. If ϕ_1 and ϕ_2 are two linearly independent solutions of the equation

$$\phi'' + f\phi = 0 ,$$

then the zeros of ϕ_1 alternate with the zeros of ϕ_2.

A particularly simple instance of Corollary 16 is provided by the equation $y'' + y/r^2 = 0$, where $r > 0$ is a constant. The solutions of this equation can all be written in the form $y(t) = b \sin(a + t/r)$. The zeros are always πr apart, so the zeros of two linearly independent solutions alternate with each other. This simple equation serves as a standard with which we can compare the Jacobi equation.

17. THEOREM (BONNET). Let M be a surface, and $\gamma\colon [0,L] \to M$ a geodesic parameterized by arclength. Let $r > 0$ be a constant.

(1) If $K(p) \le 1/r^2$ for all $p = \gamma(t)$, and γ has length $L < \pi r$, then γ contains no conjugate points.

(2) If $K(p) \ge 1/r^2$ for all $p = \gamma(t)$, and γ has length $L > \pi r$, then there is a point $\tau \in (0,L)$ conjugate to 0, and therefore γ is not of minimal length.

(3) If M is connected and complete, and $K(p) \ge 1/r^2$ for all $p \in M$, then M is actually compact, with diameter $\le \pi r$.

Proof. (1) Let Y be a unit vector field along γ with $\langle V,Y\rangle = 0$, where V is the unit vector field $V = d\gamma/dt$. The Jacobi equation for the vector field ϕY is (compare p. 312)

$$\phi''(t) + K(\gamma(t))\cdot\phi(t) = 0 .$$

The simpler equation

$$\eta''(t) + \frac{1}{r^2}\,\eta(t) = 0$$

has the solution $\eta(t) = \sin t/r$. Since $K(\gamma(t)) \leq 1/r^2$ by hypothesis, the Sturm comparison theorem shows that the first equation cannot have a solution ϕ vanishing at 0 and at $L < \pi r$, since η has no zero in $(0,L)$.

(2) Let Y be as in part 1, and consider a vector field ηY. The Jacobi equation for ηY is

$$\eta''(t) + K(\gamma(t))\cdot\eta(t) = 0 ,$$

and the simpler equation

$$\phi''(t) + \frac{1}{r^2}\,\phi(t) = 0$$

has the solution $\phi(t) = \sin t/r$ which vanishes at 0 and at πr. Since $1/r^2 \leq K(\gamma(t))$, the comparison theorem shows that any Jacobi field ηY must have a zero on the open interval $(0,\pi r) \subset (0,L)$. So if we choose any non-zero Jacobi field ηY along γ with $\eta(0) = 0$, then this Jacobi field will also vanish at some $\tau \in (0,L)$; thus τ is conjugate to 0.

(3) Any two points $p, q \in M$ can be joined by a geodesic γ of minimal length (Theorem I.9-18). Then the length of γ must be $\leq \pi r$, by part (2).

So M is bounded, with diameter $\le \pi r$. Since closed bounded sets in a complete manifold are compact, it follows that M itself is compact. ∎

I do not know whether Sturm ever saw this beautiful application of his theorem (he died in 1855, the same year that Bonnet published the result), but in his lectures he is supposed to have referred to it as the theorem "whose name I have the honor to bear."

Bonnet's Theorem fairly cries out to be generalized to higher dimensional manifolds, but a direct approach leads us into difficulties. The single normal vector field Y along γ has to be replaced by $n-1$ vector fields $Y_1,\ldots,Y_{n-1}$. Even if we choose $Y_1,\ldots,Y_{n-1}$ to be parallel, everywhere orthonormal vector fields along γ, the Jacobi equation for $\sum_i \phi_i Y_i$ reads

$$\sum_i \phi_i''(t)Y_i(t) + \sum_i \phi_i(t)\cdot R(Y_i(t),V(t))V(t) = 0 ,$$

which is equivalent to a system of ordinary differential equations

$$\phi_j''(t) + \sum_i \phi_i(t)\langle R(Y_i(t),V(t))V(t),\ Y_j(t)\rangle = 0 ,$$

and these equations do not even involve the sectional curvature directly. It is clear that we will have to approach this problem with a little more finesse.

One way to extend the results of Bonnet's theorem to higher dimensions is by an artful use of Synge's inequality (Corollary 1-7). Suppose first that $K(P) \ge 1/r^2$ for all 2-dimensional $P \subset M_{\gamma(t)}$, and that $\gamma\colon [0,L] \to M$ has length $L > \pi r$. Let Y be a parallel vector field along γ which is everywhere perpendicular to the parallel vector field $V = d\gamma/dt$, and let $S \subset M$ be a surface containing γ whose tangent space at each point $\gamma(t)$ is spanned by $V(t)$ and $Y(t)$. Then Synge's inequality shows that the Gaussian curvature

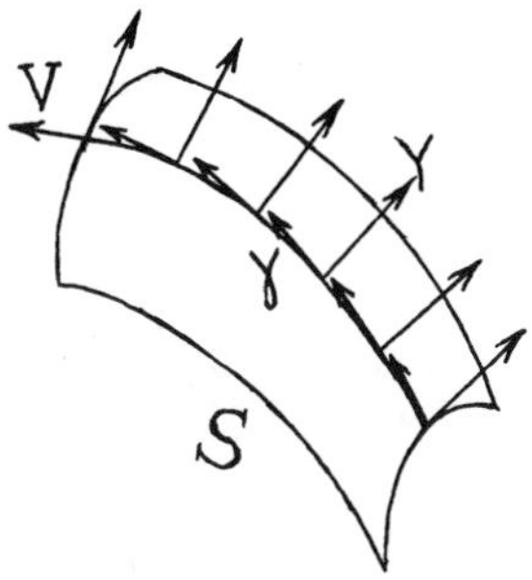

of S at $\gamma(t)$ is the same as the sectional curvature of M for the plane spanned by $V(t)$ and $Y(t)$. So the Gaussian curvature of S at $\gamma(t)$ is $\geq 1/r^2$. Bonnet's theorem then shows that there is a point $\tau \in (0,L)$ conjugate to 0 along γ. Of course, this means that τ is a conjugate value for 0 in the surface S. To conclude that there is a conjugate value in M itself, we must use the following indirect line of reasoning: Since $\tau \in (0,L)$ is a conjugate value for 0 in S, the geodesic γ is not a local minimum for length in S. Therefore it is certainly not a local minimum for length in M. Therefore, some $\sigma \in (0,L]$ must be a conjugate value for 0 in M. Applying this result to $\gamma|[0,L']$ for $\pi r < L' < L$, we see that actually some $\sigma \in (0,L'] \subset (0,L)$ is conjugate to 0 along γ in M.

In the previous paragraph, we had to choose the surface S so as not to decrease K, and then we had to show that the choice of S made no difference for the final conclusion. If we instead try to analyse the case where $K(P) \leq 1/r^2$ and γ has length $L < \pi r$, then we certainly don't care whether K is decreased, but our choice of S will be much more dependent on the desired conclusion. Suppose, then, that γ contained a conjugate point $\tau \in (0,L]$. We might as well assume that L itself is conjugate to 0 along

γ, since we can always work with $\gamma|[0,\tau]$; for the same reason, we might as well assume that L is the smallest value conjugate to 0. Then there is a variation α of γ through geodesics in M, whose variation vector field $W(t) = \partial\alpha/\partial u(0,t)$ vanishes only at 0 and L. Consider the surface

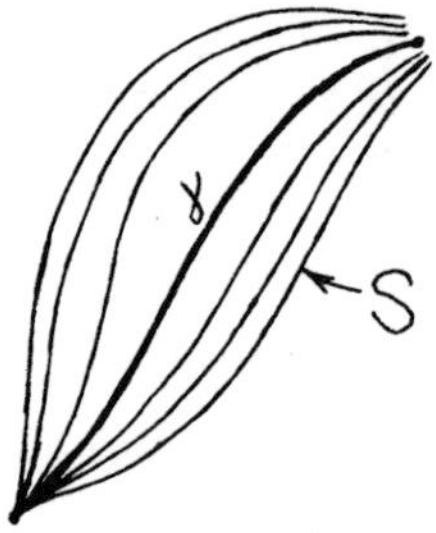

S formed by the image of α. Synge's inequality shows that the Gaussian curvature of S is $\leq 1/r^2$ along γ, and then Bonnet's theorem shows that γ cannot contain a conjugate point on S. But γ clearly does contain a conjugate point on S, because the $\bar{\alpha}(u)$ are also geodesics on S, so the variation vector field of α is also a Jacobi along γ on S. We seem to have obtained a contradiction, and thereby shown that γ cannot contain a conjugate point $\tau \in (0,L]$. The trouble with this argument is that only an excess of generosity could lead one to call S a surface, as the map α is definitely <u>not</u> an immersion at $(0,0)$ or $(0,L)$. The idea of the proof is basically sound, however, and leads to the desired result if one reasons a little more carefully (Problem 1).

We have not bestowed upon this reasoning the dignity which might accrue to it as the official proof of a theorem, because the results, and even better ones, can be obtained in a more systematic way. In fact, we shall present two new methods of generalizing Bonnet's theorem. These two methods differ significantly in their basic philosophy, but they both depend on a certain construction, similar to one used above, which is best set forth in a separate Lemma. This very generally sounding Lemma involves geodesics in two different manifolds, although in applications one of the two is always taken to be a sphere. (In the statement and proof of the Lemma, we will not use subscripts to distinguish the norms $\| \ \|$ and covariant derivatives D/dt in the two manifolds, since it should always be clear which manifold we are working in.)

18. LEMMA. Let M_1 and M_2 be two manifolds of the same dimension n, and let $\gamma_i\colon [a,b] \to M_i$ be arc-length parameterized geodesics in these two manifolds. Then there is a vector space isomorphism

$$\Phi\colon \{\text{piecewise } C^\infty \text{ vector fields along } \gamma_1\} \to \{\text{piecewise } C^\infty \text{ vector fields along } \gamma_2\}$$

such that for all $t \in [a,b]$ we have

(1) If $\dfrac{DX}{dt}$ is continuous at t, then $\dfrac{D\Phi(X)}{dt}$ is continuous at t,

(2) $\langle X(t), \gamma_1'(t)\rangle = \langle \Phi(X)(t), \gamma_2'(t)\rangle$,

(3) $\|X(t)\| = \|\Phi(X)(t)\|$,

(4) $\left\|\dfrac{DX}{dt}(t)\right\| = \left\|\dfrac{D\Phi(X)}{dt}(t)\right\|$,

it being understood that the last equation refers to left and right hand limits at discontinuity points.

Proof. Pick some fixed $t_0 \in [a,b]$. Let $\phi\colon (M_1)_{\gamma_1(t_0)} \to (M_2)_{\gamma_2(t_0)}$ be any norm preserving isomorphism with $\phi(\gamma_1'(t_0)) = \gamma_2'(t_0)$. Then we can define

$$\phi_t\colon (M_1)_{\gamma_1(t)} \longrightarrow (M_2)_{\gamma_2(t)}$$

by parallel translating a vector in $(M_1)_{\gamma_1(t)}$ along γ_1 to $\gamma_1(t_0)$, applying ϕ, and then parallel translating along γ_2 to $(M_2)_{\gamma_2(t)}$. We then define $\Phi(X)$ by

$$\Phi(X)(t) = \phi_t(X(t)) \ .$$

We can also describe $\Phi(X)$ as follows. Let $Y_1,\ldots,Y_n$ be parallel, everywhere orthonormal vector fields along γ_1 with $Y_1(t_0) = \gamma_1'(t_0)$, and let $Z_1,\ldots,Z_n$ be parallel, everywhere orthonormal vector fields along γ_2 with $Z_1(t_0) = \gamma_2'(t_0)$. If

$$X(t) = \sum_{i=1}^{n} f_i(t)Y_i(t)$$

for certain functions $f_i\colon [a,b] \longrightarrow \mathbb{R}$, then

$$\Phi(X)(t) = \sum_{i=1}^{n} f_i(t)Z_i(t) \ .$$

This shows that $\Phi(X)$ is C^∞ everywhere that X is, and that

$$<X(t), \gamma_1'(t)> = f_1(t) = <\Phi(X)(t), \gamma_2'(t)>$$

$$\|X(t)\|^2 = \sum_{i=1}^{n} [f_i(t)]^2 = \|\Phi(X)(t)\|$$

$$\left\|\frac{DX}{dt}(t)\right\| = \sum_{i=1}^{n} [f_i'(t)]^2 = \left\|\frac{D\Phi(X)}{dt}(t)\right\| \ . \qquad \blacksquare$$

In our first generalization of Bonnet's Theorem, we will consider the index of a geodesic, instead of the number of conjugate points it contains. Recall (p. 327) that the index of γ is > 0 if and only if there is some $W \in \Omega_\gamma$ with $E_{**}(W,W) < 0$.

<u>19. THEOREM</u>. Let M_1 and M_2 be two manifolds of the same dimension n, and let $\gamma_i\colon [a,b] \to M_i$ be geodesics parameterized by arclength. For each $t \in [a,b]$, suppose that for all 2-dimensional $P_i \subset (M_i)_{\gamma_i(t)}$, the curvatures K_i satisfy

$$K_1(P_1) \leq K_2(P_2) .$$

Then we have

$$\text{index } \gamma_1 \leq \text{index } \gamma_2 .$$

In particular, if $E_{**}(W_1,W_1) < 0$ for some $W_1 \in \Omega_{\gamma_1}$, then also $E_{**}(W_2,W_2) < 0$ for some $W_2 \in \Omega_{\gamma_2}$.

<u>Proof</u>. Let W be a piecewise C^∞ vector field on γ_1, and let Φ be the map in Lemma 18. The second variation formula shows that

$$(1) \qquad E_{**}(W,W) = -\int_a^b \langle R(W(t),V(t))V(t),\ W(t)\rangle\, dt$$
$$- \int_a^b \left\langle W(t),\ \frac{D^2W}{dt^2}(t)\right\rangle dt \;-\; \sum_{i=0}^{N} \left\langle W(t_i),\ \Delta_{t_i}\frac{DW}{dt}\right\rangle .$$

Now we also have

$$\frac{d}{dt}\left\langle W(t), \frac{DW}{dt}(t)\right\rangle = \left\langle \frac{DW}{dt}(t), \frac{DW}{dt}(t)\right\rangle + \left\langle W(t), \frac{D^2W}{dt^2}(t)\right\rangle .$$

Integrating this equation between t_{i-1} and t_i for each i, and adding the results, we obtain

$$-\sum_{i=0}^{N}\left\langle W(t_i), \Delta_{t_i}\frac{DW}{dt_i}\right\rangle = \int_a^b\left\langle \frac{DW}{dt}(t), \frac{DW}{dt}(t)\right\rangle dt + \int_a^b\left\langle W(t), \frac{D^2W}{dt^2}(t)\right\rangle dt .$$

So equation (1) can be written

$$E_{**}(W,W) = \int_a^b\left\{\left\langle \frac{DW}{dt}(t), \frac{DW}{dt}(t)\right\rangle - \langle R(W(t),V(t))V(t), W(t)\rangle\right\} dt .$$

From the properties of the map Φ, and the hypotheses on K, we see that

$$E_{**}(W,W) \geq E_{**}(\Phi(W),\Phi(W)) .$$

So, if $\mathcal{V} \subset \Omega_{\gamma_1}$ is a subspace on which E_{**} is negative definite, then $\Phi(\mathcal{V}) \subset \Omega_{\gamma_2}$ is a subspace of the same dimension on which E_{**} is again negative definite. Thus the index of γ_2 is certainly at least as large as the index of γ_1. ∎

<u>20. COROLLARY (THE MORSE-SCHOENBERG COMPARISON THEOREM).</u> Let M be a Riemannian manifold of dimension n, and $\gamma\colon [0,L] \to M$ a geodesic parameterized by arclength. Let $r > 0$ be a constant.

(1) If $K(P) \leq 1/r^2$ for all $P \subset M_{\gamma(t)}$, and γ has length $L < \pi r$, then the index of γ is 0, and γ contains no conjugate point. [Note that Proposition 11 is a special case.]

(2) If $K(P) \geq 1/r^2$ for all $P \subset M_{\gamma(t)}$, and γ has length $L > \pi r$, then

there is a point $\tau \in (0,L)$ conjugate to 0, and γ is not of minimal length.

Proof. (1) We apply the Theorem with $M_1 = M$ and $M_2 =$ n-sphere $S^n(r)$ of radius r, choosing γ_1 to be γ, and $\gamma_2\colon [0,L] \longrightarrow S^n(r)$ to be any geodesic parameterized by arclength. We find that

$$\text{index } \gamma \leq \text{index } \gamma_2 \ .$$

Now the index of γ_2 is certainly zero, since γ_2 contains no conjugate points, and Proposition 9 applies (all we really need is the fact that $E_{**}(W,W) \geq 0$ for $W \in \Omega_{\gamma_2}$, which follows from Theorem 8). Consequently, index $\gamma = 0$. Theorem 6 implies that no number $\tau \in (0,L)$ is conjugate to 0 along γ. We can also conclude that no number $\tau \in (0,L]$ is conjugate to 0 along γ, by extending γ to $\bar{\gamma}\colon [0,L'] \longrightarrow M$ with $L < L' < \pi r$, and applying the result to $\bar{\gamma}$.

(2) We apply the Theorem with $M_1 = S^n(r)$ and $M_2 = M$, this time choosing γ_2 to be γ. We obtain

$$\text{index } \gamma_1 \leq \text{index } \gamma \ .$$

But the index of γ_1 is at least 1, since γ_1 contains a conjugate point, and Theorem 6 applies. Consequently, index $\gamma \geq 1$. This shows that γ does contain a conjugate point $\tau \in (0,L]$ (Proposition 9 or Theorem 8 again). Applying this result to $\gamma|[0,L']$, with $\pi r < L' < L$, we see that γ contains a conjugate value $\tau \in (0,L)$. ■

For the case where $K \geq 1/r^2$, we can obtain a stronger result, involving the Ricci tensor Ric, introduced in Chapter 7.G.

21. THEOREM (MYERS). Let M be an n-dimensional Riemannian manifold, and $\gamma\colon [0,L] \to M$ a geodesic parameterized by arclength. Let $r > 0$ be a constant, and suppose that

$$-\operatorname{Ric}(\gamma'(t),\gamma'(t)) \geq \frac{n-1}{r^2} \qquad \text{for all } t,$$

and that γ has length $L > \pi r$. Then there is a point $\tau \in (0,L)$ conjugate to 0, and γ is not of minimal length.

Proof. Choose parallel, everywhere orthonormal vector fields $Y_1,\ldots,Y_n$ along γ with $Y_1 = V$. Let $W_i(t) = (\sin \pi t/L)Y_i(t)$. Then

$$\begin{aligned} E_{**}(W_i,W_i) &= -\int_0^L \left\langle W_i, \frac{D^2 W_i}{dt^2} + R(W_i,V)V \right\rangle dt \\ &= \int_0^L \left(\sin \frac{\pi t}{L}\right)^2 \left[\frac{\pi^2}{L^2} - \langle R(Y_i(t),Y_1(t))Y_1(t),\, Y_i(t)\rangle\right] dt\,. \end{aligned}$$

Summing for $i = 2,\ldots,n$, we obtain

$$\sum_{i=2}^{n} E_{**}(W_i,W_i) = \int_0^L \left(\sin \frac{\pi t}{L}\right)^2 \left[\frac{(n-1)\pi^2}{L^2} + \operatorname{Ric}(Y_1(t),Y_1(t))\right] dt\,.$$

By hypothesis, the term in brackets is < 0, so some $E_{**}(W_i,W_i)$ is < 0. Thus γ is not of minimal length, and there is $\tau \in (0,L)$ conjugate to 0 (same reasoning as in Corollary 20). ∎

Remark. If $\gamma: [0,L] \to S^n(r)$ is a geodesic parameterized by arclength, and Y is any parallel vector field along γ which is perpendicular to γ, then $W(t) = (\sin \pi t/L)Y(t)$ satisfies $E_{**}(W,W) < 0$. The vector fields W_i in the above proof come from such vector fields W by means of the map Φ of Lemma 18. This may make the proof of Myers' theorem somewhat less mysterious.

The next result reproduces the reasoning in the third part of Bonnet's theorem, together with an observation of interest only in the higher dimensional case.

22. COROLLARY. Let M be a complete connected manifold with

$$-\operatorname{Ric}(X,X) \geq \frac{n-1}{r^2}$$

for all unit vectors X, where $r > 0$ is a constant. (This hypothesis holds, in particular, if $K(P) \geq 1/r^2$ for all plane sections P.) Then M is actually compact, with diameter $\leq \pi r$. Moreover, the fundamental group of M is finite.

Proof. The proof of the first part is exactly the same as in Bonnet's theorem. To prove that the fundamental group of M is finite, we simply consider the universal covering space $\pi: \tilde{M} \to M$ of the Riemannian manifold $(M, \langle\ ,\ \rangle)$. Clearly $(\tilde{M}, \pi^*\langle\ ,\ \rangle)$ is complete, and its Ricci curvature also satisfies $-\operatorname{Ric}(X,X) \geq (n-1)/r^2$ for all unit vectors X. Therefore $\tilde{M}$ is compact. ∎

Although Theorem 19 certainly generalizes Bonnet's theorem very nicely, we do lose some information in this approach. Roughly speaking, we have generalized to higher dimensions only the first part of the Sturm comparison theorem, telling us that our Jacobi field $\Phi(W)$ must vanish somewhere on (a,b); we have not generalized the second part by comparing $\|\Phi(W)\|$ with $\|W\|$ up to the first zero of $\Phi(W)$. Such information is provided by

23. THEOREM (THE RAUCH COMPARISON THEOREM). Let M_1 and M_2 be two manifolds of the same dimension n, and let $\gamma_i\colon [a,b] \to M_i$ be geodesics parameterized by arclength such that

(1) no number $\tau \in (a,b]$ is a conjugate value of 0 along γ_1 in M_1 or along γ_2 in M_2.

Let W_i be Jacobi fields along γ_i such that

(2) $W_i(a) = 0$,

(3) $\left\|\frac{DW_1}{dt}(a)\right\| = \left\|\frac{DW_2}{dt}(a)\right\|$,

(4) W_i is perpendicular to γ_i .

For all $t \in [a,b]$, suppose that for all 2-dimensional $P_i \subset (M_i)_{\gamma_i(t)}$, the curvatures K_i satisfy

(5) $K_1(P_1) \le K_2(P_2)$.

Then

$$\|W_1(t)\| \ge \|W_2(t)\| \quad \text{for all } t \in [a,b] .$$

Proof. If $W_2 = 0$, the theorem is trivial. If W_2 is not the 0 vector field, then $W_2(t) \ne 0$ for all $t \in (a,b)$, since γ_2 has no conjugate points. Naturally, $W_1(t)$ is also non-zero for all $t \in (a,b)$. It obviously suffices to prove that

$$(1) \qquad \lim_{t\to 0} \frac{\langle W_1,W_1\rangle(t)}{\langle W_2,W_2\rangle(t)} = 1 ,$$

$$(2) \qquad \frac{d}{dt}\frac{\langle W_1,W_1\rangle(t)}{\langle W_2,W_2\rangle(t)} \geq 0 \qquad \text{for} \quad t \in (a,b) .$$

To prove (1) we note that

$$\lim_{t\to 0}\frac{\langle W_1,W_1\rangle(t)}{\langle W_2,W_2\rangle(t)} = \lim_{t\to 0}\frac{\left\langle W_1, \frac{DW_1}{dt}\right\rangle(t)}{\left\langle W_2, \frac{DW_2}{dt}\right\rangle(t)} \qquad \text{by L'Hospital's Rule}$$

$$= \lim_{t\to 0}\frac{\left\langle \frac{DW_1}{dt}, \frac{DW_1}{dt}\right\rangle(t) + \left\langle W_1, \frac{D^2W_1}{dt^2}\right\rangle(t)}{\left\langle \frac{DW_2}{dt}, \frac{DW_2}{dt}\right\rangle(t) + \left\langle W_2, \frac{D^2W_2}{dt^2}\right\rangle(t)} \qquad \text{by L'Hospital's Rule}$$

$$= 1 , \qquad \text{by hypothesis (3).}$$

(Note that the first use of L'Hospital's Rule is a genuine one, necessitated by the fact that we need to look at $\langle W_i,W_i\rangle$, rather than $\|W_i\|$. The second use, however, represents the same trivial case which occurs in the Sturm comparison theorem.)

Equation (2) is equivalent to

$$\langle W_2,W_2\rangle\cdot\left\langle W_1, \frac{DW_1}{dt}\right\rangle \geq \langle W_1,W_1\rangle\cdot\left\langle W_2, \frac{DW_2}{dt}\right\rangle ,$$

so for each $t_0 \in (a,b)$ it suffices to show that

$$(2') \qquad \left\langle W_1, \frac{DW_1}{dt}\right\rangle(t_0) \geq c^2\left\langle W_2, \frac{DW_2}{dt}\right\rangle(t_0) ,$$

$$\text{where} \quad c = \|W_1(t_0)\|/\|W_2(t_0)\| .$$

But, since W_i are Jacobi fields, and $W_i(a) = 0$, the second variation formula shows that

$$\left\langle W_i, \frac{DW_i}{dt} \right\rangle (t_0) = E_{**}(\tilde{W}_i, \tilde{W}_i) ,$$
$$\text{where} \quad \tilde{W}_i = W_i | [a, t_0] .$$

Therefore, we just have to prove that

$$(2'') \qquad E_{**}(\tilde{W}_1, \tilde{W}_1) \geq c^2 E_{**}(\tilde{W}_2, \tilde{W}_2) .$$

Consider the map Φ of Lemma 18, constructed for the geodesics $\gamma_i | [0, t_0]$. Since $W_i(t_0)$ are both non-zero, and orthogonal to γ_i, we can obviously define Φ so that

$$(3) \qquad \Phi(\tilde{W}_1)(t_0) = c\tilde{W}_2(t_0) .$$

In the first part of the proof of Theorem 19 we showed that

$$(4) \qquad E_{**}(\tilde{W}_1, \tilde{W}_1) \geq E_{**}(\Phi(\tilde{W}_1), \Phi(\tilde{W}_1)) .$$

On the other hand, we have $\tilde{W}_2(a) = \Phi(\tilde{W}_1)(a) = 0$ by hypothesis (2) and the norm-preserving property of Φ, while $c\tilde{W}_2(t_0) = \Phi(\tilde{W}_1)(t_0)$ by (3). So Corollary 10 yields

$$(5) \qquad E_{**}(\Phi(\tilde{W}_1), \Phi(\tilde{W}_1)) \geq E_{**}(c\tilde{W}_2, c\tilde{W}_2)$$
$$= c^2 E_{**}(\tilde{W}_2, \tilde{W}_2) .$$

Equations (4) and (5) together give the required equation (2"). ■

Unless you have become totally lost in these generalities, it should be clear that Theorem 23 can also be used to prove the results of Corollary 20. Rauch actually used his comparison theorem to prove a much more striking result, concerning "δ-pinched" manifolds. These are Riemannian manifolds satisfying

$$\delta A \leq K(P) \leq A$$

for all 2-dimensional subspaces $P \subset M_p$ at all points $p \in M$; here A is a constant, which we can assume is 1 if we are willing to multiply the metric $\langle\ ,\ \rangle$ by a constant. Rauch proved that if M is complete and simply connected, and δ-pinched for a certain $\delta \sim .74$, then M is homeomorphic to a sphere. Improvements by Berger and Klingenberg have established the

<u>SPHERE THEOREM</u>. Let M be a complete, simply connected Riemannian manifold of dimension n whose sectional curvatures $K(P)$ satisfy

$$\delta \leq K(P) \leq 1$$

for some constant $\delta > 1/4$. Then M is homeomorphic to S^n.

It is known that for even n this result breaks down if we allow $\delta = 1/4$. We will not go into the rather detailed proofs of this and related recent results, which together would make up a good sized monograph. A large selection is coherently presented in Gromoll, Klingenberg, Meyer {1}, and references to a few more will be found in the bibliography.

One of the most striking recent results completely clarifies the requirement in many of our theorems that the sectional curvatures should not only be positive, but also bounded away from 0. Naturally, this latter condition can fail only when the manifold M is not compact; but in this case the structure of M is completely determined:

<u>THEOREM (GROMOLL-MEYER)</u>. If M is a connected, complete, non-compact n-dimensional manifold with all sectional curvatures positive, then M is diffeomorphic to $\mathbb{R}^n$.

Although we must omit the proofs of these theorems, we can prove an older and easier, but still very striking, result about the topology of manifolds whose sectional curvatures are all positive. We begin with a lemma that will also be used later on.

<u>24. LEMMA (SYNGE)</u>. Let M be an orientable even-dimensional Riemannian manifold with all sectional curvatures positive. Let $\gamma\colon [a,b] \to M$ be a geodesic which is closed [that is, $\gamma(a) = \gamma(b)$ and $\gamma'(a) = \gamma'(b)$]. Then there is a variation $\alpha\colon (-\varepsilon,\varepsilon) \times [a,b] \to M$ of γ such that all curves $\bar{\alpha}(u)$ are closed curves with length $\bar{\alpha}(u) <$ length γ for $u \neq 0$.

<u>Proof</u>. Let $\mathcal{V} = \gamma'(0)^{\perp} \subset M_p$ be the $(n-1)$-dimensional subspace of all $X_p \in M_p$ which are perpendicular to $\gamma'(0)$. Define $\phi\colon \mathcal{V} \to \mathcal{V}$ to be the result of parallel translation around γ. Then ϕ is a norm-preserving linear transformation, with matrix A satisfying $AA^t = 1$, so $\det \phi = \pm 1$. Moreover, since M is orientable, it is easy to see that $\phi\colon \mathcal{V} \to \mathcal{V}$ must be orientation preserving, so that $\det \phi = +1$. Since the dimension of $\mathcal{V}$ is odd, the characteristic polynomial of ϕ has at least one real root, so ϕ has a real eigenvalue, which clearly must be ± 1. Moreover, the complex eigenvalues occur in conjugate pairs $\lambda, \bar{\lambda}$ with $\lambda\bar{\lambda} > 0$. The number of real eigenvalues ± 1 is therefore odd, and their product is positive, so at least one must be $+1$. Consequently, ϕ leaves some vector field fixed: $\phi(X_p) = X_p$ for some $X_p \in \mathcal{V}$. This means that parallel translation of X_p around γ produces a

vector field X along γ with $X(a) = X(b)$.

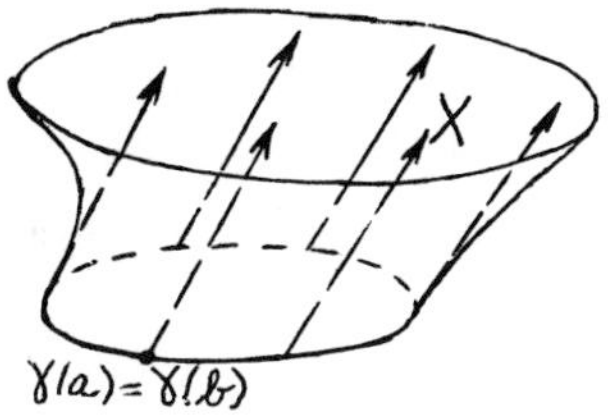

Let $\alpha\colon (-\varepsilon,\varepsilon)\times[a,b] \to M$ be a variation of γ with variation vector field $\partial\alpha/\partial u\,(0,t) = X(t)$. Since $X(a) = X(b)$, we can clearly choose α so that $\alpha(u,a) = \alpha(u,b)$ for all u, which means that each $\bar{\alpha}(u)$ is a closed curve. Applying the second variation formula, and remembering that $DX/dt = 0$, we find that

$$E_{**}(X,X) = -\int_a^b \langle X(t),\, R(X(t),V(t))V(t)\rangle\, dt$$

$$< 0, \quad \text{by the hypothesis on sectional curvatures.}$$

This means that for sufficiently small $u \neq 0$, the curves $\bar{\alpha}(u)$ have smaller energy than γ. ■

With this Lemma we can easily prove the following result, provided that we accept an "intuitively obvious" fact, whose proof will come soon afterwards. We will temporarily use the term "closed path" for a continuous path $c\colon [0,1] \to M$ with $c(0) = c(1)$; the term "smooth closed path" will be used for a smooth path $c\colon [0,1] \to M$ with $c(0) = c(1)$ and $c'(0) = c'(1)$.

<u>24. THEOREM (SYNGE).</u> Let M be a compact, connected, orientable, even-dimensional Riemannian manifold with all sectional curvatures positive. Then M is simply connected.

Proof. Pick a point $p \in M$ and suppose that $\pi_1(M,p) \neq 0$. Let $c\colon [0,1] \longrightarrow M$ be a closed path with $c(0) = c(1) = p$, representing a non-zero element of $\pi_1(M,p)$. We say that a closed path γ is in the same free homotopy class as c if c and γ are homotopic, considered simply as maps from S^1 into M.

Claim: There is a closed curve $\gamma\colon [0,1] \longrightarrow M$ in the same free homotopy class as c which has smaller length than any other closed curve in this free homotopy class.

If we accept this claim, then it is clear that γ must be a smooth closed geodesic. For every sufficiently small segment of γ must coincide with a geodesic, since geodesics are the smallest paths between sufficiently close points.

The proof is now immediate, for we obtain a contradiction by applying Synge's Lemma to γ. ■

Before we proceed with the proof of the Claim, we add a few remarks. The hypothesis that M is compact can be replaced by the hypothesis that M is complete and has sectional curvatures bounded away from 0, by Corollary 22. In fact, the Theorem on p. 352 shows that compactness can be replaced by completeness alone. The hypothesis that M is orientable is clearly necessary, as shown by the projective spaces P_n with n even. However, one can easily show (Problem 2) that if M is not orientable, then $\pi_1(M) \approx \mathbb{Z}_2$. The necessity of assuming that M is even-dimensional is shown by the projective spaces P_n with n odd. Without this assumption we must content ourselves with showing (Problem 2) that if M is a compact, connected odd-dimensional manifold with all sectional curvatures positive, then M is orientable.

We will now give two different proofs of the Claim. The first of these, the official proof, uses a few facts about covering spaces, and is generally considered to be quite elegant. The second proof is a more typical example of the sort of "direct methods" which one can sometimes use in order to establish that solutions to calculus of variation problems actually exist, instead of merely finding conditions on the presumed solution; it is similar to arguments first used by Hilbert for that sort of question (and similar arguments could be used to give an alternative demonstration that a minimal geodesic exists between any two points in a complete manifold). That, I feel, is one good reason for including it; it also turns out that this proof is no harder than the first proof if the details are handled intelligently.

25. PROPOSITION. If $(M, \langle\ ,\ \rangle)$ is a non-simply connected compact Riemannian manifold, then every free homotopy class contains a curve of minimum length.

First Proof. Let $\tilde{M}$ be the universal covering space of M and $\pi\colon \tilde{M} \to M$ the projection; the complete Riemannian metric $\pi^*\langle\ ,\ \rangle$ on $\tilde{M}$ gives an ordinary metric d on $\tilde{M}$. Recall that a homeomorphism $\phi\colon \tilde{M} \to \tilde{M}$ with $\pi\circ\phi = \pi$ is called a "covering transformation" or "deck transformation" of $\tilde{M}$. The set D of all deck transformations is in one-one correspondence with $\pi_1(M,p)$ for any $p \in M$, and d is invariant under the action of D.

Given a closed path $c\colon [0,1] \to M$, let $\tilde{c}\colon [0,1] \to \tilde{M}$ be a lifting, starting at some point $q \in \pi^{-1}(c(0))$. Then

$$\tilde{c}(1) = \delta(q) = \delta(\tilde{c}(0)) \qquad \text{for a unique } \delta \in D\ .$$

To see how this δ depends on the choice of $q \in \pi^{-1}(c(0))$, we note that any other point $q \in \pi^{-1}(c(0))$ is $\phi(q)$ for some $\phi \in D$, and that the lifting $\tilde{\tilde{c}}$ of c starting at $\phi(q)$ is just $\phi\circ\tilde{c}$. This means that

$$\tilde{\tilde{c}}(1) = \phi(\tilde{c}(1)) = \phi(\delta(q)) = \phi\circ\delta\circ\phi^{-1}(\phi(q)) = \phi\circ\delta\circ\phi^{-1}(\tilde{\tilde{c}}(0)) .$$

Thus:

(1) The conjugacy class $\{\phi\delta\phi^{-1}\colon \phi \in D\}$ does not depend on the choice of $q \in \pi^{-1}(c(0))$.

We also claim:

(2) If $c_1\colon [0,1] \to M$ is freely homotopic to c, then it determines the same conjugacy class.

For we have a map $H\colon I\times I \to M$ with

$$\begin{aligned} H(t,0) &= c(t)\\ H(t,1) &= c_1(t)\\ H(0,s) &= H(1,s) . \end{aligned}$$

Let $\tilde{H}$ be a lifting of H, and define $\tilde{c}(t) = \tilde{H}(t,0)$ and $\tilde{c}_1(t) = \tilde{H}(t,1)$. We have $\tilde{H}(1,s) = \delta(s)(\tilde{H}(0,s))$ for some $\delta(s)$, and all $\delta(s)$ must be the same δ, by continuity. Thus both c and c_1 determine the same conjugacy class $\{\phi\delta\phi^{-1}\colon \phi \in D\}$.

Finally, we claim:

(3) If $\gamma_1, \gamma_2\colon [0,1] \to \tilde{M}$ are paths with $\gamma_i(1) = \delta(\gamma_i(0))$, then $\pi\circ\gamma_1$ is freely homotopic to $\pi\circ\gamma_2$.

To prove this we let $\gamma\colon [0,1] \to \tilde{M}$ be a path from $\gamma_1(0)$ to $\gamma_2(0)$. Then

$\delta\circ\gamma$ is a path from $\gamma_1(1)$ to $\gamma_2(1)$. So we can define a continuous map $H\colon \partial([0,1]\times[0,1]) \longrightarrow \tilde{M}$ as follows:

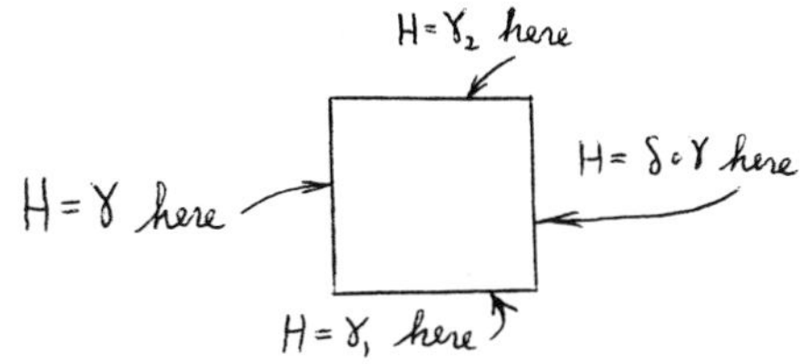

Since $\tilde{M}$ is simply connected, we can extend this to a map $H\colon [0,1]\times[0,1] \longrightarrow \tilde{M}$. Then $\pi\circ H\colon [0,1]\times[0,1] \longrightarrow M$ satisfies

$$\pi\circ H(0,s) = \pi\circ H(1,s) \qquad \text{for all } s \in [0,1] .$$

So $\pi\circ\gamma_1$ is freely homotopic to $\pi\circ\gamma_2$.

Now let $\{\phi\delta\phi^{-1}\colon \phi \in D\}$ be the conjugacy class corresponding to our given free homotopy class, and define $h_\delta\colon \tilde{M} \longrightarrow \tilde{M}$ by

$$h_\delta(q) = \inf\{d(q,\phi\delta\phi^{-1}(q))\colon \phi \in D\} ;$$

this is well-defined, since it clearly depends only on the conjugacy class of δ. Notice that for each q there is some ϕ (depending on q) such that

$$h_\delta(q) = d(q,\phi\delta\phi^{-1}(q)) ;$$

this follows from the fact that D acts discretely. It is also clear that h_δ is invariant under the action of D. It follows that h_δ takes on its minimum on $\tilde{M}$; for there is a compact set $K \subset \tilde{M}$ with $\pi(K) = M$, and consequently $D(K) = \tilde{M}$, which means that the minimum of h_δ on K is also the minimum on all of $\tilde{M}$. Say that h_δ takes on its minimum at $q_0 \in \tilde{M}$, and that

$$h_\delta(q_0) = d(q_0,\phi_0\delta\phi_0^{-1}(q_0)) .$$

Let γ be a minimal geodesic in $\tilde{M}$ from q_0 to $\phi_0\delta\phi_0^{-1}(q_0)$, with length $\gamma = h_\delta(q_0)$. Then also

$$\text{length } \pi\circ\gamma = h_\delta(q_0) \ .$$

The curve $\pi\circ\gamma$ is in the given free homotopy class, by (3). If c is any other curve in the free homotopy class, and $\tilde{c}$ is any lifting, starting at some point q, then $\tilde{c}(1)$ is $\psi\delta\psi^{-1}(q)$ for some $\psi \in D$, and consequently

$$\begin{aligned}\text{length } c = \text{length } \tilde{c} &\geq d(q,\psi\delta\psi^{-1}(q)) \\ &\geq h_\delta(q) \geq h_\delta(q_0) = \text{length } \pi\circ\gamma \ .\end{aligned}$$

<u>Second Proof</u>. Since M is compact, there is a finite open cover $U_1,\dots,U_r$ of M by geodesically convex sets. By the Lebesgue covering lemma, there is $\varepsilon > 0$ such that any set A with diameter $< \varepsilon$ lies entirely in some U_α.

A closed curve c in M will be called <u>special</u> if there is a sequence $p_0,p_1,\dots,p_N = p_0$ of points in M such that

(i) for each j, the points p_{j-1}, p_j both lie in some U_α,

(ii) c is the union of minimal geodesics c_j joining p_{j-1} to p_j.

Given an arbitrary closed curve $c\colon [0,1] \to M$, there is always a special closed curve $\bar{c}$ in the same free homotopy class as c, with length $\ell(\bar{c}) \leq \ell(c)$. To prove this, we consider the cover $\{c^{-1}(U_\alpha)\}$ of $[0,1]$. The Lebesgue covering lemma implies there is a sequence $0 = t_0 \leq \cdots \leq t_N = 1$ such that each $[t_{j-1},t_j]$ is contained in some $c^{-1}(U_\alpha)$; this means that the restriction $c|[t_{j-1},t_j]$ is contained in U_α. We can then let $\bar{c}$ be the union of the minimal geodesics in U_α joining $p_{j-1} = c(t_{j-1})$ to $p_j = c(t_j)$. It is clear that $\ell(\bar{c}) \leq \ell(c)$. Since each U_α is geodesically convex, it is

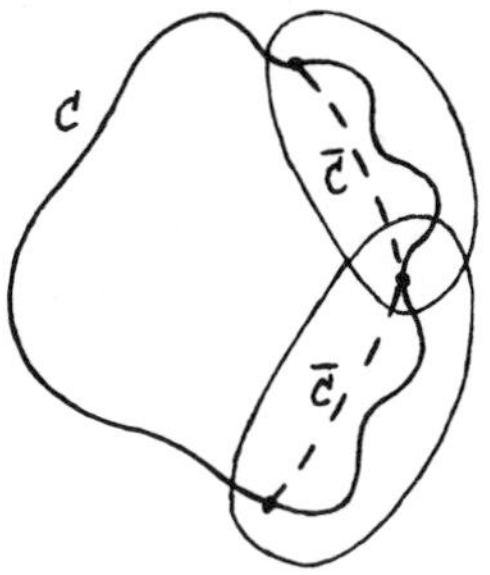

also clear that $\bar{c}$ is homotopic to c.

Now consider a particular free homotopy class of closed curves. The set of lengths of all closed curves in this free homotopy class has a greatest lower bound $\ell \geq 0$. Our aim is to find a closed curve c in this free homotopy class with length $\ell(c) = \ell$. We can certainly find a sequence $c^{(i)}$ of curves in this free homotopy class with

(1) $$\ell\left(c^{(i)}\right) \rightarrow \ell \ ;$$

we might as well assume that we also have

(2) $$\ell\left(c^{(i)}\right) < 2\ell \qquad \text{for all } i \ .$$

Finally, we can clearly assume that

(3) $$\text{each } c^{(i)} \text{ is special} \ .$$

Now in the definition of a special closed curve, no bound was placed on the number N of division points involved. However, if the N for any of our curves $c^{(i)}$ is sufficiently large, then we can always find a new $c^{(i)}$, in the same homotopy class, and with no larger length, but with a smaller N. To see why this is so, consider the [N/2] curves

$\gamma_1 = c^{(i)}$ from p_0 to p_2

$\gamma_2 = c^{(i)}$ from p_2 to p_4

$\vdots$

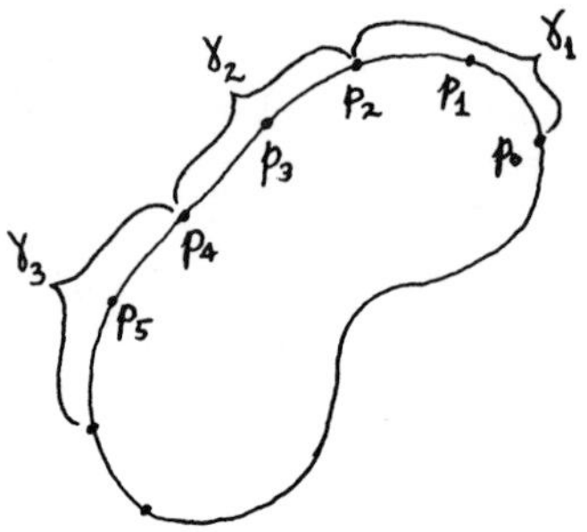

If any one of these curves has length $< \varepsilon$, then it lies entirely in some U_α, so we can replace it by a single minimal geodesic, thereby reducing N. Clearly:

$$\text{if } \ell(c) < \left[\frac{N}{2}\right]\varepsilon, \text{ then some } \gamma_\nu \text{ has length } < \varepsilon, \text{ so } N \text{ can be reduced.}$$

Using (2), we find:

$$\text{if } 2\ell < \left[\frac{N}{2}\right]\varepsilon, \text{ then some } \gamma_\nu \text{ has length } < \varepsilon, \text{ so } N \text{ can be reduced.}$$

Phrasing this slightly differently, we have:

$$\text{if } N > 2\left(\frac{2\ell}{\varepsilon} + 1\right), \text{ then } N \text{ can be reduced.}$$

Since this is true for all curves $c^{(i)}$, we can assume that all curves $c^{(i)}$ have $N \leq 2(2\ell/\varepsilon + 1)$. Since extra points can always be stuck in, we can actually assume that

(4) each $c^{(i)}$ is special with $N = N_0 = \left[2\left(\frac{2\ell}{\varepsilon} + 1\right)\right]$.

The remainder of the proof is now very simple. Let

$$p_0^{(i)}, p_1^{(i)}, \ldots, p_{N_0}^{(i)} = p_0^{(i)}$$

be the points determining $c^{(i)}$. Since M is compact, we may assume, by taking subsequences, that for each $j = 0,\ldots,N_0$ we have

$$\lim_{i\to\infty} p_j^{(i)} = p_j \in M .$$

Joining the pairs p_{j-1}, p_j by minimal geodesics, we obtain a closed curve c. Clearly

$$\begin{aligned} \ell(c) &= \sum_{j=1}^{N_0} d(p_{j-1},p_j) = \lim_{i\to\infty} \sum_{j=1}^{N_0} d(p_{j-1}^{(i)},p_j^{(i)}) \\ &= \lim_{i\to\infty} \ell\left(c^{(i)}\right) = \ell . \end{aligned}$$

To prove that c is in the same homotopy class as the $c^{(i)}$, we will carry out the construction a wee bit more carefully. We assume first that the original choice of the U_α was made so that there are geodesically convex sets $W_\alpha \supset \bar{U}_\alpha$. Now consider a fixed j. For each i, the points $p_{j-1}^{(i)}, p_j^{(i)}$ both lie in some U_α. Since there are only finitely many U_α, one of them, $U_{\alpha(j)}$ say, must contain both $p_{j-1}^{(i)}$ and $p_j^{(i)}$ for infinitely many i. By taking a subsequence, we can assume that all $p_{j-1}^{(i)}$ and $p_j^{(i)}$ are in $U_{\alpha(j)}$. There are only finitely many j to consider, so by taking subsequences we may assume that

$$\text{all } p_{j-1}^{(i)} \text{ and } p_j^{(i)} \text{ are in some } U_{\alpha(j)} , \quad j = 1,\ldots,N_0 .$$

This clearly implies that

$$p_{j-1} \text{ and } p_j \text{ are in } \bar{U}_{\alpha(j)} \subset W_{\alpha(j)} \qquad j = 1,\ldots,N_0 .$$

Using geodesic convexity of the W_α, it is easy to see that c is homotopic to any $c^{(i)}$. ■

[We now reinstate the normal terminology, and use "closed curves" for curves $c: [0,1] \to M$ with $c(0) = c(1)$ and $c'(0) = c'(1)$.]

We will end this chapter by considering a natural problem of deceptively simple appearance, which to this day remains unsolved. This problem will lead us to the study of "cut points," which are related to, but still quite different from the conjugate points which we have been considering all along. We have seen (Corollary 22) that a complete connected manifold with all sectional curvatures $\geq 1/r^2$ has diameter $\leq \pi r$. It is natural to assume that in a similar way, a complete manifold with small sectional curvatures should have large diameter -- if all $K(P) \leq 1/r^2$, then M should have diameter $\geq \pi r$. A counterexample to this conjecture is provided by projective space P_n, with constant curvature = 1, and diameter only $\pi/2$. And clearly, the larger the fundamental group, the smaller we might expect the diameter to be. An extreme case is represented by the torus, with infinite fundamental group. If we give the torus a flat metric, then $K \leq 1/r^2$ for every $r > 0$; on the other hand, we can also arrange for the diameter to be as small as we like. With the added hypothesis of simple connectivity, the conjecture still seems reasonable:

> A complete, simply connected, manifold with all $K(P) \leq 1/r^2$ should have diameter $\geq \pi r$.

One might expect to construct a proof of this conjecture along the following lines. We choose two points $p, q \in M$ at maximum distance apart, and consider a minimal geodesic $\gamma: [0,L] \to M$ which joins them. If γ has length $L < \pi r$, then we extend γ to $\bar{\gamma}: [0,L'] \to M$ with $L < L' < \pi r$, and the extended geodesic $\bar{\gamma}$ has no conjugate points, since $K(P) \leq 1/r^2$. Thus $\bar{\gamma}$ is a local minimum for length. Since p and q were already at the maximum distance apart, we might expect a contradiction to emerge from this construction. Of course, it can't, because we haven't used simple connectivity anywhere. The case of an ellipsoid shows where the problem lies. If γ is a geodesic joining the two furthest points p and q, and extending somewhat further beyond q to q', then γ is certainly not the shortest path between p and q'.

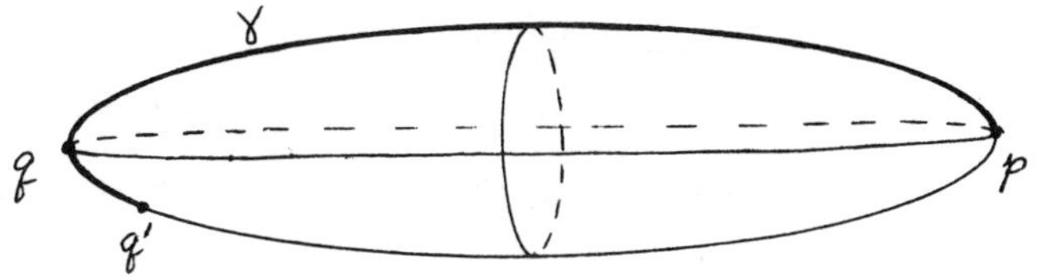

But it <u>is</u> the shortest path among nearby paths, since γ contains no conjugate points. It seems clear that there is little hope of attacking this problem if we consider only conjugate points, since they only give us information about the <u>local</u> length minimizing property of geodesics, and our problem is a global one.

The notion of a cut point was made precisely in order to deal with global minimizing properties of geodesics. For simplicity, we will deal only with the case of a complete Riemannian manifold $(M, \langle\ ,\ \rangle)$, and we will let $d\colon M \times M \to \mathbb{R}$ be the ordinary metric on M determined by the Riemannian metric $\langle\ ,\ \rangle$. Suppose we have a geodesic $\gamma\colon [0,\infty) \to M$ starting at a point $p = \gamma(0)$ in M, and parameterized by arclength. Consider the set

$$\begin{aligned} A &= \{t > 0\colon d(p,\gamma(t)) = t\} \\ &= \{t > 0\colon \gamma|[0,t] \text{ is a minimal geodesic}\}\ . \end{aligned}$$

It is clear that either $A = (0,\infty)$ or else A is a set of the form $(0,a]$.

If $A = (0,a]$, we say that $\gamma(a)$ is the cut point of p along the geodesic γ, while if $A = (0,\infty)$, we say that p has no cut point along γ. The cut locus $C(p) \subset M$ of p is then defined to be the set of all points which are cut points of p along some arclength-parameterized geodesic starting from p. We also define the cut locus $\tilde{C}(p)$ of p in M_p to be the set of all vectors $aX \in M_p$ for which X is a unit vector and $\exp aX$ is the cut point of p along the geodesic $\gamma_X(t) = \exp tX$. Thus $C(p) = \exp(\tilde{C}(p))$. On the other hand, we define the conjugate locus of p in M_p to be the set of all vectors $aX \in M_p$ for which X is a unit vector and a is the first conjugate value of 0 along γ_X. A particular ray in M_p may contain neither a point of the conjugate locus nor a point of the cut locus. But if it contains a point aX of the conjugate locus, then by Theorem 6 it must also contain a point $a'X$ of the cut locus, with $a' \le a$; briefly expressed, the cut point comes before or at the first conjugate point. Notice that if M is compact, then there is certainly a cut point along every geodesic; but there may not be any conjugate points, as is shown by the case of a compact surface of everywhere negative curvature.

Suppose now that M is a simply connected compact Riemannian manifold with all sectional curvatures $K(P) \le 1/r^2$. If there is any point $p \in M$ for which the cut locus $\tilde{C}(p)$ in M_p and the conjugate locus in M_p intersect, say at a vector $v_p \in M_p$, then the diameter of M_p must be $\ge \pi r$. For, on the one hand, the geodesic $t \mapsto \exp tv_p$ is a minimal geodesic from p to $q = \exp v_p$, so $d(p,q) = \|v_p\|$; and on the other hand, the point q is conjugate to p, so $\|v_p\| \ge \pi r$ by Corollary 20. Notice that the point q need not necessarily be the point furthest from p. For example, in the case of the ellipsoid on p. 325, the cut locus of p is the portion of the geodesic from

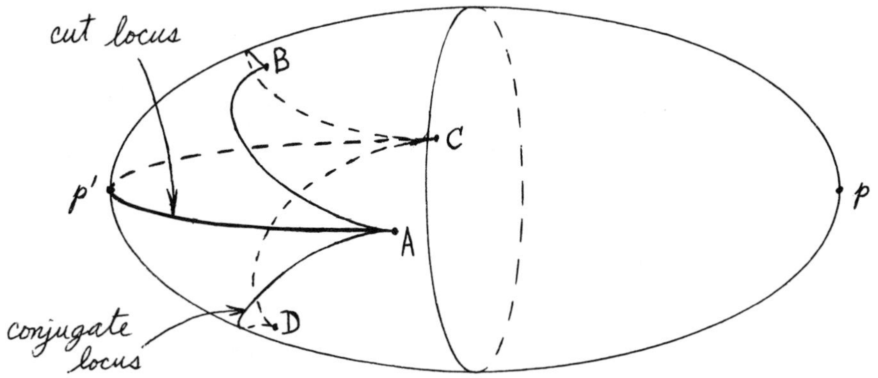

A to p' to C; so q is A or C.

Unfortunately, it is not known whether such a point p always exists on a simply connected compact manifold M. There are only partial results in this direction, and before giving one of them we will need to develop some basic properties of cut points.

One simple remark is sufficiently important to list as a separate result. Suppose that $\gamma\colon [0,\infty) \to M$ is a geodesic and $q = \gamma(t_0)$ comes <u>strictly before</u> the first cut point. Then, of course, any other geodesic ω from p to q must have length $\omega \geq t_0$. But actually the strict inequality holds:

<u>26. PROPOSITION</u>. Let M be complete, let $\gamma\colon [0,\infty) \to M$ be a geodesic parameterized by arclength, and let $\gamma(t_0)$ come strictly before the cut point $\gamma(a)$ (if there is one). Then any other geodesic ω from $p = \gamma(0)$ to $q = \gamma(t_0)$ has length $\omega > t_0$.

<u>Proof</u>. Suppose length $\omega = t_0 = \text{length } \gamma|[0,t_0]$. Choose $\varepsilon > 0$ so that $\gamma|[0,t_0+\varepsilon]$ is also minimal. Then $\gamma|[0,t_0+\varepsilon]$ has the same length as ω

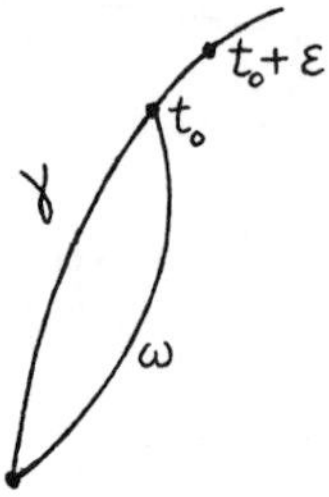

followed by $\gamma|[t_0,t_0+\varepsilon]$. But this compound curve has a corner, so it can be made shorter, and therefore $\gamma|[0,t_0+\varepsilon]$ is <u>not</u> of minimal length, a contradiction. ■

Notice that this argument does not work if $\gamma(t_0)$ is the cut point. In fact,

<u>27. PROPOSITION.</u> Let M be complete and let $\gamma\colon [0,\infty) \to M$ be a geodesic parameterized by arclength, with cut point $\gamma(a)$. Then at least one of the following holds:

(1) The number a is the first conjugate value of 0 along γ.

(2) There are at least two minimal geodesics from $p = \gamma(0)$ to $q = \gamma(a)$.

<u>Proof</u>. Choose a sequence $a_1 > a_2 > a_3 > \cdots$ with

$$\text{(1)} \qquad \lim a_i = a\ .$$

Let $b_i = d(p,\gamma(a_i)) < a_i$ and let X_i be unit vectors in M_p such that

$$t \mapsto \exp tX_i \qquad 0 \le t \le b_i$$

is a minimal geodesic from p to $\gamma(a_i)$. Naturally, all X_i are distinct from

$X = \gamma'(0)$. Then we also have

(2) $$\lim b_i = \lim d(p,\gamma(a_i)) = d(p,\gamma(a)) = a .$$

Equation (2) shows that the vectors $b_i X_i$ are contained in a compact subset of M_p. Choosing a subsequence if necessary, we can assume that

(3) $$\lim b_i X_i = aY , \qquad Y \in M_p \text{ a unit vector} .$$

Since $\exp aY = \lim \exp b_i X_i = \lim \gamma(a_i) = \gamma(a)$, the geodesic

$$t \longmapsto \exp tY \qquad 0 \leq t \leq a$$

is a minimal geodesic from p to q. So if $X \neq Y$ we have situation (2). To complete the proof we just have to show that if $X = Y$, then the number a must be conjugate to 0.

Now if $X = Y$, then $\lim b_i X_i = aY = aX = \lim a_i X$. But

(4) $$\exp(b_i X_i) = \gamma(a_i) = \exp(a_i X) .$$

So every neighborhood of aX contains infinitely many pairs $b_i X_i$, $a_i X$ on which $\exp$ has the same value; and these vectors $a_i X$ and $b_i X_i$ are definitely distinct (since the X_i are different from X, or, just as conclusively, since $b_i < a < a_i$). So aX must be a critical point of $\exp$. Thus Theorem 7 shows that a is a conjugate value of 0 along γ. ■

Proposition 27 can be used to derive several other facts about cut points. First of all, we have

28. PROPOSITION. In a complete manifold M, if q is the cut point of p along a geodesic γ from p to q, then p is the cut point of q along the geodesic $\bar{\gamma}$ obtained by traversing γ in the opposite direction.

Proof. The hypothesis implies that γ is a minimal geodesic from p to q. So $\bar{\gamma}$ is minimal from q to p; consequently, the cut point of $\bar{\gamma}$, if there is one, occurs past or at p. Now γ must satisfy one of the two alternatives

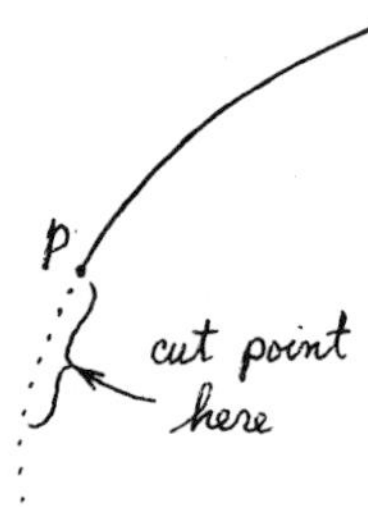

in Proposition 27.

(1) If q is conjugate to p along γ, then of course p is conjugate to q along $\bar{\gamma}$. The cut point must then occur before or at p. So it must occur at p.

(2) If there is another minimizing geodesic from q to p, then again p must be the cut point, since Proposition 26 shows that there cannot be another minimal geodesic to a point strictly before the cut point. ∎

Consider now the "sphere bundle" $S(M)$ of M, consisting of all unit tangent vectors at all points of M; this is a submanifold of the tangent bundle TM. Let $\mathbb{R}^* = \mathbb{R} \cup \{\infty\}$ be the real numbers together with some other set "∞". The ordering $<$ on $\mathbb{R}$ can be extended to $\mathbb{R}^*$ by defining $a < \infty$ for all $a \in \mathbb{R}$. We give $\mathbb{R}^*$ the order topology (a basis consists of all sets of the form $(a,b) \subset \mathbb{R}$, together with all sets of the form

$(a,\infty] = (a,\infty) \cup \{\infty\}$.) We now define a function $\mu\colon S(M) \to \mathbb{R}^*$ by

$$\mu(X) = \begin{cases} a > 0 & \text{if } aX \text{ is the cut point of } p \text{ along the} \\ & \text{geodesic } \gamma_X(t) = \exp tX \\ \infty & \text{if } \gamma_X \text{ has no cut point.} \end{cases}$$

29. THEOREM. If M is a complete manifold, then the function $\mu\colon S(M) \to \mathbb{R}^*$ is continuous.

Proof. Let $X_1, X_2, X_3, \ldots$ be a sequence of unit vectors in $S(M)$ converging to a unit vector $X \in M_p$, and suppose that $a_i = \mu(X_i)$ did not converge to $a = \mu(X)$. Since the values of μ lie in the compact set $\{\alpha \in \mathbb{R}^*\colon \alpha \geq 0\}$, we can assume, by choosing a subsequence, that a_i converges to some $\alpha \in \mathbb{R}^*$ with $\alpha \neq a$. Suppose for the moment that α is in $\mathbb{R}$ (and consequently all but finitely many a_i are in $\mathbb{R}$). Then a_iX_i converges to αX. Now it is clear from the definition of μ that

$$d(p, \exp a_iX_i) = a_i .$$

So

$$\begin{aligned} d(p, \exp \alpha X) &= d(p, \lim \exp a_iX_i) \\ &= \lim d(p, \exp a_iX_i) \\ &= \lim a_i = \alpha . \end{aligned}$$

This shows that the geodesic $t \mapsto \exp tX$ is minimizing on $[0,\alpha]$, and consequently $a = \mu(X) \geq \alpha$. If $\alpha = \infty$, it is easy to see that we must again have $a \geq \alpha$. So in order to derive a contradiction from the assumption that $a \neq \alpha$, we can assume that $a > \alpha$. Thus we are assuming that the vectors $a_i X_i$ approach the vector αX with $\alpha < a$. This means, in particular, that $\exp_*$ is not singular at αX, since a conjugate point cannot come before a cut point.

By choosing a subsequence, we can assume that either each $\gamma_i(t) = \exp(ta_i X_i)$ satisfies (1) of Proposition 27, or else that each γ_i satisfies (2). If each γ satisfies (1), then $\exp_*$ is singular at each $a_i X_i$. Hence $\exp_*$ is singular at $\alpha X = \lim a_i X_i$, a contradiction.

If each γ_i satisfies (2), then there are unit vectors $Y_i \neq X_i$ such that $\exp(a_i Y_i) = \exp(a_i X_i)$. Since $\exp$ is a diffeomorphism on some open neighborhood U of αX, these vectors $a_i Y_i$ must lie outside U. By choosing a subsequence, we can assume that Y_i approach a unit vector Y at p. Clearly Y also lies outside U, so $Y \neq X$. But

$$\begin{aligned} d(p, \exp \alpha Y) &= \lim d(p, \exp a_i Y_i) \\ &= \lim d(p, \exp a_i X_i) \\ &= \lim a_i = \alpha \,. \end{aligned}$$

This shows that $t \mapsto \exp tY$ is another minimal geodesic from p to $\exp(\alpha X)$. Since $\exp(\alpha X)$ comes before the cut point, this is impossible, by Proposition 26. ■

As a particular consequence of Theorem 29, the map $\mu\colon M_p \to \mathbb{R}^*$ is continuous for each $p \in M$. Therefore the set

$$E(p) = \{tv\colon v \in M_p \text{ is a unit vector and } 0 \leq t < \mu(v)\}$$

is clearly homeomorphic to an open n-dimensional cell.

30. THEOREM. Let M be complete. Then $\exp\colon M_p \to M$ maps $E(p)$ diffeomorphically onto an open subset of M, and M is the disjoint union of $\exp E(p)$ and $C(p)$.

Proof. Clearly $\exp_*$ is one-one on $E(p)$, since there are no vectors $w \in E(p)$ with $\exp w$ conjugate to p. To see that $\exp$ is one-one on $E(p)$, consider $w_1, w_2 \in E(p)$, with $\|w_1\| \leq \|w_2\|$, say. If we had $\exp w_1 = \exp w_2 = q$, then the geodesic $\omega(t) = \exp tw_1$ would have length from p to q less than or equal to that of the geodesic $\gamma(t) = \exp tw_2$. This contradicts Proposition 26, since q comes before the cut point of γ.

We next claim that $\exp E(p)$ and $C(p)$ are disjoint. If not, then there is $w \in E(p)$ and $u \in \tilde{C}(p)$ with $\exp w = \exp u = q$. If $\|u\| \leq \|w\|$, we have the same contradiction as before. If $\|w\| < \|u\|$ we still have a contradiction, for then $t \mapsto \exp tw$ would be a geodesic from p to q shorter than the geodesic $t \mapsto \exp tu$, which is minimal since $u \in \tilde{C}(p)$.

Finally, let q be any point of M. Then there is an arclength parameterized minimal geodesic $\gamma(t) = \exp tv$ from $p = \gamma(0)$ to $q = \gamma(a)$. Clearly $a \leq \mu(v)$. So $av \in E(p)$ or $av \in \tilde{C}(p)$. ∎

31. COROLLARY. If M is complete, and $p \in M$, then M is compact if and only if every geodesic through p has a cut point. In particular, if every geodesic through M has a conjugate point, then M is compact.

Proof. We already know that if M is compact, then every geodesic through p has a cut point. On the other hand, if every such geodesic has a cut point, then $\tilde{C}(p) \subset M_p$ is homeomorphic to S^{n-1}, and $E(p) \cup \tilde{C}(p)$ is a compact set.

So $M = \exp\big(E(p) \cup \tilde{C}(p)\big)$ is also compact. ■

One reason that the cut locus $C(p)$ is so important is that most of the topological properties of M are concentrated in $C(p)$. For it is easy to see that there is a deformation retraction of $M - \{p\}$ into $C(p)$ -- we just push points of $\exp E(p) - \{p\}$ along geodesics through p until they hit $C(p)$. Thus the homotopy groups $\pi_p(C(p))$ and singular homology $H_k(C(p))$ groups are isomorphic to $\pi_k(M - \{p\})$ and $H_k(M - \{p\})$, respectively; and there are well-known relations between these groups and $\pi_k(M)$ and $H_k(M)$.

Another simple consequence of Theorem 29 is:

<u>32. COROLLARY</u>. If M is complete, then the distance $d(p, C(p))$ between p and its cut locus is a continuous function of p.

We are now beginning to approach our goal, although it may not look like it. We first prove the following important lemma, which improves on Proposition 27 when $\gamma(a)$ is a special point in $C(p)$.

<u>33. LEMMA</u>. Let p be a point in a complete manifold M and let q be a point of $C(p)$ closest to p. Then at least one of the following holds:

(1) The point q is conjugate to p along some minimal geodesic from p to q.

(2) There are exactly two minimal geodesics from p to q, and their tangent vectors at q are negatives of each other, so that together they give a geodesic beginning and ending at p.

Proof. Suppose (1) does not hold. Then by Proposition 27 there are at least two minimal geodesics γ_1 and γ_2 from p to q. We will show that the tangent vectors of any two such γ_1 and γ_2 are negatives of each other at q; this clearly implies in addition that there is not a third minimal geodesic γ_3 from p to q.

Let K_1 be a "cone" formed by the points on all geodesics of length $d(p,q)$ whose tangent vectors lie in a neighborhood of γ_1' at p; and define K_2 similarly. The set $\mathcal{E}(K_1)$ of all the endpoints of the geodesics making up K_1 is a hypersurface containing q. If the tangent vectors of γ_1 and γ_2 are not negatives of each other at q, then $\mathcal{E}(K_1)$ crosses the corresponding

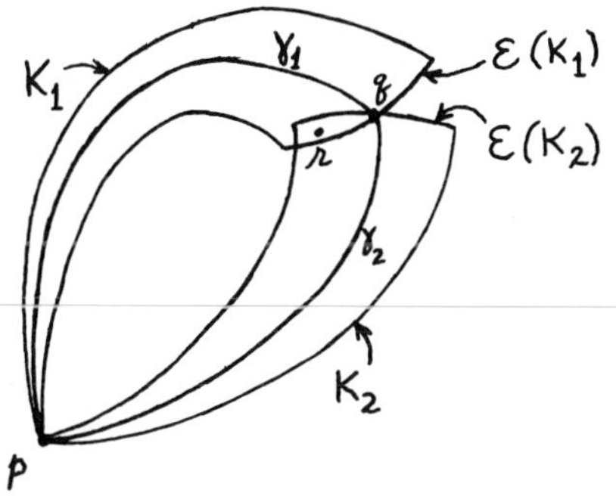

hypersurface $\mathcal{E}(K_2)$. It follows that there is a point r with

$$r \in [K_1 - \mathcal{E}(K_1)] \cap [K_2 - \mathcal{E}(K_2)] .$$

Now r is joined to p by a geodesic $\bar{\gamma}_1$ lying in K_1, and a geodesic $\bar{\gamma}_2$ lying in K_2. Since q is the point of C(p) closest to p, the point r must come strictly before the first conjugate point on both γ_1 and γ_2. But this is impossible by Proposition 26. ■

When the point p of Lemma 33 is very special we can say even more.

34. LEMMA. Let p be a point in a complete manifold M for which $d(p,C(p))$ is smallest, and let q be a point of $C(p)$ closest to p. Suppose that q is not conjugate to p along a minimal geodesic from p to q. Then there is a closed geodesic made up of two minimal geodesics from p to q.

Proof. Since q is a point of $C(p)$ closest to p, there are, by Lemma 33, exactly two minimal geodesics γ_1 and γ_2 from p to q, and their tangent vectors are negatives of each other at q. But our hypotheses imply also that p is a point of $C(q)$ closest to q. So there are also exactly two minimal geodesics from q to p, namely γ_1 and γ_2 again, and their tangent vectors are negatives of each other at p. ∎

35. THEOREM (KLINGENBERG). Let M be a compact simply connected even-dimensional Riemannian manifold whose sectional curvatures satisfy $0 < K(P)$ for all 2-dimensional $P \subset M_q$, for all $q \in M$. Then for some point $p \in M$, the cut locus $\tilde{C}(p)$ in M_p and the conjugate locus in M_p intersect.

Consequently, if we also have $K(P) \leq 1/r^2$ for some $r > 0$, then M has diameter $\geq \pi r$.

Proof. Let p be a point for which $d(p,C(p))$ has the smallest value, L say, and let $q \in C(p)$ be a point closest to p. We claim that q is conjugate to p along a minimal geodesic. Suppose it were not. Then by Lemma 34, there is a closed geodesic $\gamma\colon [0,1] \to M$, of length $2L$, made up of two minimal geodesics from p to q. By Synge's Lemma, there is a variation

$\alpha\colon [0,\varepsilon)\times[0,1]\to M$ of α such that all $\bar{\alpha}(u)$ are closed curves of length $< 2L$ for $u > 0$. This means that for each $u > 0$, the set of points $\{\alpha(u,t)\}$ is the image under $\exp_{\alpha(u,0)}$ of a set $\mathcal{S}_u$ in $E(\alpha(u,0))$; this set $\mathcal{S}_u$ is a closed curve, since $\exp_{\alpha(u,0)}$ is a diffeomorphism on $E(\alpha(u,0))$. But the points of γ are <u>not</u> all in $\exp_p E(p)$. Instead, the set $\{\gamma(t)\} - \{q\}$ is the image of a set in $E(p)$; this set consists of two open rays from $0 \in M_p$ to two vectors $v, -v \in M_p$. We will show that such a situation cannot arise.

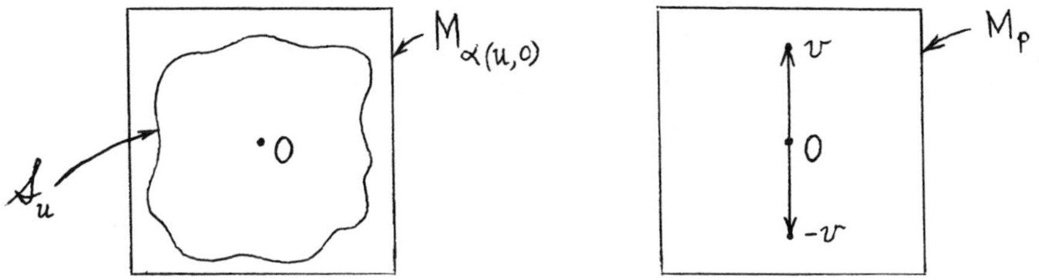

Let $\mathcal{S} = \bigcup_{u>0} \mathcal{S}_u$, and consider the set C of all points in $\mathcal{S}$ corresponding to points of the form $\alpha(u, 1/2)$ for $u > 0$. We claim that C is

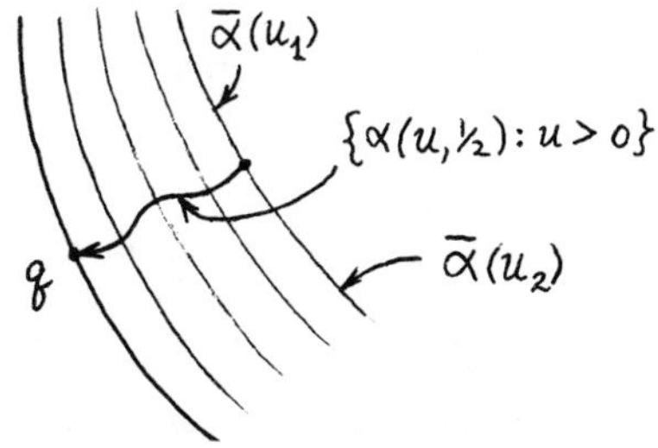

connected. This is because the map

$$u \mapsto \exp_{\alpha(u,0)}{}^{-1}(\alpha(u, 1/2))$$

is continuous, where $\exp_{\alpha(u,0)}{}^{-1}$ denotes the inverse of the map $\exp_{\alpha(u,0)}: E(\alpha(u,0)) \longrightarrow M$. Similarly, if C_n is the set of all points in $\mathcal{S}$ corresponding to points of the form $\alpha(u, 1/2)$ for $0 < u < 1/n$, then C_n is also connected.

Now consider the set B_n of points in $\mathcal{S}$ corresponding to points of the form

$$\alpha(u,t) \quad \text{for} \quad t \in \left(\frac{1}{2}-\frac{1}{n}, \frac{1}{2}+\frac{1}{n}\right) \quad \text{and} \quad 0 < u < \frac{1}{n} .$$

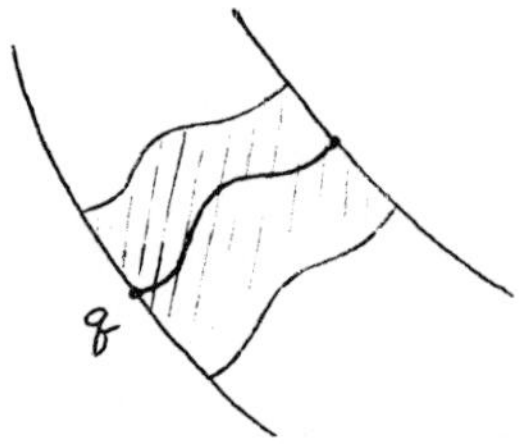

The set B_n is also connected: it consists of the union of connected sets in $\mathcal{S}_u$ for $0 < u < 1/n$, and each of these contains a point of the connected set C_n.

Finally, consider the set

$$B = \bigcap_n \bar{B}_n .$$

As a decreasing intersection of compact, connected sets, it is also connected. It is clear that it is completely contained in M_p, and that it contains both v and $-v$. Therefore it must contain some other vector $w \in M_p$. But then it is easy to see that $t \longmapsto \exp tw$ is another minimal geodesic from p to q, contradicting Lemma 33. ∎

PROBLEMS

1\. (a) Prove the following "delicate Sturm Comparison theorem": Let f and h be two continuous functions $f \leq h$ on an open interval (a,b), and let ϕ and η be two functions satisfying

$$(1) \qquad \phi'' + f\phi = 0$$

$$(2) \qquad \eta'' + h\eta = 0$$

on (a,b). Assume that $\phi(t) \neq 0$ for $t \in (a,b)$, and that

$$\lim_{t\to a+} \phi(t) = \lim_{t\to b-} \phi(t) = 0.$$

Then η must have a zero on (a,b), unless $f = h$ everywhere on (a,b) and η is a constant multiple of ϕ on (a,b).

(b) In the situation considered on pp. 339-340, let $V = d\gamma/dt$, and let Y be the unit vector field along $\gamma|(0,L)$ which is perpendicular to V and tangent to image α along $\gamma|(0,L)$. Let $W = fV + \phi Y$ be the decomposition of Proposition 3 for image α; note that f and ϕ have (left- and right-hand) limits 0 at 0 and L. Conclude that $f = 0$ and that ϕ satisfies the hypotheses of part (a). Thus obtain a contradiction, demonstrating that we cannot have $L < \pi r$.

2\. (a) Let M be a non-orientable C^∞ manifold. Let $\tilde{M}$ be the set of all orientations μ_p for M_p, for all $p \in M$, and define $\pi \colon \tilde{M} \longrightarrow M$ to be the map which takes each of the two orientations of M_p into p. Show that $\tilde{M}$ has a natural C^∞ structure that makes $\pi\colon \tilde{M} \to M$ a 2-fold covering space of M, and that $\tilde{M}$ is orientable.

(b) If M is a compact, connected, non-orientable, even-dimensional Riemannian manifold with all sectional curvatures positive, then $\pi_1(M) \approx \mathbb{Z}_2$.

(c) Let $c, \gamma\colon [0,1] \to M$ be freely homotopic closed curves, and let $\tilde{c}, \tilde{\gamma}\colon [0,1] \to \tilde{M}$ be curves with $\pi \circ \tilde{c} = c$ and $\pi \circ \tilde{\gamma} = \gamma$. Then $\tilde{c}(0) = \tilde{c}(1)$ if and only if $\tilde{\gamma}(0) = \tilde{\gamma}(1)$. Hence if M is non-orientable, then there is a closed curve $c\colon [0,1] \to M$ of minimal length such that $\tilde{c}(0) \neq \tilde{c}(1)$.

(d) If M is a compact, connected, odd-dimensional Riemannian manifold with all sectional curvatures positive, then M is orientable.

Chapter 9. Variations of Length, Area, and Volume

The classical calculus of variations was extended, quite soon after its inception, to deal with problems in several variables. In this Chapter we will use these methods to study n-dimensional submanifolds $M \subset (N^m, <\ ,\ >)$ with minimal n-dimensional volume. Thus the material of this Chapter may be regarded as a generalization of the study of geodesics which was carried out in Chapter I.9 and in Chapter 8 of this Volume. One big difference, aside from the greater difficulties to be encountered, is the fact that our results are truly extrinsic -- all our theorems will be about the submanifolds of N, not about the structure of N itself.

When we look for curves which have the shortest length among all curves between 2 fixed endpoints, we find that the only possible candidates are geodesics (provided that we parameterize all curves proportionally to arclength). For the 2-dimensional analogue of this situation, we replace the 2 fixed endpoints in our Riemannian manifold $(N, <\ ,\ >)$ by a compact 1-dimensional manifold M_0 (diffeomorphic to a finite union of circles). We then consider all immersed compact 2-dimensional manifolds-with-boundary M satisfying $\partial M = M_0$. Among these, we seek one which has minimum area; by the <u>area</u> of an

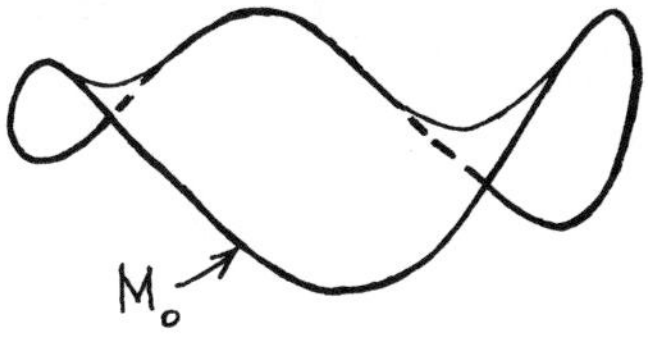

immersed surface $f\colon M \to N$ we mean the integral over M of the (2-dimensional) volume element dA determined by the induced metric $f^*<\ ,\ >$ (when M is oriented, we can consider dA to be a 2-form). Our approach to this problem will be similar to our approach in the analogous 1-dimensional case;

we will find "critical points" for the area function. One important difference is that no particular parameterization of M will play a favored role.

Before we try to find a general formula for the "variation of area," we will first investigate the case of surfaces in $\mathbb{R}^3$, which leads to an extraordinarily rich theory, of a very special sort. At first, we will not even consider general immersed surfaces-with-boundary, but only immersions $f: D \to \mathbb{R}^3$, where $D \subset \mathbb{R}^2$ is a compact 2-dimensional manifold-with boundary. By a <u>variation</u> α of f we will mean a C^∞ function $\alpha: (-\varepsilon,\varepsilon)\times D \to \mathbb{R}^3$ with $\alpha(0,p) = f(p)$ for $p \in D$; for each $u \in (-\varepsilon,\varepsilon)$, we then define the function $\bar{\alpha}(u): D \to \mathbb{R}^3$ by $\bar{\alpha}(u)(p) = \alpha(u,p)$. Since $f = \bar{\alpha}(0)$ is an immersion, the same must be true of $\bar{\alpha}(u)$ for sufficiently small u (one needs compactness of D to prove this), so with no loss of generality we can assume that all $\bar{\alpha}(u)$ are immersions. As in the previous Chapter, we define the <u>variation vector field</u> W by

$$W(p) = \frac{\partial \alpha}{\partial u}(0,p) \;;$$

notice that $W(p) \in \mathbb{R}^3{}_{f(p)}$, so that W is a "vector field along f."

In almost every differential geometry book under the sun, the only variations considered are those of the form

$$(1) \quad \alpha(u,t_1,t_2) = f(t_1,t_2) + u\cdot\phi(t_1,t_2)\cdot n(t_1,t_2) \;,$$

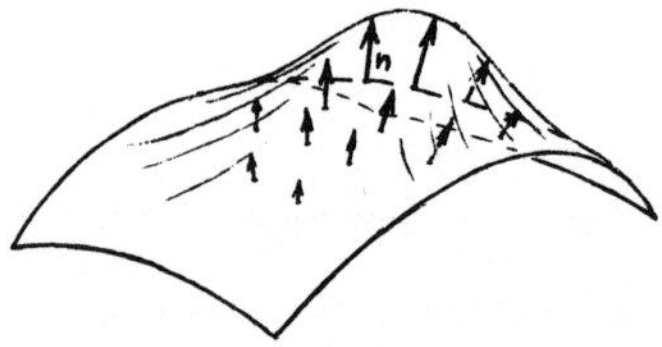

where $n(t_1,t_2)$ is the unit normal at $f(t_1,t_2)$, and ϕ is some C^∞ function. Thus α is a very special sort of "normal variation" -- each curve $u \mapsto \alpha(u,t_1,t_2)$ is a straight line normal to the surface f, and the variation vector field W is just

$$W(t_1,t_2) = \phi(t_1,t_2)\cdot n(t_1,t_2) .$$

The decision to ignore more general variations is partially justified by the following observations. In the first place, if we are given a variation $\alpha\colon (-\varepsilon,\varepsilon)\times D \to \mathbb{R}^3$ of f, then we can usually find a new variation $\beta\colon (-\varepsilon',\varepsilon')\times D \to \mathbb{R}^3$ of f such that

(a) $\dfrac{\partial\beta}{\partial u}(0,p)$ is perpendicular to $f(D)$,

(b) the surfaces $\bar{\alpha}(u)(D)$ and $\bar{\beta}(u)(D)$ are always the same, even though the parameterizations $\bar{\alpha}(u)$ and $\bar{\beta}(u)$ may be different.

To do this, we assume that the surfaces $\bar{\alpha}(u)(D)$ are all disjoint, and we consider the curves, parameterized by arclength, which are orthogonal to all the surfaces $\bar{\alpha}(u)(D)$. Then we let $\beta(u,p)$ be the unique point of $\bar{\alpha}(u)(D)$

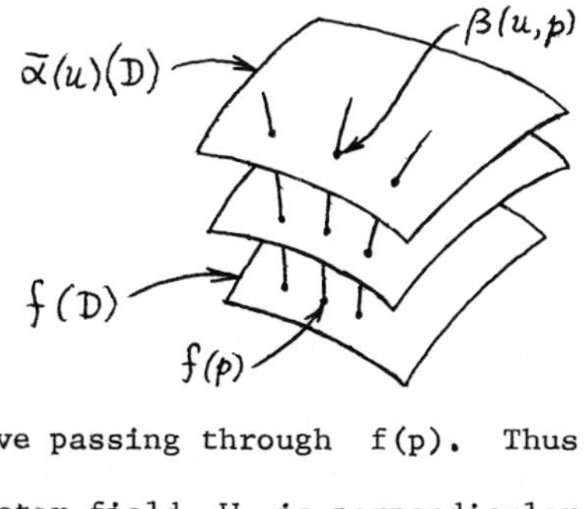

which lies on the curve passing through $f(p)$. Thus we can usually assume that our variation vector field W is perpendicular to $f(D)$. In the second place, when we take the derivative at 0 of the areas of the surfaces $\bar{\alpha}(u)(D)$, we naturally expect that the answer will depend only on W, exactly as in the case of arclength. If this expectation is correct, then we can even assume that α is of the form (1). This line of argument, intuitively reasonable as it may be, is perhaps not very satisfying. But we will have adequate opportunity to consider more general variations later on, when we re-examine surfaces immersed in an arbitrary Riemannian manifold. So for the time being, let us endulge in the classical simplification, which makes the calculations so much more manageable.

For the special variation given by (1) we have

$$(2) \quad \frac{\partial\alpha}{\partial t_i}(u,t_1,t_2) = \frac{\partial f}{\partial t_i} + u\cdot\frac{\partial\phi}{\partial t_i}\cdot n + u\cdot\phi\cdot\frac{\partial n}{\partial t_i} \qquad \text{[all partials on the right evaluated at } (t_1,t_2)].$$

Let

$$(3) \qquad g_{ij}(u)(t_1,t_2) = \left\langle \frac{\partial\alpha}{\partial t_i}(u,t_1,t_2), \frac{\partial\alpha}{\partial t_j}(u,t_1,t_2) \right\rangle ,$$

so that the functions $g_{ij}(u)$ are the components of $\bar{\alpha}(u)^*\langle\ ,\ \rangle$; in particular, then, $g_{ij} = g_{ij}(0)$ are the components of $f^*\langle\ ,\ \rangle$. Since

$$\left\langle \frac{\partial f}{\partial t_i}, \frac{\partial n}{\partial t_j} \right\rangle = -\ell_{ij} , \qquad \left\langle \frac{\partial f}{\partial t_i}, n \right\rangle = 0 ,$$

equations (2) and (3) give

$$g_{ij}(u) = g_{ij} - 2u\phi\ell_{ij} + u^2 a_{ij}(u) ,$$

where $(u,t_1,t_2) \mapsto a_{ij}(u)(t_1,t_2)$ is some continuous function. From this we obtain

$$\begin{aligned} \det g_{ij}(u) &= \det g_{ij} - 2u\phi[g_{11}\ell_{22} + g_{22}\ell_{11} - 2g_{12}\ell_{12}] + u^2 b(u) \\ &= (\det g_{ij})[1 - 4u\phi H] + u^2 b(u) \text{ , by formula (B) of Chapter 3;} \end{aligned}$$

in this equation, $b(u)$ is a function having the same property as the $a_{ij}(u)$. It is now easy to see that

$$\left.\frac{\partial}{\partial u}\right|_{u=0} \det g_{ij}(u) = -\,4\phi H \det g_{ij} ,$$

from which we obtain

$$\left.\frac{\partial}{\partial u}\right|_{u=0} \sqrt{\det g_{ij}(u)} = -\,2\phi H\sqrt{\det g_{ij}} .$$

Let us denote by $A(\bar{\alpha}(u))$ the area of the immersed surface $\bar{\alpha}(u)\colon D \to \mathbb{R}^3$. Then

$$(*) \qquad \left.\frac{dA(\bar{\alpha}(u))}{du}\right|_{u=0} = \left.\frac{d}{du}\right|_{u=0} \int_D \sqrt{\det g_{ij}(u)}\; dt_1 dt_2$$

$$= \int_D \left.\frac{\partial}{\partial u}\right|_{u=0} \sqrt{\det g_{ij}(u)}\; dt_1 dt_2$$

$$= -\int_D 2\phi H\sqrt{\det g_{ij}}\; dt_1 dt_2$$

$$= -\int_D 2\phi H\; dA , \qquad dA = \text{volume element on } D \text{ for the metric } f^*\langle\ ,\ \rangle.$$

We are now ready to draw a conclusion.

1. PROPOSITION. Let M be a compact 2-dimensional manifold-with-boundary, and $f\colon M \to \mathbb{R}^3$ an immersion such that $f(\partial M)$ is a given compact 1-manifold $M_0 \subset \mathbb{R}^3$. If M is a critical point for the area function, among all such immersions, then M must be a minimal surface ($H = 0$ everywhere). In particular, if M has the minimum area among all such surfaces, then M is a minimal surface.

Proof. Suppose that $H(p) \neq 0$ for some $p \in M$, say $H(p) > 0$. Choose a neighborhood $\mathscr{D}$ of p so small that $H(q) > 0$ for all $q \in \mathscr{D}$. We can assume that $f(\mathscr{D})$ is also the image $g(D)$ for some immersion $g\colon D \to \mathbb{R}^3$ of a compact 2-dimensional manifold-with-boundary $D \subset \mathbb{R}^2$. Let $\phi\colon D \to \mathbb{R}$

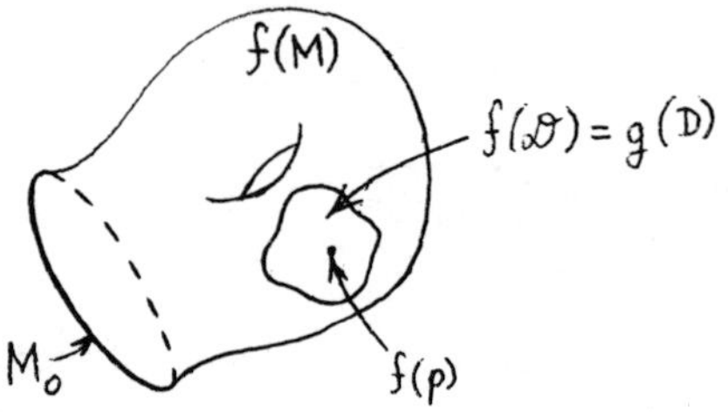

be a C^∞ function which is ≥ 0 on D and $= 0$ in a neighborhood of ∂D. We can then define a variation α of f by letting

$$\alpha(u,p) = f(p) \qquad p \notin \mathscr{D}$$

$$\alpha(u,p) = f(p) + u\cdot\phi(\bar{p})\cdot n(\bar{p}) \,, \qquad \text{for } \bar{p} = g^{-1}(f(p)),\ p \in \mathscr{D} \,.$$

Formula (*) shows that

$$\left.\frac{dA(\bar{\alpha}(u))}{du}\right|_{u=0} = -\int_D 2\phi\tilde{H}\, dA \,,$$

where $\tilde{H}(t_1,t_2)$ is the mean curvature H at $f^{-1}(g(t_1,t_2))$. Since $\tilde{H} > 0$ everywhere on D, and since ϕ is ≥ 0 on D, but is not identically 0, the integral is positive; this is a contradiction. ■

In the statement of Proposition 1 we have deliberately not claimed that a minimal surface actually is a critical point for the area function. We found that $H = 0$ is a necessary condition for a critical point by considering variations α which, first of all, vanish outside a small region, and, second of all, are normal to the surface. It is conceivable (well, just barely) that if we considered arbitrary variations, we would obtain another condition more stringent than $H = 0$. So we will have to wait a bit before we can assert with assurance that minimal surfaces are precisely the critical points for the area function. On the other hand, the second part of Proposition 1 is already the best we can hope for: among those surfaces with boundary M_0, the one with minimum area must be a minimal surface; but we would not expect every minimal surface to have this property, any more than we expect every geodesic to be the shortest length between its endpoints.

The next result only begins to suggest how special minimal surfaces are.

2. PROPOSITION. Let M be an immersed surface in $\mathbb{R}^3$ with normal map $n\colon M \to S^2$. If M is minimal, then n is conformal (angle-preserving) at all points where $K \neq 0$. Conversely, if n is conformal, and M is connected, then either M is a minimal surface, with $K < 0$ everywhere, or M is part of a sphere.

Proof. Recall (Lemma II.7-20) that the map n is conformal at p if and only if there is $\mu(p) \neq 0$ such that

$$(1) \qquad \langle n_* X_p, n_* Y_p \rangle = \mu(p) \langle X_p, Y_p \rangle \qquad X_p, Y_p \in M_p .$$

We will make use of the third fundamental form III of M, which was defined in Chapter 2:

$$\begin{aligned} \mathrm{III}(p)(X_p, Y_p) &= \langle n_* X_p, n_* Y_p \rangle \\ &= \langle {n_*}^2 (X_p), Y_p \rangle , \qquad \text{for } X_p, Y_p \in M_p . \end{aligned}$$

By Proposition 2-6 we have

$$\text{III} - 2H\cdot\text{II} + K\cdot\text{I} = 0 . \tag{2}$$

Suppose first that M is minimal. Then (2) gives $\text{III} = -K\cdot\text{I}$, which shows that (1) holds with $\mu(p) = -K(p)$; hence n is conformal when $K(p) \neq 0$.

Conversely, suppose that n is conformal, so that it satisfies (1) for some function μ which is non-zero, and hence obviously positive. Then (2) gives

$$(K+\mu)\cdot\text{I} - 2H\cdot\text{II} = 0 .$$

At a point p with $H(p) \neq 0$ we can therefore write $\text{II}(p)$ as a multiple of $\text{I}(p)$, which means that p is an umbilic. At a point p with $H(p) = 0$, we have $K(p) = -\mu(p) < 0$, so p cannot be an umbilic. In short,

$$p \text{ is an umbilic if and only if } H(p) \neq 0 .$$

The set of umbilics is thus open. But it is also closed. So either: no points are umbilics, and $H = 0$ everywhere; or all points p are umbilics, and these umbilics are not flat points (since $H(p) \neq 0$), so M is part of a sphere. ∎

Back in Volume II, p. 334, we mentioned that every 2-dimensional Riemannian manifold M is locally conformally equivalent to the plane: around each point p we can choose an "isothermal" coordinate system for which we have $g_{ij} = \mu\delta_{ij}$. Addendum 1 contains a proof of this result for general Riemannian 2-manifolds. On the other hand, Proposition 2 provides an easy way of introducing isothermal coordinates around any non flat point p of a minimal surface M. We need only find a conformal map $\sigma\colon S^2 - \{\text{point}\} \to \mathbb{R}^2$, and then $\sigma\circ n$ will be the required isothermal coordinate system in a neighborhood of p. But we already know such a conformal map σ, namely stereographic projection. It will be convenient to use the second version of stereographic projection, given on p. 158. Recall that

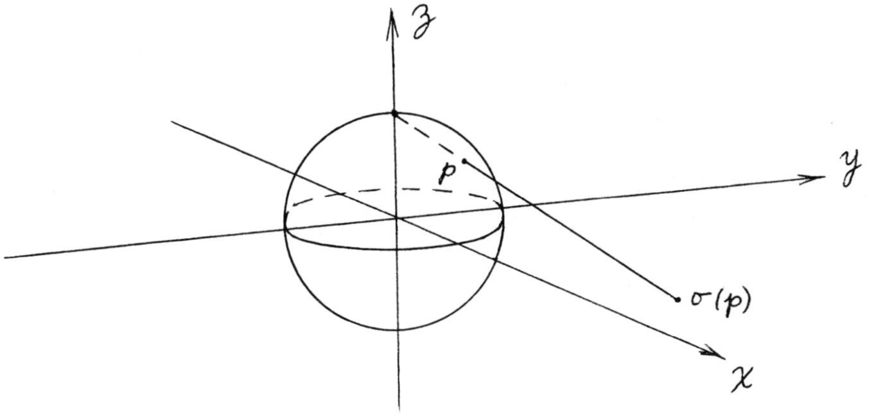

$$\sigma(a,b,c) = \left(\frac{a}{1-c}, \frac{b}{1-c}\right)$$

$$\sigma^{-1}(x,y) = \left(\frac{2x}{x^2+y^2+1}, \frac{2y}{x^2+y^2+1}, \frac{x^2+y^2-1}{x^2+y^2+1}\right).$$

Naturally, we can find a conformal map $S^2 - \{p\} \to \mathbb{R}^2$ for any other point p merely by first rotating S^2 so that p goes to $(0,0,1)$.

Unfortunately, this method does not work at a flat point. To include such points we can, of course, appeal to the result of Addendum 1, valid for all surfaces. However, for minimal surfaces there is a considerably easier argument that still works at all points.

3. PROPOSITION. Isothermal coordinates can be introduced around any point of a minimal surface $M \subset \mathbb{R}^3$.

Proof. We can assume that M is the graph of a function $h: U \to \mathbb{R}$, for $U \subset \mathbb{R}^2$, so that M is the image of the map $f(x,y) = (x,y,h(x,y))$. Introducing the classical notation

$$p = \frac{\partial h}{\partial x}, \qquad q = \frac{\partial h}{\partial y}$$

$$r = \frac{\partial^2 h}{\partial x^2}, \qquad s = \frac{\partial^2 h}{\partial x \partial y}, \qquad t = \frac{\partial^2 h}{\partial y^2},$$

and using equation (B') on p. III.201, we have

$$(1) \qquad (1+q^2)r - 2pqs + (1+p^2)t = 0 .$$

Setting

$$W = \sqrt{1+p^2+q^2},$$

we note that

$$\frac{\partial}{\partial x}\left(\frac{1+q^2}{W}\right) - \frac{\partial}{\partial y}\left(\frac{pq}{W}\right) = -\frac{p}{W^3}\left[(1+q^2)r - 2pqs + (1+p^2)t\right]$$

$$= 0 \qquad \text{by (1)},$$

and similarly

$$\frac{\partial}{\partial x}\left(\frac{pq}{W}\right) - \frac{\partial}{\partial y}\left(\frac{1+p^2}{W}\right) = 0 .$$

This means that we can locally find functions α and β with

$$(2) \qquad \begin{array}{llll} \text{(a)} & \dfrac{\partial\alpha}{\partial x} = \dfrac{1+p^2}{W} & \text{(c)} & \dfrac{\partial\beta}{\partial x} = \dfrac{pq}{W} \\[2ex] \text{(b)} & \dfrac{\partial\alpha}{\partial y} = \dfrac{pq}{W} & \text{(d)} & \dfrac{\partial\beta}{\partial y} = \dfrac{1+q^2}{W} . \end{array}$$

Consider the transformation of Lewy:

$$T(x,y) = \big(x+\alpha(x,y),\ y+\beta(x,y)\big) \ .$$

Its Jacobian is

$$J(T)(x,y) = \begin{pmatrix} 1+\frac{1+p^2}{W} & \frac{pq}{W} \\ \frac{pq}{W} & 1+\frac{1+q^2}{W} \end{pmatrix} ,$$

with determinant

$$2 + \frac{2+p^2+q^2}{W} \geq 2 \ .$$

So T has an inverse locally, and

$$J(T^{-1})(T(x,y)) = [J(T)(x,y)]^{-1} = \frac{1}{\det J(T)(x,y)} \begin{pmatrix} 1+\frac{1+q^2}{W} & -\frac{pq}{W} \\ -\frac{pq}{W} & 1+\frac{1+p^2}{W} \end{pmatrix}$$

$$= C \begin{pmatrix} 1+W+q^2 & -pq \\ -pq & 1+W+p^2 \end{pmatrix} \qquad \text{for some } C \ .$$

So

$$J(f\circ T^{-1})(T(x,y)) = J(f)(x,y)\cdot J(T^{-1})(T(x,y))$$

$$= C\cdot \begin{pmatrix} 1 & 0 \\ 0 & 1 \\ p & q \end{pmatrix} \begin{pmatrix} 1+W+q^2 & -pq \\ -pq & 1+W+p^2 \end{pmatrix}$$

$$= C \cdot \begin{pmatrix} 1+W+q^2 & -pq \\ -pq & 1+W+p^2 \\ p+pW & q+qW \end{pmatrix} .$$

It is easy to check that the two column vectors in this matrix are orthogonal, and that they have the same squared length

$$(1+p^2+q^2)(2W+2+p^2+q^2) .$$

Thus $f \circ T^{-1}$ is conformal, and its inverse is the desired isothermal coordinate system. ■

The reader has probably noticed the similarity between this proof and the proof of Jörgen's Theorem (7-45). As a matter of fact, that proof of Jörgen's Theorem was motivated by manipulations with the minimal surface equation, and the original application of Jörgen's Theorem itself had been to reprove a result about minimal surfaces:

<u>4. THEOREM (BERNSTEIN)</u>. Planes are the only minimal surfaces in $\mathbb{R}^3$ which are the graph of a function $h: \mathbb{R}^2 \to \mathbb{R}$.

<u>Proof</u>. Suppose we have a function $h: \mathbb{R}^2 \to \mathbb{R}$ satisfying equation (1) in the previous proof. Then the functions α and β of equation (2) are defined on all of $\mathbb{R}^2$ (since $\mathbb{R}^2$ is simply connected). From (b) and (c) of equation (2) we see that there is a function $\phi: \mathbb{R}^2 \to \mathbb{R}$ with

$$\phi_x = \alpha \qquad \text{and} \qquad \phi_y = \beta .$$

Together with (a) and (d), we then have

$$\phi_{xx} = \frac{1+p^2}{W}, \qquad \phi_{xy} = \frac{pq}{W}, \qquad \phi_{yy} = \frac{1+q^2}{W},$$

which implies that

$$\phi_{xx}\phi_{yy} - (\phi_{xy})^2 = 1 .$$

Jörgen's theorem implies that

$$\frac{1+p^2}{W}, \qquad \frac{pq}{W}, \qquad \frac{1+q^2}{W}$$

are constants. A simple exercise shows that p and q must be constants. ∎

The manipulations of the past few pages were undoubtedly unpleasant (not to say, slightly unmotivated), but they were really worth the trouble, because isothermal coordinates play such a vital role in the study of minimal surfaces.

5. PROPOSITION. If $f\colon M \to \mathbb{R}^3$ is a minimal immersion, and (u^1, u^2) is an isothermal coordinate system on M, then

$$\frac{\partial^2 f^i}{\partial u^1 \partial u^1} + \frac{\partial^2 f^i}{\partial u^2 \partial u^2} = 0 \qquad i = 1,2,3 .$$

Conversely, if this equation holds for a collection of isothermal coordinate systems covering M, then f is a minimal immersion.

Proof. By equation (7) on p. 198 we have

$$\Delta f = 2Hn ,$$

where n is the normal map, and Δ is the Laplacian. Therefore f is minimal if and only if $\Delta f^i = 0$ for $i = 1,2,3$. Since our coordinate system (u^1,u^2) is isothermal, Problem 7-23 shows that

$$\Delta f^i = \frac{1}{E}\left(\frac{\partial^2 f^i}{\partial u^1 \partial u^1} + \frac{\partial^2 f^i}{\partial u^2 \partial u^2}\right) . \blacksquare$$

Let us rephrase Proposition 5 just slightly. If $u = (u^1,u^2)\colon U \to V \subset \mathbb{R}^2$ is an isothermal coordinate system on $U \subset M$, and $f\colon M \to \mathbb{R}^3$ is a minimal immersion, then each real-valued function $g^i = f^i \circ u^{-1}\colon V \to \mathbb{R}$ satisfies "Laplace's equation"

$$\frac{\partial^2 g^i}{\partial x^2} + \frac{\partial^2 g^i}{\partial y^2} = 0 ,$$

where $\partial/\partial x$ and $\partial/\partial y$ denote the ordinary partial derivatives in $\mathbb{R}^2$. Now at this point complex analysis comes rushing in, waving its hands excitedly in its eagerness to enlighten us. It is a well-known result that locally any such function is the real part of a complex analytic function; we recall the argument briefly. Suppose that g satisfies Laplace's equation

$$\frac{\partial^2 g}{\partial x^2} + \frac{\partial^2 g}{\partial y^2} = 0 ,$$

which we can also write as

$$\frac{\partial\left(\frac{\partial g}{\partial x}\right)}{\partial x} = \frac{\partial\left(-\frac{\partial g}{\partial y}\right)}{\partial y} .$$

According to Proposition I.6-0, there is locally a function h such that

$$\frac{\partial h}{\partial x} = -\frac{\partial g}{\partial y} \qquad \frac{\partial h}{\partial y} = \frac{\partial g}{\partial x} .$$

But these are just the Cauchy-Riemann equations for $g + ih$, showing that this function is complex-analytic, with real part $\text{Re}(g + ih) = g$. The converse is even easier: If g is the real part of a complex analytic function $g + ih$, then the Cauchy-Riemann equations immediately lead to Laplace's equation for g.

A minimal surface M can thus be represented locally by

$$(x,y) \longmapsto \Phi(x,y) = (\text{Re } \phi_1(x+iy), \text{ Re } \phi_2(x+iy), \text{ Re } \phi_3(x+iy)) \in \mathbb{R}^3 ,$$

where the ϕ_i are complex analytic functions, and Φ itself is the inverse of an isothermal coordinate system. As one consequence of this representation, we see that every minimal surface in $\mathbb{R}^3$ is <u>automatically</u> real analytic (C^ω).

The fact that Φ^{-1} is an isothermal coordinate system can just as well be expressed by saying that Φ is conformal, and hence by the following two equations for the vectors $\partial\Phi/\partial x$, $\partial\Phi/\partial y \in \mathbb{R}^3$:

$$\left\langle \frac{\partial\Phi}{\partial x}, \frac{\partial\Phi}{\partial x} \right\rangle = \left\langle \frac{\partial\Phi}{\partial y}, \frac{\partial\Phi}{\partial y} \right\rangle \qquad \left\langle \frac{\partial\Phi}{\partial x}, \frac{\partial\Phi}{\partial y} \right\rangle = 0 .$$

Since the complex derivative ϕ_k' is given by

$$\begin{aligned} \phi_k'(x+iy) &= \frac{\partial \text{ Re } \phi_k}{\partial x} + i\frac{\partial \text{ Im } \phi_k}{\partial x} \\ &= \frac{\partial \text{ Re } \phi_k}{\partial x} - i\frac{\partial \text{ Re } \phi_k}{\partial y} \\ &= \frac{\partial \Phi^k}{\partial x} - i\frac{\partial \Phi^k}{\partial y} , \end{aligned}$$

our pair of equations for Φ is equivalent to the one complex equation $\Sigma(\phi_k')^2 = 0$; in terms of the functions $\psi_k = \phi_k'$ we can thus write our conditions as

$$\psi_1^2 + \psi_2^2 + \psi_3^2 = 0 .$$

Now we can describe the solutions of this equation explicitly.

6. LEMMA. Let $V \subset \mathbb{C}$ be open, let g be meromorphic in V, and let f be analytic in V with a zero of order at least $2m$ at each point where g has a pole of order m. Then the functions

$$\psi_1 = \tfrac{1}{2}f(1-g^2) , \qquad \psi_2 = \tfrac{i}{2}f(1+g^2) , \qquad \psi_3 = fg$$

are analytic in V and satisfy $\psi_1^2 + \psi_2^2 + \psi_3^2 = 0$. Conversely, every triple ψ_1, ψ_2, ψ_3 of analytic functions satisfying $\psi_1^2 + \psi_2^2 + \psi_3^2 = 0$ on V can be represented this way.

Proof. The first half of the Lemma is a direct calculation. Suppose, conversely, that we are given functions ψ_i satisfying the equation $\psi_1^2 + \psi_2^2 + \psi_3^2 = 0$, which we can also write in the form

$$(1) \qquad (\psi_1 - i\psi_2)(\psi_1 + i\psi_2) = -\psi_3^2 .$$

If ψ_3 is the 0 function, we choose $g = 0$ and $f = 2\psi_1$. If ψ_3 is not the 0 function, then $\psi_1 - i\psi_2$ is also not the 0 function, so we can define

$$(2) \qquad f = \psi_1 - i\psi_2 , \qquad g = \frac{\psi_3}{\psi_1 - i\psi_2} ,$$

with f analytic and g meromorphic. Then equation (1) gives

(3) $$\psi_1 + i\psi_2 = \frac{-\psi_3{}^2}{\psi_1 - i\psi_2} = -fg^2 .$$

Equation (3) together with the definition of f in equation (2) shows that the ψ_i have the desired form. Equation (3) also shows that fg^2 is analytic, so f must have a zero of order at least $2m$ at each point where g has a pole of order m. ■

It is now a simple matter to give a representation of minimal surfaces, due to Enneper and Weierstrass, which plays a major role in the theory.

7. THEOREM. Every point of a minimal surface $M \subset \mathbb{R}^3$ is in the image of some conformal map $\Phi\colon V \to M \subset \mathbb{R}^3$, where $V \subset \mathbb{C}$ is a simply connected open set. Each such conformal map Φ is of the form $\Phi = \Phi_{(f,g)}$, where

$$\Phi_{(f,g)}{}^1(x,y) = \operatorname{Re} \int \tfrac{1}{2} f(w)(1 - g(w)^2)\,dw + c_1$$

$$\Phi_{(f,g)}{}^2(x,y) = \operatorname{Re} \int \tfrac{i}{2} f(w)(1 + g(w)^2)\,dw + c_2$$

$$\Phi_{(f,g)}{}^3(x,y) = \operatorname{Re} \int f(w)g(w)\,dw + c_3 .$$

In these equations, the c_i are real numbers, g is meromorphic on V, and f is an analytic function on V vanishing precisely at the poles of g, the order of the zero being exactly twice the order of the pole; the integrals are taken along any path from a fixed point $x_0 + iy_0 \in V$ to the point $x + iy$.

Conversely, every such $\Phi_{(f,g)}$ is a conformal map into a minimal surface.

Proof. We have already seen that there is a conformal map $\Phi\colon V \longrightarrow M \subset \mathbb{R}^3$ given by

$$\Phi^k(x,y) = \mathrm{Re}\ \phi_k(x+iy)\ , \tag{1}$$

for complex analytic functions ϕ_k satisfying

$$\Sigma(\phi_k{}')^2 = 0\ . \tag{2}$$

By Lemma 6 we have

$$\phi_1{}' = \tfrac{1}{2}f(1-g^2)\ , \qquad \phi_2{}' = \tfrac{i}{2}f(1+g^2)\ , \qquad \phi_3{}' = fg\ , \tag{3}$$

where f has a zero of order at least $2m$ at each point where g has a pole of order m. We just have to show that the order of f is exactly $2m$ at such a pole. Now if we had $\phi_1{}'(x+iy) = \phi_2{}'(x+iy) = 0$, then we would also have $\phi_3{}'(x+iy) = 0$ by (2). Since

$$\phi_k{}'(x+iy) = \frac{\partial\Phi^k}{\partial x} - i\frac{\partial\Phi^k}{\partial y}\ , \tag{4}$$

this would mean that $\partial\Phi/\partial x = \partial\Phi/\partial y = 0$ at (x,y), contradicting the fact that Φ is conformal (and hence an immersion). So $\phi_k{}'(x+iy) \neq 0$ for $k = 1$ or 2 (or both). Then equation (3) implies that the order of f is at most $2m$ at a pole of g of order m.

Conversely, consider $\Phi = \Phi_{(f,g)}$ where f and g have the stated properties. Then we have equation (1), where the ϕ_k are given by (3), and hence satisfy (2). It follows from (2) and (4) that

$$\left\langle\frac{\partial\Phi}{\partial x},\frac{\partial\Phi}{\partial x}\right\rangle = \left\langle\frac{\partial\Phi}{\partial y},\frac{\partial\Phi}{\partial y}\right\rangle \qquad \left\langle\frac{\partial\Phi}{\partial x},\frac{\partial\Phi}{\partial y}\right\rangle = 0\ . \tag{5}$$

Now our hypotheses on f and g imply [by (3)] that ϕ_1' and ϕ_2' are nowhere zero, and thus that $\partial\Phi/\partial x$ and $\partial\Phi/\partial y$ are nowhere zero. Since they are also orthogonal, by (5), they are linearly independent, so the map Φ is an immersion, and thus a conformal immersion into its image. Since the Φ^k are the real parts of complex analytic functions, they satisfy Laplace's equation, so Φ is also a minimal immersion, by Proposition 5. ■

In order to connect this with the differential geometric properties of minimal surfaces, we need the following additional information, which will also explain the significance of the poles of g.

8. PROPOSITION. For the immersion $\Phi = \Phi_{(f,g)}$ of Theorem 7, the metric $\Phi^*\langle\ ,\ \rangle$ on V has components $g_{ij} = \mu\delta_{ij}$, where

$$\mu(z) = \left[\frac{|f(z)|(1+|g(z)|^2)}{2}\right]^2$$

[this expression will approach some limit at z if z is a pole of g].

If n is the normal map of Φ, then

$$n(z) = \left(\frac{2\operatorname{Re} g(z)}{|g(z)|^2+1}, \frac{2\operatorname{Im} g(z)}{|g(z)|^2+1}, \frac{|g(z)|^2-1}{|g(z)|^2+1}\right) \in S^2$$

[$= (0,0,1) \in S^2$ if z is a pole of g].

Proof. Since Φ is conformal, we have $g_{ij} = \mu\delta_{ij}$, where

$$\mu = \left\langle\frac{\partial\Phi}{\partial x},\frac{\partial\Phi}{\partial x}\right\rangle = \left\langle\frac{\partial\Phi}{\partial y},\frac{\partial\Phi}{\partial y}\right\rangle .$$

Using

$$\phi_k'(x+iy) = \frac{\partial\Phi^k}{\partial x} - i\frac{\partial\Phi^k}{\partial y}$$

and equation (3) of the previous proof, this gives

$$\mu(z) = \frac{1}{2}\sum_k |\phi_k'(z)|^2 = \left[\frac{|f(z)|(1+|g(z)|^2)}{2}\right]^2 .$$

We also see that at points where g does not have a pole, we have

$$\begin{aligned}\frac{\partial\Phi}{\partial x} \times \frac{\partial\Phi}{\partial y} &= (\text{Re } \phi_1', \text{Re } \phi_2', \text{Re } \phi_3') \times -(\text{Im } \phi_1', \text{Im } \phi_2', \text{Im } \phi_3') \\ &= (\text{Re } \phi_3' \text{ Im } \phi_2' - \text{Re } \phi_2' \text{ Im } \phi_3', \ldots) \\ &= (\text{Im } \phi_2\bar{\phi}_3, \text{Im } \phi_3\bar{\phi}_1, \text{Im } \phi_1\bar{\phi}_2) \\ &= \frac{|f|^2(1+|g|^2)}{4}(2 \text{ Re } g, 2 \text{ Im } g, |g|^2-1) .\end{aligned}$$

From this we compute that

$$\left|\frac{\partial\Phi}{\partial x} \times \frac{\partial\Phi}{\partial y}\right| = \left[\frac{|f|(1+|g|^2)}{2}\right]^2 = \mu ,$$

which we should have known anyway, and finally get

$$\frac{\dfrac{\partial\Phi}{\partial x} \times \dfrac{\partial\Phi}{\partial y}}{\left|\dfrac{\partial\Phi}{\partial x} \times \dfrac{\partial\Phi}{\partial y}\right|} = \left(\frac{2 \text{ Re } g}{|g|^2+1}, \frac{2 \text{ Im } g}{|g|^2+1}, \frac{|g|^2-1}{|g|^2+1}\right) .$$

As we approach a pole, this clearly approaches $(0,0,1)$, since $g \to \infty$. ■

The representation in Theorem 7 is not unique, because there are many different conformal maps $\Phi\colon V \to M$. If $\Phi_i\colon V_i \to M$ are two conformal maps, then the map

$$\alpha = \Phi_2^{-1}\circ\Phi_1\colon U \to \mathbb{R}^2 \qquad\qquad U = \Phi_1^{-1}(\Phi_2(V_2)) ,$$

from the open set $U \subset \mathbb{R}^2$ into $\mathbb{R}^2$, is conformal with respect to the usual Riemannian metric on $\mathbb{R}^2$. It is easy to see (Problem 4-9) that such conformal maps α are precisely the one-one complex analytic maps α and their conjugates. Conversely, if we are given $\Phi_{(f,g)}\colon V \to M$ in Theorem 7, and a one-one analytic or conjugate analytic map $\alpha\colon W \to V$, then $\Phi_{(f,g)}\circ\alpha\colon W \to M$ is another conformal map into the same minimal surface, and it must have the same form, with different f and g. One can obtain the new f and g by making the substitution $w = \alpha(u)$ in the integrals of Theorem 7.

The non-uniqueness in Theorem 7 is not really much of a problem, for we have already seen that there is practically a canonical way to select a conformal map $\Phi\colon V \to M$ which covers a given point p of an imbedded minimal surface $M \subset \mathbb{R}^3$. We only have to assume that p is not a flat point, and also that $\nu(p) \neq (0,0,1) \in S^2$. Then $\nu(U) \subset S^2 - \{(0,0,1)\}$ for some neighborhood U of p, and $\sigma\circ\nu\colon U \to V \subset \mathbb{C}$ is conformal, where $\sigma\colon S^2 - \{(0,0,1)\} \to \mathbb{C}$ is stereographic projection. Hence we can choose $\nu^{-1}\circ\sigma^{-1}\colon V \to \mathbb{R}^3$ as our conformal map, and Theorem 7 shows that there are f and g with

$$\nu^{-1}\circ\sigma^{-1} = \Phi_{(f,g)} , \qquad \text{or} \qquad n = \nu\circ\Phi_{(f,g)} = \sigma^{-1} .$$

But the formula in Proposition 8, together with the formula for σ^{-1} on p. 387, shows that $n = \sigma^{-1}$ precisely when $g(z) = z$ for all z. We therefore have a representation of M in the following form (traditionally written with omission of the constants c_i):

$$(*)\qquad \begin{aligned} \Phi^1 &= \operatorname{Re}\int \tfrac{1}{2}F(w)(1-w^2)\,dw \\ \Phi^2 &= \operatorname{Re}\int \tfrac{i}{2}F(w)(1+w^2)\,dw \\ \Phi^3 &= \operatorname{Re}\int F(w)w\,dw \end{aligned} \qquad (F \text{ nowhere } 0)\ .$$

We could also have obtained this representation in a different way, by beginning with the formulas for $\Phi_{(f,g)}$ in Theorem 7 and then making the substitution $w = g^{-1}(u)$; in other words, we could find the formulas for $\Phi_{(f,g)}\circ g^{-1}$. Notice that a local inverse g^{-1} exists around z precisely when z is not a pole of g and $g'(z) \neq 0$; the first condition is equivalent to $\nu(\Phi(z)) \neq (0,0,1)$, and it is easy to see that the second condition is equivalent to ν_* being one-one at $\Phi(z)$.

The representation (*) is especially nice to work with. Problem 1 gives the choices of F which lead to the helicoid, the catenoid, and Scherk's minimal surface; if we take the simplest case of all, $F(w) = 1$, we obtain Enneper's surface, which seemed so mysterious when it was first introduced in Chapter 3. Naturally, the geometric information given by Proposition 8 now simplifies considerably. If Φ_F is given by (*), then

$$(**)\qquad \begin{aligned} n &= \nu\circ\Phi_F = \sigma^{-1} \\ \Phi_F{}^*\langle\ ,\ \rangle &= \mu(dx\otimes dx + dy\otimes dy)\ , \\ &\qquad \text{where } \mu(z) = \frac{|F(z)|^2(1+|z|^2)^2}{4}\ . \end{aligned}$$

Notice in particular, that for real θ, the minimal surfaces $\Phi = \Phi_{e^{-i\theta}F}$,

$$\Phi^1 = \operatorname{Re} e^{-i\theta} \int \tfrac{1}{2} F(w)(1-w^2)\,dw$$

$$\Phi^2 = \operatorname{Re} e^{-i\theta} \int \tfrac{i}{2} F(w)(1+w^2)\,dw$$

$$\Phi^3 = \operatorname{Re} e^{-i\theta} \int F(w)w\,dw ,$$

are all locally isometric, the isometry being given by

$$\Phi_{e^{-i\theta}F}(z) \longmapsto \Phi_{e^{-i\phi}F}(z) .$$

In general, we call two connected minimal surfaces associated if they have this representation for the same F and real θ and ϕ. It suffices to have this for some small piece of each surface, since minimal surfaces are analytic. We also define two planes to be associated surfaces (these are the only minimal surfaces where the representation (*) cannot be achieved [except at isolated points]). Associated minimal surfaces are not only locally isometric, but can also clearly be made part of a continuous family of isometric surfaces. With the proper choice of F we obtain (Problem 1) the continuous family of isometric surfaces between the catenoid and helicoid which is pictured on pp.III.248-249.

At first consideration, it seems to be a pure stroke of luck that the catenoid and helicoid are not only isometric, but also associated. However, the following result dispels that misconception.

9. THEOREM (H. SCHWARZ). If two minimal surfaces are isometric, then one of them is congruent to an associated surface of the other.

Proof. If one of the surfaces is a plane, the other must be also; for $H = 0$

and $K = 0$ implies that both principal curvatures are 0. So we will assume neither is a plane. We can then represent them as

$$f = \Phi_F \colon V \longrightarrow \mathbb{R}^3$$

$$g = \Phi_G \colon W \longrightarrow \mathbb{R}^3 .$$

By hypothesis, there is a map $\alpha \colon V \longrightarrow W$ such that the correspondence $\Phi_F(z) \longmapsto \Phi_G(\alpha(z))$ is an isometry. We want to show that after changing the second minimal surface by a congruence we will actually have $\alpha = \text{identity}$. Then relations (**) will show that $|F(z)| = |G(z)|$, and the maximum modulus principle applied to G/F will imply that we have $G = e^{-i\theta}F$ for some real θ.

The third fundamental form will play a role. Since the surfaces f and $g \circ \alpha$ are minimal, Proposition 2-6 gives

$$\mathrm{III}_f = -(K \circ f)\mathrm{I}_f$$

$$\mathrm{III}_{g\circ\alpha} = -(K \circ g \circ \alpha)\mathrm{I}_{g\circ\alpha} .$$

On the other hand, $\mathrm{I}_f = \mathrm{I}_{g\circ\alpha}$ by hypothesis, and therefore $K \circ f = K \circ g \circ \alpha$ by the Theorema Egregium. So

$$\mathrm{III}_f = \mathrm{III}_{g\circ\alpha} .$$

But by Proposition 2-7 we have

$$\mathrm{III}_f = \mathrm{I}_{n_1} = -\mathrm{II}_{n_1} \qquad n_1 = \text{normal map of } f$$

$$\mathrm{III}_{g\circ\alpha} = \mathrm{I}_{n_2} = -\mathrm{II}_{n_2} \qquad n_2 = \text{normal map of } g\circ\alpha .$$

We thus find that

$$I_{n_1} = I_{n_2} \qquad \text{and} \qquad II_{n_1} = II_{n_2} \ .$$

The Fundamental Theorem of Surface Theory then implies that n_1 and n_2 are the same up to a congruence. So if we change our second surface by a congruence we can assume that $n_1 = n_2$. But then (**) gives

$$\sigma^{-1}(z) = \sigma^{-1}(\alpha(z)) \ .$$

So we must have $\alpha(z) = z$. ■

We conclude with one curious phenomenon concerning the representation (*). This representation was supposed to depend only on the imbedded minimal surface M, but this is not exactly the case, for it also depends on the choice of the normal map ν, or equivalently, on the choice of an orientation for M. So while ν gives rise to the map Φ_F with

(1) $$\nu \circ \Phi_F = \sigma^{-1} \qquad \text{defined on some } V \subset \mathbb{C} \ ,$$

the map $-\nu$ will give rise to a map $\Phi_{\tilde{F}}$ with

(2) $$-\nu \circ \Phi_{\tilde{F}} = \sigma^{-1} \qquad \text{defined on some } W \subset \mathbb{C} \ .$$

Since Φ_F and $\Phi_{\tilde{F}}$ are conformal maps into M, inducing opposite orientations, there must be a conjugate analytic map $\alpha\colon W \longrightarrow V$ such that

(3) $$\Phi_F(z) = \Phi_{\tilde{F}}(\alpha(z)) \qquad z \in W \ .$$

This means that for all $z \in W$ we have

$$\begin{aligned}\sigma^{-1}(\alpha(z)) &= -\nu(\Phi_{\tilde{F}}(\alpha(z)) && \text{by (2)}\\ &= -\nu(\Phi_F(z)) && \text{by (3)}\\ &= -\sigma^{-1}(z) && \text{by (1)}.\end{aligned}$$

Thus we must have

$$\begin{aligned}\alpha(z) &= \sigma(-\sigma^{-1}(z))\\ &= -\frac{1}{\bar{z}}.\end{aligned}$$

Writing equation (3) in terms of (*), we thus obtain, for example,

$$\begin{aligned}\mathrm{Re}\int^{z} F(w)(1-w^2)dw \text{ (+ constant)} &= \mathrm{Re}\int^{-1/\bar{z}} \tilde{F}(w)(1-w^2)\,dw\\ &= \mathrm{Re}\overline{\left(\int^{-1/\bar{z}} \tilde{F}(w)(1-w^2)\,dw\right)}\\ &= \mathrm{Re}\int^{-1/z} \overline{\tilde{F}(\bar{w})(1-\bar{w}^2)}\,dw\\ &= \mathrm{Re}\int^{-1/z} \overline{\tilde{F}(\bar{w})}(1-w^2)\,dw\\ &= \mathrm{Re}\int^{z} \overline{\tilde{F}\left(-\frac{1}{\bar{w}}\right)}\left(1-\frac{1}{w^2}\,\frac{1}{w^2}\right)dw && \text{(by substitution)},\end{aligned}$$

as well as two other equations. As one would certainly hope, these equations all lead to the same relation:

$$\tilde{F}(z) = -\frac{1}{z^4}\,\overline{F\left(-\frac{1}{\bar{z}}\right)}.$$

This $\tilde{F}$ gives the exact same surface as F, but it induces the opposite orientation on M.

Now the interesting thing is, that there are functions F which equal $\tilde{F}$, the simplest example being

$$F(z) = 1 - \frac{1}{z^4} = \frac{z^4 - 1}{z^4} .$$

This choice of F leads to Henneberg's minimal surface

$$\Phi^1 = \mathrm{Re} \int \frac{1}{2} \frac{w^4 - 1}{w^4}(1 - w^2)\, dw$$

$$\Phi^2 = \mathrm{Re} \int \frac{i}{2} \frac{w^4 - 1}{w^4}(1 + w^2)\, dw$$

$$\Phi^3 = \mathrm{Re} \int \frac{w^4 - 1}{w^3}\, dw .$$

The map Φ can be defined on all of $\mathbb{C} - \{0\}$ (we don't even have to restrict ourselves to a simply connected domain, since all integrands have residue 0 at 0, so the integrals are independent of the path); however, Φ is not an immersion at ± 1, $\pm i$, the points where F is zero. Using stereographic projection, we can identify $\mathbb{C} - \{0, \pm 1, \pm i\}$ with S^2 minus three pairs of antipodal points, the points ± 1, $\pm i$ occurring on the equator of S^2. Since

$$\Phi_F(z) = \Phi_{\tilde{F}}(\alpha(z)) = \phi_{\tilde{F}}(\sigma(-\sigma^{-1}(z))) = \Phi_F(\sigma(-\sigma^{-1}(z))) ,$$

the map $\Phi_F \circ \sigma^{-1} \colon S^2 \to \mathbb{R}^3$ is invariant under the antipodal map, so our surface is the image of the projective plane punctured at three points. The figure below shows the image of a symmetric strip around the equator of S^2.

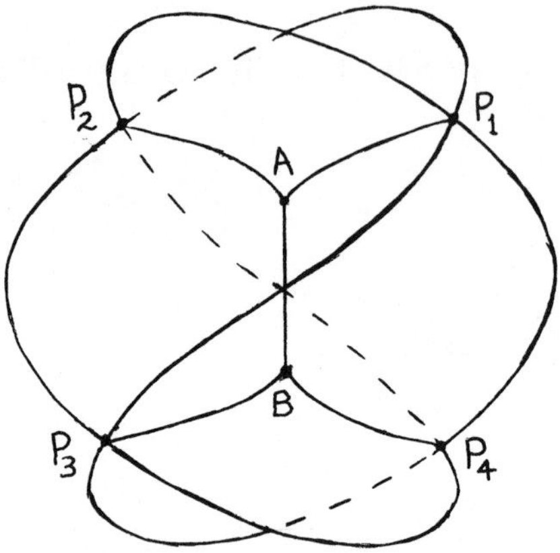

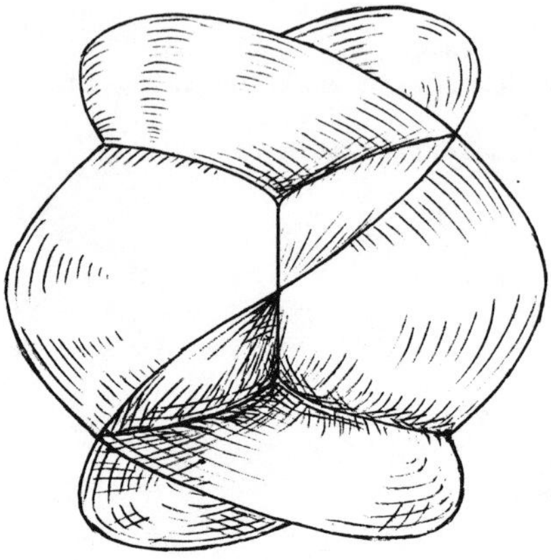

The equator maps onto the vertical segment AB, with the pair of points corresponding to $\pm i$ mapping onto the upper endpoint, and the pair ± 1 onto the lower. The boundary circles of the strip each map into the closed curve which intersects itself at points $P_1, \ldots, P_4$. The points of the segment AB are all double points of the immersion, but the surface crosses itself in a funny way along this line -- it contains two congruent helicoid-like surfaces,

with AB common to both. The final figure below shows an imbedded Möbius strip inside the immersed surface.

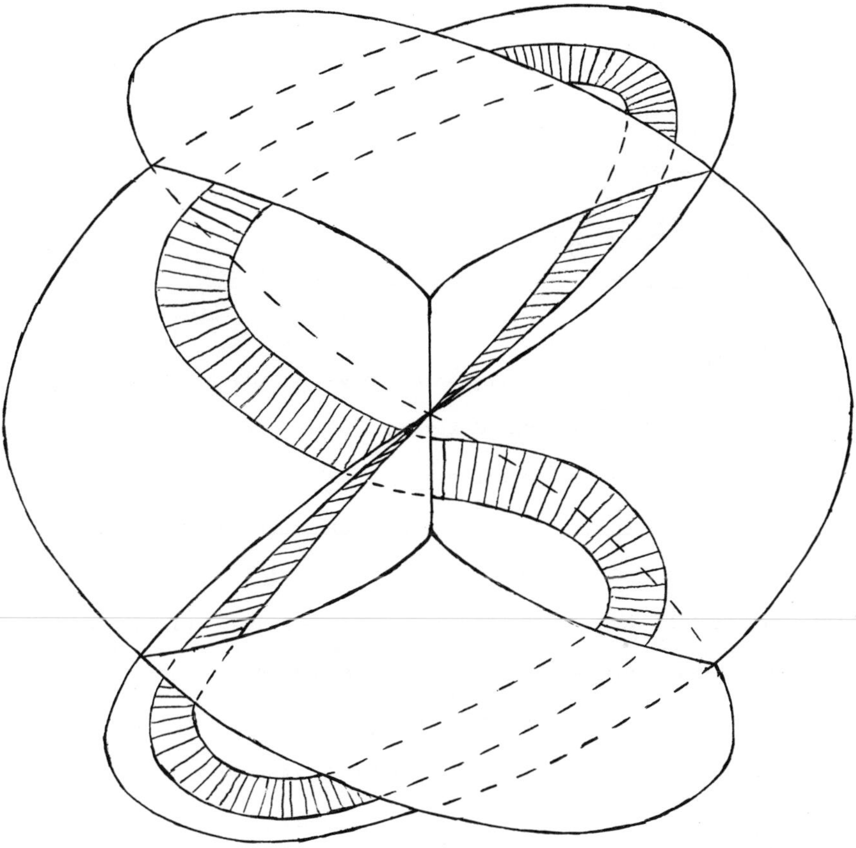

No physicist, by the way, would be surprised to learn that there are minimal surfaces in the shape of a Möbius strip. If one dips an appropriately bent piece of wire into a soap solution, then a soap film will be formed in the shape of this surface (actually, one always obtains the sort of film pictured on the left; after the middle sheet is pierced, the soap film snaps back

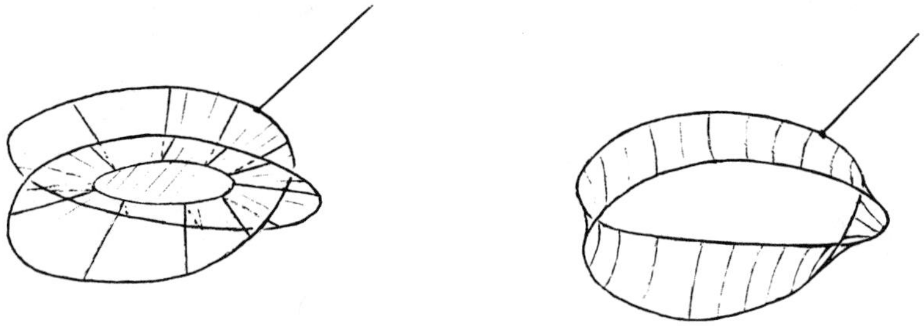

into the Möbius strip). If one neglects the slight effect of gravity, then any soap film ought to be a minimal surface, since the surface tension makes the film contract as much as possible.

Considerations of this sort were first introduced by the blind experimental physicist Plateau, who gave a much more elaborate discussion of the problem, taking into account the thickness of the films. His writings gave rise to the Plateau problem, to prove that every imbedded circle in $\mathbb{R}^3$ is the boundary of an immersed disc which has minimum area among all such immersed discs; this very difficult problem was first solved by Jesse Douglas and Tibor Rado. Douglas' methods work just as well for higher dimensions, and his work won him the Field's medal in 1936. We will not even enter into a discussion of this work, which is almost purely analytic in nature, but descriptions of the methods used may be found in several references in the bibliography. There are many questions related to Plateau's problem, some of which have led to the invention of powerful new techniques. Notice, for example, that Plateau's problem is in some ways not even the natural question to ask, since it is concerned only with surfaces homeomorphic to a disc. Thus the solution of the Plateau problem for the curve pictured above will not be the Möbius strip, but a surface like the one shown below. Probably the simplest

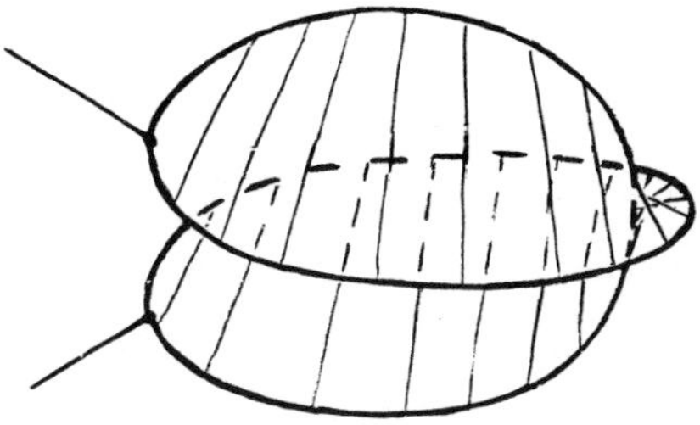

way to picture this surface is to make a piece of wire in the right shape and dip it into a bubble solution (the 2 loops should be rather further apart than in the previous picture). It is fairly easy to find a shape that gives both a Möbius strip and a disc, depending on how it is dipped in. Since the two different soap films have unequal areas, this shows us that we should slightly revise our criterion for the shape of a soap film spanned by a given wire loop. The film need not have a minimum area -- a local minimum should suffice. A surface which is a critical value, but not a local minimum, would presumably correspond to position of unstable equilibrium -- the slightest disturbance would cause the soap film to change shape; presumably such films could never occur in practice (in addition, of course, all sorts of physical considerations might rule out other surfaces on practical grounds).

If the wire loop is equipped with a pair of handles, then by gently pulling the two parts of the loop apart one can see the film jump from a Möbius strip to a disc, presumably at the point where the Möbius strip is no longer in stable equilibrium. Even for those who are willing to get involved in all the analysis necessary for the Plateau problem, experiments like this can be as instructive as they are fascinating, and provide convincing evidence for assertions that are still not mathematically provable; the interested reader should consult Courant [1]. And even if you are not eager to get your hands all soaped up, there is one description of simple experiments that you simply cannot afford to miss. This is a series of lectures by Boys {1} which treats soap films and soap bubbles, the mathematical correlates of which we will study a little later on. They were given to an audience of children in the good old Victorian days, and are probably the best science writing ever produced. I seriously suggest that you put down the silly stuff you are presently reading, rush right out to purchase Boys' little gem of a book, and get high on physics for a while.

* * *

Returning to purely mathematical questions, we now seek a formula for the variation of area when we are dealing with an arbitrary variation of an immersed surface $f\colon M \to N$, in a general Riemannian manifold $(N, \langle\ ,\ \rangle)$. We would even like to find the variation of n-dimensional volume for an immersion $f\colon M^n \to N^m$ (but at least we will not worry about maps and variations which are only piecewise C^∞). As a start in this direction, we consider a simple general problem from the classical calculus of variations in several variables. Suppose we are given a (suitably differentiable) function

$$F\colon \mathbb{R}^n \times \mathbb{R} \times \mathbb{R}^n \to \mathbb{R}$$

and a compact n-dimensional manifold-with-boundary $D \subset \mathbb{R}^n$. We seek, among all functions $g\colon D \to \mathbb{R}$ with prescribed values on ∂D, one which will maximize (or minimize) the quantity

$$J(g) = \int_D F(t_1,\dots,t_n,\ g(t_1,\dots,t_n),\ D_1g(t_1,\dots,t_n),\dots,D_ng(t_1,\dots,t_n))dt_1\cdots dt_n$$

$$= \int_D F(t,g(t),Dg(t))dt_1\cdots dt_n\ , \qquad \text{in abbreviated form.}$$

This is a direct generalization of the problem considered on p. I.429. For any variation $\alpha\colon (-\varepsilon,\varepsilon)\times D \longrightarrow \mathbb{R}$ of g, we compute the variation of J as follows. It will be convenient to denote a typical point in the domain of F by

$$(t_1,\dots,t_n,\ x,\ y_1,\dots,y_n) \qquad \text{or, even more briefly, by} \qquad (t,x,y)\ .$$

Then

$$\left.\frac{dJ(\bar{\alpha}(u))}{du}\right|_{u=0} = \left.\frac{d}{du}\right|_{u=0} \int_D F(t,\ \alpha(u,t),\ \frac{\partial\alpha}{\partial t}(u,t))dt_1\cdots dt_n$$

$$\left[\frac{\partial\alpha}{\partial t} \text{ stands for } \frac{\partial\alpha}{\partial t_1},\dots,\frac{\partial\alpha}{\partial t_n}\right]$$

$$= \int_D \left[\left.\frac{d}{du}\right|_{u=0} F(t,\ \alpha(u,t),\ \frac{\partial\alpha}{\partial t}(u,t))\right]dt_1\cdots dt_n$$

$$= \int_D \left[\frac{\partial\alpha}{\partial u}(0,t)\cdot\frac{\partial F}{\partial x}(\bullet) + \sum_{i=1}^{n} \frac{\partial^2\alpha}{\partial u\partial t_i}(0,t)\cdot\frac{\partial F}{\partial y_i}(\bullet)\right]dt_1\cdots dt_n\ ,$$

where

$$(1)\quad \bullet = (t_1,\dots,t_n,\ g(t_1,\dots,t_n),\ \frac{\partial g}{\partial t_1}(t_1,\dots,t_n),\dots,\frac{\partial g}{\partial t_n}(t_1,\dots,t_n))\ .$$

Introducing the abbreviations

$$w(t) = \frac{\partial\alpha}{\partial u}(0,t)$$

$$(2)\qquad A(t) = \frac{\partial F}{\partial x}(\bullet) \qquad \text{(all of these are functions on } D)$$

$$B_i(t) = \frac{\partial F}{\partial y_i}(\bullet)\ ,$$

we can write

$$\left.\frac{dJ(\bar{\alpha}(u))}{du}\right|_{u=0} = \int_D \left[w\cdot A + \sum_{i=1}^{n} \frac{\partial w}{\partial t_1}\cdot B_i\right] dt_1 \wedge \dots \wedge dt_n \ .$$

We now have to pull an integration-by-parts-type trick on the second term in the integrand. We do this by considering the $(n-1)$-form ϖ defined by

$$(3) \qquad \varpi = \sum_{i=1}^{n} (-1)^{i+1} (w\cdot B_i)\, dt_1 \wedge \dots \wedge \widehat{dt_i} \wedge \dots \wedge dt_n \ .$$

Since

$$d\varpi = \left[w \cdot \sum_{i=1}^{n} \frac{\partial B_i}{\partial t_i}\right] dt_1 \wedge \dots \wedge dt_n + \left[\sum_{i=1}^{n} \frac{\partial w}{\partial t_i} B_i\right] dt_1 \wedge \dots \wedge dt^n \ ,$$

we have

$$\left.\frac{dJ(\bar{\alpha}(u))}{du}\right|_{u=0} = \int_D \left[w \cdot \left(A - \sum_{i=1}^{n} \frac{\partial B_i}{\partial t_i}\right)\right] dt_1 \wedge \dots \wedge dt_n + \int_D d\varpi$$

$$= \int_D \left[w \cdot \left(A - \sum_{i=1}^{n} \frac{\partial B_i}{\partial t_i}\right)\right] dt_1 \wedge \dots \wedge dt_n + \int_{\partial D} \varpi \ .$$

From the definition of ϖ, we see that $\varpi = 0$ on ∂D if α is a variation keeping the boundary fixed. So g is a critical point for J if and only if

$$0 = \left.\frac{dJ(\bar{\alpha}(u))}{du}\right|_{u=0} = \int_D \left[\frac{\partial\alpha}{\partial u}(0,t)\cdot\left(A - \sum_{i=1}^{n} \frac{\partial B_i}{\partial t_i}\right)\right] dt_1 \wedge \dots \wedge dt_n$$

for all variations α keeping the boundary fixed. From this we easily see that g must satisfy the equation

$$A - \sum_{i=1}^{n} \frac{\partial B_i}{\partial t_i} = 0 ,$$

that is,

$$(*) \qquad \frac{\partial F}{\partial x}(\bullet) - \sum_{i=1}^{n} \frac{\partial^2 F}{\partial t_i \partial y_i}(\bullet) = 0 , \qquad \text{where } \bullet \text{ is given by (1).}$$

This is the classical analogue of Euler's equation (Theorem I.9-8).

As a particular example, we take $n = 2$, and let

$$(4) \qquad F(t_1,t_2,x,y_1,y_2) = \sqrt{1+{y_1}^2+{y_2}^2} ,$$

so that

$$J(g) = \int_D \sqrt{1+{g_1}^2+{g_2}^2}\, dt_1 dt_2 \qquad [g_i = \frac{\partial g}{\partial t_i}]$$

$$= \int_D \sqrt{EG-F^2}\, dt_1 dt_2 \qquad \text{by formulas (A') on p. III.200.}$$

Thus $J(g)$ is the area of the imbedded surface

$$f(t_1,t_2) = (t_1,t_2,g(t_1,t_2)) .$$

A variation $\alpha\colon (-\varepsilon,\varepsilon)\times D \longrightarrow \mathbb{R}$ of g gives rise to a variation $\beta\colon (-\varepsilon,\varepsilon)\times D \longrightarrow \mathbb{R}^3$ of the imbedding f, defined by

$$\beta(u,t_1,t_2) = (t_1,t_2,\alpha(u,t_1,t_2)) .$$

This variation β is perpendicular to the (t_1,t_2) plane, instead of being perpendicular to the surface $M = f(D)$; it has variation vector

$$W(t_1,t_2) = (0,0,w(t_1,t_2))_{f(t_1,t_2)} .$$

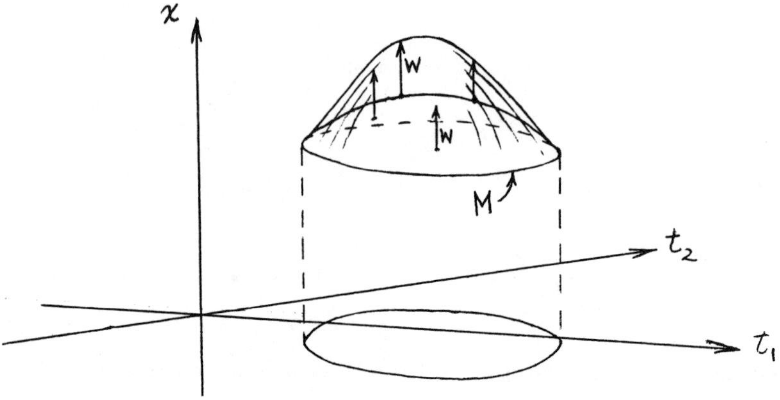

This is the one other kind of variation sometimes encountered in differential geometry books, and the kind which is always used in books on the calculus of variations. Indeed, this particular example was chosen by Lagrange to illustrate the general methods which he had developed (1760) for the calculus of variations in several variables. In this case, equation (4) gives

$$\frac{\partial F}{\partial x} = 0 , \qquad \frac{\partial F}{\partial y_i} = \frac{y_i}{\sqrt{1+{y_1}^2+{y_2}^2}} ,$$

so equation (*) becomes

$$\frac{\partial}{\partial t_1}\left(\frac{g_1}{R}\right) + \frac{\partial}{\partial t_2}\left(\frac{g_2}{R}\right) = 0 , \qquad R = \sqrt{1+{g_1}^2+{g_2}^2} ,$$

which boils down to exactly the equation

$$(1+{g_1}^2)g_{11} - 2g_1g_2g_{12} + (1+{g_2}^2)g_{22} = 0$$

which we found in the proof of Proposition 3; it was only in 1776 that Meusnier interpreted this equation in terms of the mean curvature of f.

We will also be interested in the 1-form ϖ which we obtain in this case; from (2) and (3) we see that

$$\varpi = \frac{w \cdot g_1}{R} dt_2 - \frac{w \cdot g_2}{R} dt_1 \ .$$

We will express ϖ in terms of the form ω on $M = f(D)$ with $\varpi = f^*\omega$. We have

$$\omega((1,0,g_1)) = \varpi\left(f_*\left(\frac{\partial}{\partial t_1}\right)\right) = \left\langle (1,0,g_1), \left(-\frac{w \cdot g_2}{R}, -\frac{w \cdot g_1}{R}, 0\right)\right\rangle$$

$$\omega((0,1,g_2)) = \varpi\left(f_*\left(\frac{\partial}{\partial t_2}\right)\right) = \left\langle (0,1,g_2), \left(-\frac{w \cdot g_2}{R}, -\frac{w \cdot g_1}{R}, 0\right)\right\rangle .$$

But

$$\left(-\frac{w \cdot g_2}{R}, -\frac{w \cdot g_1}{R}, 0\right) = \left(-\frac{g_1}{R}, -\frac{g_2}{R}, \frac{1}{R}\right) \times (0,0,w)$$

$$= \nu \times W ,$$

where ν is the normal vector. So for all $X \in M_p$ we have

$$\begin{aligned} \omega(p)(X) &= \langle \nu(p) \times W(p), X\rangle \\ &= \langle W(p) \times X, \nu(p)\rangle \\ &= \langle \mathbf{T}W(p) \times X, \nu(p)\rangle , \qquad \mathbf{T}W(p) = \text{tangential component of } W(p). \end{aligned}$$

If dA is the 2-dimensional volume form on M, then we have

$$\omega(X) = dA(\mathbf{T}W, X) \ .$$

Using the notation introduced on p. I.309, we can thus write

$$\omega = \mathbf{T}W \lrcorner\, dA \ .$$

Without going through the calculations, we merely state that if we take an arbitrary n and let

$$F(t_1,\ldots,t_n, x, y_1,\ldots,y_n) = \sqrt{1+\sum_i {y_i}^2} ,$$

so that $J(g)$ represents the n-dimensional volume of the imbedded n-manifold $\{f(t_1,\ldots,t_n, g(t_1,\ldots,t_n))\}$, then the $(n-1)$-form ϖ is $f^*\omega$, where the $(n-1)$-form ω on M is defined by

$$\omega = \mathbf{T}W \lrcorner dV \qquad dV = \text{volument element on } M .$$

This $(n-1)$-form ω will be very important when we look for an invariant description of the variation of n-dimensional volume for an immersion $f: M^n \to N$. We have always expressed length or area as an integral involving a coordinate system, and calculated the derivative with respect to the variation parameter u by using "Leibniz' Rule" to bring the derivative inside the integral sign. Before we go any further, we will need an invariant description of this procedure.

Suppose we have a C^∞ 1-parameter family of k-forms on an n-manifold (-with-boundary) M; thus, for each $u \in (-\varepsilon,\varepsilon)$, we have a k-form $\Gamma(u)$ on M. For each $p \in M$, the map $u \mapsto \Gamma(u)(p) \in \Omega^k(M_p)$ into the vector space $\Omega^k(M_p)$ then has a derivative, which at each u is again an element $\dot\Gamma(u)(p) \in \Omega^k(M_p)$. Thus a C^∞ 1-parameter family of k-forms $u \mapsto \Gamma(u)$ on M gives rise to a new C^∞ 1-parameter family of k-forms $u \mapsto \dot\Gamma(u)$ on M.

<u>10. PROPOSITION (LEIBNIZ' RULE)</u>. Let M be a compact oriented n-dimensional manifold-with-boundary and $u \mapsto \Gamma(u)$ a C^∞ 1-parameter family of n-forms on M. Then

$$\left.\frac{d}{du}\right|_{u=u_0} \int_M \Gamma(u) = \int_M \dot{\Gamma}(u_0) .$$

Proof. Let $\mathcal{O}$ be a finite cover of M by open sets V each contained in $c([0,1]^n)$ for some orientation-preserving singular n-cube $c\colon [0,1]^n \to M$. Let $\Phi = \{\phi_V\}$ be a partition of unity subordinate to this cover. Then

$$\int_M \phi_V \cdot \Gamma(u) = \int_{[0,1]^n} (\phi_V \circ c) \cdot c^*\Gamma(u) .$$

It is easy to see that the ordinary Leibniz' Rule implies that

$$\left.\frac{d}{du}\right|_{u=u_0} \int_M \phi_V \cdot \Gamma(u) = \int_{[0,1]^n} (\phi_V \circ c) \cdot c^*\dot{\Gamma}(u_0) = \int_M \phi_V \dot{\Gamma}(u_0) .$$

Since

$$\int_M \Gamma(u) = \sum_{\{\phi_V\}} \int_M \phi_V \cdot \Gamma(u) ,$$

and similarly for $\int_M \dot{\Gamma}(u_0)$, the result follows. ∎

Now consider a compact oriented n-dimensional manifold-with-boundary M, and a C^∞ map $\alpha\colon (-\varepsilon,\varepsilon) \times M \to N$, where $(N, \langle\ ,\ \rangle)$ is a Riemannian manifold. We will assume that each $\bar{\alpha}(u)\colon M \to N$ is an immersion. Then the metric $\bar{\alpha}(u)^*\langle\ ,\ \rangle$ on M determines a volume element $\Gamma(u)$ on M; using the given orientation of M, we can consider this to be an n-form on M, which we call the volume form. What we want to determine is

$$\left.\frac{d}{du}\right|_{u=0} \int_M \Gamma(u) .$$

According to Proposition 10, it suffices to determine $\dot{\Gamma}(0)$. For this we do not even need M to be compact.

11. THEOREM (VARIATION OF VOLUME FORMULA). Let $f\colon M \to N$ be an immersion of an oriented n-dimensional manifold (-with-boundary) M into a Riemannian manifold $(N^m, \langle\ ,\ \rangle)$ and let $\alpha\colon (-\varepsilon,\varepsilon)\times M \to N$ be a variation of f through immersions, with variation vector field W. If $\Gamma(u)$ is the volume form of M determined by the metric $\bar{\alpha}(u)^*\langle\ ,\ \rangle$ and the given orientation of M, then

$$\dot{\Gamma}(0) = -\langle W, n\cdot\eta\rangle\cdot\Gamma(0) + d(\mathbf{T}W \lrcorner\, \Gamma(0))$$

where η is the mean curvature normal of the immersion f. [Notice that there is a slight abuse of notation here; at each $p \in M$, the vector $\mathbf{T}W$ really denotes the unique vector $X \in M_p$ with $f_*(X) = \mathbf{T}W$ at $f(p)$.]

Proof. The theorem involves two n-forms on M which we have to prove are equal at all points of M. Let us first consider a point $p_0 \in M$ where $W(p_0)$ is not tangent to $f(M)$. By choosing a sufficiently small neighborhood V of p_0, and decreasing ε if necessary, we can then assume that $\alpha\colon (-\varepsilon,\varepsilon)\times V \to N$ is an imbedding.

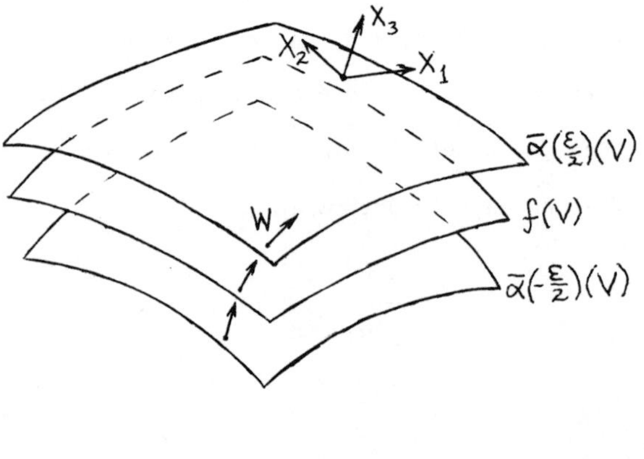

It will be convenient to identify V with $f(V)$, so that $f = \bar{\alpha}(0)$ is just the inclusion map $i\colon V \to N$. On some open set containing image α, we can choose an orthonormal moving frame $X_1,\dots,X_n,X_{n+1},\dots,X_m$ such that

(1) $X_j(\alpha(u,p))$ is tangent to the submanifold $\bar{\alpha}(u)(V)$ $\quad 1 \le j \le n$,

(2) $X_r(\alpha(u,p))$ is normal to the submanifold $\bar{\alpha}(u)(V)$ $\quad n+1 \le r \le m$.

If $\phi^1,\dots,\phi^n,\phi^{n+1},\dots,\phi^m$ are the dual 1-forms, then clearly

(1') $\bar{\alpha}(u)^*(\phi^1 \wedge \dots \wedge \phi^n) = \Gamma(u)$

(2') $\bar{\alpha}(u)^*(\phi^r) = 0 \qquad n+1 \le r \le m$.

Now the variation vector field W, defined along V, is the restriction of the vector field $\tilde{W} = \partial\alpha/\partial u$ defined along all of image α. We can further extend $\tilde{W}$ to a vector field defined on some open set containing image α; we will use the same symbol $\tilde{W}$ for this extension. Associated to this vector field $\tilde{W}$ is a certain local 1-parameter group of local diffeomorphisms $\{\rho_u\}$; recall (Chapter I.5) that $\rho_u(q)$ is the result of following for time u the integral curve of $\tilde{W}$ that starts at q. Clearly the integral curve of $\tilde{W}$ that starts at a point $p \in V$ is just $u \mapsto \alpha(u,p)$. So

$$\rho_u(p) = \alpha(u,p) = \bar{\alpha}(u)(p), \qquad p \in V.$$

It is therefore clear that if Y is a tangent vector of V, then

(3) $$\rho_{u*}(i_*Y) = \bar{\alpha}(u)_*(Y).$$

Now let us recall the Lie derivate (pp. I.207, 238, 318): if ω is a k-form on N, then $L_{\tilde{W}}\omega$ is another k-form defined by

$$L_{\tilde{W}}\omega(Z_1,\ldots,Z_k) = \lim_{h\to 0}\frac{1}{h}\Big[\omega(\rho_{h*}Z_1,\ldots,\rho_{h*}Z_k) - \omega(Z_1,\ldots,Z_k)\Big] .$$

We claim that

$$\dot{\Gamma}(0) = i^*\{L_{\tilde{W}}(\phi^1\wedge\ldots\wedge\phi^n)\} . \tag{4}$$

The proof of this will be quite straightforward. We adopt the abbreviation $\Phi = \phi^1\wedge\ldots\wedge\phi^n$. If $Y_1,\ldots,Y_n$ are tangent vectors of V, then we have

$$\begin{aligned}
\dot{\Gamma}(0)(Y_1,\ldots,Y_n) &= \lim_{h\to 0}\frac{1}{h}\Big[\Gamma(h)(Y_1,\ldots,Y_n) - \Gamma(0)(Y_1,\ldots,Y_n)\Big] \\
&= \lim_{h\to 0}\frac{1}{h}\Big[\bar{\alpha}(h)^*\Phi(Y_1,\ldots,Y_n) - i^*\Phi(Y_1,\ldots,Y_n)\Big] && \text{by (1')} \\
&= \lim_{h\to 0}\frac{1}{h}\Big[\Phi(\bar{\alpha}(h)_*Y_1,\ldots,\bar{\alpha}(h)_*Y_n) - \Phi(i_*Y_1,\ldots,i_*Y_n)\Big] \\
&= \lim_{h\to 0}\frac{1}{h}\Big[\Phi(\rho_{h*}i_*Y_1,\ldots,\rho_{h*}i_*Y_n) - \Phi(i_*Y_1,\ldots,i_*Y_n)\Big] && \text{by (3)} \\
&= L_{\tilde{W}}\Phi(i_*Y_1,\ldots,i_*Y_n) ,
\end{aligned}$$

which proves (4).

The reason for bringing in the Lie derivative is that we have some useful formulas for it. In particular (p. I.319), we have

$$L_{\tilde{W}}\omega = \tilde{W} \lrcorner\, d\omega + d(\tilde{W} \lrcorner\, \omega) .$$

Substituting this into (4) we obtain

$$\dot{\Gamma}(0) = i^*\{\tilde{W} \lrcorner\, d\Phi\} + d(i^*\{\tilde{W} \lrcorner\, \Phi\}) . \tag{5}$$

We will show that the two terms on the right are precisely the terms appearing in the statement of the theorem.

We first compute $d\Phi = d(\phi^1 \wedge \dots \wedge \phi^n)$ by using the first structural equation for N, which will bring in the connection forms ψ^α_β $(1 \le \alpha, \beta \le m)$ for N associated to $\phi^1, \dots, \phi^m$:

$$(6) \qquad d\Phi = d(\phi^1 \wedge \dots \wedge \phi^n) = \sum_{j=1}^{n} (-1)^{j+1} \phi^1 \wedge \dots \wedge d\phi^j \wedge \dots \wedge \phi^n$$

$$= \sum_{j=1}^{n} (-1)^{j+1} \phi^1 \wedge \dots \wedge \Big(-\sum_{\alpha=1}^{m} \psi^j_\alpha \wedge \phi^\alpha\Big) \wedge \dots \wedge \phi^n$$

$$= \sum_{j=1}^{n} \sum_{r=n+1}^{m} \phi^r \wedge \phi^1 \wedge \dots \wedge \psi^j_r \wedge \dots \wedge \phi^n .$$

So if $Y_1, \dots, Y_n$ are the tangent vectors of V with $i_* Y_j = X_j$ along V, then

$$(7) \quad i^*\{\tilde{W} \lrcorner\, d\Phi\}(Y_1, \dots, Y_n) = d\Phi(W, X_1, \dots, X_n)$$

$$= \sum_{j=1}^{n} \sum_{r=n+1}^{m} (\phi^r \wedge \phi^1 \wedge \dots \wedge \psi^j_r \wedge \dots \wedge \phi^n)(W, X_1, \dots, X_n)$$

$$= \sum_{j=1}^{n} \sum_{r=n+1}^{m} \phi^r(W) \psi^j_r(X_j) ,$$

since (2') says that $\phi^r(X_j) = 0$ for $i \le n < r$. On the other hand, we have

$$\nabla'_{X_j} X_j = \sum_{\alpha=1}^{m} \psi^\alpha_j(X_j) \cdot X_\alpha = -\sum_{\alpha=1}^{m} \psi^j_\alpha(X_j) \cdot X_\alpha ,$$

so

$$n\cdot\eta = \mathbf{\perp}\left(\sum_{j=1}^{n} \nabla'_{X_j} X_j\right) = \sum_{j=1}^{n} \left(- \sum_{r=n+1}^{m} \psi_r^j(X_j) X_r\right) ,$$

and hence

$$(8) \qquad - \langle W, n\cdot\eta\rangle = \sum_{j=1}^{n} \sum_{r=n+1}^{m} \phi^r(W)\psi_r^j(X_j) .$$

Equations (7) and (8) thus give

$$(9) \qquad i^*\{\tilde{W} \lrcorner d\Phi\} = - \langle W, n\cdot\eta\rangle\cdot\Gamma(0) .$$

As for the other term in (5), if $Y_1,\ldots,Y_{n-1}$ are tangent vectors of V, then we have

$$\begin{aligned} i^*\{\tilde{W} \lrcorner \Phi\}(Y_1,\ldots,Y_{n-1}) &= \Phi(W, i_*Y_1,\ldots,i_*Y_{n-1}) \\ &= (\phi^1 \wedge \ldots \wedge \phi^n)(W, i_*Y_1,\ldots,i_*Y_{n-1}) \\ &= (\phi^1 \wedge \ldots \wedge \phi^n)(\mathbf{T}W, i_*Y_1,\ldots,i_*Y_{n-1}) \\ &\qquad \text{since each } \phi^j(\mathbf{\perp}W) = 0 \\ &= [\mathbf{T}W \lrcorner i^*(\phi^1 \wedge \ldots \wedge \phi^n)](Y_1,\ldots,Y_{n-1}) . \end{aligned}$$

Thus

$$(10) \qquad i^*\{\tilde{W} \lrcorner \Phi\} = \mathbf{T}W \lrcorner \Gamma(0) .$$

This completes the proof of the theorem at any point p_0 where $W(p_0)$ is not tangent to $f(M)$.

The general case can be disposed of by a technical trick. Let $\mathbf{N} = N\times\mathbb{R}$, with the product Riemannian metric, which we also denote by $\langle\ ,\ \rangle$, and define $\boldsymbol{\alpha}\colon (-\varepsilon,\varepsilon)\times M \longrightarrow \mathbf{N}$ by

$$\boldsymbol{\alpha}(u,p) = (\alpha(u,p),u) .$$

The new variation vector field $\mathbf{W}$ is

$$\mathbf{W}(p) = (W(p), \mathbf{1}) ,$$

where $\mathbf{1}$ denotes the unit vector field on $\mathbb{R}$. Clearly $\mathbf{W}$ is not tangent to $\bar{\boldsymbol{\alpha}}(0)(M) \subset N \times \{0\}$, so the theorem holds for $\boldsymbol{\alpha}$. On the other hand, it is easy to see that the new mean curvature normal $\boldsymbol{\eta}$ is just

$$\boldsymbol{\eta}(p) = (\eta, 0) ,$$

so that $<\mathbf{W}, \boldsymbol{\eta}> = <W,\eta>$; thus the result for $\boldsymbol{\alpha}$ implies the result for α. ∎

<u>12. COROLLARY</u>. Let $\alpha\colon (-\varepsilon,\varepsilon) \times M \to N$ be a variation of an immersion $f\colon M \to N$ of a compact oriented n-dimensional manifold-with-boundary M into a Riemannian manifold $(N, < , >)$. If $V(\bar{\alpha}(u))$ is the n-dimensional volume of M determined by the metric $\bar{\alpha}(u)^*< , >$ and the given orientation of M, then

$$\left.\frac{dV(\bar{\alpha}(u))}{du}\right|_{u=0} = - \int_M <W, n\cdot\eta> \, dV + \int_{\partial M} \omega ,$$

where dV is the volume element determined by $f^*< , >$ and $\omega = W \lrcorner \, dV$. In particular, if α is a variation keeping ∂M fixed, then

$$\left.\frac{dV(\bar{\alpha}(u))}{du}\right|_{u=0} = - \int_M <W, n\cdot\eta> \, dV .$$

The immersion f is a critical point for V, among all immersions $g\colon M \to N$ with $g = f$ on ∂M, if and only if $\eta = 0$ everywhere.

Proof. The first statement follows from Theorem 11, Leibniz' Rule, and Stokes' Theorem. If α keeps ∂M fixed, then $W = 0$ on ∂M, so also $\omega = 0$ on ∂M; this proves the second statement. To prove the third, we can choose $W = \phi \cdot \eta$, where ϕ is a C^∞ function on M which is positive on $M - \partial M$ and 0 on ∂M. ■

Notice that in the expression

$$-\int_M \langle W, n\cdot\eta\rangle \, dV + \int_{\partial M} \omega ,$$

the first term depends only on the normal component $\perp W$ of W; for $\langle W, n\cdot\eta\rangle = \langle \perp W, n\cdot\eta\rangle$, since η is perpendicular to $f(M)$. This partially confirms our suspicion that we need work only with normal variations. On the other hand, in the term $\int_{\partial M} \omega$, only the tangential component $\mathbf{T}W$ enters; roughly speaking, the integral measures how much the volume of M is changing because of the way that the variation is expanding its boundary. In particular, we see that $\int_{\partial M} \omega$ is 0 not only when the variation keeps the boundary fixed, but also when W is normal to M along the boundary. Consequently, if $\eta = 0$ on M, then $dV(\bar{\alpha}(u))/du\big|_{u=0}$ will be 0 for every variation which is perpendicular on the boundary of M, not merely for those variations which keep ∂M fixed. Back in our original equation (*) on p. 383 we didn't have any term involving an integral over ∂D precisely because we were dealing only with normal variations. This leads to an interesting phenomenon in the case of minimal surfaces $M \subset \mathbb{R}^3$. If ν is the unit normal vector on M, then we can define a variation α of the inclusion $i\colon M \to \mathbb{R}^3$ by

$$\alpha(u,p) = p + u\cdot\nu(p) .$$

The various surfaces $\{\alpha(u,p): p \in M\}$ are called the parallel surfaces of M.

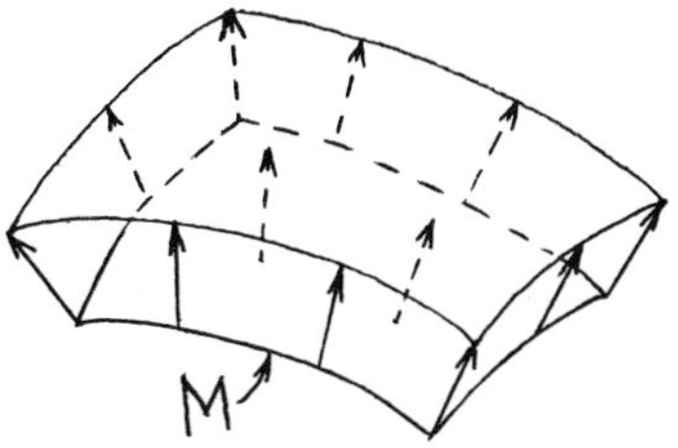

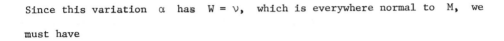

Since this variation α has $W = \nu$, which is everywhere normal to M, we must have

$$\left.\frac{dA(\bar{\alpha}(u))}{du}\right|_{u=0} = 0 .$$

But this equality does not necessarily mean that $A(\bar{\alpha}(0))$ is a minimum. Indeed, as Problem 3-12 shows, each parallel surface has smaller area than M, so actually $A(\bar{\alpha}(0))$ is a maximum! Something quite similar happens in the case of geodesics on a surface of positive curvature. For example, on S^2, a portion of a great circle is longer than a "parallel" curve. The phenomenon

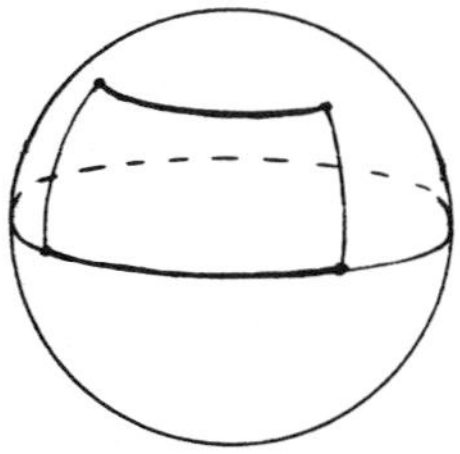

for minimum surfaces is analysed in greater detail in Addendum 4, which considers the second variation of volume.

Although we have derived the fundamental formula for the variation of volume in all dimensions, we will not proceed to discuss the analogues of minimal surfaces in higher dimensions, although this topic has generated much interest in recent years. We should also mention that the study of minimal hypersurfaces in spheres has also attracted much attention, and differs greatly from the theory for Euclidean spaces. For example (Lawson [1]), every compact orientable surface can be imbedded as a minimal surface in S^3.

For the remainder of this Chapter, we will discuss a few other topics involving the variation of volume. We will often digress quite a bit from purely differential-geometric matters, and unfortunately our remarks will not form a coherent subject like the study of minimal surfaces.

Two special cases of Corollary 12 will form the starting point of our considerations. Suppose first that M is simply a compact manifold with no boundary. Then we have

$$\text{(I)} \qquad \left.\frac{dV(\bar{\alpha}(u))}{du}\right|_{u=0} = -\int_M \langle W, n\cdot\eta\rangle \, dV .$$

We can also apply Corollary 12 when M and N have the same dimension n, so that $M \subset N$ is a compact n-dimensional manifold-with-boundary in the n-dimensional manifold N. In this case, $M_p = N_p$ for all $p \in M$, so $\top\colon N_p \to M_p$ is the identity, while $\perp\colon N_p \to N_p$ is the 0 map. Consequently, η is automatically 0, and we have only the boundary term left,

$$\left.\frac{dV(\bar{\alpha}(u))}{du}\right|_{u=0} = \int_{\partial M} \omega .$$

It is easily checked that this can be written

$$\text{(II)} \qquad \left.\frac{dV(\bar{\alpha}(u))}{du}\right|_{u=0} = \int_{\partial M} \langle W, \nu\rangle \, dV_{n-1}$$

where ν is the outward-pointing normal on ∂M and dV_{n-1} is the $(n-1)$-dimensional volume element on ∂M. This formula is certainly reasonable, for when we move each point p on ∂M a distance $\phi(p)$ along $\nu(p)$, we add on a narrow band whose volume is approximately $\int_{\partial M} \phi \, dV_{n-1}$.

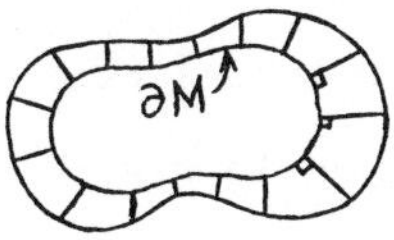

Both formulas (I) and (II) are important for a discussion of the <u>isoperimetric problem</u>. The classical isoperimetric problem was to find the curve of fixed length L which encloses the largest area; one naturally expects the answer to be a circle. One can also seek the curve of smallest length which encloses a fixed area; presumably the answer to this "dual" problem is also a circle. We should also mention the <u>problem of Dido</u>, to find the curve of fixed length between two points P and Q which, together with the straight line between P and Q, encloses the largest area; the expected answer is an arc of a circle. These classical problems have given rise to a whole class of problems in the calculus of variations, known generically as "isoperimetric problems." To illustrate this sort of problem we will, for simplicity, stay in dimension 1. Consider two functions

$$F\colon \mathbb{R}\times\mathbb{R}\times\mathbb{R} \to \mathbb{R} \qquad \text{and} \qquad G\colon \mathbb{R}\times\mathbb{R}\times\mathbb{R} \to \mathbb{R} .$$

For a function $f\colon [a,b] \to \mathbb{R}$ we define

$$J(f) = \int_a^b F(t,f(t),f'(t))\,dt$$

$$K(f) = \int_a^b G(t,f(t),f'(t))dt \ .$$

Among all functions $f\colon [a,b] \to \mathbb{R}$ with fixed values at a and b, and a fixed value $J(f) = C$, we seek the one which maximizes or minimizes $K(f)$. The "dual" problem is to find that f with fixed values at a and b, and fixed $K(f) = C'$, which minimizes or maximizes $J(f)$. This problem is approached by generalizing the methods which work for the corresponding problem in ordinary calculus, a review of which is now in order.

Suppose we are given two differentiable functions $j, k\colon \mathbb{R}^n \to \mathbb{R}$, and we seek the maximum or minimum of j on the set $k^{-1}(C)$. The method of "Lagrangian multipliers" states that if j attains its maximum or minimum on $k^{-1}(C)$ at the point p, and p is not a critical point of k, then there is a number λ such that

$$(1) \qquad \frac{\partial j}{\partial x_i}(p) = \lambda \frac{\partial k}{\partial x_i}(p) \qquad i = 1,\ldots,n \ .$$

The proof of this assertion has already been outlined in Problem 3-3 , but it is so crucial to the present discussion that it will be repeated here. We note that the hypotheses on k imply that in a neighborhood of p, the set $k^{-1}(C) \subset \mathbb{R}^n$ is a hypersurface M, and that $k_*(X_p) = 0$ for $X_p \in \mathbb{R}^n{}_p$ precisely when $X_p \in M_p$. Every such X_p is $c'(0)$ for some curve c in M.

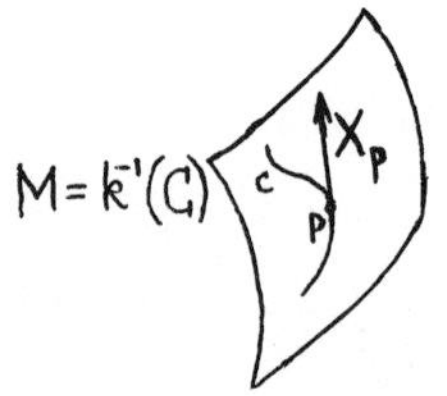

It follows that $j(c(t))$ has a maximum or minimum at $t = 0$, which means that

$j_*(X_p) = 0$. Thus the two linear functions $j_*, k_*\colon \mathbb{R}^n{}_p \to \mathbb{R}$ have the property that $\ker k_* \subset \ker j_*$. This implies that $j_* = \lambda k_*$ for some λ, which is equivalent to equation (1).

Notice that if k attains its maximum or minimum on $g^{-1}(C')$ at q, and q is not a critical point of j, then there is a number μ such that

$$(2) \qquad \frac{\partial k}{\partial x_i}(q) = \mu \frac{\partial j}{\partial x_i}(q) .$$

Equations (1) and (2) are equivalent, since $\lambda, \mu \neq 0$ (as p and q are not critical points). Thus, if p is a maximum point of g on $k^{-1}(C)$ and we set $C' = j(p)$, then p is at least one of the candidates for the minimum point of k on $j^{-1}(C')$. If we simply look for critical points for j on $k^{-1}(C)$ and for k on $j^{-1}(C')$, then these two "dual" problems are completely equivalent.

Let us apply these ideas to our two functions J and K. Suppose that the maximum or minimum of J on $K^{-1}(C)$ occurs at a C^2 function f which is not a critical point of K. Consider any variation $\alpha\colon (-\varepsilon,\varepsilon)\times[a,b] \to \mathbb{R}$ of f which keeps endpoints fixed. We know from formula (**) on p. I.433 that $dJ(\bar\alpha(u))/du\big|_{u=0}$ depends only on the function $\partial\alpha/\partial u\,(0,t)$ on $[a,b]$. For any C^2 function W on $[a,b]$ with $W(a) = W(b) = 0$, we define

$$J_{f*}(W) = \left.\frac{dJ(\bar\alpha(u))}{du}\right|_{u=0} \qquad \text{for any variation } \alpha \text{ of } f \text{ with } \frac{\partial\alpha}{\partial u}(0,t) = W(t) .$$

Notice that there always is a variation α with this property, for example,

$$\alpha(u,t) = f(t) + uW(t) .$$

The same W can be used to give a variation of any function f, so the "f" in the symbol $J_{f*}(W)$ is important. Nevertheless, since we will be considering only one f, we will usually write simply $J_*(W)$ for convenience. We define $K_*(W)$ in precisely the same way. We thus have functions $J_*, K_*: \mathcal{V} \to \mathbb{R}$, where $\mathcal{V}$ is the vector space of all C^2 functions W on $[a,b]$ with $W(a) = W(b) = 0$. We claim that J_* (and likewise K_*) is linear. To see this we choose two variations α_1 and α_2 with

$$\frac{\partial \alpha_i}{\partial u}(0,t) = W_i(t) ,$$

and define the variation α by

$$\alpha(u,t) = \alpha_1(u,t) + \alpha_2(u,t) .$$

Then

$$\frac{\partial \alpha}{\partial u}(0,t) = W_1(t) + W_2(t) ,$$

so

$$\begin{aligned} J_*(W_1 + W_2) &= \left.\frac{dJ(\bar{\alpha}(u))}{du}\right|_{u=0} \\ &= \left.\frac{dJ(\bar{\alpha}_1(u))}{du}\right|_{u=0} + \left.\frac{dJ(\bar{\alpha}_2(u))}{du}\right|_{u=0} , \\ &\qquad \text{from formula (**) on p. I.433} \\ &= J_*(W_1) + J_*(W_2) . \end{aligned}$$

Homogeneity is proved similarly.

We now make the following

Claim: If $K_*(W) = 0$, then $W = \partial\alpha/\partial u\ (0,t)$ for some variation α with the property that each $\bar{\alpha}(u)$ is in $K^{-1}(C)$.

Remember that, by hypothesis, f is not a critical point of K. From a modern point of view, our claim seems very reasonable, for the set of all C^2 functions $\phi\colon [a,b] \to \mathbb{R}$, with given values at a and b, forms an infinite dimensional manifold, and in a neighborhood of f the set $K^{-1}(C)$ should be a sub-manifold of codimension 1; each "tangent vector" W at f with $K_*(W) = 0$ is a tangent vector to the submanifold $K^{-1}(C)$ and should therefore come from a "curve" α in $K^{-1}(C)$. The classical argument runs as follows.

13. LEMMA. If $K(f) = C$, where the C^2 function f is not a critical point of K, and $K_*(W) = K_{f*}(W) = 0$, then $W = \partial\alpha/\partial u\ (0,t)$ for some variation α with the property that each $\bar{\alpha}(u)$ is in $K^{-1}(C)$.

Proof. Since f is not a critical point, there is W_1 with $K_*(W_1) \neq 0$. Let $L\colon \mathbb{R}^2 \to \mathbb{R}$ be

$$L(r,s) = K(f + rW + sW_1)\ .$$

If we define

$$\beta(u,t) = f(t) + uW_1(t)\ ,$$

then β is a variation of f with $\partial\beta/\partial u\ (0,t) = W_1(t)$ and $\bar{\beta}(u) = f + uW_1$.

So

$$K_*(W_1) = \lim_{u\to 0} \frac{K(f+uW_1) - K(f)}{u} = \frac{\partial L}{\partial s}(0,0) \ .$$

Similarly,

$$K_*(W) = \frac{\partial L}{\partial r}(0,0) \ .$$

Since

$$\begin{cases} L(0,0) = K(f) = C \\ \frac{\partial L}{\partial s}(0,0) = K_*(W_1) \neq 0 \ , \end{cases}$$

the implicit function theorem shows that there is a C^2 function $r \longmapsto s(r)$, from a neighborhood of 0 in $\mathbb{R}$ to a neighborhood of 0 in $\mathbb{R}$, such that

(1) $$C = L(r,s(r)) = K(f + rW + s(r)W_1) \qquad \text{for small } r \ .$$

Notice that the first part of the equation gives, upon differentiating with respect to r,

$$0 = \frac{\partial L}{\partial r}(0,0) + \frac{\partial L}{\partial s}(0,0)s'(0) = K_*(W) + K_*(W_1)s'(0) = K_*(W_1)s'(0) \ ,$$

and hence

$$s'(0) = 0 \ .$$

Thus, if we define the variation α by

$$\alpha(u,t) = f(t) + uW(t) + s(u)W_1(t) \ ,$$

then each $\bar{\alpha}(u) = f + uW + s(u)W_1$ is in $K^{-1}(C)$ by (1), and also

$$\frac{\partial\alpha}{\partial u}(0,t) = W(t) + s'(0)W_1(t) = W(t) . \blacksquare$$

14. THEOREM (EULER'S RULE). If the maximum or minimum of J on $K^{-1}(C)$ occurs at a C^2 function f which is not a critical point of K, then there is a number λ such that f is a critical point of $J - \lambda K$ (and consequently the Euler equations for $J - \lambda K$ hold for f).

Proof. Consider the two linear functions $J_*, K_*: \mathcal{V} \to \mathbb{R}$. If $K_*(W) = 0$, let α be the variation given by Proposition 13, with all $\bar{\alpha}(u)$ in $K^{-1}(C)$. Since the maximum or minimum of J on $K^{-1}(C)$ occurs at f, the function $u \mapsto J(\bar{\alpha}(u))$ has a maximum or minimum at 0, and consequently

$$J_*(W) = \left.\frac{dJ(\bar{\alpha}(u))}{du}\right|_{u=0} = 0 .$$

Thus $\ker K_* \subset \ker J_*$. The vector space $\mathcal{V}$ is infinite dimensional, but it still follows (Problem 3-2) that there is a number λ with $J_* = \lambda K_*$, which is equivalent to the assertion that f is a critical point of $J - \lambda K$. $\blacksquare$

In Problem I.9-19, we showed that the Euler equations actually make sense and hold for a critical function of J which is only known to be C^1. A similar result holds for Euler's Rule; because this strengthened form of the rule will be so important for us, the details of the proof will be given here.

Let f be a C^1 function on $[a,b]$, and let W be a C^1 function with $W(a) = W(b) = 0$. Since we no longer have equation (**) on p. I.433 , we can no longer define $J_{f*}(W)$ quite as before. Instead we define

$$J_*(W) = J_{f*}(W) = \left.\frac{dJ(\bar{\alpha}(u))}{du}\right|_{u=0} ,$$

where α is the particular variation

$$\alpha(u,t) = f(t) + uW(t) .$$

The formula in Problem I.9-19 shows that

$$J_*(W) = \int_a^b W'(t)\left[\frac{\partial F}{\partial y}(t,f(t),f'(t)) - \int_a^t \frac{\partial F}{\partial x}(t,f(t),f'(t))dt\right] dt .$$

We define $K_*(W)$ similarly. It is clear that J_* and K_* are linear. Notice that if f is a critical point of K, in the sense that $dK(\bar{\alpha}(u))/du\big|_{u=0} = 0$ for <u>all</u> variations α keeping end-points fixed, then surely $K_*(W) = 0$ for all W. Conversely, suppose that $K_*(W) = 0$ for all W. Then Du Bois Reymond's Lemma (see Problem I.9-19) shows that

$$\frac{\partial G}{\partial y}(t,f(t),f'(t)) - \int_a^t \frac{\partial G}{\partial x}(t,f(t),f'(t))dt = c ,$$

for some constant c. So for any variation α keeping endpoints fixed we have (p. I.480)

$$\left.\frac{dK(\bar{\alpha}(u))}{du}\right|_{u=0} = c\int_a^b \frac{\partial^2\alpha}{\partial u\partial t}(0,t)dt = c\left[\frac{\partial\alpha}{\partial u}(0,b) - \frac{\partial\alpha}{\partial u}(0,a)\right]$$

$$= 0 - 0 .$$

Thus f is a critical point for K if and only if $K_{f*}(W) = 0$ for all W.

13'. LEMMA. If $K(f) = C$, where the C^1 function f is not a critical point of K, and $K_*(W) = 0$, then $W = \partial\alpha/\partial u\ (0,t)$ for some variation α [not of the special sort considered above] with the property that each $\bar{\alpha}(u)$ is in $K^{-1}(C)$.

Proof. The proof of Proposition 13 goes through unchanged; all variations constructed in the proof are of the special sort considered above, except for the final variation α. ∎

14'. THEOREM (EULER'S RULE FOR C^1 FUNCTIONS). If the maximum or minimum of J on $K^{-1}(C)$ occurs at a C^1 function f which is not a critical point of K, then there is a number λ such that f is a critical point of $J - \lambda K$ (and consequently the Euler equations for $J - \lambda K$ make sense and hold for f, by Problem I.9-19).

Proof. Let $\mathscr{W}$ be the vector space of all C^1 functions W on $[a,b]$ with $W(a) = W(b) = 0$, and consider the two linear functions $J_*, K_*\colon \mathscr{W} \to \mathbb{R}$. If $K_*(W) = 0$, let α be the variation given by Proposition 13'. Then the function $u \mapsto J(\bar{\alpha}(u))$ has a maximum or minimum at 0, and consequently

$$0 = \left.\frac{dJ(\bar{\alpha}(u))}{du}\right|_{u=0} = \int_a^b W'(t)\left[\frac{\partial F}{\partial y}(t,f(t),f'(t)) - \int_a^t \frac{\partial F}{\partial x}(t,f(t),f'(t))dt\right] dt$$

(by Problem I.9-19)

$$= J_*(W) \ .$$

Thus $\ker K_* \subset \ker J_*$. So there is a number λ with $J_* = \lambda K_*$ on $\mathscr{W}$. This means that

$$\int_a^b W'(t)\left[\frac{\partial F}{\partial y}(t,f(t),f'(t)) - \int_a^t \frac{\partial F}{\partial x}(t,f(t),f'(t))dt\right] dt$$

$$= \int_a^b W'(t)\lambda\left[\frac{\partial G}{\partial y}(t,f(t),f'(t)) - \int_a^t \frac{\partial G}{\partial x}(t,f(t),f'(t))dt\right] dt ,$$

for all $W \in \mathcal{W}$. DuBois Reymond's Lemma then implies, as in the argument preceeding Lemma 13', that f is a critical point of $J - \lambda K$. ∎

Notice that, as in the simpler case of functions on $\mathbb{R}^n$, the dual problem has exactly the same critical points as the original.

Given a certain amount of trust, that similar results hold for functions $f: [a,b] \to \mathbb{R}^m$, we can finally tackle the classical isoperimetric problem. Consider an imbedding $f: S^1 \to \mathbb{R}^2$, and let $\alpha: (-\varepsilon,\varepsilon) \times S^1 \to \mathbb{R}^2$ be a variation of f through imbeddings. For the length $L(\bar{\alpha}(u))$ of $\bar{\alpha}(u)(S^1)$ we have, by formula (I) on p. 426,

$$\left.\frac{dL(\bar{\alpha}(u))}{du}\right|_{u=0} = -\int_{S^1} \langle W,\eta\rangle \, ds$$

$$= -\int_{S^1} \langle W, \mathbf{n}\rangle\cdot\kappa \, ds ,$$

where $\mathbf{n}$ is the principal normal of f and κ is the curvature of f. For the area $A(\bar{\alpha}(u))$ bounded by $\bar{\alpha}(u)(S^1)$ we easily derive, from formula (II) on p. 426,

$$\left.\frac{dA(\bar{\alpha}(u))}{du}\right|_{u=0} = \int_{S^1} \langle W, \mathbf{n}\rangle \, ds .$$

We want to find the imbedding $f: S^1 \to \mathbb{R}^2$ which maximizes A for fixed L.

Since $f(S^1)$ cannot lie on a straight line, f is not a critical point for L. Therefore Euler's Rule shows that there is some λ with

$$0 = \int_{S^1} \langle W, \mathbf{n} \rangle \, ds + \lambda \int_{S^1} \langle W, \mathbf{n} \rangle \kappa \, ds$$

$$= \int_{S^1} \langle W, \mathbf{n} \rangle [1 + \lambda\kappa] \, ds ,$$

for all variations W. It clearly follows that we must have $1 + \lambda\kappa = 0$, so κ must be a constant, $-1/\lambda$, and f must be an imbedding as a circle.

[It is perhaps worth pointing out that for this problem one can give an elementary proof that if $L_*(W) = 0$ for some W, then there is a variation $\alpha: (-\varepsilon, \varepsilon) \times S^1 \to \mathbb{R}^2$ of f with $\partial\alpha/\partial u\,(0,t) = W$, for which each $\bar{\alpha}(u)(S^1)$ has length $L(0)$. In fact, if β is any variation with $\partial\beta/\partial u\,(0,t) = W(t)$, then we can set

$$\alpha(u,t) = \frac{L(0)}{L(u)} \cdot \beta(u,t) \in \mathbb{R}^2 , \qquad L(u) = \text{length of } \bar{\beta}(u)(S^1) .$$

We have

$$\frac{\partial\alpha}{\partial u}(0,t) = 1 \cdot \frac{\partial\beta}{\partial u}(0,t) - \frac{L'(0)}{L(0)^2} \cdot \beta(0,t)$$

$$= \frac{\partial\beta}{\partial u}(0,t) , \qquad \text{since } L'(0) = L_*(W) = 0 ;$$

and

$$\text{length of } \bar{\alpha}(u)(S^1) = \frac{L(0)}{L(u)} \cdot \text{length of } \bar{\beta}(u)(S^1)$$

$$= \frac{L(0)}{L(u)} \cdot L(u) = L(0) .]$$

We can also apply Euler's Rule to the dual problem of finding the imbedding $f\colon S^1 \to \mathbb{R}^2$ which minimizes L for fixed A. Since no f can be a critical point for A, we find once again that $f(S^1)$ must be a circle. Finally, consider the problem of Dido, to join two fixed points P and Q by a curve c of fixed length $L > d(P,Q)$ so that the area enclosed by c and the line segment $\overline{PQ}$ is a maximum. We consider an imbedding $f\colon [a,b] \to \mathbb{R}^2$ with $f(a) = P$ and $f(b) = Q$, and let $\alpha\colon (-\varepsilon,\varepsilon) \times [a,b] \to \mathbb{R}^2$ be a variation of f through imbeddings. For the length $L(\bar{\alpha}(u))$ of $\bar{\alpha}(u)([a,b])$ we have, by formula (I) on p. 426 ,

$$\left.\frac{dL(\bar{\alpha}(u))}{du}\right|_{u=0} = -\int_a^b \langle W, \mathbf{n} \rangle \cdot \kappa \, ds ,$$

while for the area $A(\bar{\alpha}(u)$ bounded by $\bar{\alpha}(u)([a,b])$ and $\overline{PQ}$, formula (II) on p. 426 reduces to

$$\left.\frac{dA(\bar{\alpha}(u))}{du}\right|_{u=0} = \int_a^b \langle W, \mathbf{n} \rangle \, ds .$$

Euler's Rule shows, once again, that f must have constant curvature, so that $f([a,b])$ must be an arc of a circle. We find the same result for the dual problem.

There are, unfortunately, two difficulties with our solution of the isoperimetric problem. We have been working with C^1 curves, and we could have obtained similar results for piecewise C^1 curves with a little more effort. But the obvious class of curves to consider for the isoperimetric problem is the class of rectifiable curves, the curves with finite length (defined as the least upper bound of the lengths of inscribed polygonal curves). Moreover, we have merely found that the circle is the solution of the isoperimetric problem *if a solution exists*; we have not proved that the circle actually is a solution.

Although this will lead us astray from the righteous path of differential geometry, at this point I cannot resist the impulse to mention one of the extremely clever solutions of the isoperimetric problem, involving no assumptions about differentiability, which was given by the great geometer Steiner. Note first that we might as well restrict our attention to convex curves, because the convex hull C^* of a nonconvex curve C has smaller length and encloses a larger area -- a suitable region C^{**} similar to C^* will then have the same length as C, and yet still larger area.

Let us therefore consider a convex curve C which is not a circle. We will show that it cannot be a solution to the isoperimetric problem. Choose two points A and B on C which divide C into two curves C_1 and C_2 of equal length, and let R_i be the region bounded by C_i and the line segment AB. We can assume that area $R_1 \geq$ area R_2; we claim that we actually have area R_1 = area R_2. To prove this, we reflect region R_1 in the line AB, obtaining a region R_1' on the opposite side. Then $R_1 \cup R_1'$ has

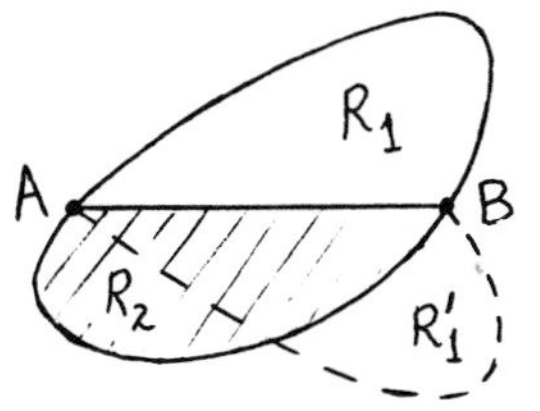

area $\geq$ the area of $R_1 \cup R_2$, while its circumference is the same. If C is a solution to the isoperimetric problem, then we must actually have area $R_1 \cup R_1' =$ area $R_1 \cup R_2$, so we have area $R_1 =$ area $R_1' =$ area R_2.

Now since C is not a circle, we can choose A and B so that neither C_1 nor C_2 is a semi-circle. Since area $R_1 =$ area R_2, the region $R_1 \cup R_1'$ with boundary $C_1 \cup C_1'$ will be another solution to the isoperimetric problem, and it will also not be a circle. In other words, we can assume that C is symmetric with respect to AB.

Now there is a point P on C_1 such that $\angle APB$ is not a right angle; let Q be the symmetric point on C_2. The region inside C is made up of the quadrilateral APBQ together with 4 regions $T_1, \ldots, T_4$ as shown in part (a) of the figure below. In part (b) of this figure we have drawn a quadrilateral

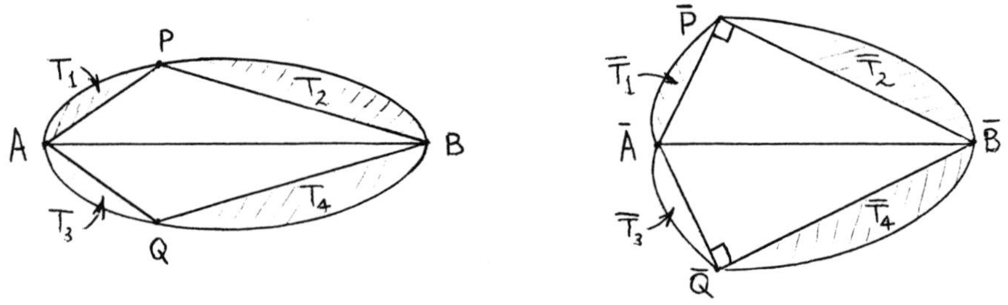

$\bar{A}\bar{P}\bar{B}\bar{Q}$ with $AP = AQ = \bar{A}\bar{P} = \bar{A}\bar{Q}$ and $BP = BQ = \bar{B}\bar{P} = \bar{B}\bar{Q}$, but with $\angle\bar{A}\bar{P}\bar{B}$ and $\angle\bar{A}\bar{Q}\bar{B}$ both right angles. Then on $\bar{A}\bar{P}$ we have drawn a region $\bar{T}_1$ congruent to the region T_1 in part (a), and regions $\bar{T}_2, \ldots, \bar{T}_4$ have been drawn similarly. The new figure clearly has the same circumference as the original curve C. On the other hand, it has larger area, since the quadrilateral $\bar{A}\bar{P}\bar{B}\bar{Q}$ clearly has larger area than APBQ. Thus C could not be a solution to the isoperimetric problem. This completes the proof that a circle is the only curve which can be a solution to the isoperimetric problem.

This ingenious proof, although it assumes absolutely nothing about the differentiability of C, still has a defect, which, to be sure, Steiner would never have worried about. This proof, like our previous one, shows only that the circle is the solution of the isoperimetric problem, if a solution exists. In Blaschke {1}, {2}, one can find many rigorous solutions of the isoperimetric problem which avoid this pitfall by showing that for a closed curve of length L, enclosing a region of area A, we always have $L^2 - 4\pi A \geq 0$, with equality only when the curve is a circle. These proofs exhibit various degrees of ingenuity and elegance, but there is also a straightforward, if somewhat lengthy, direct proof of existence, which will be useful for us to examine.

Let (X,d) be a bounded metric space, and let $\mathcal{C}(X)$ be the set of all non-empty closed subsets of X. The distance $d(x,C)$ from a point $x \in X$ to a closed set $C \in \mathcal{C}(X)$ is defined as

$$d(x,C) = \min_{y \in C} d(x,y) ,$$

and we define the ε-neighborhood $V_\varepsilon(C)$ of C as

$$V_\varepsilon(C) = \{x\colon d(x,C) < \varepsilon\} .$$

Given $C_1, C_2 \in \mathcal{C}(X)$, we then define

$$\rho(C_1,C_2) = \inf\{\varepsilon > 0\colon C_1 \subset V_\varepsilon(C_2) \text{ and } C_2 \subset V_\varepsilon(C_1)\} .$$

It is easy to check that ρ is a metric, the Hausdorff metric, on $\mathcal{C}(X)$. When X is compact, the corresponding topology on $\mathcal{C}(X)$ depends only on the topology of X, not on the given metric d, since any neighborhood of $C \in \mathcal{C}(X)$ contains an ε-neighborhood.

15. PROPOSITION. If (X,d) is compact, then so is $(\mathcal{C}(X),\rho)$.

Proof. Given $\varepsilon > 0$, choose a finite number of sets $A_1,\ldots,A_n$ of diameter $< \varepsilon$ which cover X. For each finite set $F \subset \{1,\ldots,n\}$ let

$$\mathcal{A}_F = \{C \in \mathcal{C}(X)\colon C \cap A_j \neq \emptyset \iff j \in F\} .$$

Then the sets $\mathcal{A}_F$ cover $\mathcal{C}(X)$ and have diameter $\leq 2\varepsilon$. This shows that $(\mathcal{C}(X),\rho)$ is totally bounded.

Now let $C_1,C_2,\ldots$ be a Cauchy sequence in $(\mathcal{C}(X),\rho)$. Let C be the set of all $x \in X$ such that every neighborhood of x contains points from infinitely many C_n. The set C is non-empty, for if $x_n \in C_n$ and x is an accumulation point of the sequence $\{x_n\}$, then $x \in C$. It is also clear that C is closed. We claim that $C = \lim C_n$. Given $\varepsilon > 0$, we first show that the C_n are eventually in the open ε-neighborhood $V_\varepsilon(C)$ of C. For suppose that an infinite sequence $C_{i_1},C_{i_2},\ldots$ intersected the compact set $X - V_\varepsilon(C)$. Then we could choose $x_{i_n} \in C_{i_n} \cap [X - V_\varepsilon(C)]$; some point $x \in X - V_\varepsilon(C)$ would be an accumulation point of the sequence $\{x_{i_n}\}$, hence $x \in C$, a contradiction. We also claim that C is in $V_\varepsilon(C_n)$ for sufficiently large n. In fact, since C_n is a Cauchy sequence, there is some N such that $C_m \subset V_{\varepsilon/2}(C_n)$ for all $m, n > N$; this implies that $V_{\varepsilon/2}(C_m) \subset V_\varepsilon(C_n)$ for all $m, n > N$. So if $n > N$ and C is not contained in $V_\varepsilon(C_n)$, then also C contains a point which is not in $V_{\varepsilon/2}(C_m)$ for all $m > N$, which is clearly impossible. ■

We will apply this result to the case where X is a closed disc in $\mathbb{R}^2$. The set $\mathrm{Con}(X) \subset \mathcal{C}(X)$ consisting of all non-empty closed <u>convex</u> subsets of X is easily seen to be a closed, and hence compact, subset of $\mathcal{C}(X)$. If $A\colon \mathcal{C}(X) \to \mathbb{R}$ is the function $A(C)$ = area of C (= Lebesgue measure of A, say), then A is clearly continuous. Define $L\colon \mathrm{Con}(X) \to \mathbb{R}$ by $L(C)$ = length of boundary C.

<u>16. PROPOSITION</u>. The function $L\colon \mathrm{Con}(X) \to \mathbb{R}$ is continuous.

<u>Proof</u>. If $\rho(C_1, C_2) < \varepsilon$, then $C_1 \subset (1+\varepsilon)\cdot C_2$ and $C_2 \subset (1+\varepsilon)\cdot C_1$, so the result follows from

<u>17. LEMMA</u>. If γ_1 and γ_2 are convex curves with γ_1 contained inside γ_2, then length $\gamma_1 \leq$ length γ_2.

<u>Proof</u>. The following picture shows that if P is a polygonal arc inscribed

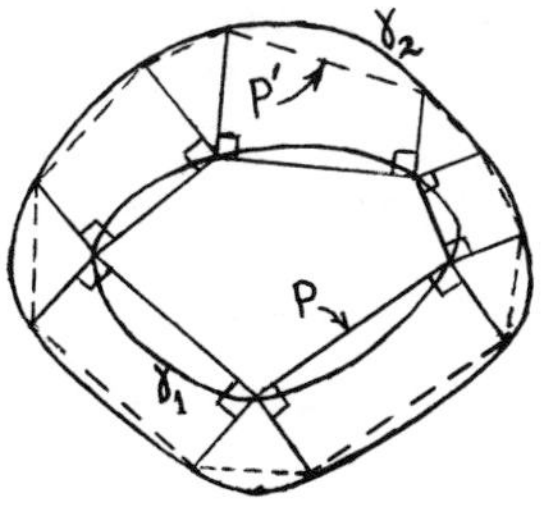

in γ_1, then P is shorter than some polygonal arc P' inscribed in γ_2. ■

It is now an easy matter to prove the existence of a (convex) curve, with fixed length L_0, of maximum area: We can clearly restrict our attention to convex sets contained within a closed disc X of radius L_0; then the set $L^{-1}(L_0) \subset \text{Con}(X)$ is a closed subset of the compact space Con(X), so the continuous function A takes on its maximum somewhere on the set. This proof of existence, together with Steiner's argument, rigorously solves the isoperimetric problem. The dual problem can be handled similarly. Its solution is also contained in our solution of the original problem, for we now know that the relation $L^2 - 4\pi A \geq 0$ always holds, with equality only for circles, and this proves that the circle is also the solution of the dual problem. It is also easy to derive this fact from the solution of the original problem by using the similarities of the plane. Finally we mention that the problem of Dido can be settled by similar methods; for instance, we can consider the space of all closed convex sets which have a given line segment PQ as part of their boundary.

I would now like to discuss briefly a line of argument which could be used if Steiner's argument were not available, and we had to rely solely on Euler's Rule. Clearly the only problem is to show that the solution of the isoperimetric problem (whose existence we can prove) must be a C^1 curve. The first step would be to show that the solution curve has a tangent line everywhere. Now it is well-known (Problem 2) that a convex function always has left- and right-hand derivatives, so we just have to show that our convex curve has no corners. If our curve actually contained

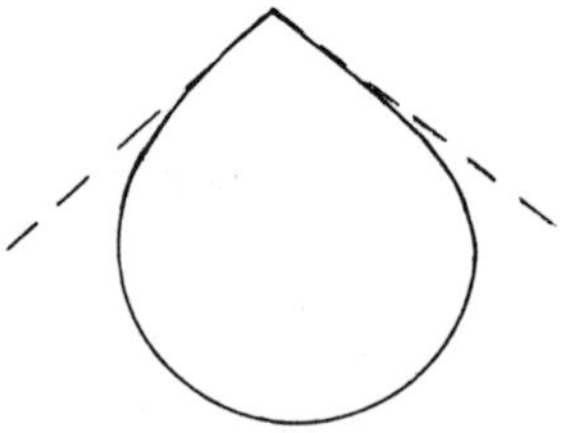

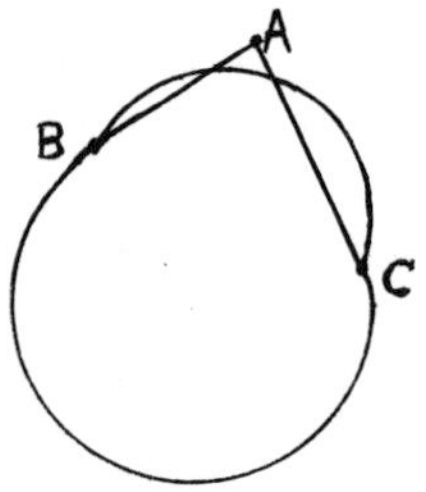

two straight line segments AB and AC meeting at an angle at A, then it would be easy to show that it is not a solution to the isoperimetric problem. For the two segments could be replaced by an arc of a circle with equal length, but enclosing larger area, since such an arc is a solution to the problem of Dido. One doesn't really need the whole solution to the problem of Dido to reach this conclusion, however, for a simple calculation will show that the appropriate arc together with line BC encloses more area than triangle ABC. (If we had worked out the calculus of variations argument for piecewise C^1 curves we would have another way of seeing that the two segments can be replaced by some nearly curve of the same length, but enclosing larger area.) In the general case, the same idea can be made to work by an approximation argument.

Now it is also easy to see (Problem 2) that if a convex function is everywhere differentiable, then its derivative is <u>automatically</u> continuous. This shows that the solution to the isoperimetric problem must be a C^1 curve; Euler's Rule then leads to the conclusion that it must be a circle.

As differential geometers, we naturally think of generalizing the isoperimetric problem to an arbitrary surface M. Given a variation $\alpha\colon (-\varepsilon,\varepsilon)\times S^1 \to M$ of a map $f\colon S^1 \to M$ we now have

$$\left.\frac{dA(\bar{\alpha}(u))}{du}\right|_{u=0} = \int_{S^1} \langle W, \mathbf{u}\rangle\, ds \,,$$

$$\left.\frac{dL(\bar{\alpha}(u))}{du}\right|_{u=0} = -\int_{S^1} \langle W, \eta\rangle\, ds$$

$$= -\int_{S^1} \langle W, \mathbf{u}\rangle \kappa_g \, ds \,,$$

where $\mathbf{u}$ is the second member of the Darboux frame for f, and κ_g is the geodesic curvature of f. These formulas, together with a few ruthlessly suppressed details which are necessary to transfer Euler's Rule from $\mathbb{R}^m$ to manifolds, show that if f maximizes A for fixed L, then there is a constant λ such that

$$\begin{aligned} 0 &= \int_{S^1} \langle W, \mathbf{u} \rangle \, ds + \lambda \int_{S^1} \langle W, \mathbf{u} \rangle \kappa_g \, ds \\ &= \int_{S^1} \langle W, \mathbf{u} \rangle [1 + \lambda \kappa_g] \, ds \end{aligned}$$

for all variations W. This implies that f has <u>constant geodesic curvature</u>. The geodesic curvature was first invented by Minding, in 1830, when he obtained this solution (for the problem of Dido, rather than the isoperimetric problem). Minding dealt with surfaces in $\mathbb{R}^3$, and defined κ_g extrinsically, but he then showed that it was a bending invariant; its present name was given it by Bonnet, in 1848.

A rigorous discussion of the isoperimetric problem on an arbitrary surface M is considerably more complicated than for the plane, if for no other reason than because the problem itself is more involved. First of all, Euler's Rule is not always applicable, because there might be closed curves which are geodesics, and consequently critical points for L. For example, on the surface M shown below, the equator C of the smaller spherical part is <u>not</u> a critical

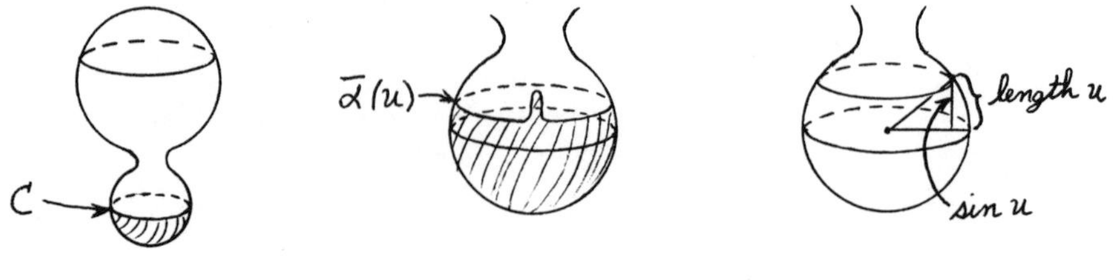

point for area among all curves with length equal to L(C). We can obtain a variation α of C by moving C up distance u along geodesics perpendicular to C, and then adding on a bulge to bring the length up to L(C). Then A(u) - A(0) is greater than the area of M enclosed between two parallel planes at distance sin u (see the third picture above), so

$$A(u) - A(0) > \sin u$$

and consequently $A'(0) \neq 0$. The figure below shows a curve C' on M, with L(C') = L(C), which is a critical point for A among curves with this length.

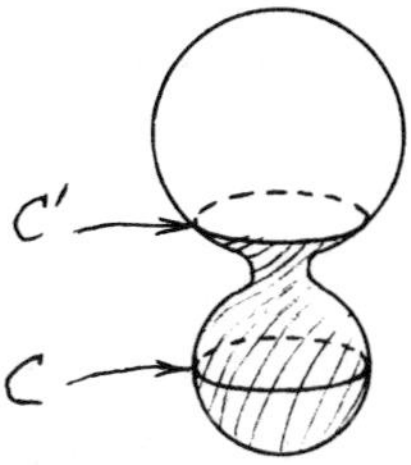

If we accept the fact that a circle is a solution to the isoperimetric problem on the sphere, then C' must be a solution to the isoperimetric problem on M. Of course, we really have to decide which of the two regions of M bounded by C' should be maximized; if we take the top region, then C' actually minimizes. The necessity of making this decision correctly is further illustrated by the fact that there is another curve C" higher up with length L that is also a critical point for A among curves of this length. In fact, if we make the wrong decision we might be led to say that there are curves of length L bounding regions with area arbitrarily close to that of M. This becomes

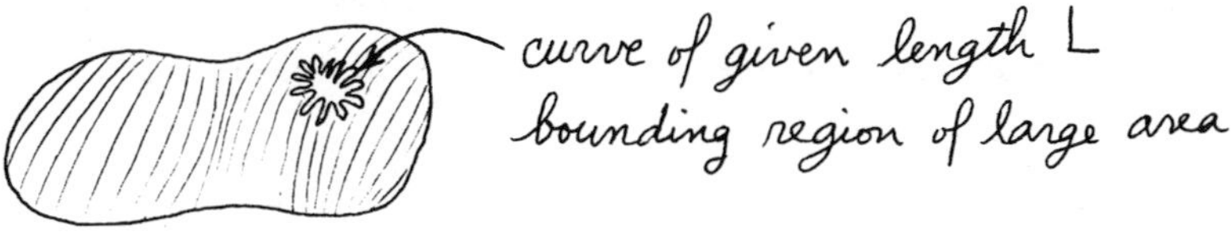

quite critical if there is a closed curve of length L which divides M into two pieces with the same area, as may happen for example on a sphere.

I also suspect that in some cases the solution of the isoperimetric problem will have to be a curve which intersects itself, as in the following picture;

notice that we want the curve to go around as much as possible of the part of the surface with large curvature.

Finally, we point out that on most surfaces there will be just a few solutions of the isoperimetric problem, and that they may be completely different curves. In this respect the problem of Dido is more natural on a general surface; given a geodesic segment γ from P to Q, we would expect that among all curves c from P to Q with given length $L > d(P,Q)$ there is just one on each side of γ which maximizes the area enclosed by γ and c.

I think that a reasonable approach to the isoperimetric problem on a compact surface M is to consider only lengths L so small that a closed curve of length L must be contained in a geodesically convex set. It is then clear that our solution must be the boundary of a geodesically convex set, and there is no problem deciding which region it bounds. All our previous considerations can be suitably modified to show that a solution of the isoperimetric problem exists and is C^1, so that it must have constant geodesic curvature. This proves, in particular, that there are <u>closed</u> curves of constant geodesic curvature; proving this result directly seems almost hopeless. By the way, it is a classical theorem that if every curve of constant geodesic curvature is closed, then M must have constant curvature (Blaschke {1}).

In this connection, an interesting experiment can be performed with a soap film on a wire loop. If a small loop of thread held between two thin sticks is dipped into the soap solution, it can then be thrust into the soap film without breaking it. If the part of the soap film inside the thread is

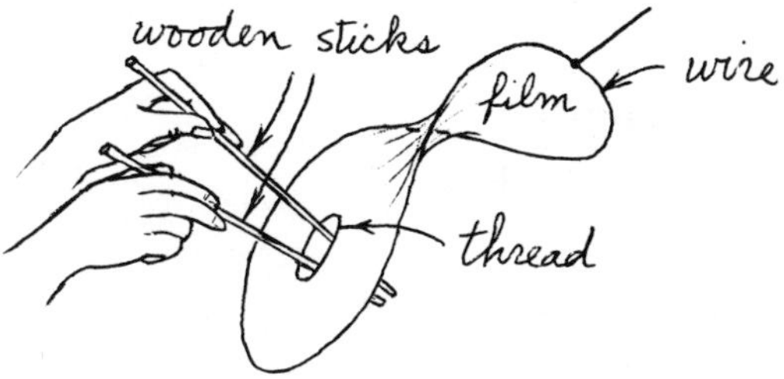

then broken, and the sticks are removed, the thread should take a form which is a solution to the isoperimetric problem on the soap film. When one tries this experiment it turns out that, unless the wire loop is very flat, the string always rushes off toward the wire loop, no matter where it is placed. I take this to mean that there are no curves of constant geodesic curvature on a non-flat minimal surface, but I haven't the slightest idea how one would prove it. [Actually, as Osserman pointed out to me, the experiment involves a rather more complicated problem, since the shape of the surface changes as the thread moves.]

For our next application of Euler's Rule we will work only in $\mathbb{R}^3$, and consider the 3-dimensional isoperimetric problem, to find the surface of fixed area A which encloses the greatest volume. Consider an imbedding $f\colon S^2 \to \mathbb{R}^3$, and let $\alpha\colon (-\varepsilon,\varepsilon) \times S^2 \to \mathbb{R}^3$ be a variation of f through imbeddings. For the area $A(\bar{\alpha}(u))$ of $\bar{\alpha}(u)(S^2)$ we have, by formula (I) on

p. 426 ,

$$\left.\frac{dA(\bar{\alpha}(u))}{du}\right|_{u=0} = -\int_{S^2} \langle W,\eta\rangle \, dA$$

$$= -\int_{S^2} 2H\langle W,\nu\rangle \, dA ,$$

where ν is the normal of $f(S^2)$ and H is the mean curvature. For the volume $V(\bar{\alpha}(u))$ enclosed by $\bar{\alpha}(u)(S^2)$ we have, by formula (II) on p. 426 ,

$$\left.\frac{dV(\bar{\alpha}(u))}{du}\right|_{u=0} = \int_{S^2} \langle W,\nu\rangle \, dA .$$

We want to find the imbedding $f\colon S^2 \to \mathbb{R}^3$ which maximizes V for fixed A. The compact surface $f(S^2)$ cannot have $H = 0$ everywhere (Corollary 7-31), so f is not a critical point for A. Therefore Euler's Rule shows that there is some λ with

$$0 = \int_{S^2} \langle W,\nu\rangle \, dA + \lambda \int_{S^2} 2H\langle W,\nu\rangle \, dA = \int_{S^2} \langle W,\nu\rangle[1+2\lambda H] \, dA$$

for all variations W. It clearly follows that $f(S^2)$ must have constant mean curvature.

At this point we encounter new difficulties, for we first have to find all the surfaces of constant mean curvature. This particular problem is interesting of itself, quite apart from any connection with the isoperimetric problem. For one thing, such surfaces are the possible shape for soap bubbles -- the increased air pressure within the bubble naturally makes it take a form which maximizes the enclosed volume. We already know (Theorem 5-3) that a convex surface with constant mean curvature must be a standard sphere. H. Hopf [1] proved that an immersed surface homeomorphic to S^2 with constant mean curvature must be a standard sphere. The proof of this is deferred to Addendum 2, since it uses the existence of isothermal parameters, which is proved in Addendum 1. Alexandrov [1] proved that any imbedded compact hypersurface of $\mathbb{R}^{n+1}$ with constant mean curvature must be a standard sphere; this proof is presented in Addendum 3. Alexandrov's theorem holds just as well for hypersurfaces in the hyperbolic space H^{n+1} or in an open hemisphere of S^{n+1}. It definitely fails even for surfaces with $H = 0$ in the sphere S^3, as we mentioned on p. 426 . It is still unknown (Hopf's Problem) whether every immersed compact hypersurface with constant mean curvature is a standard sphere, even for the case of $\mathbb{R}^3$; for all we know, some one may some day blow a soap bubble in the shape of an immersed torus.

At first sight the isoperimetric problem seems easier, since it seems that a solution ought to be convex. Proving this directly seems almost hopeless, however, for the boundary of the convex hull C^* of a set C in $\mathbb{R}^3$ may well have larger surface area than the boundary of C. Of course, the

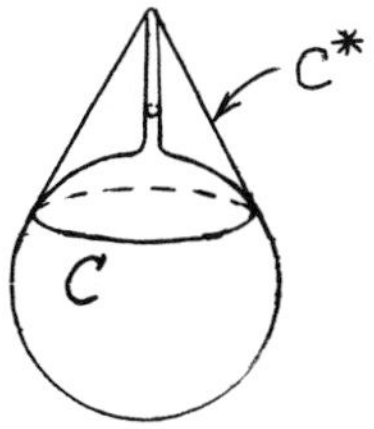

volume of C^* is also larger than that of C -- the big question is whether it is larger by enough. In Blaschke {2 } there is a proof that the sphere is the solution to the isoperimetric problem provided that we restrict our attention to convex sets. In the general case there is such an overwhelming multitude of problems, not least of which is the difficulty of defining surface area, that we will say no more about the problem, merely refering the interested reader to the bibliography.

To conclude this rather disconnected series of remarks, we shall very briefly discuss a problem which requires for its solution even more elaborate machinery than any yet mentioned, but which is of much greater interest to differential geometry. In his investigations of the "three body problem", Poincare was led to consider simple closed geodesics on a compact convex surface $M \subset \mathbb{R}^3$. Poincare gave a rather long proof that at least one simple closed geodesic exists on M, and then outlined a much more direct argument for the same conclusion. Although many (probably hopelessly difficult) subsidiary results would be required to make this argument into a complete proof, it is nevertheless an interesting application of Euler's rule for isoperimetric problems. We notice first that if c is a simple closed geodesic on M, then Theorem 6-5 implies that $\nu \circ c$ divides S^2 into two regions each of area 2π. To establish the existence of such a geodesic, we will consider the set of all simple closed curves γ on M such that $\nu \circ \gamma$ divides S^2 into two regions of equal area, and then among these choose one, $c\colon S^1 \to M$, of shortest length. We claim that c must be a geodesic. To prove this we consider a variation $\alpha\colon (-\varepsilon,\varepsilon) \times S^1 \to M$ of c. For the length $L(\bar{\alpha}(u))$ of $\bar{\alpha}(u)(S^1)$ we have, by the formula on p. 445,

$$(1) \qquad \left.\frac{dL(\bar{\alpha}(u))}{du}\right|_{u=0} = -\int_{S^1} \langle W, \mathbf{u} \rangle \cdot \kappa_g \, ds \,.$$

Now extend f to a map $f\colon D \to M$, of the unit disc into M, so that $f(D)$

is one of the regions bounded by $f(S^1)$; extend α to a map $\alpha\colon (-\varepsilon,\varepsilon)\times D \longrightarrow M$ similarly. Let $A(\bar\alpha(u))$ be the area of the image $\nu(\bar\alpha(u)(D)) \subset S^2$. Then

$$A(\bar\alpha(u)) = \int_{\bar\alpha(u)(D)} K\, dA ,$$

where dA is the volume element of M and K is the Gaussian curvature of M. It certainly seems reasonable that we should have

$$\text{(2)} \qquad \left.\frac{dA(\bar\alpha(u))}{du}\right|_{u=0} = \int_{S^1} (K\circ f)\langle W, \mathbf{u}\rangle\, ds ,$$

for $A(\bar\alpha(h)) - A(\bar\alpha(0))$ is the integral of $K\,dA$ over a small band around $f(S^1)$ whose width is given approximately by the function $\langle W, \mathbf{u}\rangle$. To prove this rigorously, we write $A(\bar\alpha(u))$ as

$$A(\bar\alpha(u)) = \int_D [K\circ\bar\alpha(u)]\cdot\Gamma(u) , \qquad \Gamma(u) = \bar\alpha(u)^* dA .$$

Then

$$\begin{aligned}
\left.\frac{dA(\bar\alpha(u))}{du}\right|_{u=0} &= \left.\frac{d}{du}\right|_{u=0} \int_D [K\circ\bar\alpha(u)]\cdot\Gamma(u) \\
&= \int_D \left[\left.\frac{d}{du}\right|_{u=0} K\circ\bar\alpha(u)\right]\cdot\Gamma(0) + \int_D (K\circ f)\cdot\dot\Gamma(0) \\
&\qquad\qquad \text{by Leibnitz's Rule} \\
&= \int_D \left[\left.\frac{d}{du}\right|_{u=0} K\circ\bar\alpha(u)\right]\cdot\Gamma(0) - \int_D (K\circ f)\langle W, \mathbf{u}\rangle\, \Gamma(0) \\
&\qquad + \int_{S^1} (K\circ f)(W \lrcorner\, \Gamma(0)) \qquad \text{by Theorem 11}
\end{aligned}$$

$$= \int_{S^1} (K\circ f)\cdot(W \lrcorner \Gamma(0))$$

$$= \int_{S^1} (K\circ f)\cdot\langle W, \mathbf{u}\rangle \, ds .$$

Now if our curve c is a solution to the isoperimetric problem of minimizing L for fixed $A = 2\pi$, then Euler's rule says that there is a constant λ such that

$$0 = \int_{S^1} \langle W, \mathbf{u}\rangle[\lambda(K\circ f) - \kappa_g] \, ds$$

for all variations W. This implies that $\kappa_g = \lambda(K\circ f)$. On the other hand, applying Theorem 6-5 to $f(D) \subset M$, we obtain

$$-\int_{S^1} \kappa_g \, ds + 2\pi = \int_{f(D)} K \, dA = 2\pi ,$$

and thus

$$0 = \int_{S^1} \kappa_g \, ds = \lambda \int_{S^1} (K\circ f) \, ds .$$

So if M has $K > 0$ everywhere, then we must have $\lambda = 0$, and thus $\kappa_g = 0$; consequently, c is a geodesic.

In Blaschke {1; pp. 211-212} there is a further argument, due to Herglotz, to show that M actually contains at least 3 closed geodesics. Nowadays, all such results are proved by quite different, rigorous methods of far greater generality. Although much is known about closed geodesics on compact manifolds (see the Bibliography for references) the conjecture that every compact manifold always has at least 3 closed geodesics remains unproved.

Addendum 1. Isothermal Coordinates

As we mentioned in Volume II, the existence of isothermal coordinates on any surface was first proved by Gauss, who resorted to a trick that works only in the analytic case. Although we will treat the more general case also, Gauss' proof will be given first, as it is interesting in its own right. First we need to review some facts about differential equations. The equation

$$y'(x) = f(x,y(x))$$

is written classically as

$$\frac{dy}{dx} = f(x,y) ,$$

or even as

$$dy - f(x,y)dx = 0 .$$

Most elementary differential equations courses indicate that one method of solving this equation is to find an "integrating factor" for it, that is, a nowhere zero function λ such that $\lambda(dy - fdx)$ is exact, say

$$\lambda(dy - fdx) = dg .$$

Then the solutions of the original equation are the same as the solutions of $dg = 0$, i.e. the curves $g = \text{constant}$. For example, to solve the equation

$$\begin{aligned} 0 &= (x^2y + x)dy + (xy^2 - y)dx \\ &= x\,dy - y\,dx + xy(x\,dy + y\,dx) , \end{aligned}$$

we multiply by $1/xy$, to obtain

$$0 = \frac{dy}{y} - \frac{dx}{x} + d(xy) ,$$

with the solution

$$\log y(x) - \log x + x \cdot y(x) = \text{constant} .$$

As a more interesting example, we consider the general first order linear equation

$$\frac{dy}{dx} + \phi(x)y = \psi(x) ,$$

which we write as

$$[\phi(x)y - \psi(x)]dx + dy = 0 .$$

In order for

$$[\lambda(x)\phi(x)y - \lambda(x)\psi(x)]dx + \lambda(x)dy$$

to be exact, we need

$$\frac{\partial}{\partial y}[\lambda(x)\phi(x)y - \lambda(x)\psi(x)] = \frac{d\lambda}{dx}$$

or

$$\begin{aligned} \lambda(x)\phi(x) = \frac{d\lambda}{dx} &\implies \frac{d\lambda}{\lambda} = \phi(x)dx \\ &\implies \log \lambda = \int \phi \\ &\implies \lambda = e^{\int\phi} . \end{aligned}$$

So we write our original equation as

$$e^{\int\phi}\frac{dy}{dx} + e^{\int\phi}\phi(x)y = \psi(x)e^{\int\phi} ,$$

which gives

$$\frac{d}{dx}\left(e^{\int\phi} y\right) = \psi(x)e^{\int\phi} ,$$

$$e^{\int\phi} y = \int e^{\int\phi}\psi + C ,$$

$$y = e^{-\int\phi}\left(\int e^{\int\phi}\psi + C\right) .$$

Of course, only in the most fortuitous cases can one find an integrating factor by inspection. What is theoretically more interesting is the observation that for any 1-form

$$(*) \qquad \omega = \alpha\, dx + \beta\, dy$$

on $\mathbb{R}^2$ with $\alpha(p_0)$, $\beta(p_0) \neq 0$, an integrating factor exists in a neighborhood of p_0. To prove this, we consider the differential equation

$$(**) \qquad y'(x) = -\frac{\beta}{\alpha}(x,y(x)) .$$

Since $-(\beta/\alpha)(p_0) \neq 0$, the integral curves of this differential equation form a foliation in a neighborhood of p_0 and there is a diffeomorphism h from a

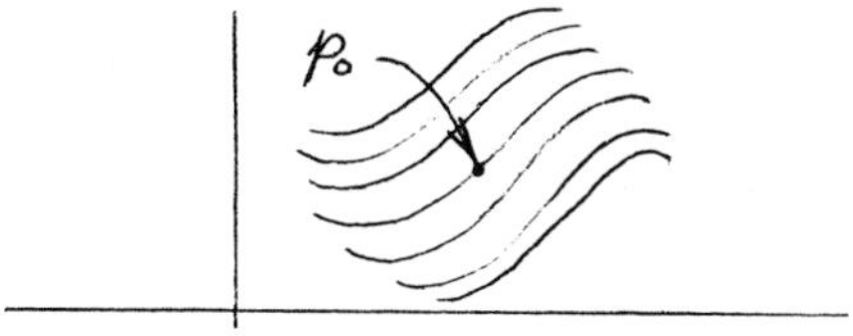

neighborhood of p_0 to $\mathbb{R}^2$ such that the integral curves go into the sets with 2nd coordinate constant. Let

$$g(p) = 2^{\text{nd}} \text{ coordinate } h(p) .$$

Then

$$\ker dg(p) = \text{tangent space at } p \text{ of the solution curve of } (**) \text{ going through } p$$
$$= \ker \omega(p) .$$

This proves that

$$dg(p) = \lambda(p)\cdot\omega(p)$$

for some $\lambda(p) \neq 0$.

In problem I.6-9 we showed that the differential equation

$$y'(z) = f(z,y(z)) \qquad (' = \text{complex derivative})$$

can always be solved if $f\colon \mathbb{C}\times\mathbb{C} \to \mathbb{C}$ is complex analytic. From this we easily conclude, by modifying the preceeding argument, that if α and β are complex-valued functions on $\mathbb{R}^2$ which are the restrictions of complex analytic functions on $\mathbb{C}^2$, and $\alpha(p_0)$, $\beta(p_0) \neq 0$, then there is a complex-valued function λ in a neighborhood of p_0 such that

$$\lambda(\alpha\, dx + \beta\, dy) = dg$$

for some complex-valued function g; both λ and g are the restrictions of complex analytic functions on $\mathbb{C}^2$. Now we can prove

<u>18. THEOREM</u>. Let $\langle\ ,\ \rangle$ be a Riemannian metric on a neighborhood V of $0 \in \mathbb{R}^2$ whose components g_{ij} with respect to the standard coordinate system on $\mathbb{R}^2$ are C^ω (= real analytic). Then there exists a C^ω isothermal coordinate system for $\langle\ ,\ \rangle$ in a neighborhood of 0.

Proof. Let X_1, X_2 be a C^ω orthonormal moving frame in a neighborhood of 0, with dual 1-forms θ^1, θ^2. Then

$$< , > = \theta^1 \otimes \theta^1 + \theta^2 \otimes \theta^2 ,$$

and consequently the corresponding quadratic function $\| \ \|^2$ can be written as

$$\| \ \|^2 = \theta^1 \cdot \theta^1 + \theta^2 \cdot \theta^2 .$$

Let ϕ be the complex-valued differential form

$$\phi = \theta^1 + i\theta^2 , \qquad \text{with} \qquad \bar{\phi} = \theta^1 - i\theta^2 .$$

Then

$$\| \ \|^2 = \phi \cdot \bar{\phi} .$$

If we construct X_1, X_2 explicitly by applying the Gram-Schmidt orthogonalization process to $\partial/\partial x^1$, $\partial/\partial x^2$, then the coefficients of X_1, X_2 will appear as algebraic combinations of the g_{ij}. The same is thus true of θ^1, θ^2. Since the g_{ij} are C^ω, and hence the restrictions of complex analytic functions on $\mathbb{C}^2$, the same is true for θ^1, θ^2. So by the remark preceeding the theorem, there is a complex-valued function λ such that

$$\lambda\phi = dg \implies \bar{\lambda}\bar{\phi} = d\bar{g}$$

for some complex-valued function g. This implies that

$$\lambda\bar{\lambda}\| \ \|^2 = \lambda\bar{\lambda}\phi \cdot \bar{\phi} = dg \cdot d\bar{g} ,$$

so that

$$\| \ \|^2 = \frac{1}{\lambda\bar{\lambda}} \, dg \cdot d\bar{g} \ .$$

If we write $g = u + iv$ for real-valued u and v, then the Jacobian of $(u,v) : \mathbb{R}^2 \to \mathbb{R}^2$ is not zero, for if it were, then dg would be zero, and hence $\| \ \|^2$ would be zero. Now

$$dg \cdot d\bar{g} = du \cdot du + dv \cdot dv \ ,$$

so

$$\| \ \|^2 = \frac{1}{\lambda\bar{\lambda}} (du \cdot du + dv \cdot dv) \ .$$

By polarization,

$$\langle \ , \ \rangle = \frac{1}{\lambda\bar{\lambda}} (du \otimes du + dv \otimes dv) \ .$$

The functions u and v are C^ω since g is the restriction of a complex analytic function on $\mathbb{C}^2$. Thus (u,v) is the required C^ω isothermal coordinate system. ∎

The proof of Theorem 18 when the g_{ij} are not C^ω will be much more involved. First we introduce some new classes of functions. A function $f : \mathbb{R}^n \to \mathbb{R}^m$ is said to satisfy a Hölder condition of order α $(0 < \alpha < 1)$ on $U \subset \mathbb{R}^n$ if there is a constant K such that

$$|f(p) - f(q)| \leq K \cdot |p - q|^\alpha \qquad \text{for all } p, q \in U \ .$$

Such functions are called C^α functions, and a function f is $C^{n+\alpha}$ if all mixed n^{th} order derivatives of f exist and are C^α. We will eventually show that if the g_{ij} in Theorem 18 are C^{α}, then there is a $C^{1+\alpha}$ isothermal coordinate system in a neighborhood of 0. We will also show that if the g_{ij} are $C^{n+\alpha}$, then this same coordinate system is $C^{n+1+\alpha}$; in particular, if the g_{ij} are C^{∞}, so is the coordinate system. There need not be a C^1 isothermal coordinate system when the g_{ij} are merely C^0 (= continuous).

The condition that (u,v) be isothermal is

$$\Sigma\, g_{ij}\, dx^i \otimes dx^j = \langle\ ,\ \rangle = \lambda(du\otimes du + dv\otimes dv)\ , \qquad \text{some } \lambda > 0\ .$$

To derive explicit equations for u and v, it is easiest to consider the dual metric $\langle\ ,\ \rangle^*$ on $T^*\mathbb{R}^2$, which must satisfy

$$\Sigma\, g^{ij}\left(\frac{\partial}{\partial x^i}\right)^{**} \otimes \left(\frac{\partial}{\partial x^j}\right)^{**} = \langle\ ,\ \rangle^* = \frac{1}{\lambda}\left[\left(\frac{\partial}{\partial u}\right)^{**}\otimes\left(\frac{\partial}{\partial u}\right)^{**} + \left(\frac{\partial}{\partial v}\right)^{**}\otimes\left(\frac{\partial}{\partial v}\right)^{**}\right]\ .$$

Denoting (x^1,x^2) by (x,y), setting

$$(g^{ij}) = \begin{pmatrix} a & b \\ b & c \end{pmatrix}\ ,$$

and applying our equation to the pairs (du,du), (dv,dv), and (du,dv), we obtain

$$(1) \qquad a{u_x}^2 + 2bu_xu_y + c{u_y}^2 = \frac{1}{\lambda} = a{v_x}^2 + 2bv_xv_y + c{v_y}^2$$

$$(2) \qquad au_xv_x + b(u_xv_y + u_yv_x) + cu_yv_y = 0\ .$$

Equation (2) can be written

$$u_x(av_x + bv_y) + u_y(bv_x + cv_y) = 0\ ,$$

which implies that there is a function ρ with

$$u_x = \rho(bv_x + cv_y)$$

$$u_y = -\rho(av_x + bv_y) .$$

Substituting into (1) we find that

$$\rho^2(ac - b^2) = 1 .$$

We thus have the

<u>Beltrami equations</u>

$$u_x = \frac{bv_x + cv_y}{\sqrt{ac-b^2}} , \qquad u_y = -\frac{av_x + bv_y}{\sqrt{ac-b^2}} , \qquad \begin{pmatrix} a & b \\ b & c \end{pmatrix} = (g^{ij})$$

as necessary and sufficient conditions that (u,v) be isothermal coordinates for the metric $\langle\ ,\ \rangle = \sum_{i,j=1} g_{ij}\, dx^i \otimes dx^j$.

At this point it becomes extremely convenient to introduce the notation of formal complex derivatives. We will often denote a typical point of $\mathbb{C} = \mathbb{R}^2$ by z, and z will also be used to denote the identity map $z\colon \mathbb{C} \to \mathbb{C}$. We have already used $x, y\colon \mathbb{R}^2 \to \mathbb{R}$ for the coordinate functions on $\mathbb{R}^2$, so the equation $z = x + iy$ is a (true) equation concerning the three complex-valued functions x, y, z on $\mathbb{R}^2$. Because of this equation we have

$$dz = dx + i\, dy , \qquad d\bar{z} = dx - i\, dy .$$

Since any complex-valued differential on $\mathbb{R}^2$ can be written in terms of dx and dy, it can also be written in terms of dz and $d\bar{z}$. So for any complex-valued function w on $\mathbb{R}^2$, there are unique functions $w_z = \partial w/\partial z$ and $w_{\bar{z}} = \partial w/\partial \bar{z}$ with

$$dw = w_z dz + w_{\bar{z}} d\bar{z} .$$

Substituting from the above equations, we have

$$dw = (w_z + w_{\bar{z}})dx + i(w_z - w_{\bar{z}})dy ,$$

so that

$$w_z + w_{\bar{z}} = \frac{\partial w}{\partial x} = w_x$$

$$i(w_z - w_{\bar{z}}) = \frac{\partial w}{\partial y} = w_y ,$$

which gives

$$\boxed{\begin{aligned} w_z &= \tfrac{1}{2}(w_x - iw_y) \\ w_{\bar{z}} &= \tfrac{1}{2}(w_x + iw_y) \end{aligned}} \quad \text{or} \quad \boxed{\begin{aligned} w_x &= w_z + w_{\bar{z}} \\ w_y &= \frac{w_{\bar{z}} - w_z}{i} \end{aligned}} .$$

The usual differentiation rules apply to the operators $\partial/\partial z$ and $\partial/\partial \bar{z}$, and we have

$$\frac{\partial}{\partial z}(z) = 1 , \qquad \frac{\partial}{\partial \bar{z}}(z) = 0$$

$$\frac{\partial}{\partial \bar{z}}(z) = 0 , \qquad \frac{\partial}{\partial \bar{z}}(\bar{z}) = 1 .$$

It is also easy to check that we always have

$$w_{z\bar{z}} = w_{\bar{z}z} .$$

Another easily checked result is

$$\bar{w}_{\bar{z}} = \overline{(w_z)} .$$

The chain rule becomes

$$(w \circ \zeta)_z = (w_z \circ \zeta) \cdot \zeta_z + (w_{\bar{z}} \circ \zeta) \cdot \bar{\zeta}_z$$

$$(w \circ \zeta)_{\bar{z}} = (w_z \circ \zeta) \cdot \zeta_{\bar{z}} + (w_{\bar{z}} \circ \zeta) \cdot \bar{\zeta}_{\bar{z}} .$$

[This looks a little nicer if we agree to write $w_z \circ \zeta = w_\zeta$ and $w_{\bar{z}} \circ \zeta = w_{\bar{\zeta}}$, so that we have

$$(w \circ \zeta)_z = w_\zeta \cdot \zeta_z + w_{\bar{\zeta}} \cdot \bar{\zeta}_z$$

$$(w \circ \zeta)_{\bar{z}} = w_\zeta \cdot \zeta_{\bar{z}} + w_{\bar{\zeta}} \cdot \bar{\zeta}_{\bar{z}} .]$$

Finally, we note that if $w = u + iv$ for real-valued u and v, then the condition

$$0 = w_{\bar{z}} = \frac{1}{2}(u_x + iv_x + i[u_y + iv_y])$$

is equivalent to the Cauchy-Riemann equations

$$u_x = v_y , \qquad u_y = - v_x .$$

So $w_{\bar{z}} = 0$ if and only if w is complex analytic on U; in this case it is also easy to see that

$$w_z = w' , \qquad \text{the complex derivative} .$$

Now suppose that u, v satisfy the Beltrami equations. If we set

$$w = u + iv ,$$

we find that

$$2w_{\bar{z}}\sqrt{ac - b^2} = \left(b - ia + i\sqrt{ac - b^2}\right)v_x + \left(c - ib - \sqrt{ac - b^2}\right)v_y ,$$

$$2w_z\sqrt{ac - b^2} = \left(b + ia + i\sqrt{ac - b^2}\right)v_x + \left(c - ib - \sqrt{ac - b^2}\right)v_y .$$

A short calculation shows that the coefficients of v_x and v_y on the right hand sides of these two equations are proportional, and we have

$$\frac{w_{\bar{z}}}{w_z} = \frac{c - a - 2ib}{c + a + 2\sqrt{ac - b^2}}$$

or

$$(*) \qquad w_{\bar{z}} = \mu w_z , \qquad \mu = \frac{c - a - 2ib}{c + a + 2\sqrt{ac - b^2}} .$$

Conversely, it is easy to see that the Beltrami equations follow from (*). Notice that if the g_{ij} are $C^{n+\alpha}$, then so are a, b, c and hence μ. Moreover, $|\mu| < 1$. Notice also that we always have

$$u_x v_y - u_y v_x = |w_z|^2 - |w_{\bar{z}}|^2 .$$

So if w satisfies (*), then

$$u_x v_y - u_y v_x = |w_z|^2(1 - |\mu|^2) .$$

Since $|\mu| < 1$, it follows that (u,v) has non-zero Jacobian at any point where $w_z \neq 0$.

The first major step on the road to our final result will be to prove that if μ is C^{α} and $|\mu(0)| < 1$, then equation (*) has a $C^{1+\alpha}$ solution w in a neighborhood of 0, with $w_z(0) \neq 0$. In outline our proof will go as follows. We will let $D(R)$ denote the open disc of radius $R > 0$. Suppose that f is C^{α} in $D(R)$. For all $z_0 \in D(R)$, define

$$F(z_0) = -\frac{1}{\pi} \iint\limits_{D(R)} \frac{f(z)}{z - z_0} \, dxdy \qquad (z = x + iy) .$$

We will show that

$$\text{(A)} \qquad F_{\bar{z}}(z_0) = f(z_0) .$$

We thus have a way of producing a function F with $F_{\bar{z}} = f$.

Now suppose for the moment that we have a function w satisfying (*). If we define

$$F(z_0) = -\frac{1}{\pi} \iint\limits_{D(R)} \frac{\mu(z) w_z(z)}{z - z_0} \, dxdy \qquad z_0 \in D(R) ,$$

then (A) gives

$$F_{\bar{z}}(z_0) = \mu(z_0) w_z(z_0) = w_{\bar{z}}(z_0) .$$

But this means that $(w - F)_{\bar{z}} = 0$, so $w - F$ is complex analytic. Thus we have

$$w(z_0) = -\frac{1}{\pi} \iint\limits_{D(R)} \frac{\mu(z) w_z(z)}{z - z_0} \, dxdy + g(z_0) ,$$

for some complex analytic function g. Conversely, if w satisfies this integral equation for some complex analytic function g, then (A) shows that w satisfies (*), since $g_{\bar{z}} = 0$. We will solve (*) by showing that the equivalent integral equation always has a solution.

In order to get to the proof of (A), we need a succession of simple lemmas.

19. LEMMA (GENERALIZED CAUCHY INTEGRAL THEOREM). Let $D \subset \mathbb{R}^2$ be a compact 2-dimensional manifold-with-boundary, and let $f\colon D \to \mathbb{C}$ be C^1. Then

$$\int_{\partial D} f\,dz = 2i \iint_D f_{\bar{z}}\,dxdy .$$

Proof. If $f = u + iv$ for C^1 functions $u, v\colon D \to \mathbb{R}$, then

$$\int_{\partial D} f\,dz = \int_{\partial D} (u+iv)(dx+i\,dy) = \int_{\partial D} u\,dx - v\,dy \quad + \quad i \int_{\partial D} v\,dx + u\,dy ,$$

while

$$2i \iint_D f_{\bar{z}}\,dxdy = 2i \iint_D \tfrac{1}{2}(f_x + i\,f_y)dxdy = \iint_D (-\,f_y + i\,f_x)dxdy$$

$$= \iint_D (-\,u_y - v_x)dxdy + i \iint_D (u_x - v_y)dxdy .$$

The real and imaginary parts of these two expressions are equal by Stokes' Theorem. ■

Remark. We define the line integral $\int_c f\,d\bar{z}$ as

$$\int_c f\, d\bar{z} = \int_c f\cdot(dx - i\, dy) .$$

It is easy to check that this definition is equivalent to the one usually adopted in complex analysis books,

$$\int_c f\, d\bar{z} = \overline{\left(\int_c \bar{f}\, dz\right)} .$$

Since

$$\bar{f}_{\bar{z}} = \overline{(f_z)} ,$$

Lemma 19 gives

$$\int_{\partial D} f\, d\bar{z} = \overline{\left(\int_{\partial D} \bar{f}\, dz\right)} = -2i\overline{\left(\iint_D \bar{f}_{\bar{z}}\, dxdy\right)}$$

$$= -2i \iint_D f_z\, dxdy .$$

20. <u>LEMMA (GENERALIZED CAUCHY INTEGRAL FORMULA)</u>. For D and f as in Lemma 19, and $z_0 \in$ interior D, we have

$$f(z_0) = \frac{1}{2\pi i} \int_{\partial D} \frac{f(z)}{z - z_0}\, dz - \frac{1}{\pi} \iint_D \frac{f_{\bar{z}}(z)}{z - z_0}\, dxdy .$$

<u>Proof</u>. Let $B(\varepsilon) \subset D$ be a disc of radius ε around z_0. Applying Lemma 19 to the function

$$z \mapsto \frac{f(z)}{z - z_0} \qquad \text{on} \qquad D - \text{interior } B(\varepsilon) ,$$

we have

$$2i \iint_{D - \text{int } B(\varepsilon)} \frac{f_{\bar{z}}(z)}{z - z_0}\, dxdy = \int_{\partial D} \frac{f(z)}{z - z_0}\, dz + \int_{\partial B(\varepsilon)} \frac{f(z)}{z - z_0}\, dz .$$

Taking the limit as $\varepsilon \to 0$, we find that

$$2i \iint_{D} \frac{f_{\bar{z}}(z)}{z - z_0}\, dxdy = \int_{\partial D} \frac{f(z)}{z - z_0}\, dz + 2\pi i\, f(z_0) . \blacksquare$$

21. LEMMA. If $z_0 \in D(R)$, then

$$\bar{z}_0 = -\frac{1}{\pi} \iint_{D(R)} \frac{1}{z - z_0}\, dxdy .$$

Proof. Let $\overline{D(R)}$ be the closure of $D(R) \subset \mathbb{C}$. Applying Lemma 20 to $\bar{z}: \overline{D(R)} \to \mathbb{C}$ we have

$$\bar{z}_0 = \frac{1}{2\pi i} \int_{\partial \overline{D(R)}} \frac{\bar{z}}{z - z_0}\, dz - \frac{1}{\pi} \iint_{D(R)} \frac{1}{z - z_0}\, dxdy ,$$

and

$$\int_{\partial \overline{D(R)}} \frac{\bar{z}}{z - z_0}\, dz = \int_{\partial \overline{D(R)}} \frac{R^2}{z(z - z_0)}\, dz = 0 ,$$

since the sum of the residues of $R^2/z(z - z_0)$ inside $\partial\overline{D(R)}$ is 0. $\blacksquare$

22. LEMMA. If $z_0 \in D(R)$, then

$$|z_0|^2 = -\frac{1}{\pi}\iint_{D(R)} \frac{z}{z-z_0}\,dxdy + R^2 .$$

Proof. Since $|z|^2 = z\bar{z}$, so that

$$\frac{\partial |z|^2}{\partial \bar{z}} = z ,$$

Lemma 20 now gives

$$|z_0|^2 = \frac{R^2}{2\pi i}\int_{\overline{\partial D(R)}} \frac{1}{z-z_0}\,dz - \frac{1}{\pi}\int_{D(R)} \frac{z}{z-z_0}\,dxdy ,$$

and

$$\frac{R^2}{2\pi i}\int_{\overline{\partial D(R)}} \frac{1}{z-z_0}\,dz = R^2 . \quad \blacksquare$$

And now one somewhat more technical lemma.

23. LEMMA. Let $0 < \varepsilon_1, \varepsilon_2 \le 1$, with $\varepsilon_1 + \varepsilon_2 \ne 2$. Then there is a constant $c(\varepsilon_1, \varepsilon_2)$, not depending on R, such that

$$\iint_{D(R)} \frac{dxdy}{|z-z_1|^{2-\varepsilon_1}\cdot|z-z_2|^{2-\varepsilon_2}} \le c(\varepsilon_1,\varepsilon_2)\cdot\frac{1}{|z_1-z_2|^{2-\varepsilon_1-\varepsilon_2}}$$

for all $z_1, z_2 \in D(R)$ with $z_1 \ne z_2$.

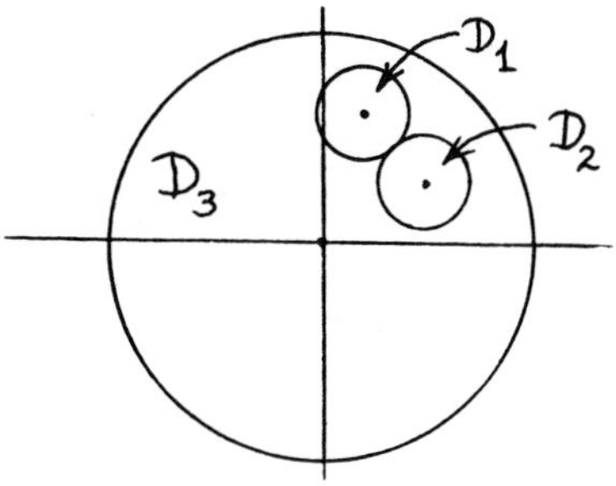

Proof. Let $|z_1 - z_2| = 2\delta$ and define

D_1 = disc of radius δ around z_1

D_2 = disc of radius δ around z_2

$D_3 = D(R) - (D_1 \cup D_2)$

Clearly

$$\iint_{D_1} \frac{dx\,dy}{|z-z_1|^{2-\varepsilon_1} \cdot |z-z_2|^{2-\varepsilon_2}} \le \frac{1}{\delta^{2-\varepsilon_2}} \iint_{D_1} \frac{dx\,dy}{|z-z_1|^{2-\varepsilon_1}}$$

$$= \frac{1}{\delta^{2-\varepsilon_2}} \int_0^{2\pi}\int_0^{\delta} \frac{1}{r^{2-\varepsilon_1}} \cdot r\,dr d\theta \qquad \text{using polar coordinates around } z_1$$

$$= \frac{1}{\delta^{2-\varepsilon_2}} \cdot 2\pi \cdot \int_0^{\delta} r^{\varepsilon_1 - 1}\,dr$$

$$= \frac{1}{\delta^{2-\varepsilon_2}} \cdot 2\pi \cdot \frac{\delta^{\varepsilon_1}}{\varepsilon_1} = \frac{2\pi}{\varepsilon_1} \cdot \frac{1}{\delta^{2-\varepsilon_1-\varepsilon_2}} .$$

Similarly, the integral over D_2 is

$$\le \frac{2\pi}{\varepsilon_2} \cdot \frac{1}{\delta^{2-\varepsilon_1-\varepsilon_2}} .$$

These bounds both have the desired form

$$c(\varepsilon_1,\varepsilon_2) \cdot \frac{1}{|z_1 - z_2|^{2-\varepsilon_1-\varepsilon_2}} .$$

Now we always have

$$|z - z_1| \leq |z - z_2| + |z_1 - z_2| = |z - z_2| + 2\delta ,$$

and consequently

$$\frac{|z - z_1|}{|z - z_2|} \leq 1 + \frac{2\delta}{|z - z_2|} \qquad z \neq z_2 .$$

So on D_3 (in fact on $\mathbb{R}^2 - D_2$) we have

$$\frac{|z - z_1|}{|z - z_2|} \leq 1 + \frac{2\delta}{\delta} = 3 .$$

So

$$\iint_{D_3} \frac{dx\, dy}{|z - z_1|^{2-\varepsilon_1} \cdot |z - z_2|^{2-\varepsilon_2}} = \iint_{D_3} \left|\frac{z - z_1}{z - z_2}\right|^{2-\varepsilon_2} \cdot \frac{dxdy}{|z - z_1|^{4-\varepsilon_1-\varepsilon_2}}$$

$$\leq 3^{2-\varepsilon_2} \iint_{D_3} \frac{dxdy}{|z - z_1|^{4-\varepsilon_1-\varepsilon_2}}$$

$$\leq 3^{2-\varepsilon_2} \iint_{\mathbb{R}^2 - D_1} \frac{dxdy}{|z - z_1|^{4-\varepsilon_1-\varepsilon_2}}$$

$$\leq 3^{2-\varepsilon_2} \int_0^{2\pi} \int_\delta^\infty r^{\varepsilon_1+\varepsilon_2-3}\, dr \qquad \text{using polar coordinates around } z_1$$

$$= 3^{2-\varepsilon_2} \cdot 2\pi \cdot \frac{1}{2 - \varepsilon_1 - \varepsilon_2} \cdot \delta^{\varepsilon_1+\varepsilon_2-2} ,$$

which is again of the desired form. ■

We are now ready to give the precise formulation of (A), which includes three inequalities that are essential for proving that the equivalent integral equation can be solved.

24. PROPOSITION. Let $f\colon D(R) \longrightarrow \mathbb{C}$ satisfy

$$|f(z)| \le M \qquad z \in D(R)$$

$$|f(z_1) - f(z_2)| \le K|z_1 - z_2|^\alpha \qquad z_1, z_2 \in D(R) .$$

Define

$$F(z_0) = -\frac{1}{\pi}\iint_{D(R)} \frac{f(z)}{z - z_0}\,dxdy , \qquad z_0 \in D(R) .$$

Then

(a) $$F_{\bar z}(z_0) = f(z_0)$$

(b) $$F_z(z_0) = -\frac{1}{\pi}\iint_{D(R)} \frac{f(z) - f(z_0)}{(z - z_0)^2}\,dxdy .$$

Moreover for all $z_0, z_1, z_2 \in D(R)$ we have

(c) $$|F(z_0)| \le 4RM$$

(d) $$|F_z(z_0)| \le \frac{2^{\alpha+1}}{\alpha} R^\alpha K$$

(e) $$|F_z(z_1) - F_z(z_2)| \le CK|z_1 - z_2|^\alpha ,$$

where C is a constant that <u>does not depend on</u> R, <u>or on the function</u> f.

<u>Proof</u>. For fixed z_0, let

$$\tilde{F}(z) = F(z) - f(z_0)\bar{z} ,$$

so that by Lemma 21

$$\tilde{F}(z') = -\frac{1}{\pi}\iint_{D(R)} \frac{f(z) - f(z_0)}{z - z'}\, dxdy .$$

We claim that the complex derivative $\tilde{F}'(z_0)$ exists and that in fact

$$\tilde{F}'(z_0) = -\frac{1}{\pi}\iint_{D(R)} \frac{f(z) - f(z_0)}{(z - z_0)^2}\, dxdy .$$

To prove this we have to show that as $h \to 0$, the same is true of

$$\left|\frac{\tilde{F}(z_0 + h) - \tilde{F}(z_0)}{h} + \frac{1}{\pi}\iint_{D(R)} \frac{f(z) - f(z_0)}{(z - z_0)^2}\, dxdy\right|$$

$$= \left|-\frac{1}{\pi h}\iint_{D(R)} [f(z) - f(z_0)]\cdot\left\{\frac{1}{z - z_0 - h} - \frac{1}{z - z_0}\right\} dxdy + \frac{1}{\pi}\iint_{D(R)} \frac{f(z) - f(z_0)}{(z - z_0)^2}\, dxdy\right|$$

$$= \left|-\frac{1}{\pi}\iint_{D(R)} \frac{f(z) - f(z_0)}{(z - z_0 - h)(z - z_0)} + \frac{1}{\pi}\iint_{D(R)} \frac{f(z) - f(z_0)}{(z - z_0)^2}\, dxdy\right|$$

$$= \frac{1}{\pi}\left|\iint_{D(R)} \frac{f(z) - f(z_0)}{z - z_0}\left\{\frac{1}{z - z_0 - h} - \frac{1}{z - z_0}\right\} dxdy\right|$$

$$= \frac{|h|}{\pi}\left|\iint_{D(R)} \frac{f(z) - f(z_0)}{(z - z_0)^2(z - z_0 - h)}\, dxdy\right|$$

$$\leq \frac{|h|}{\pi} \iint_{D(R)} \frac{|f(z) - f(z_0)|}{|z - z_0|^2 |z - z_0 - h|} \, dxdy$$

$$\leq \frac{K|h|}{\pi} \iint_{D(R)} \frac{|z - z_0|^{\alpha}}{|z - z_0|^2 |z - z_0 - h|} \, dxdy$$

$$= \frac{K|h|}{\pi} \iint_{D(R)} \frac{dx\,dy}{|z - z_0|^{2-\alpha} |z - z_0 - h|}$$

$$\leq \frac{K}{\pi} |h| \; c(\alpha,1) |h|^{\alpha+1-2} \qquad \text{by Lemma 23}$$

$$= \frac{K}{\pi} c(\alpha,1) |h|^{\alpha} .$$

This indeed approaches 0 as $h \to 0$.

Now since the complex derivative $\tilde{F}'(z_0)$ exists for all $z_0 \in D(R)$, the ordinary partials F_x, F_y exist, and hence $\tilde{F}_z$ and $\tilde{F}_{\bar{z}}$ exist. Moreover, since $\tilde{F}$ is complex analytic, from the definition of $\tilde{F}$ we obtain

$$0 = \tilde{F}_{\bar{z}}(z_0) = F_{\bar{z}}(z_0) - f(z_0) \cdot 1 ,$$

which proves (a). Furthermore

$$\tilde{F}'(z_0) = \tilde{F}_z(z_0) = F_z(z_0) - 0 ,$$

which proves (b).

To prove (c), we note that

$$|F(z_0)| \leq \frac{M}{\pi} \iint_{D(R)} \frac{1}{|z - z_0|} \, dxdy$$

$$\leq \frac{M}{\pi} \iint_{D} \frac{1}{|z - z_0|} \, dxdy , \qquad \text{where } D \supset D(R) \text{ is the disc of radius } 2R \text{ about } z_0$$

$$= \frac{M}{\pi}\int_0^{2\pi}\int_0^{2R} \frac{1}{r}\cdot r\,drd\theta\ , \qquad \text{using polar coordinates around } z_0$$

$$= 4RM\ .$$

Similarly, for (d) we have

$$|F_z(z_0)| = \left|\frac{1}{\pi}\iint_{D(R)} \frac{|f(z)-f(z_0)|}{|z-z_0|^2}\,dxdy\right| \qquad \text{by (b)}$$

$$\leq \frac{K}{\pi}\iint_{D(R)} \frac{1}{|z-z_0|^{2-\alpha}}\,dxdy$$

$$\leq \frac{K}{\pi}\iint_{D} \frac{1}{|z-z_0|^{2-\alpha}}\,dxdy$$

$$= \frac{K}{\pi}\int_0^{2\pi}\int_0^{2R} \frac{1}{r^{2-\alpha}}\,r\,drd\theta$$

$$= \frac{2K(2R)^{\alpha}}{\alpha}\ .$$

To prove (e) let z_1, z_2 be fixed, and define

$$\begin{cases} B = \dfrac{f(z_1)-f(z_2)}{z_1-z_2} \\ \tilde{F}(z) = F(z) - Bz\bar{z}\ . \end{cases}$$

If we set

$$\tilde{f}(z) = f(z) - Bz\ ,$$

then by Lemma 22 we have

$$\tilde{F}(z_0) = -\frac{1}{\pi}\iint_{D(R)} \frac{\tilde{f}(z)}{z-z_0}\,dxdy - BR^2\ .$$

So by (b) we have

$$\tilde{F}_z(z_0) = -\frac{1}{\pi}\iint_{D(R)} \frac{\tilde{f}(z) - \tilde{f}(z_0)}{(z-z_0)^2}\,dxdy \,.$$

Thus

$$\tilde{F}_z(z_1) - \tilde{F}_z(z_2) = -\frac{1}{\pi}\iint_{D(R)} \left\{\frac{\tilde{f}(z) - \tilde{f}(z_1)}{(z-z_1)^2} - \frac{\tilde{f}(z) - \tilde{f}(z_2)}{(z-z_2)^2}\right\} dxdy \,.$$

But we easily check that

$$\tilde{f}(z_1) = \tilde{f}(z_2) \,.$$

Therefore

$$\begin{aligned}
\tilde{F}_z(z_1) - \tilde{F}_z(z_2) &= -\frac{1}{\pi}\iint_{D(R)} [\tilde{f}(z) - \tilde{f}(z_1)]\cdot\left\{\frac{1}{(z-z_1)^2} - \frac{1}{(z-z_2)^2}\right\} dxdy \\
&= -\frac{1}{\pi}\iint_{D(R)} \frac{[\tilde{f}(z) - \tilde{f}(z_1)]\cdot(z_1 - z_2)(2z - z_1 - z_2)}{(z-z_1)^2(z-z_2)^2}\,dxdy \\
&= -\frac{1}{\pi}\iint_{D(R)} \frac{[\tilde{f}(z) - \tilde{f}(z_1)]\cdot(z_1 - z_2)[(z-z_1) + (z-z_2)]}{(z-z_1)^2(z-z_2)^2}\,dxdy \,.
\end{aligned}$$

Now since

$$\begin{aligned}
\tilde{f}(z) - \tilde{f}(z_1) &= f(z) - f(z_1) - B(z-z_1) \\
&\parallel \\
\tilde{f}(z) - \tilde{f}(z_2) &= f(z) - f(z_2) - B(z-z_2) \,,
\end{aligned}$$

we have

$$[\tilde{f}(z)-\tilde{f}(z_1)][(z-z_1)+(z-z_2)] = [f(z)-f(z_1)-B(z-z_1)](z-z_2)$$
$$+ [f(z)-f(z_2)-B(z-z_2)](z-z_1)$$
$$= [f(z)-f(z_1)](z-z_2) + [f(z)-f(z_2)](z-z_1)$$
$$- 2B(z-z_1)(z-z_2) .$$

So we get

$$\tilde{F}_z(z_1) - \tilde{F}_z(z_2) = -\frac{(z_1-z_2)}{\pi}\iint_{D(R)} \frac{f(z)-f(z_1)}{(z-z_1)^2(z-z_2)}\,dxdy$$
$$-\frac{(z_1-z_2)}{\pi}\iint_{D(R)} \frac{f(z)-f(z_2)}{(z-z_1)(z-z_2)^2}\,dxdy$$
$$+\frac{2B}{\pi}\iint_{D(R)} \frac{(z_1-z_2)}{(z-z_1)(z-z_2)}\,dxdy$$
$$= I_1 + I_2 + I_3 \qquad \text{say} .$$

Now

$$|I_1| \le \frac{|z_1-z_2|}{\pi}\iint_{D(R)} \frac{K}{|z-z_1|^{2-\alpha}\cdot|z-z_2|}\,dxdy$$
$$\le \frac{|z_1-z_2|}{\pi}\,\frac{K\,c(\alpha,1)}{|z_1-z_2|^{1-\alpha}} \qquad \text{by Lemma 23}$$
$$= \frac{Kc(\alpha,1)}{\pi}|z_1-z_2|^{\alpha} .$$

Similarly,

$$|I_2| \le \frac{Kc(\alpha,1)}{\pi}|z_1-z_2|^{\alpha} .$$

Finally,

$$I_3 = \frac{2B}{\pi} \iint_{D(R)} \left(\frac{1}{z - z_1} - \frac{1}{z - z_2} \right) dxdy$$

$$= 2B(\bar{z}_1 - \bar{z}_2) \qquad \text{by Lemma 21},$$

so

$$|I_3| \le 2|B| \cdot |z_1 - z_2| .$$

We have

$$|B| = \frac{|f(z_1) - f(z_2)|}{|z_1 - z_2|} \le K|z_1 - z_2|^{\alpha-1} .$$

Therefore

$$|I_3| \le 2K|z_1 - z_2|^{\alpha} .$$

Thus

$$|\tilde{F}_z(z_1) - \tilde{F}_z(z_2)| \le |I_1| + |I_2| + |I_3|$$

$$\le \text{const} \cdot K \cdot |z_1 - z_2|^{\alpha} .$$

From the definition of $\tilde{F}$ we have

$$\tilde{F}_z(z) = F_z(z) - B\bar{z} ,$$

so we have

$$|F_z(z_1) - F_z(z_2)| \le \text{const} \cdot K \cdot |z_1 - z_2|^{\alpha} + |B| \cdot |\bar{z}_1 - \bar{z}_2|$$

$$\le \text{const} \cdot K \cdot |z_1 - z_2|^{\alpha} + K|z_1 - z_2|^{\alpha-1} \cdot |z_1 - z_2| ,$$

by the estimate for $|B|$ above

$$\le CK|z_1 - z_2|^{\alpha} . \blacksquare$$

Instead of solving the equation

$$(*) \qquad w_{\bar{z}} = \mu w_z ,$$

or the equivalent integral equation, for reasons that will appear later we will instead solve the more general equation

$$(**) \qquad w_{\bar{z}} = \mu w_z + \gamma w + \delta ,$$

where μ, γ, δ are C^{α} and $|\mu(0)| < 1$; moreover, we will show that solutions exist with any given values for $w(0)$ and $w_z(0)$. There is no loss of generality in assuming that $\mu(0) = 0$:

25. LEMMACHEN. If the equation

$$(**) \qquad w_{\bar{z}} = \mu w_z + \gamma w + \delta$$

has a $C^{1+\alpha}$ solution in a neighborhood of 0, with arbitrary values for $w(0)$ and $w_z(0)$, for all C^{α} functions μ, γ, δ with $\mu(0) = 0$, then it also has such $C^{1+\alpha}$ solutions for all C^{α} functions μ, γ, δ with $|\mu(0)| < 1$.

Proof. For any function w, define $\tilde{w}$ by

$$\tilde{w}(z) = w(z - \mu(0)\bar{z}) ,$$

so that

$$w(z) = \tilde{w}(z + \mu(0)\bar{z}) .$$

The chain rule on p. 464 gives

(1)
$$w_z(z) = \tilde{w}_z + \tilde{w}_{\bar{z}}\cdot\overline{\mu(0)} ,$$
$$w_{\bar{z}}(z) = \tilde{w}_z\cdot\mu(0) + \tilde{w}_{\bar{z}} ,$$

where $\tilde{w}_z$, $\tilde{w}_{\bar{z}}$ are to be evaluated at $z+\mu(0)\bar{z}$. Therefore

$$(**) \qquad w_{\bar{z}} = \mu w_z + \gamma w + \delta$$

if and only if

$$\mu(0)\cdot\tilde{w}_z + \tilde{w}_{\bar{z}} = \mu(z)[\tilde{w}_z + \overline{\mu(0)}\tilde{w}_{\bar{z}}] + \gamma(z)w(z) + \delta(z) ,$$

or

$$\tilde{w}_{\bar{z}} = \left(\frac{\mu(z) - \mu(0)}{1-\overline{\mu(0)}\mu(z)}\right)\tilde{w}_z + \frac{\gamma(z)}{1-\overline{\mu(0)}\mu(z)}\, w(z) + \frac{\delta(z)}{1-\overline{\mu(0)}\mu(z)}$$
$$= \rho(z)\tilde{w}_{\bar{z}} + \sigma(z)w(z) + \tau(z) , \qquad \text{say} ,$$

where $\rho(0) = 0$. In this equation $\tilde{w}_{\bar{z}}$, $\tilde{w}_z$ are evaluated at $z+\overline{\mu(0)}z$. Replacing z by $z-\overline{\mu(0)}z$, we get the equivalent equation

$$(\tilde{**}) \quad \tilde{w}_{\bar{z}}(z) = \rho(z-\overline{\mu(0)}z)\tilde{w}_{\bar{z}}(z) + \sigma(z-\overline{\mu(0)}z)\tilde{w}(z) + \tau(z-\overline{\mu(0)}z) ,$$

which is of the same form as (**), with the coefficient of $\tilde{w}_{\bar{z}}$ being 0 at 0. So by hypothesis we can solve for a $C^{1+\alpha}$ function $\tilde{w}$ with any desired initial values

$$\tilde{w}(0) = \tilde{a} , \qquad \tilde{w}_z(0) = \tilde{b} .$$

This gives

$$w(0) = \tilde{w}(0) = \tilde{a} ,$$

while by equation (1)

$$w_z(0) = \tilde{w}_z(0) + \overline{\mu(0)}\tilde{w}_{\bar{z}}(0) .$$

Using equation $(\widetilde{**})$, we have

$$\tilde{w}_{\bar{z}}(0) = \sigma(0)\tilde{w}(0) + \tau(0) = \sigma(0)\cdot\tilde{a} + \tau(0) ,$$

so

$$w_z(0) = \tilde{b} + \overline{\mu(0)}[\sigma(0)\cdot\tilde{a} + \tau(0)] .$$

So to solve (**) for

$$w(0) = a , \qquad w_z(0) = b ,$$

we just solve (**) for

$$\tilde{w}(0) = a$$

$$\tilde{w}_z(0) = b - \overline{\mu(0)}[\sigma(0)\tilde{a} + \tau(0)] . \qquad \blacksquare$$

Since we will be solving the general equation

$$(**) \qquad w_{\bar{z}} = \mu w_z + \gamma w + \delta \qquad \mu(0) = 0 ,$$

we first want to find an integral equation equivalent to it. To do this we form

$$F(z_0) = -\frac{1}{\pi}\iint_{D(R)} \frac{\mu(z)w_z(z) + \gamma(z)w(z) + \delta(z)}{z - z_0}\, dxdy .$$

Proposition 24 gives

$$F_{\bar z} = \mu w_z + \gamma w + \delta = w_{\bar z} \qquad \text{if } w \text{ satisfies } (**) ,$$

and hence $(w - F)_{\bar z} = 0$, so that

$$w(z_0) = -\frac{1}{\pi}\iint_{D(R)} \frac{\mu(z)w_z(z)}{z - z_0}\,dxdy - \frac{1}{\pi}\iint_{D(R)} \frac{\gamma(z)w(z)}{z - z_0}\,dxdy - \frac{1}{\pi}\iint_{D(R)} \frac{\delta(z)}{z - z_0}\,dxdy$$
$$+ g(z_0)$$

for some complex analytic function g. Conversely, of course, if w satisfies this equation for a complex analytic g, then it satisfies (**). By complicating our integral equation, we can arrange that $w(0) = g(0)$; clearly we just have to add

$$\frac{1}{\pi}\iint_{D(R)} \frac{\mu(z)w_z(z)}{z}\,dxdy + \cdots$$

to the right hand side. Similarly, if we add

$$z\cdot\left\{\frac{1}{\pi}\iint_{D(R)} \frac{\mu(z)w_z(z)}{z^2} + \cdots\right\}$$

to the right hand side, we will have $w_z(0) = g'(0)$; this follows from Proposition 24 (and the fact that $\mu(0) = 0$). So we see that we can solve (**) for w with any given values of $w(0)$, $w_z(0)$ provided that we can solve the following equation for w, where g is any complex analytic function (actually it would suffice to solve it for functions of the form $g(z) = \tilde{a}z + \tilde{b}$):

$$w(z_0) = -\frac{1}{\pi}\iint_{D(R)} \frac{\mu(z)w_z(z)}{z - z_0}\,dxdy - \frac{1}{\pi}\iint_{D(R)} \frac{\gamma(z)w(z)}{z - z_0}\,dxdy - \frac{1}{\pi}\iint_{D(R)} \frac{\delta(z)}{z - z_0}\,dxdy$$

$$+ \frac{1}{\pi}\iint_{D(R)} \frac{\mu(z)w_z(z)}{z}\,dxdy + \frac{1}{\pi}\iint_{D(R)} \frac{\gamma(z)w(z)}{z}\,dxdy + \frac{1}{\pi}\iint_{D(R)} \frac{\delta(z)}{z}\,dxdy$$

$$+ z_0\left\{\frac{1}{\pi}\iint_{D(R)} \frac{\mu(z)w_z(w)}{z^2}\,dxdy + \frac{1}{\pi}\iint_{D(R)} \frac{\gamma(z)w(z)}{z^2}\,dxdy + \frac{1}{\pi}\iint_{D(R)} \frac{\delta(z)}{z^2}\,dxdy\right\}$$

$$+ g(z_0)\ .$$

Now the first integral involving δ is a $C^{1+\alpha}$ function Δ, for Proposition 24 shows that

$\Delta_{\bar{z}} = \delta$ which is C^{α} by assumption,

Δ_z is C^{α} by part (e) of Proposition 24 .

The other two integrals involving δ are just constants. So it certainly suffices to show that we can solve the following equation for C^{α} functions μ, γ with $\mu(0) = 0$ and any $C^{1+\alpha}$ function h:

$$\text{(I)}\qquad w(z_0) = -\frac{1}{\pi}\iint_{D(R)} \frac{\mu(z)w_z(z)}{z - z_0}\,dxdy - \frac{1}{\pi}\iint_{D(R)} \frac{\gamma(z)w(z)}{z - z_0}\,dxdy$$

$$+ \frac{1}{\pi}\iint_{D(R)} \frac{\mu(z)w_z(z)}{z}\,dxdy + \frac{1}{\pi}\iint_{D(R)} \frac{\gamma(z)w(z)}{z}\,dxdy$$

$$+ z_0\left\{\frac{1}{\pi}\iint_{D(R)} \frac{\mu(z)w_z(z)}{z^2}\,dxdy + \frac{1}{\pi}\iint_{D(R)} \frac{\gamma(z)w(z)}{z^2}\,dxdy\right\}$$

$$+ h(z_0)\ .$$

The integral equation (I) will be solved by the only method available to us, namely, the method of successive approximation, which we have always formulated in terms of the Contraction Lemma (I.5-1). First we need to concoct the right space of functions to work with. Consider first the set

$$H(R,\alpha) = \{C^\alpha \text{ functions } w\colon D(R) \longrightarrow \mathbb{C}\} .$$

For $w \in H(R,\alpha)$ we define

$$\|w\|_R = \sup_{z \in D(R)} |w(z)| + R^\alpha \sup_{\substack{z_1, z_2 \in D(R) \\ z_1 \neq z_2}} \frac{|w(z_1) - w(z_2)|}{|z_1 - z_2|^\alpha} .$$

The first term $\sup|w(z)|$ insures that $w_n \to 0$ uniformly if $\|w_n\|_R \to 0$. The term $\sup|w(z_1) - w(z_2)|/|z_1 - z_2|^\alpha$ is simply the "best" constant K in the definition of w being C^α; the fudge factor R^α is reasonable, for it insures that

$$\|w\|_R = \|\tilde{w}\|_1 \qquad \text{where} \qquad \tilde{w}(z) = w(Rz) .$$

It is easy to check that

$$\|w\|_R > 0 \qquad \text{for} \qquad w \neq 0 ,$$

$$\|\lambda w\|_R = |\lambda| \cdot \|w\|_R \qquad \lambda \in \mathbb{R} ,$$

$$\|w_1 + w_2\|_R \leq \|w_1\|_R + \|w_2\|_R .$$

So we obtain a metric on $H(R,\alpha)$ by defining

$$\text{distance from } w_1 \text{ to } w_2 = \|w_1 - w_2\|_R ,$$

and it is easy to see that $H(R,\alpha)$ is complete in this metric. Finally, it is easily checked that if $w_1, w_2 \in H(R,\alpha)$, then

$$\|w_1 w_2\|_R \le \|w_1\|_R \cdot \|w_2\|_R .$$

Next consider the set

$$H(R, \alpha+1) = \{C^{\alpha+1} \text{ functions } w: D(R) \longrightarrow \mathbb{C}\} .$$

For $w \in H(R, \alpha+1)$ we define

$$|||w|||_R = \sup_{z \in D(R)} |w(z)| + R \cdot \|w_z\|_R + R \cdot \|w_{\bar{z}}\|_R .$$

It is once again easy to check that

$$|||w|||_R > 0 \quad \text{for} \quad w \neq 0$$

$$|||\lambda w|||_R = |\lambda| \cdot |||w|||_R$$

$$|||w_1 + w_2|||_R \le |||w_1|||_R + |||w_2|||_R ,$$

and that $H(R, \alpha+1)$ is a complete metric space with respect to the metric

$$\text{distance from } w_1 \text{ to } w_2 = |||w_1 - w_2|||_R .$$

Consider, for the moment, a function $f: (-R,R) \longrightarrow \mathbb{R}$, and suppose that

$$R^{\alpha+1} \cdot \frac{|f'(x_1) - f'(x_2)|}{|x_1 - x_2|^\alpha} \le K .$$

Defining

$$g(s) = f(x_1 + s(x_2 - x_1)) - f(x_2 + s(x_1 - x_2)) ,$$

we have

$$g(0) = f(x_1) - f(x_2) , \qquad g(1) = f(x_2) - f(x_1) .$$

So the mean value theorem gives

$$2[f(x_2) - f(x_1)] = \frac{g(1) - g(0)}{1 - 0} = g'(\xi) \qquad \xi \in (0,1)$$

$$= (x_2 - x_1)\cdot[f'(\eta_1) - f'(\eta_2)] \qquad \eta_1, \eta_2 \in (x_1,x_2) .$$

Thus

$$R^\alpha|f(x_1) - f(x_2)| \le \frac{R^\alpha|x_1 - x_2|}{2} \cdot \frac{K|\eta_1 - \eta_2|^\alpha}{R^{\alpha+1}}$$

$$= \frac{K}{2}\frac{|x_1 - x_2|}{R}\cdot |\eta_1 - \eta_2|^\alpha$$

$$\le K\cdot |x_1 - x_2|^\alpha ,$$

or finally

$$R^\alpha \frac{|f(x_1) - f(x_2)|}{|x_1 - x_2|^\alpha} \le K .$$

For functions $w: D(R) \longrightarrow \mathbb{C}$ there is a similar argument, using Taylor's formula to estimate $|g(1) - g(0)|$. The answer involves the derivative Dw, which can be expressed in terms of w_z and $w_{\bar{z}}$. From this argument we easily see that there is an inequality of the form

$$\|w\|_R \le \text{constant} \cdot |||w|||_R .$$

26. PROPOSITION. Let μ, γ be C^{α} functions in a neighborhood of 0 with $\mu(0) = 0$, and let h be $C^{\alpha+1}$ in a neighborhood of 0. Then for sufficiently small $R > 0$ there is a $C^{\alpha+1}$ function $w: D(R) \longrightarrow \mathbb{C}$ satisfying (I) for all $z_0 \in D(R)$.

Proof. We suppose that μ, γ are C^{α} in $D(R_0)$ for some $R_0 \leq 1$, and we will henceforth consider only $R \leq R_0$. For $w \in H(R, \alpha+1)$, define the function Sw on $D(R)$ by setting $(Sw)(z_0)$ equal to the right side of (I) without the $h(z_0)$: we will abbreviate this expression by

$$\begin{aligned}(Sw)(z_0) = \quad & I_1(z_0) \quad + I_2(z_0) \\ & + I_3(z_0) \quad + I_4(z_0) \\ & + z_0\{I_5(z_0) + I_6(z_0)\} .\end{aligned}$$

We make the crucial

Claim: There is a constant C', depending only on α, and not on R, such that

$$|||Sw|||_R \leq C' \cdot R^{\alpha} \cdot |||w|||_R$$

for all $w \in H(R, \alpha+1)$.

Assuming this Claim for the moment, the remainder of the proof goes as follows. Since $R^{\alpha} \longrightarrow 0$ as $R \longrightarrow 0$, there is clearly some R_* such that for all $R \leq R_*$ we have

$$|||Sw|||_R \leq C'' \cdot |||w|||_R ,$$

where C'' is a constant with

$$C'' < 1, \frac{|||h|||_R}{3} .$$

Define $T: H(R, \alpha+1) \longrightarrow H(R, \alpha+1)$ by

$$Tw = Sw + h .$$

If $R \leq R_*$, then for all w with

$$|||w|||_R \leq \frac{3}{2}|||h|||_R$$

we have

$$\begin{aligned} |||Tw|||_R = |||Sw+h|||_R &\leq |||Sw|||_R + |||h|||_R \\ &\leq \frac{|||h|||_R}{3} \cdot |||w|||_R + |||h|||_R \\ &\leq \frac{1}{2}|||h|||_R + |||h|||_R \\ &= \frac{3}{2}|||h|||_R . \end{aligned}$$

Thus, for $R \leq R_*$, the map T takes the complete metric space

$$M = \left\{ w \in H(R, \alpha+1): |||w|||_R \leq \frac{3}{2}|||h|||_R \right\}$$

into itself. Moreover, the map $T: M \longrightarrow M$ is a contraction, for

$$\begin{aligned} |||Tw_1 - Tw_2|||_R &= |||Sw_1 - Sw_2|||_R \\ &= |||S(w_1 - w_2)|||_R \leq C''|||w_1 - w_2|||_R . \end{aligned}$$

By the contraction Lemma, there is some $w \in M$ with

$$w = Tw = Sw + h ,$$

which is precisely the equation we want.

To prove the Claim we will use all the information in Lemma 24. First we want to show that

$$|||I_1|||_R \leq \text{constant} \cdot R^{\alpha} \cdot |||w|||_R ,$$

where the constant is independent of R. It clearly suffices to prove the same inequality for each of

$$\sup|I_1(z)| , \qquad R \cdot \|(I_1)_z\|_R , \qquad R \cdot \|(I_1)_{\bar{z}}\|_R .$$

Let L be a number such that

$$|\mu(z_1) - \mu(z_2)| \leq L \cdot |z_1 - z_2|^{\alpha} \qquad z_1, z_2 \in D(R_0) .$$

Since $\mu(0) = 0$, it follows that

$$|\mu(z)| \leq LR^{\alpha} , \qquad z \in D(R) , \quad R \leq R_0$$

and therefore that

$$\|\mu\|_R \leq 2LR^{\alpha} .$$

Thus for all $z, z_1, z_2 \in D(R)$ we have

(1) $$|\mu(z)w_z(z)| \leq \|\mu w_z\|_R \leq \|\mu\|_R \cdot \|w_z\|_R$$

$$\leq 2LR^{\alpha} \cdot \frac{|||w|||_R}{R}$$

$$= 2LR^{\alpha-1} \cdot |||w|||_R ,$$

(2) $$\frac{|\mu(z_1)w_z(z_1) - \mu(z_2)w_z(z_2)|}{|z_1 - z_2|^{\alpha}} \leq \frac{\|\mu w_z\|_R}{R^{\alpha}} \leq \frac{\|\mu\|_R \cdot \|w_z\|_R}{R^{\alpha}}$$

$$\leq \frac{2LR^{\alpha} \cdot |||w|||_R}{R^{\alpha} \cdot R}$$

$$= \frac{2L}{R} \cdot |||w|||_R .$$

We can now apply the inequalities of Proposition 24. Inequality (c) gives

(3) $$|I_1(z)| \leq 4R \cdot 2LR^{\alpha-1} \cdot |||w|||_R$$

$$= 8L \cdot R^{\alpha} \cdot |||w|||_R ,$$

which is the desired inequality for $\sup|I_1(z)|$. Inequalities (d) and (e) give

(4) $$R \cdot |(I_1)_z(z)| \leq R\, \frac{2^{\alpha+1}}{\alpha} R^{\alpha} \cdot \frac{2L}{R}|||w|||_R$$

$$= \frac{2^{\alpha+2}}{\alpha} L \cdot R^{\alpha} \cdot |||w|||_R$$

(5) $$R^{\alpha+1} \cdot \frac{|(I_1)_z(z_1) - (I_1)_z(z_2)|}{|z_1 - z_2|^{\alpha}} \leq R^{\alpha+1} \cdot C \cdot \frac{2L}{R}|||w|||_R$$

$$= 2CL \cdot R^{\alpha} \cdot |||w|||_R ;$$

these give the desired inequality for $R\cdot\|(I_1)_z\|_R$. Finally, since $(I_1)_{\bar{z}} = \mu z$, the necessary inequalities for $R\cdot\|(I_1)_{\bar{z}}\|_R$ follow immediately from (1), (2). We have therefore shown that

$$|||I_1|||_R \le \text{constant} \cdot R^{\alpha} \cdot |||w|||_R \ .$$

Now consider I_2. We first note that for $z \in D(R)$ we have

$$\begin{aligned} |\gamma(z)w(z)| &\le \|\gamma w\|_R \le \|\gamma\|_R \cdot \|w\|_R \\ &\le \|\gamma\|_{R_0} \cdot \text{constant} \cdot |||w|||_R \end{aligned}$$ (see p. 487).

This is a <u>better</u> inequality than (1): since $0 < R \le 1$ and $0 < \alpha < 1$ we have $1 \le R^{\alpha-1}$, so we can write

$$(1') \qquad |\gamma(z)w(z)| \le \text{constant} \cdot R^{\alpha-1} \cdot |||w|||_R \ .$$

Similarly, if $z_1, z_2 \in D(R)$, then

$$\begin{aligned} (2') \qquad \frac{|\gamma(z_1)w(z_1) - \gamma(z_2)w(z_2)|}{|z_1 - z_2|^{\alpha}} &\le \frac{\|\gamma w\|_R}{R^{\alpha}} \le \frac{\|\gamma\|_R \cdot \|w\|_R}{R^{\alpha}} \\ &\le \frac{\|\gamma\|_{R_0}}{R^{\alpha}} \cdot \text{constant} \cdot |||w|||_R \\ &\le \frac{\text{constant}}{R} \cdot |||w|||_R \ . \end{aligned}$$

Now (1'), (2') give the inequality

$$|||I_2|||_R \le \text{constant} \cdot R^{\alpha} \cdot |||w|||_R$$

in the same way that (1), (2) gave the inequality for $|||I_1|||_R$.

Since I_3 is just a constant, $I_3(z) = I_1(0)$, we have

$$|||I_3|||_R = |||I_1(0)|||_R = |I_1(0)| \le \sup_{z \in D(R)} |I_1(z)|$$

$$\le |||I_1|||_R \le \text{constant} \cdot R^\alpha \cdot |||w|||_R \,.$$

Similarly for I_4.

As for the term $zI_5(z) = z(I_1)_z(0)$, we have

$$|z(I_1)_z(0)| \le |z| \cdot |(I_1)_z(0)|$$

$$\le R \cdot \frac{|||I_1|||_R}{R} = |||I_1|||_R$$

$$\le \text{constant} \cdot R^\alpha \cdot |||w|||_R \,,$$

$$R \cdot \|\{z \cdot (I_1)_z(0)\}_z\|_R = R \cdot \|(I_1)_z(0)\|_R$$

$$= R \cdot |(I_1)_z(0)|$$

$$\le \text{constant} \cdot R^\alpha \cdot |||w|||_R \,, \qquad \text{as above.}$$

Thus $|||z(I_1)_z(0)|||_R \le \text{constant} \cdot R^\alpha \cdot |||w|||_R$, and the term $zI_6(z)$ works exactly the same. ■

27. <u>COROLLARY</u>. If μ, γ, δ are C^α functions in a neighborhood of 0, with $|\mu(0)| < 1$, and $a, b \in \mathbb{C}$ are any two complex numbers, then there is a $C^{\alpha+1}$ function w in a neighborhood of 0 such that

$$(**) \qquad \begin{aligned} w_{\bar z} &= \mu w_z + \gamma w + \delta , \\ w(0) &= a \\ w_z(0) &= b . \end{aligned}$$

In particular, there is a $C^{1+\alpha}$ isothermal coordinate system around any point of a surface with a C^{α} metric.

Proof. Proposition 26 and Lemmachen 25. ∎

Our next task is to show that if μ, γ, δ in Corollary 27 are $C^{n+\alpha}$, then there is a solution w of (**) which is $C^{n+1+\alpha}$. First a technical lemma.

28. LEMMA. If f is $C^{n+\alpha}$ $(n \geq 1)$ on $D(R)$ and we define

$$F(z_0) = -\frac{1}{\pi} \iint_{D(R)} \frac{f(z)}{z - z_0}\, dxdy , \qquad z_0 \in D(R) ,$$

then F is $C^{n+1+\alpha}$.

Proof. Induction on n. Consider first the case $n = 1$. By Proposition 24 we have $F_{\bar z} = f$, so $F_{\bar z}$ is $C^{1+\alpha}$. We just have to show that F_z is $C^{1+\alpha}$, since this then implies that F_x, F_y are $C^{1+\alpha}$, so that F is $C^{2+\alpha}$. Now we easily check that

$$\frac{\partial}{\partial z} \log |z - z_0|^2 = \frac{1}{z - z_0} ,$$

and therefore

$$F(z_0) = -\frac{1}{\pi}\iint\limits_{D(R)} \frac{\partial}{\partial z}(f \log|z-z_0|^2)\, dxdy + \frac{1}{\pi}\iint\limits_{D(R)} f_z \log|z-z_0|^2\, dxdy .$$

Using the Remark after Lemma 19, we write this as

$$F(z_0) = \frac{1}{2\pi i}\int\limits_{\partial D(R)} f \log|z-z_0|^2\, d\bar{z} + \frac{1}{\pi}\iint\limits_{D(R)} f_z \log|z-z_0|^2\, dxdy .$$

We can now differentiate under the integral signs to obtain

$$(1) \qquad F_z(z_0) = \frac{1}{2\pi i}\int\limits_{\partial \overline{D(R)}} \frac{f(z)}{z-z_0}\, d\bar{z} + \frac{1}{\pi}\iint\limits_{D(R)} \frac{f_z(z)}{z-z_0}\, dxdy .$$

The first integral is C^∞ (since we can keep differentiating under the integral sign); the second is $C^{1+\alpha}$ by Proposition 24.

Now suppose the result holds for $C^{n+\alpha}$ functions, and let f be $C^{n+1+\alpha}$. We still have $F_{\bar{z}} = f$, so that $F_{\bar{z}}$ is $C^{n+1+\alpha}$, and we also have equation (1), in which the first integral is C^∞. Now f_z is $C^{n+\alpha}$, so by the induction assumption, the second integral is $C^{n+1+\alpha}$. Thus F_z is $C^{n+1+\alpha}$, so F is $C^{n+2+\alpha}$. ■

<u>29. PROPOSITION.</u> If μ, γ, δ are $C^{n+\alpha}$ functions in a neighborhood of 0, with $|\mu(0)| < 1$, and $a, b \in \mathbb{C}$ are any two complex numbers, then there is a $C^{n+1+\alpha}$ function w in a neighborhood of 0 such that

(**) $$w_{\bar{z}} = \mu w_z + \gamma w + \delta$$
$$w(0) = a$$
$$w_z(0) = b\ .$$

In particular, there is a $C^{n+1+\alpha}$ isothermal coordinate system around any point of a surface with a $C^{n+\alpha}$ metric.

<u>Proof</u>. Induction on n. The case $n = 0$ is Corollary 26. Now suppose the result is true for n, and let μ, γ, δ be $C^{n+1+\alpha}$.

<u>Case 1</u>. $\gamma = 0$. The motivation for the proof is the following. If w satisfies

(1) $$w_{\bar{z}} = \mu w_z + \delta\ ,$$

then we should have

$$(w_z)_{\bar{z}} = w_{\bar{z}z} = \mu(w_z)_z + \mu_z w_z + \delta_z\ .$$

So we first solve this equation for w_z. To be precise, we note that μ, μ_z, δ_z are $C^{n+\alpha}$, so since the result is assumed true for n, there is a function f satisfying

(2) $$f_{\bar{z}} = \mu f_z + \mu_z f + \delta_z$$

in some disc $D(R)$; moreover, we can obtain any desired values for $f(0)$ and $f_z(0)$. [Notice that equation (2) contains f explicitly even though equation (1) does not contain w explicitly.] Define W by

$$\overline{W}(z_0) = -\frac{1}{\pi}\iint_{D(R)} \frac{\overline{f}(z)}{z - z_0}\,dxdy\ .$$

Then W is $C^{n+2+\alpha}$ by Lemma 28, and by Proposition 24 we have

$$\bar{f}(z_0) = \bar{W}_{\bar{z}}(z_0) = \overline{W_z(z_0)} \implies f(z_0) = W_z(z_0) .$$

So

$$(W_{\bar{z}})_z = W_{z\bar{z}} = f_{\bar{z}} = \mu f_z + \mu_z f + \delta_z \qquad \text{by (1)}$$

$$= (\mu f)_z + \delta_z = (\mu W_z)_z + \delta_z .$$

Hence $(W_{\bar{z}} - \mu W_z - \delta)_z = 0$. This means that we can write

$$W_{\bar{z}}(z) - \mu(z)W_z(z) - \delta(z) = g(\bar{z}) , \tag{3}$$

where g is complex analytic. Let G be a complex analytic function with $G_{\bar{z}}(\bar{z}) = g(\bar{z})$, and let

$$w(z) = W(z) - G(\bar{z}) .$$

Then

$$w_z = W_z - 0$$

$$w_{\bar{z}}(z) = W_{\bar{z}}(z) - g(\bar{z}) = \mu(z)W_z(z) + \delta(z) \qquad \text{by (3)}$$

$$= \mu(z)w_z(z) + \delta(z) .$$

Thus w is a $C^{n+2+\alpha}$ solution of our equation. We also have

$$w(0) = W(0) - G(0)$$

$$w_z(0) = W_z(0) = f(0) .$$

So we obtain the condition $w_z(0) = b$ by choosing a solution f of (2) with

$f(0) = b$. We can obtain $w(0) = a$ as G is only determined up to a constant.

Case 2. General case. We look for a solution of the form $w = e^{\lambda}\sigma$. We find that the equation

$$(4) \qquad w_{\bar{z}} = \mu w_z + \gamma w + \delta$$

is equivalent to

$$\sigma_{\bar{z}} + \lambda_{\bar{z}}\sigma = \mu\sigma_z + \mu\lambda_z\sigma + \gamma\sigma + e^{-\lambda}\delta ,$$

or

$$\sigma(\lambda_{\bar{z}} - \mu\lambda_z - \gamma) + \sigma_{\bar{z}} = \mu\sigma_z + e^{-\lambda}\delta .$$

By Case 1, there are $C^{n+2+\alpha}$ functions λ, σ satisfying

$$\lambda_{\bar{z}} = \mu\lambda_z + \gamma ; \qquad \lambda(0) = 0 , \quad \lambda_z(0) = 0$$

$$\sigma_{\bar{z}} = \mu\sigma_z + e^{-\lambda}\delta ; \qquad \sigma(0) = a , \quad \sigma_z(0) = b .$$

Then $w = e^{\lambda}\sigma$ is $C^{n+2+\alpha}$ and satisfies (4), and $w(0) = a$, $w_z(0) = b$. ■

Notice that Proposition 29 does not give a C^{∞} isothermal coordinate system in the C^{∞} case; for although the equation $w_{\bar{z}} = \mu w_z$ will have $C^{n+1+\alpha}$ solutions for all n, these solutions might be defined on smaller and smaller neighborhoods of 0. But this is now easy to take care of. First let us note that if (u,v) is an isothermal coordinate system, and $f\colon \mathbb{C} \to \mathbb{C}$ is complex analytic, with f' never 0, then $f\circ(u,v)$ is also an isothermal coordinate system, since f is angle preserving. We can also prove this from our equation

$w_{\bar z} = \mu w_z$, for since $f_{\bar z} = 0$, the chain rule gives

$$(f\circ w)_z = (f_z\circ w)\cdot w_z$$

$$(f\circ w)_{\bar z} = (f_z\circ w)\cdot w_{\bar z} ,$$

and hence we have $(f\circ w)_{\bar z} = \mu\cdot(f\circ w)_z$. This argument can also be reversed, allowing us to prove

30. PROPOSITION. If μ is a $C^{n+\alpha}$ function with $|\mu| < 1$, and w is <u>any</u> solution of

$$(*) \qquad w_{\bar z} = \mu w_z ,$$

then w is $C^{n+1+\alpha}$. So if μ is C^∞, any solution w is also C^∞.

In particular, there is a C^∞ isothermal coordinate system around any point of a surface with a C^∞ metric.

<u>Proof</u>. We know that around any point there is <u>some</u> $C^{n+1+\alpha}$ solution σ of $(*)$ which has an inverse around that point. So we can write

$$(1) \qquad w = f\circ\sigma$$

for some f. Then the chain rule gives

$$w_z = (f_z\circ\sigma)\cdot\sigma_z + (f_{\bar z}\circ\sigma)\bar\sigma_z$$

$$w_{\bar z} = (f_z\circ\sigma)\cdot\sigma_{\bar z} + (f_{\bar z}\circ\sigma)\bar\sigma_{\bar z} .$$

Since w is a solution of $(*)$, we have

$$(f_z \circ \sigma)\sigma_{\bar z} + (f_{\bar z} \circ \sigma)\bar\sigma_{\bar z} = \mu[(f_z \circ \sigma)\sigma_z + (f_{\bar z} \circ \sigma)\bar\sigma_z] .$$

Since σ is a solution, this leads to

$$(2) \qquad (f_{\bar z} \circ \sigma)[\bar\sigma_{\bar z} - \mu\bar\sigma_z] = 0 .$$

Since $\sigma_{\bar z} = \mu\sigma_z$ implies that

$$\bar\sigma_z = \overline{(\sigma_{\bar z})} = \bar\mu\overline{(\sigma_z)} = \bar\mu\bar\sigma_{\bar z} ,$$

we see that

$$\begin{aligned} \bar\sigma_{\bar z} - \mu\bar\sigma_z &= \bar\sigma_{\bar z} - \mu\bar\mu\bar\sigma_{\bar z} \\ &= \bar\sigma_{\bar z}(1 - |\mu|^2) . \end{aligned}$$

This is non-zero, since $|\mu| < 1$, and σ has non-zero Jacobian at the point in question. It follows from (2) that $f_{\bar z} = 0$, i.e. f is analytic. Then (1) shows that w must be $C^{n+1+\alpha}$ too. ■

Addendum 2. Immersed Spheres with Constant Mean Curvature

Let $f\colon U \to M$ be an immersion (for $U \subset \mathbb{R}^2$ open) which is conformal, so that the components E, F, G of I_f satisfy

$$E = G \qquad F = 0 \;;$$

such immersions always exist by the results* of Addendum 1. From equation (B) on p. III.198 we have

$$K = k_1 k_2 = \frac{\ell n - m^2}{E^2} \tag{1}$$

$$H = \frac{1}{2}(k_1 + k_2) = \frac{\ell + n}{2E} \,. \tag{2}$$

A little calculation shows that the Codazzi-Mainardi equations (p. III.79) become

$$\ell_y - m_x = \frac{E_y}{2E}(\ell + n) = E_y H$$

$$m_y - n_x = -\frac{E_x}{2E}(\ell + n) = -E_x H \,.$$

But

$$EH = \frac{\ell + n}{2} \implies \begin{cases} E_y H = -EH_y + \dfrac{\ell_y}{2} + \dfrac{n_y}{2} \\ E_x H = -EH_x + \dfrac{\ell_x}{2} + \dfrac{n_x}{2} \,, \end{cases}$$

so the Codazzi-Mainardi equations can be written

---

*At present we need Proposition 29 or 30, but we could make do with the much simpler Theorem 18, since it follows from (hard) theorems on partial differential equations that a surface of constant mean curvature must be analytic (see pp. V.213-214).

$$(3) \qquad \begin{aligned} \left(\frac{\ell - n}{2}\right)_x + m_y &= EH_x \\ \left(\frac{\ell - n}{2}\right)_y - m_x &= -EH_y . \end{aligned}$$

If we define the function $\Phi\colon U \to \mathbb{C}$ by

$$(4) \qquad \Phi = \frac{\ell - n}{2} - im ,$$

then

$$\begin{aligned} |\Phi|^2 &= \frac{(\ell - n)^2}{4} + m^2 = \frac{(\ell + n)^2}{4} + m^2 - \ell n \\ &= E^2(H^2 - K) \qquad \text{by (1) and (2)} \\ &= E^2(k_1 - k_2)^2/4 . \end{aligned}$$

Thus the umbilics on $f(U)$ are the image of the zeros of Φ. Notice that if H is constant, so that $H_x = H_y = 0$, then equations (3) are precisely the Cauchy-Riemann equations for Φ; thus Φ is complex analytic. So we immediately have

31. LEMMA. If M is a connected surface immersed in $\mathbb{R}^3$ with constant mean curvature, then either all points of M are umbilics, or else the umbilics are isolated.

Proof. Since the analytic function Φ is identically zero if its zeros have an accumulation point, we see that for every $p \in M$ one of two possibilities must hold:

(1) p has a neighborhood with no umbilics, except perhaps p,

(2) p has a neighborhood all of whose points are umbilics.

But the set of points p satisfying (1) is open, and so is the set of points p satisfying (2). Since M is connected, either (1) holds everywhere, or (2) holds everywhere. ■

Now consider the lines of curvature on M, or rather their images in U under the map f^{-1}. Formula (D) on p.III.199 says that a vector $\mathbf{v} = (a_1, a_2)$ is tangent to one of these curves if and only if

$$0 = \det \begin{pmatrix} {a_2}^2 & -a_1a_2 & {a_1}^2 \\ E & 0 & E \\ \ell & m & n \end{pmatrix}$$

$$= -E[-m{a_1}^2 + (\ell - n)a_1a_2 + m{a_2}^2]$$

{ℓ, m, n evaluated at the point where $\mathbf{v}$ is considered to be a tangent vector}.

Thus $\mathbf{v}$ is tangent to [f^{-1} of] a line of curvature if and only if

$$-m\{dx(\mathbf{v})\}^2 + (\ell - n)dx(\mathbf{v})dy(\mathbf{v}) + m\{dy(\mathbf{v})\}^2 = 0 .$$

We can write the left side of this equation as the imaginary part of a complex number, namely

$$\operatorname{Im} [\tfrac{\ell - n}{2} - im]\cdot[\{dx(\mathbf{v})\}^2 - \{dy(\mathbf{v})\}^2 + 2i\, dx(\mathbf{v})dy(\mathbf{v})]$$

$$= \operatorname{Im} \Phi \cdot [\{dx(\mathbf{v})\}^2 - dy(\mathbf{v})^{\ 2} + 2i\, dx(\mathbf{v})dy(\mathbf{v})] .$$

Introducing the complex-valued 1-form dz, as on p. 462 , we can thus write our equation as

$$\text{Im } \Phi\cdot\{dz(\mathbf{v})\}^2 = 0 .$$

For any complex number $w \neq 0$, we let $\arg w$ be some angle between the x-axis and the ray from 0 through w, so that $w = |w|e^{\arg w}$. Then the above equation holds if and only if there is an integer m with

$$\begin{aligned} m\pi &= \arg \Phi\cdot\{dz(\mathbf{v})\}^2 \\ &= \arg \Phi + 2 \arg dz(\mathbf{v}) , \end{aligned}$$

or

$$(*) \qquad \arg dz(\mathbf{v}) = -\frac{1}{2} \arg \Phi + \frac{m\pi}{2} \qquad \text{for some integer } m .$$

In a neighborhood of an isolated umbilic of our surface M with constant H we consider the 1-dimensional distribution Δ formed by the multiples of the principle vectors with the larger principal curvature, say. The index of this distribution was defined in Addendum 2 to Chapter 4. We can now compute it in terms of Φ.

32. PROPOSITION. Let $f: U \to M$ be a conformal immersion into a surface M of constant mean curvature H, with corresponding analytic function Φ. Suppose that $p = f(0)$ is an isolated umbilic, so that $\Phi(0) = 0$, and consequently

$$\Phi(z) = a_n z^n + \cdots \qquad a_n \neq 0 , \quad n \geq 1 .$$

Then the index of Δ at p is $-n/2$.

Proof. We consider the distribution on U which is f^{-1} of Δ. Let $c\colon [0,1] \longrightarrow U$ be a small circle around 0. To compute the index of the distribution at 0, we must choose a continuous function $\theta\colon [0,1] \longrightarrow \mathbb{R}$ such that $\theta(t)$ is an angle between the x-axis and the direction of the distribution at $c(t)$; then the index is $[\theta(1) - \theta(0)]/2\pi$. First choose a continuous function $\phi\colon [0,1] \longrightarrow \mathbb{R}$ such that $\phi(t)$ is an argument for $\Phi(c(t))$. Then equation (*) shows that we must have

$$\theta(t) = -\frac{1}{2}\,\phi(t) + \frac{m\pi}{2}\,,$$

where the integer m must be constant, by continuity. So the index in question is

$$\frac{1}{2\pi}[\theta(1) - \theta(0)] = -\frac{1}{2}\cdot\frac{1}{2\pi}[\phi(1) - \phi(0)]\ .$$

But standard complex analysis results say that $\phi(1) - \phi(0) = 2\pi n$. [Here is a direct proof. Clearly $[\phi(1) - \phi(0)]/2\pi$ is just the degree of the map α from S^1 to $\mathbb{C} - \{0\}$ defined by

$$\begin{aligned}\alpha(t) &= \frac{1}{a_n}\,\Phi(c(t)) = c(t)^n[1 + \cdots]\\ &= c(t)^n[1 + d(t)]\ ,\end{aligned}$$

where we have

$$|d(t)| < 1 \qquad \text{for a sufficiently small circle } c\ .$$

Now

$$|\alpha(t) - c(t)^n| = |c(t)^n d(t)| < |c(t)|^n\ .$$

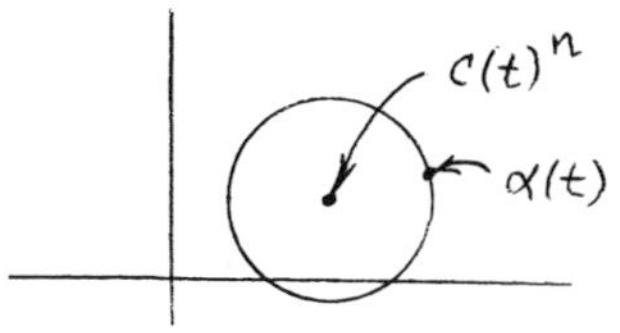

So the line segment from $c(t)^n$ to $\alpha(t)$ does not contain 0. This means that α and $t \mapsto c(t)^n$ are homotopic as maps from S^1 to $\mathbb{C} - \{0\}$. So they have the same degree. But the degree of $t \mapsto c(t)^n$ is n.] ■

All of this leads up to

<u>33. THEOREM (H. HOPF)</u>. If M is an immersed sphere in $\mathbb{R}^3$ with constant mean curvature H, then M is a standard sphere.

<u>Proof</u>. If all points of M were not umbilics, then by Lemma 31 there would be only finitely many umbilics. By Proposition 32, the index of Δ at each umbilic would be negative. This contradicts Theorem 4-21, since $\chi(M) = 2 > 0$. ■

Addendum 3. Imbedded Surfaces with Constant Mean Curvature

In this Addendum we will prove that a compact imbedded surface $M \subset \mathbb{R}^3$ with constant mean curvature H_0 must be a standard sphere. Essentially the same proof works for imbedded hypersurfaces in $\mathbb{R}^{n+1}$, H^{n+1}, or an open hemisphere of S^{n+1}. The proof depends on a simple ingenious geometric construction, together with analytic results (Theorems 10-17 and 10-20) from Addendum 2 to Chapter 10; the proofs of these theorems can be read right now, for they do not depend on any material from Chapter 10 proper. These analytic results are applied to the present situation as follows.

Consider a surface given as the graph of a function $h\colon \mathbb{R}^2 \to \mathbb{R}$, and introduce the standard abbreviations

$$p = \frac{\partial h}{\partial x} \qquad q = \frac{\partial h}{\partial y}$$

$$r = \frac{\partial^2 h}{\partial x^2} \qquad s = \frac{\partial^2 h}{\partial x \partial y} \qquad t = \frac{\partial^2 h}{\partial y^2}\,.$$

For the condition that the surface has constant mean curvature H_0 we find, from (B') on p. III.201, the equation

$$(*) \qquad 0 = (1+q^2)r - 2pqs + (1+p^2)t - 2H_0(1+p^2+q^2)^{3/2}$$
$$= F(p,q,r,s,t)\,.$$

Now let h_1 and h_2 be two solutions of (*), with corresponding partials $p_1,\dots,t_1$ and $p_2,\dots,t_2$. If we denote the partial derivatives of F with respect to its 5 arguments as $F_p,\dots,F_t$, then at all points of $\mathbb{R}^2$ we have

$$0 = F(p_1,q_1,r_1,s_1,t_1) - F(p_2,q_2,r_2,s_2,t_2)$$

$$= \int_0^1 \frac{d}{d\tau}F(\tau p_1 + (1-\tau)p_2,\ldots,\tau t_1 + (1-\tau)t_2)\ d\tau$$

$$= \int_0^1 (p_1 - p_2)F_p(\bullet) + \cdots + (t_1 - t_2)F_t(\bullet)\ d\tau$$

$$\text{where} \quad \bullet = (\tau p_1 + (1-\tau)p_2,\ldots,\tau t_1 + (1-\tau)t_2)$$

$$= A\cdot(p_1 - p_2) + B\cdot(q_1 - q_2) + C\cdot(r_1 - r_2)$$

$$+ D\cdot(s_1 - s_2) + E\cdot(t_1 - t_2)\ , \qquad \text{say.}$$

Setting $u = h_1 - h_2$, and letting $p,q,\ldots,t$ now denote the partials of u, we see that u satisfies the equation

$$(**) \qquad A\cdot p + B\cdot q + C\cdot r + D\cdot s + E\cdot t = 0\ .$$

<u>34. LEMMA</u>. Let h_1 and h_2 be two functions whose graphs are surfaces of the same constant mean curvature H_0, both functions being defined either in a neighborhood of 0 in $\mathbb{R}^2$, or in a neighborhood of 0 in the closed half-plane $\{(x,y)\colon y \geq 0\}$. Suppose that $h_1 \geq h_2$ in this domain, and that $h_1(0) = h_2(0)$. If h_1 and h_2 are defined only in the half-plane, assume also that $\partial h_1/\partial x(0) = \partial h_2/\partial x(0)$. Then $h_1 = h_2$ in a neighborhood of 0, or in a neighborhood of 0 in $\{(x,y)\colon y \geq 0\}$.

<u>Proof</u>. Notice that for all $(\lambda,\mu) \neq (0,0)$ we have

$$F_r\lambda^2 + F_s\lambda\mu + F_t\mu^2 = (1+q^2)\lambda^2 - 2pq\lambda\mu + (1+p^2)\mu^2$$

$$= \lambda^2 + \mu^2 + (q\lambda - p\mu)^2 > 0\ ,$$

where F_r, F_s, F_t are evaluated at any point of $\mathbb{R}^5$. So we also have

$$C\lambda^2 + D\lambda\mu + E\mu^2 = \int_0^1 F_r(\bullet)\lambda^2 + F_s(\bullet)\lambda\mu + F_t(\bullet)\mu^2 \, d\tau$$

$$> 0 .$$

Thus Theorems 10-17 and 10-20 apply to the solution $u = h_1 - h_2$ of equation (**). ■

For the geometric part of the proof, we first note that the standard spheres are the only compact surfaces which have a plane of symmetry in every direction. In fact,

35. LEMMA. If $A \subset \mathbb{R}^3$ is bounded and has a plane of symmetry in every direction, then A is invariant under all rotations about some point * (hence A is a union of concentric spheres around *).

Proof. Choose 3 mutually orthogonal planes P_1, P_2, P_3 which are planes of symmetry for A, and let * be the unique point in $P_1 \cap P_2 \cap P_3$. Let P be any other plane of symmetry. It is easy to see that if P does not go through *, then suitable compositions of the reflections through P_1, P_2, P_3, and P will take any given point in A to points arbitrarily far from *. So if A is bounded, then we must have $* \in P$. Thus A is invariant under reflection through every plane through *. This implies that A is invariant under all rotations about *. ■

It is this symmetry property of spheres which we will establish for any surface of constant mean curvature.

36. THEOREM (ALEKSANDROV). Let M be a compact surface imbedded in $\mathbb{R}^3$ with constant mean curvature H_0. Then M has a plane of symmetry in every direction, so M is a standard sphere.

Proof. We know (Theorem I.11-14) that M is the boundary of some closed domain D. We can assume that our direction is the z-axis, and that M is placed so that it lies in the region where $z \geq 0$, and touches the plane $z = 0$. For each $a > 0$, let P_a be the plane $z = a$. The set of points of M which lie below P_a is a "hump" H_a. Let $\tilde{H}_a$ be the reflection of H_a in P_a. For

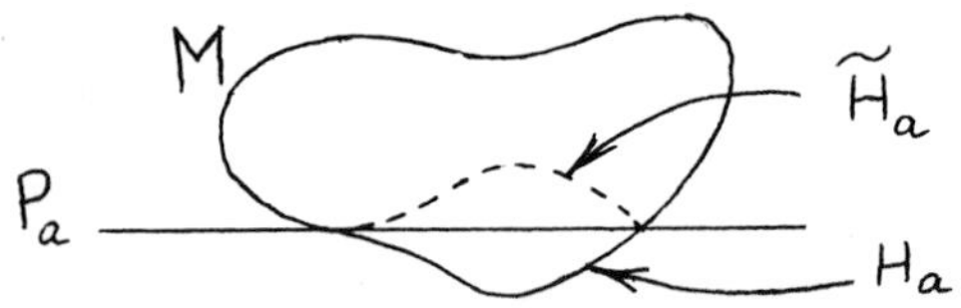

sufficiently small $a > 0$, the set $\tilde{H}_a$ will lie inside D. Consider the set of all $b > 0$ such that $\tilde{H}_a$ lies in D for $0 \leq a \leq b$. This set clearly has a largest element c. There are then two possible cases, as illustrated below.

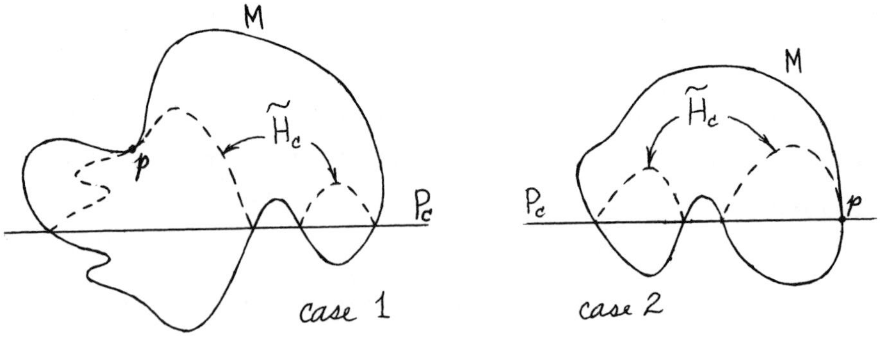

In the first case, there is a point $p \in \tilde{H}_c \cap M$ which is not on P_c. From the definition of c, it is easy to see that near p the surfaces M and $\tilde{H}_c$ are the graphs of two functions h_1, h_2 with $h_1 \geq h_2$. Then Lemma 35 shows that M and $\tilde{H}_c$ coincide in a neighborhood of p.

If there is no point $p \in \tilde{H}_c \cap M - P_c$, then we must have the situation shown in the second figure: for some point $p \in P_c$, the surface M has a vertical tangent plane at p. The part of M which lies above or on P_c is a surface-with-boundary, and near p it can be represented as a function h_1 on a closed half-plane perpendicular to the (x,y)-plane. Similarly, $\tilde{H}_c$ can be represented as a function h_2 on the same closed half-plane. This time we have $h_2 \geq h_1$. Lemma 35 shows that $\tilde{H}_c$ coincides near p with the part of M which lies above or on P_c.

Now let $\tilde{K}$ be the component of $\tilde{H}_c$ which contains the point p (in either case 1 or case 2). This component $\tilde{K}$ is the reflection in P_c of a component $K \subset H_c$. The argument of the above two paragraphs, together with a simple connectedness argument, shows that $\tilde{K} \subset M$. But $\tilde{K} \cup K \subset M$ is already a compact manifold. So we must have $\tilde{K} \cup K = M$. Thus M is symmetric with respect to the plane P_c. ■

The most interesting aspect of this proof is the fact that constancy of the mean curvature was used in such a weak way. There are numerous other conditions which can be treated similarly; Alexandrov has a whole series of papers on this subject. A somewhat later paper by Alexandrov [2] generalizes Theorem 36 so as to allow many types of self-intersections of M. For example, if $M \subset \mathbb{R}^3$ is a compact surface, bounding a domain D, and $f\colon M \to \mathbb{R}^3$ is an immersion which can be extended to an immersion of D into $\mathbb{R}^3$, then $f(M)$

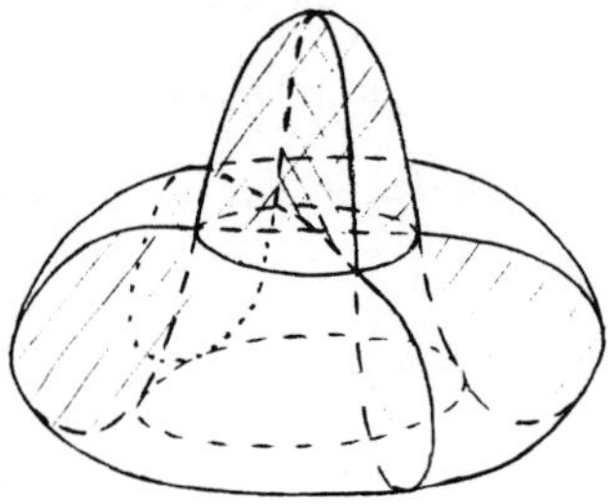

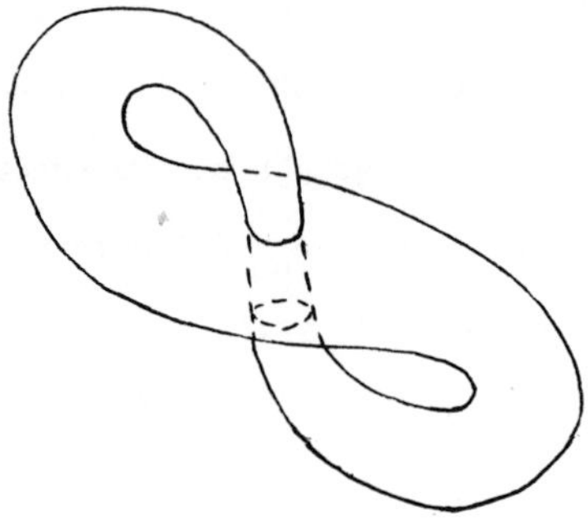

does not have constant mean curvature unless it is a standard sphere. Not every immersion $f: M \to \mathbb{R}^3$ has this property, however, even if $M = S^2$. For example, let $f: S^2 \to \mathbb{R}^3 - \{0\}$ be an immersion such that $r \circ f: S^2 \to S^2$ has degree > 1, where $r: \mathbb{R}^3 - \{0\} \to S^2$ is the obvious retraction. Then f cannot be extended to an immersion of the 3-dimensional disc.

Addendum 4. The Second Variation of Volume

In this Addendum we will derive the formula for the second variation of volume, and give some applications. The calculation itself is a real bitch, and even the final formula is quite involved, so some preliminaries will be required.

1. For a submanifold $M \subset N$ and a vector $\xi \in M_p{}^{\perp}$ we have the map $A_\xi \colon M_p \to M_p$ with

$$\langle A_\xi(X), Y\rangle = \langle s(X,Y), \xi\rangle .$$

Since A_ξ is symmetric, it has n real eigenvalues $\lambda_1, \ldots, \lambda_n$. We will let

$$\Sigma_2(\xi) = \sum_{i=1}^{n} \lambda_i{}^2 = \operatorname{trace} A_\xi{}^2 .$$

If ξ denotes, instead, a section of the normal bundle Nor M, then $\Sigma_2(\xi)$ is a function on M.

2. Given $\xi \in M_p{}^{\perp}$, we define the "partial Ricci Tensor"

$$\operatorname{Ric}_M(\xi) = -\sum_{i=1}^{n} \langle R'(\xi, X_i)X_i, \xi\rangle ,$$

where $X_1, \ldots, X_n$ is an orthonormal basis of M_p; it is easily seen that this does not depend on the choice of $X_1, \ldots, X_n$. Naturally, $\operatorname{Ric}_M(\xi)$ denotes a function on M if ξ denotes a section of the normal bundle Nor M.

3. Recall from Addendum 1 of Chapter 7 that if W is a vector field tangent along the manifold M, then div W is the function on M defined by

$$(\mathrm{div}\ W)(p) = \text{trace of } X_p \longmapsto \nabla_{X_p} W = \sum_{i=1}^{n} \langle \nabla_{X_i} W, X_i \rangle ,$$

where $X_1,\ldots,X_n$ is any orthonormal basis of M_p.

4. We also recall from this same Addendum that we have defined the Laplacian $\Delta\psi$ for a section ψ of a vector bundle over a Riemannian manifold M; to define this we needed a connection on the bundle. For a submanifold $M \subset N$ of a Riemannian manifold N, we have the induced metric on M, and a connection D on the normal bundle defined by

$$D_X\psi = \perp\nabla'_X\psi , \qquad \nabla' = \text{the covariant derivative in } N.$$

Thus if ψ is a section of the normal bundle, we have

$$\Delta\psi(p) = \sum_{j=1}^{n} \perp\nabla'_{X_j(p)}(\perp\nabla'_{X_j}\psi) ,$$

where $X_1,\ldots,X_n$ is an orthonormal moving frame with

$$\nabla_{X_i}X_j(p) = 0 , \qquad \nabla = \text{covariant derivative in } M .$$

5. We will require the following properties of contractions $X \lrcorner\, \omega$ and Lie derivatives $L_Z\omega$:

(a) $Z \lrcorner (\phi \wedge \eta) = (Z \lrcorner\, \phi) \wedge \eta + (-1)^k \phi \wedge (Z \lrcorner\, \eta)$ for ϕ a k-form [Problem I.7-4]

(b) $L_Z(\phi \wedge \eta) = L_Z\phi \wedge \eta + \phi \wedge L_Z\eta$

(c) $L_Z\omega = Z \lrcorner\, d\omega + d(Z \lrcorner\, \omega)$

(d) $L_Z d\omega = dL_Z\omega$

[Problem I.7-18] (for (b)–(d))

(e) $L_{Y+Z}\omega = L_Y\omega + L_Z\omega$ [Problem I.5-14]

(f) $L_Z(Y \lrcorner \omega) = [Z,Y] \lrcorner \omega + Y \lrcorner L_Z\omega$ [an exercise, using Problem I.7-18(c)] .

6. Finally, there is one important way that the second variation formula for volume will differ from the second variation formula for energy: If $\alpha\colon (-\varepsilon,\varepsilon)\times M \to N$ is a variation of $\bar{\alpha}(0) = f\colon M \to N$, and $W(p) = \partial\alpha/\partial u\,(0,p)$ is the variation vector field, then our formula will involve not merely W, but also $\tilde{W} = \partial\alpha/\partial u$. We will define vector fields $\mathbf{T}\tilde{W}$ and $\perp\tilde{W}$ along α by writing $\tilde{W}(u,p) = \mathbf{T}\tilde{W}(u,p) + \perp\tilde{W}(u,p)$, where $\mathbf{T}\tilde{W}(u,p)$ is tangent to $\bar{\alpha}(u)(M)$ at $u(p)$ and $\perp\tilde{W}(u,p)$ is orthogonal to $\bar{\alpha}(u)(M)$ at $u(p)$.

<u>37. THEOREM</u>. Let $f\colon M \to N$ be a minimal immersion of an oriented n-dimensional manifold (-with-boundary) M into a Riemannian manifold $(N^m, \langle\ ,\ \rangle)$, and let $\alpha\colon (-\varepsilon,\varepsilon)\times M \to N$ be a variation of f through immersions. Let W be the variation vector field, and let $\tilde{W} = \partial\alpha/\partial u$. If $\Gamma(u)$ is the volume form on M determined by the metric $\bar{\alpha}(u)^*\langle\ ,\ \rangle$ and the given orientation of M, then

$$\ddot{\Gamma}(0) = [\mathrm{Ric}_M(\perp W) - \Sigma_2(\perp W) - \langle \perp W, \Delta(\perp W)\rangle]\cdot\Gamma(0)$$
$$+ d\Big(\mathrm{div}\,\mathbf{T}W\cdot(\mathbf{T}W \lrcorner \Gamma(0)) + \mathbf{T}[\perp\tilde{W},\mathbf{T}\tilde{W}] \lrcorner \Gamma(0)\Big) .$$

<u>Proof</u>. We will regard this proof as a continuation of the proof of Theorem 11; we will refer to equations (1)-(10) in that proof, and therefore commence our numbering of equations with (11). Once again we first consider a point $p_0 \in M$ where $W(p_0)$ is not tangent to $f(M)$. We choose V as before, and assume that f is just the inclusion $i\colon V \to N$. We will use all the notation

introduced in the proof of Theorem 11, and we will also introduce the abbreviations

$$(11) \qquad \theta^j = i^*\phi^j \qquad 1 \le j \le n .$$

Since our immersion is minimal ($\eta = 0$), equation (9) shows that

$$(12) \qquad i^*\{Z \lrcorner\, d\Phi\} = 0 \qquad \text{for \underline{all} vector fields } Z \text{ along } V .$$

Using (6), we can write $d\Phi$ as

$$(13) \quad d\Phi = \sum_{r=n+1}^{m} \phi^r \wedge \mu_r , \qquad \text{for } \mu_r = \sum_{j=1}^{n} \phi^1 \wedge \dots \wedge \psi_r^j \wedge \dots \wedge \phi^n .$$

Then (12) becomes

$$\sum_{r=n+1}^{m} \theta^r(Z) i^*\mu_r = 0 , \qquad \text{using (a) on p.514 and } i^*\phi^r = 0 .$$

Since this is true for arbitrary Z, we have

$$(14) \qquad i^*\mu_r = 0 , \qquad \text{and hence} \quad \sum_{j=1}^{n} \psi_r^j(X_j) = 0 \quad \text{along } V .$$

Now let us apply equation (5) to all u, not just $u = 0$. We obtain

$$\dot\Gamma(u) = \bar\alpha(u)^*(\tilde W \lrcorner\, d\Phi) + \bar\alpha(u)^* d(\tilde W \lrcorner\, \Phi)) .$$

As before, this implies that

$$\begin{aligned}(15) \qquad \ddot\Gamma(0) &= i^*\{L_{\tilde W}(\tilde W \lrcorner\, d\Phi)\} + i^*\{L_{\tilde W} d(\tilde W \lrcorner\, \Phi)\} \\ &= i^*\{L_{\tilde W}(\tilde W \lrcorner\, d\Phi)\} + d(i^*\{L_{\tilde W}(\tilde W \lrcorner\, \Phi)\}) \qquad \text{by (d)} .\end{aligned}$$

Once again we will show that the two terms on the right are precisely the terms appearing in the statement of the theorem.

The first term is the one that will give us all the difficulties, and we will use some preliminary tricks to make the calculations manageable. First of all, we want to be more particular in our choice of the moving frame $X_1,\dots,X_n,X_{n+1},\dots,X_m$. We will assume that $X_1(p_0),\dots,X_n(p_0)$ is a basis of eigenvectors for $A_{\perp W(p_0)}$, with corresponding eigenvalues $\lambda_1,\dots,\lambda_n$. This means that

$$\begin{aligned}
\lambda_j\delta_{jk} &= \langle A_{\perp W(p_0)}X_j(p_0),\ X_k(p_0)\rangle \\
&= \langle s(X_k(p_0),X_j(p_0)),\ \perp W(p_0)\rangle \\
&= \langle \perp\nabla'_{X_k}X_j,\ \perp W\rangle \quad \text{at } p_0 \\
&= \left\langle \sum_{r=n+1}^{m} \psi_j^r(X_k)X_r,\ \perp W\right\rangle \quad \text{at } p_0 \\
&= -\sum_{r=n+1}^{m} \phi^r(W)\cdot\psi_r^j(X_k) \quad \text{at } p_0 ,
\end{aligned}$$

and consequently,

$$(16)\qquad \sum_{r=n+1}^{m} \phi^r(W)i^*\psi_r^j = -\lambda_j\theta^j \quad \text{at } p_0 .$$

We still have considerable leeway in the choice of our moving frame $X_1,\dots,X_m$. We can replace it with a new moving frame $\bar X_1,\dots,\bar X_m$ defined by

$$\bar X_\alpha = \sum_{\beta=1}^{m} M_\alpha^\beta X_\beta ,$$

where (M_α^β) is a matrix of functions such that

(i) (M_α^β) is always orthogonal, $(M_\alpha^\beta)^{-1} = (M_\beta^\alpha)$,

(ii) $M_r^j = M_j^r = 0 \quad 1 \le j \le n, \quad n+1 \le r \le m$,

(iii) $(M_\alpha^\beta(p_0)) = I$.

Condition (i) means that the new moving frame is orthonormal, and condition (ii) implies that (1) and (2) still hold, so the $\bar{X}_j$ are tangent to the $\bar{\alpha}(u)(V)$, while the $\bar{X}_r$ are orthogonal. Condition (iii) means that the frame is not changed at p_0, so that equation (16) still holds. The dual 1-forms $\bar{\phi}^\alpha$ are related to the ϕ^β by

$$\bar{\phi}^\alpha = \sum_{\beta=1}^m M_\alpha^\beta \phi^\beta , \qquad \bar{\phi}^\alpha(p_0) = \phi^\alpha(p_0) ,$$

so the corresponding connection forms $\bar{\psi}_\beta^\alpha$ satisfy

$$\begin{aligned}
-\sum_{\beta=1}^m \bar{\psi}_\beta^\alpha \wedge \bar{\phi}^\beta = d\bar{\phi}^\alpha &= \sum_{\beta=1}^m dM_\alpha^\beta \wedge \phi^\beta + \sum_{\beta=1}^m M_\alpha^\beta \wedge d\phi^\beta \\
&= \sum_{\beta=1}^m dM_\alpha^\beta \wedge \phi^\beta - \sum_{\beta,\gamma=1}^m M_\alpha^\beta \psi_\gamma^\beta \wedge \phi^\gamma \\
&= \sum_{\beta=1}^m \sum_{\delta=1}^m M_\beta^\delta \, dM_\alpha^\beta \wedge \bar{\phi}^\delta - \sum_{\beta,\gamma,\delta=1}^m M_\alpha^\beta M_\delta^\gamma \psi_\gamma^\beta \wedge \bar{\phi}^\delta \\
&= -\sum_{\beta=1}^m \left[\sum_{\gamma,\delta=1}^m M_\alpha^\delta M_\beta^\gamma \psi_\gamma^\delta - \sum_{\delta=1}^m M_\delta^\beta \, dM_\alpha^\delta \right] \wedge \bar{\phi}^\beta .
\end{aligned}$$

Now $\sum_{\gamma,\delta} M_\alpha^\delta M_\beta^\gamma \psi_\gamma^\delta$ is easily seen to be skew-symmetric with respect to α and β, since $\psi_\gamma^\delta = -\psi_\delta^\gamma$. Since (M_α^β) is orthogonal and $M(p_0) = I$, we also have skew-symmetry for $\sum_{\delta} M^{\beta}_{\delta}\, dM^{\delta}_{\alpha}$ at p_0. So Proposition II.7-4 (which is really a result about forms on a single vector space) shows that at p_0 we have

$$\text{(iv)} \qquad \bar{\psi}^{\alpha}_{\beta}(p_o) = \psi^{\alpha}_{\beta}(p_0) - dM^{\beta}_{\alpha}(p_0) \ .$$

Now we claim that it is possible to choose M^{β}_{α} so that

$$\text{(v)} \qquad \begin{cases} dM^{j}_{k}(p_0) = \psi^{j}_{k}(p_0) & 1 \le j,\ k \le n \\ dM^{r}_{s}(p_0) = \psi^{r}_{s}(p_0) & n+1 \le r,\ s \le m \end{cases}$$

In fact, for every unit tangent vector X at p_0 we can define

$$\begin{aligned} M^{j}_{k}(\exp tX) &= \exp(t\psi^{j}_{k}(X)) & 1 \le j,\ k \le n \\ M^{r}_{s}(\exp tX) &= \exp(t\psi^{r}_{s}(X)) & n+1 \le r,\ s \le m \\ M^{j}_{r} &= M^{r}_{j} = 0\ , \end{aligned}$$

where the exp on the right is the ordinary exponential of matrices. Then the matrices (M^{α}_{β}) satisfy (v); they also satisfy (i)-(iii), the matrices (M^{j}_{k}) and (M^{r}_{s}) being orthogonal since they are exp of skew-symmetric matrices. In connection with (iv), we thus see that our moving frame can be picked so that it satisfies not only (16), but also

$$\text{(17)} \qquad \begin{cases} \psi^{j}_{k}(p_0) = 0 & 1 \le j,\ k \le n \\ \psi^{r}_{s}(p_0) = 0 & n+1 \le r,\ s \le m\ . \end{cases}$$

In addition to this special choice of the moving frame, we require another preliminary move. We are trying to show that $i^*\{L_{\tilde{W}}(\tilde{W} \lrcorner d\Phi)\}$ is the first term in the formula of the theorem. We notice that this term involves only the perpendicular component $\perp W$ of W. We can ease the strain of the calculations by first proving that the same is true of the expression which we have to work with. In the following lemma, $\top \tilde{W}$ and $\perp \tilde{W}$ actually denote extensions of these vector fields to a neighborhood of image α.

<u>38. LEMMA</u>. For a minimal immersion we have

$$i^*\{L_{\tilde{W}}(\tilde{W} \lrcorner d\Phi)\} = i^*\{L_{\perp \tilde{W}}(\perp \tilde{W} \lrcorner d\Phi)\} .$$

<u>Proof</u>. By property (c), which we stated before the theorem, we have

$$i^*\{L_{\tilde{W}}(\tilde{W} \lrcorner d\Phi)\} = i^*\{\tilde{W} \lrcorner d(\tilde{W} \lrcorner d\Phi)\} .$$

For vector fields Y and Z in N, define

$$\mathscr{S}(Y,Z) = i^*\{Y \lrcorner d(Z \lrcorner d\Phi)\} ,$$

so that

$$i^*\{L_{\tilde{W}}(\tilde{W} \lrcorner d\Phi)\} = \mathscr{S}(\tilde{W},\tilde{W}) .$$

It is clear that $\mathscr{S}$ is bilinear over $\mathbb{R}$. We will show that $\mathscr{S}(Y,Z) = 0$ if either Y or Z is tangent along M. The lemma then follows by writing $\tilde{W} = \top \tilde{W} + \perp \tilde{W}$.

Suppose first that Y is tangent along M. Then

$$\begin{aligned} \mathscr{S}(Y,Z) &= i^*\{Y \lrcorner d(Z \lrcorner d\Phi)\} \\ &= i^*\{L_Y(Z \lrcorner d\Phi) - d(Y \lrcorner Z \lrcorner d\Phi)\} \qquad \text{by (c)} . \end{aligned}$$

Now equation (12) tells us that $Z \lrcorner d\Phi$ gives 0 when applied to tangent vectors of M. The same is clearly true for $Y \lrcorner Z \lrcorner d\Phi$, since Y is itself tangent along M. So

$$i^*\{Y \lrcorner Z \lrcorner d\Phi\} = 0 , \quad \text{and hence} \quad i^*\{d(Y \lrcorner Z \lrcorner d\Phi)\} = 0 .$$

On the other hand, since Y is tangent to M we clearly also have

$$i^*\{L_Y(Z \lrcorner d\Phi)\} = 0 .$$

Thus $\mathcal{S}(Y,Z) = 0$.

Now suppose that Z is tangent along M. We have

$$\begin{aligned}
\mathcal{S}(Y,Z) &= i^*\{Y \lrcorner d(Z \lrcorner d\Phi)\} \\
&= i^*\{Y \lrcorner L_Z d\Phi\} && \text{by (c)} \\
&= i^*\{[Y,Z] \lrcorner d\Phi\} + i^*\{L_Z(Y \lrcorner d\Phi)\} && \text{by (f)} \\
&= i^*\{[Y,Z] \lrcorner d\Phi\} + i^*\{Z \lrcorner d(Y \lrcorner d\Phi)\} + i^*\{d(Z \lrcorner Y \lrcorner d\Phi)\} && \text{by (c)} \\
&= i^*\{[Y,Z] \lrcorner d\Phi\} + \mathcal{S}(Z,Y) + i^*\{d(Z \lrcorner Y \lrcorner d\Phi)\} .
\end{aligned}$$

The first term is 0 by (12). The second term is 0 by what we have already proved. The third term is 0 for the same reason that $i^*\{d(Y \lrcorner Z \lrcorner d\Phi)\}$ was 0 before. Q.E.D.

We are now finally ready to carry out the computation.

Step 1. We claim that

$$(18) \qquad i^*\{L_{\tilde{W}}\phi^r\} = 0 = i^*\{L_{\perp \tilde{W}}\phi^r\} \qquad n+1 \le r \le m .$$

To see this, choose Y to be a vector field tangent to V and let $i_*Y = X$. Then

$$\begin{aligned} i^*\{L_{\tilde{W}}\phi^r\}(Y) &= L_{\tilde{W}}\phi^r(X) \\ &= d(W \lrcorner\, \phi^r)(X) + (W \lrcorner\, d\phi^r)(X) && \text{by (c)} \\ &= X(\phi^r(W)) + d\phi^r(W,X) \\ &= X(\phi^r(W)) + [W(\phi^r(X)) - X(\phi^r(W)) - \phi^r([W,X])] && \text{by p. I.292} \\ &= -\,\phi^r([W,X]) \ . \end{aligned}$$

But if $t^1,\ldots,t^n$ is a coordinate system around p_0 in V, then X is a linear combination of $\partial\alpha/\partial t^1,\ldots,\partial\alpha/\partial t^n$,

$$X = \sum_{j=1}^{n} a_j \frac{\partial\alpha}{\partial t^j} \ .$$

We have

$$\left[W, \frac{\partial\alpha}{\partial t^j}\right] = \left[\frac{\partial\alpha}{\partial u}, \frac{\partial\alpha}{\partial t^j}\right] = \alpha_*\left(\left[\frac{\partial}{\partial u}, \frac{\partial}{\partial t^j}\right]\right) = 0 \ ,$$

so

$$[W,X] = \left[W, \sum_{j=1}^{n} a_j \frac{\partial\alpha}{\partial t^j}\right] = -\sum_{j=1}^{n} W(a_j)\frac{\partial\alpha}{\partial t^j} \qquad \text{by p. I.215.}$$

Thus $[W,X]$ is also tangent to V, so $\phi^r([W,X]) = 0$. This show that $i^*\{L_{\tilde{W}}\phi^r\} = 0$.

We also have

$$i^*\{L_{\mathbf{T}\tilde{W}}\phi^r\} = -\,\phi^r([\mathbf{T}W,X]) = 0 \ ,$$

since $[\mathbf{T}W,X]$ is tangent to V. Hence $i^*\{L_{\perp\tilde{W}}\phi^r\} = 0$ also.

Step 2. For $1 \leq j \leq n$, we have

$$i^*\{L_{\tilde{W}}\phi^j\} = i^*\{d(\tilde{W} \lrcorner \phi^j)\} + i^*\{\tilde{W} \lrcorner d\phi^j\} \qquad \text{by (c)}$$

$$= d(\phi^j(W)) - i^*\{\tilde{W} \lrcorner \sum_{\alpha=1}^{m} \psi^j_\alpha \wedge \phi^\alpha\}$$

$$= d(\phi^j(W)) - \sum_{k=1}^{n} \psi^j_k(W)\theta^k + \sum_{\alpha=1}^{m} \phi^\alpha(W) i^*\psi^j_\alpha \qquad \text{by (a)} .$$

Using (16) and (17) we see that

$$(19) \qquad i^*\{L_{\tilde{W}}\phi^j\} = d(\phi^j(W)) - \lambda_j\theta^j \qquad \text{at } p_0 .$$

Similarly, we find that

$$(20) \qquad i^*\{L_{\perp\tilde{W}}\phi^j\} = -\lambda_j\theta^j \qquad \text{at } p_0 .$$

Step 3. Using the second structural equation to express $d\psi^j_r$ in terms of the curvature forms Ψ^j_r, we have

$$i^*\{L_{\perp\tilde{W}}\psi^j_r\} = i^*\{\perp\tilde{W} \lrcorner d\psi^j_r\} + i^*\{d(\perp\tilde{W} \lrcorner \psi^j_r)\} \qquad \text{by (c)}$$

$$= -\, i^*\{\perp\tilde{W} \lrcorner \sum_{\gamma=1}^{m} \psi^j_\gamma \wedge \psi^\gamma_r\} + i^*\{\perp\tilde{W} \lrcorner \Psi^j_r\} + d(\psi^j_r(\perp W)) .$$

Because of equation (17), each term $\psi^j_\gamma \wedge \psi^\gamma_r$ always has one factor equal to 0 at p_0, so we obtain

$$(21) \qquad i^*\{L_{\perp\tilde{W}}\psi^j_r\} = i^*\{\perp\tilde{W} \lrcorner \Psi^j_r\} + d(\psi^j_r(\perp W)) \qquad \text{at } p_0 .$$

Step 4. Referring to (13) for the definition of μ_r, we now compute

$$i^*\{L_{\perp\tilde{W}}\mu_r\} = i^*\{L_{\perp\tilde{W}} \sum_{j=1}^{n} \phi^1 \wedge \dots \wedge \psi_r^j \wedge \dots \wedge \phi^n\}$$

$$= \sum_{j=1}^{n} \theta^1 \wedge \dots \wedge i^*\{L_{\perp\tilde{W}}\psi_r^j\} \wedge \dots \wedge \theta^n$$

$$+ \sum_{j=1}^{n} [\sum_{k\neq j} \theta^1 \wedge \dots \wedge i^*\{L_{\perp\tilde{W}}\phi^k\} \wedge \dots \wedge i^*\psi_r^j \wedge \dots \wedge \theta^n] .$$

Substituting from (20) and (21), and rearranging slightly, we have

$$i^*\{L_{\perp\tilde{W}}\mu_r\} = \sum_{j=1}^{n} \theta^1 \wedge \dots \wedge i^*\{\perp\tilde{W} \lrcorner \Psi_r^j\} \wedge \dots \wedge \theta^n$$

$$+ \sum_{j=1}^{n} [\sum_{k\neq j} \theta^1 \wedge \dots \wedge -\lambda_k\theta^k \wedge \dots \wedge i^*\psi_r^j \wedge \dots \wedge \theta^n]$$

$$+ \sum_{j=1}^{n} \theta^1 \wedge \dots \wedge d(\psi_r^j(\perp W)) \wedge \dots \wedge \theta^n \qquad \text{at } p_0 .$$

Notice that

$$\sum_{j=1}^{n} [\sum_{k\neq j} \theta^1 \wedge \dots \wedge -\lambda_k\theta^k \wedge \dots \wedge i^*\psi_r^j \wedge \dots \wedge \theta^n]$$

$$= \sum_{j=1}^{n} (\sum_{k\neq j} -\lambda_k)\theta^1 \wedge \dots \wedge i^*\psi_r^j \wedge \dots \wedge \theta^n .$$

But $\sum_{k=1}^{n} \lambda_k = 0$, since our immersion is minimal; so $\sum_{k\neq j} -\lambda_k = \lambda_j$. Thus

$$(22)\qquad \begin{aligned} i^*\{L_{\perp\tilde{W}}\mu_r\} &= \sum_{j=1}^{n} \theta^1 \wedge \dots \wedge i^*\{\perp\tilde{W} \lrcorner \psi_r^j\} \wedge \dots \wedge \theta^n \\ &+ \sum_{j=1}^{n} \lambda_j \theta^1 \wedge \dots \wedge i^*\psi_r^j \wedge \dots \wedge \theta^n \\ &+ \sum_{j=1}^{n} \theta^1 \wedge \dots \wedge d(\psi_r^j(\perp W)) \wedge \dots \wedge \theta^n \qquad \text{at } p_0 . \end{aligned}$$

Step 5.

$$\begin{aligned} i^*\{L_{\tilde{W}}(\tilde{W} \lrcorner d\Phi)\} &= i^*\{L_{\perp\tilde{W}}(\perp\tilde{W} \lrcorner d\Phi)\} && \text{by Lemma 38} \\ &= i^*\{\perp\tilde{W} \lrcorner d(\perp\tilde{W} \lrcorner d\Phi)\} && \text{by (c)} \\ &= i^*\{\perp\tilde{W} \lrcorner L_{\perp\tilde{W}} d\Phi\} && \text{by (c) again} \\ &= \sum_{r=n+1}^{m} i^*\{\perp\tilde{W} \lrcorner L_{\perp\tilde{W}}(\phi^r \wedge \mu_r)\} && \text{by (13)} \\ &= \sum_{r=n+1}^{m} i^*\{\perp\tilde{W} \lrcorner (L_{\perp\tilde{W}}\phi^r \wedge \mu_r)\} \\ &\quad + \sum_{r=n+1}^{m} i^*\{\perp\tilde{W} \lrcorner (\phi^r \wedge L_{\perp\tilde{W}}\mu_r)\} && \text{by (b)} . \end{aligned}$$

When we expand the first of these sums by (a), we obtain two terms, one involving $i^*\mu_r$ and one involving $i^*\{L_{\perp\tilde{W}}\phi^r\}$. These will both be 0, by (14) and (18), so we obtain

$$\begin{aligned} i^*\{L_{\tilde{W}}(\tilde{W} \lrcorner d\Phi)\} &= \sum_{r=n+1}^{m} i^*\{\perp\tilde{W} \lrcorner (\phi^r \wedge L_{\perp\tilde{W}}\mu_r)\} \\ &= \sum_{r=n+1}^{m} \phi^r(W) i^*\{L_{\perp\tilde{W}}\mu_r\} \qquad \text{by (a)} . \end{aligned}$$

Substituting in from (22), we obtain

$$(23) \quad i^*\{L_{\tilde{W}}(\tilde{W} \lrcorner d\Phi)\} = \sum_{r=n+1}^{m} \phi^r(W) \sum_{j=1}^{n} \theta^1 \wedge \dots \wedge i^*\{\perp \tilde{W} \lrcorner \Psi_r^j\} \wedge \dots \wedge \theta^n$$

$$+ \sum_{r=n+1}^{m} \phi^r(W) \sum_{j=1}^{n} \lambda_j \theta^1 \wedge \dots \wedge i^*\psi_r^j \wedge \dots \wedge \theta^n$$

$$+ \sum_{r=n+1}^{m} \phi^r(W) \sum_{j=1}^{n} \theta^1 \wedge \dots \wedge d(\psi_r^j(\perp W)) \wedge \dots \wedge \theta^n \quad \text{at } p_0$$

$$= S_1 + S_2 + S_3, \quad \text{say.}$$

<u>Step 6</u>. We will see what each of these sums gives when applied to the n-tuple of vectors $X_1(p_0), \dots, X_n(p_0)$.

Recall that

$$\Psi_r^j(X_\alpha, X_\beta) = \langle R'(X_\alpha, X_\beta)X_r,\ X_j\rangle .$$

Thus we have

$$\begin{aligned} S_1(X_1,\dots,X_n) &= \sum_{r=n+1}^{m} \phi^r(W) \sum_{j=1}^{n} \Psi_r^j(\perp W, X_j) && \text{at } p_0 \\ &= \sum_{r=n+1}^{m} \phi^r(W) \sum_{j=1}^{n} \langle R'(\perp W, X_j)X_r,\ X_j\rangle && \text{at } p_0 \\ &= -\sum_{r=n+1}^{m} \phi^r(W)\langle R'(\perp W, X_j)X_j,\ X_r\rangle && \text{at } p_0 \\ &= -\langle R'(\perp W, X_j)X_j, \perp W\rangle = \mathrm{Ric}_M(\perp W) && \text{at } p_0 . \end{aligned}$$

Hence

$$(24) \qquad S_1 = \mathrm{Ric}_M(\perp W)\cdot\Gamma(0) \qquad \text{at } p_0 .$$

Next we have

$$S_2(X_1,\dots,X_n) = \sum_{r=n+1}^{m} \phi^r(W) \sum_{j=1}^{n} \lambda_j \psi_r^j(X_j) \qquad \text{at } p_0$$

$$= \sum_{j=1}^{n} \lambda_j \sum_{r=n+1}^{m} \phi^r(W)\psi_r^j(X_j) \qquad \text{at } p_0$$

$$= - \sum_{j=1}^{n} \lambda_j{}^2 \qquad \text{by (16) .}$$

Hence

$$(25) \qquad S_2 = - \Sigma_2(\perp W)\cdot\Gamma(0) \qquad \text{at } p_0 .$$

To evaluate S_3, we note that $d(\psi_r^j(\perp W)) = \sum_i d(\psi_r^j(\perp W))(X_i)\cdot\theta^i$. So we obtain

$$(26) \qquad S_3(X_1,\ldots,X_n) = \sum_{r=n+1}^{m} \phi^r(W) \sum_{j=1}^{n} d(\psi_r^j(\perp W))(X_j) \qquad \text{at } p_0$$

$$= \sum_{j=1}^{n} \sum_{r=n+1}^{m} \phi^r(W)X_j(\psi_r^j(\perp W)) \qquad \text{at } p_0 .$$

Step 7. The coefficient of $\Gamma(0)$ in the statement of the theorem will clearly be completely accounted for as soon as we show that

$$(27) \qquad S_3(X_1,\ldots,X_n) = - <\perp W, \Delta(\perp W)> \qquad \text{at } p_0 .$$

Equation (17) implies that $<\nabla'_{X_\ell} X_k, X_j> = 0$ at p_0, and hence that $\nabla_{X_\ell} X_k = 0$ at p_0, where ∇ is the covariant derivative in V. So

$$\Delta(\perp W) = \sum_{j=1}^{n} \perp\nabla'_{X_j}(\perp\nabla'_{X_j}\perp W) .$$

Now

$$\nabla'_{X_j}\perp W = \nabla'_{X_j}\Big(\sum_{r=n+1}^{m} \phi^r(W)X_r\Big)$$

$$= \sum_{r=n+1}^{m} X_j(\phi^r(W))X_r + \sum_{r=n+1}^{m} \phi^r(W)\nabla'_{X_j}X_r$$

$$= \sum_{r=n+1}^{m} X_j(\phi^r(W))X_r + \sum_{r=n+1}^{m} \phi^r(W) \sum_{\alpha=1}^{m} \psi_r^\alpha(X_j)X_\alpha ,$$

so

$$\perp(\nabla'_{X_j} \perp W) = \sum_{r=n+1}^{m} [X_j(\phi^r(W)) + \sum_{s=n+1}^{m} \phi^s(W)\psi_s^r(X_j)]X_r .$$

Hence

$$\nabla'_{X_j}(\perp\nabla'_{X_j} \perp W) = \sum_{r=n+1}^{m} X_j\Big(X_j(\phi^r(W)) + \sum_{s=n+1}^{m} \phi^s(W)\psi_s^r(X_j)\Big)\cdot X_r$$

$$+ \sum_{r=n+1}^{m} [X_j(\phi^r(W)) + \sum_{s=n+1}^{m} \phi^s(W)\psi_s^r(X_j)]\cdot \sum_{\alpha=1}^{m} \psi_r^\alpha(X_j)X_\alpha .$$

Using (17), we obtain*

$$\perp\nabla'_{X_j}(\perp\nabla'_{X_j} \perp W) = \sum_{r=n+1}^{m} [X_j(X_j(\phi^r(W))) + \sum_{s=n+1}^{m} \phi^s(W)X_j(\psi_s^r(X_j))]X_r \qquad \text{at } p_0 ,$$

and therefore

$$(28) \quad \langle \perp W, \Delta(\perp W)\rangle = \sum_{j=1}^{n} \sum_{r=n+1}^{m} \phi^r(W)\cdot[X_j(X_j(\phi^r(W))) + \sum_{s=n+1}^{m} \phi^s(W)X_j(\psi_s^r(X_j))] \qquad \text{at } p_0$$

$$= \sum_{j=1}^{n} \sum_{r=n+1}^{m} \phi^r(W)X_j(X_j(\phi^r(W)) \qquad \text{at } p_0 ,$$

since $\psi_s^r = -\psi_r^s$.

We can find out something about the $X_j(\phi^r(W))$ by writing (18) in the form

---

*Note that $X_j(\psi_s^r(X_j))$ need not be zero at p_0, even though $\psi_s^r(X_j) = 0$ at p_0.

$$0 = i^*\{L_{\perp\tilde{W}}\phi^r\} = i^*\{d(\perp\tilde{W} \lrcorner \phi^r)\} + i^*\{\perp\tilde{W} \lrcorner d\phi^r\} \qquad \text{by (c)}$$

$$= d(\phi^r(W)) - i^*\{\perp\tilde{W} \lrcorner \sum_{\alpha=1}^{m} \psi^r_\alpha \wedge \phi^\alpha\}$$

$$= d(\phi^r(W)) - \sum_{k=1}^{n} \psi^r_k(\perp W)\theta^k + \sum_{s=n+1}^{m} \phi^s(W) i^*\psi^r_s \qquad \text{by (a) .}$$

This gives us

$$X_j(\phi^r(W)) = -\psi^j_r(\perp W) - \sum_{s=n+1}^{m} \phi^s(W)\psi^r_s(X_j) \ ;$$

using (17) we obtain

$$X_j(X_j(\phi^r(W)) = -X_j(\psi^j_r(\perp W)) - \sum_{s=n+1}^{m} \phi^s(W) X_j(\psi^r_s(X_j)) \qquad \text{at } p_0 \ .$$

Substituting into (28), we get

$$\langle \perp W, \Delta(\perp W)\rangle = -\sum_{j=1}^{n} \sum_{r=n+1}^{m} \phi^r(W) X_j(\psi^j_r(\perp W)) - \sum_{j=1}^{n} \sum_{r,s=n+1}^{m} \phi^r(W)\phi^s(W) X_j(\psi^r_s(X_j))$$

$$= -\sum_{j=1}^{n} \sum_{r=n+1}^{m} \phi^r(W) X_j(\psi^j_r(\perp W)) \qquad \text{at } p_0, \text{ since } \psi^r_s = -\psi^s_r \ .$$

This proves (27), and completes our calculation of the first term in (15).

The second term in (15) will not be nearly so bad. We have

$$(29) \qquad i^*\{L_{\tilde{W}}(\tilde{W} \lrcorner \Phi)\} = i^*\{\tilde{W} \lrcorner d(\tilde{W} \lrcorner \Phi)\} \qquad \text{by (c)}$$

$$= i^*\{\tilde{W} \lrcorner L_{\tilde{W}}\Phi\} \qquad \text{by (c)}$$

$$= i^*\{\tilde{W} \lrcorner \sum_{j=1}^{n} \phi^1 \wedge \ldots \wedge L_{\tilde{W}}\phi^j \wedge \ldots \wedge \phi^n\} \qquad \text{by (b) .}$$

To show that

$$(30)\qquad i^*\{L_{\tilde W}(\tilde W \lrcorner \Phi)\} = \operatorname{div} \mathbf{T}W\cdot(\mathbf{T}W \lrcorner \Gamma(0)) \;+\; \mathbf{T}[\bot\tilde W, \mathbf{T}\tilde W] \lrcorner \Gamma(0)\ ,$$

it obviously suffices to show that both sides give the same result when applied to any $(n-1)$-tuple $(X_1,\ldots,\hat X_\ell,\ldots,X_n)$ at p_0. Since the X_i enter symmetrically, we can assume, by renumbering, that $\ell = 1$. Now (29) gives

$$\begin{aligned} i^*\{L_{\tilde W}(\tilde W \lrcorner \Phi)\}(X_2,\ldots,X_n) &= \Big(\sum_{j=1}^n \phi^1 \wedge \ldots \wedge L_{\tilde W}\phi^j \wedge \ldots \wedge \phi^n\Big)(W,X_2,\ldots,X_n) \qquad \text{at } p_0 \\ &= (L_{\tilde W}\phi^1 \wedge \ldots \wedge \phi^n)(W,X_2,\ldots,X_n) \\ &\quad + \sum_{j=2}^n (\phi^1 \wedge \ldots \wedge L_{\tilde W}\phi^j \wedge \ldots \wedge \phi^n)(W,X_2,\ldots,X_n) \qquad \text{at } p_0. \end{aligned}$$

In computing the first term, the only permutations of $(W,X_2,\ldots,X_n)$ that do not give zero are interchanges of W with one X_j; in the second sum only the given order $(W,X_2,\ldots,X_n)$ produces a non-zero result. So

$$\begin{aligned} &i^*\{L_{\tilde W}(\tilde W \lrcorner \Phi)\}(X_2,\ldots,X_n) \\ &= \Big[(L_{\tilde W}\phi^1)(W) - \sum_{j=2}^n \phi^j(W)(L_{\tilde W}\phi^1)(X_j)\Big] + \sum_{j=2}^n \phi^1(W)(L_{\tilde W}\phi^j)(X_j) \qquad \text{at } p_0 \\ &= (L_{\tilde W}\phi^1)(W) - \sum_{j=1}^n \phi^j(W)(L_{\tilde W}\phi^1)(X_j) + \phi^1(W)\sum_{j=1}^n (L_{\tilde W}\phi^j)(X_j) \qquad \text{at } p_0 \\ &\qquad\text{[since } j = 1 \text{ gives the same new term in each sum]} \\ &= (L_{\tilde W}\phi^1)(W) - (L_{\tilde W}\phi^1)\Big(\sum_{j=1}^n \phi^j(W)X_j\Big) + \phi^1(W)\sum_{j=1}^n (L_{\tilde W}\phi^j)(X_j) \qquad \text{at } p_0 \\ &= (L_{\tilde W}\phi^1)(\bot W) + \phi^1(W)\sum_{j=1}^n [X_j(\phi^j(W)) - \lambda_j] \qquad \text{at } p_0 \\ &\qquad\text{[by (19)]} \end{aligned}$$

$$= (L_{\tilde{W}}\phi^1)(\mathbf{\perp} W) + \phi^1(W) \sum_{j=1}^{n} X_j(\phi^j(\mathbf{T}W)) \qquad \text{at } p_0$$

$$\left[\text{since } \sum_{j=1}^{n} \lambda_j = 0\right]$$

$$= (W \lrcorner\, d\phi^1)(\mathbf{\perp} W) + d(\tilde{W} \lrcorner\, \phi^1)(\mathbf{\perp} W) + \phi^1(W) \sum_{j=1}^{n} X_j(\langle \mathbf{T}W, X_j\rangle) \qquad \text{at } p_0$$

$$= d\phi^1(W, \mathbf{\perp} W) + \mathbf{\perp} W(\phi^1(\tilde{W})) + \phi^1(W) \sum_{j=1}^{n} [\langle \nabla_{X_j} \mathbf{T} W, X_j\rangle + \langle \mathbf{T}W, \nabla_{X_j} X_j\rangle] \qquad \text{at } p_0$$

$$= [W(\phi^1(\mathbf{\perp}\tilde{W})) - \mathbf{\perp} W(\phi^1(\tilde{W})) - \phi^1([\tilde{W}, \mathbf{\perp}\tilde{W}])] + \mathbf{\perp} W(\phi^1(\tilde{W}))$$

$$+ \phi^1(W) \sum_{j=1}^{n} \langle \nabla_{X_j} \mathbf{T} W, X_j\rangle \qquad \text{at } p_0 \quad \left[\text{since } \sum \nabla_{X_j} X_j = \eta = 0\right]$$

$$= -\phi^1([\tilde{W}, \mathbf{\perp}\tilde{W}]) + \phi^1(W) \sum_{j=1}^{n} \langle \nabla_{X_j} \mathbf{T} W, X_j\rangle$$

$$= \langle \mathbf{T}[\mathbf{\perp}\tilde{W}, \mathbf{T}\tilde{W}], X_1\rangle + \langle \mathbf{T}W, X_1\rangle \sum_{j=1}^{n} \langle \nabla_{X_j} \mathbf{T} W, X_j\rangle \; .$$

This is exactly the value of the right side of (30) on $X_2,\ldots,X_n$; we have thus completed the calculation of the second term in (15).

Finally, we again dispose of the general case, where $W(p_0)$ may be tangent to V, by considering $\mathbf{N} = N \times \mathbb{R}$, with the product metric, and the map $\boldsymbol{\alpha}\colon (-\varepsilon,\varepsilon) \times M \longrightarrow \mathbf{N}$ defined by

$$\boldsymbol{\alpha}(u,p) = (\alpha(u,p), u) \; .$$

We recall that

$$\mathbf{W}(p) = (W(p), \mathbf{1}) \qquad \text{and} \qquad \boldsymbol{\eta}(p) = (\eta, 0) \; .$$

So $\bar{\boldsymbol{\alpha}}(0)$ is minimal if $\bar{\alpha}(0)$ is. If $\mathbf{R}'$ is the curvature tensor in $\mathbf{N}$, then we have

$$\mathrm{Ric}_{\mathbf{M}}(\perp \mathbf{W}) = \mathrm{Ric}_{\mathbf{M}}((\perp W, \mathbf{1}))$$
$$= -\sum_{i=1}^{n} \langle \mathbf{R}'(\perp W, X_i)X_i, \perp W\rangle - \sum_{i=1}^{n} \langle \mathbf{R}'(\mathbf{1}, X_i)X_i, \mathbf{1}\rangle ,$$

for $X_1,\ldots,X_n$ an orthonormal basis of M. Using the fact that we have a product metric, we easily find that

$$\mathrm{Ric}_{\mathbf{M}}(\perp \mathbf{W}) = -\sum_{i=1}^{n} \langle R'(\perp W, X_i)X_i, \perp W\rangle$$
$$= \mathrm{Ric}_M(\perp W) .$$

The map $\mathbf{s}(p)\colon \mathbf{M}_p \times \mathbf{M}_p \longrightarrow \mathbf{M}_p{}^{\perp}$ is obviously given by

$$\mathbf{s}(p)(X,Y) = (s(X,Y),0) , \qquad X, Y \in M_p ,$$

so the map $\mathbf{A}_{\perp \mathbf{W}}$ is given by

$$\langle \mathbf{A}_{\perp \mathbf{W}}(X),Y\rangle = \langle (s(X,Y),0), (\perp W, \mathbf{1})\rangle = \langle s(X,Y), \perp W\rangle = \langle A_{\perp W}(X),Y\rangle .$$

Consequently,

$$\Sigma_2(\perp \mathbf{W}) = \Sigma_2(\perp W) .$$

We also have

$$\Delta(\perp \mathbf{W}) = \Delta((\perp W, \mathbf{1})) = (\Delta(\perp W),0) ,$$

and hence

$$\langle \perp \mathbf{W}, \Delta(\perp \mathbf{W})\rangle = \langle (\perp W,\mathbf{1}),(\Delta(\perp W),0)\rangle = \langle \perp W,\Delta(\perp W)\rangle .$$

Since we obviously have $\mathbf{T W} = \mathbf{T} W$, we have

$$\operatorname{div} \mathbf{T}\mathbf{W}\cdot(\mathbf{T}\mathbf{W} \lrcorner \Gamma(0)) = \operatorname{div} \mathbf{T}W\cdot(\mathbf{T}W \lrcorner \Gamma(0)) .$$

Finally,

$$\begin{aligned}[\boldsymbol{\perp}\tilde{\mathbf{W}}, \mathbf{T}\tilde{\mathbf{W}}] &= [(\boldsymbol{\perp}\tilde{W},\mathbf{1}),(\mathbf{T}\tilde{W},0)] \\ &= [\boldsymbol{\perp}\tilde{W}, \mathbf{T}\tilde{W}] + [\mathbf{1}, \mathbf{T}\tilde{W}] \\ &= [\boldsymbol{\perp}\tilde{W}, \mathbf{T}\tilde{W}] ,\end{aligned}$$

the second bracket vanishing since there is clearly a coordinate system $x^1,\dots,x^m,\tau$ on $\mathbf{N}$ with $\partial/\partial x^1 = \mathbf{T}\tilde{W}$ and $\partial/\partial\tau = \mathbf{1}$. Thus the result for $\boldsymbol{\alpha}$ implies the result for α. ■

To integrate the result of Theorem 37 succinctly, we introduce the <u>outward pointing unit normal of ∂M along f</u>; for each $q \in \partial M$, we define $\mu(q) \in N_{f(q)}$ to be the unit vector tangent to $f(M)$, perpendicular to $f(\partial N)$, and outward pointing. Recall (p. I.352) that the orientation for ∂M is chosen so

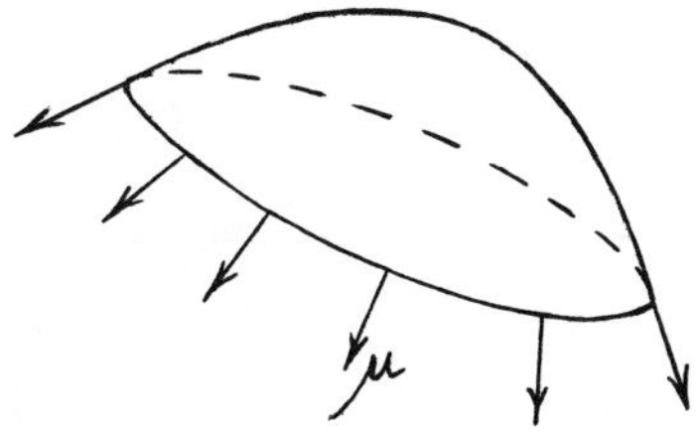

that $v_1,\dots,v_{n-1}$ is positively oriented at q if and only if $\mu(q),v_1,\dots,v_{n-1}$ is positively oriented on M.

<u>39. COROLLARY.</u> Let $f\colon M \to N$ be a minimal immersion of a compact oriented n-dimensional manifold-with-boundary M into a Riemannian manifold $(N^m, \langle\ ,\ \rangle)$, and let $\alpha\colon (-\varepsilon,\varepsilon)\times M \to N$ be a variation of f through immersions. Let W

be the variation vector field, let $\tilde{W} = \partial\alpha/\partial u$, and let μ be the outward pointing unit normal of ∂M along f. If $V(\bar{\alpha}(u))$ is the n-dimensional volume of M determined by the metric $\bar{\alpha}(u)^*\langle\ ,\ \rangle$ and the given orientation of M, then

$$\left.\frac{d^2V(\bar{\alpha}(u))}{du^2}\right|_{u=0} = \int_M [\mathrm{Ric}_M(\perp W) - \Sigma_2(\perp W) - \langle\perp W, \Delta(\perp W)\rangle]\, dV$$
$$+ (-1)^{n+1}\int_{\partial M}[\mathrm{div}\,\top W\cdot\langle\top W,\mu\rangle + \langle\top[\perp\tilde{W},\top\tilde{W}],\mu\rangle]\, dV^{n-1},$$

where dV is the volume element determined by $f^*\langle\ ,\ \rangle$, and dV^{n-1} is the induced volume element on ∂M. In particular, if α is a variation keeping ∂M fixed, then

$$\left.\frac{d^2V(\bar{\alpha}(u))}{du^2}\right|_{u=0} = \int_M [\mathrm{Ric}_M(\perp W) - \Sigma_2(\perp W) - \langle\perp W, \Delta(\perp W)\rangle]\, dV .$$

Proof. Left to the reader. ■

Problem 3 shows what our formula reduces to in the case of a geodesic $\gamma\colon [a,b] \longrightarrow N$. Here we will consider the case of a hypersurface $M \subset N$, with $i\colon M \longrightarrow N$ the inclusion map. Then $\perp W = h\nu$ for some function h, where ν is a unit normal vector field. Since

$$\perp\nabla'_{X_j(p)}(\perp\nabla'_{X_j} h\nu) = \perp\nabla'_{X_j(p)}(X_j(h)\nu)$$
$$= X_j(p)(X_j(h))\cdot\nu ,$$

we see that

$$\langle \perp W, \Delta(\perp W) \rangle = h\Delta h \ ,$$

where Δ now denotes the Laplacian on functions, computed by means of the induced metric $i^*\langle\ ,\ \rangle$ on M. So if Σ_2 denotes the sum of the squares of the eigenvalues of the symmetric map $II\colon M_p \times M_p \to \mathbb{R}$, then our integral becomes

$$\int_M [h^2 \mathrm{Ric}_M(\nu) - h^2\Sigma_2 - h\Delta h]\, dV \ .$$

Suppose in particular, that we consider the variation by parallel surfaces, $\alpha(u,p) = \exp_p u\cdot\nu(p)$. Then $h = 1$ and (Problem 3-12) $\tilde{W}(u,p)$ is always perpendicular to $\bar{\alpha}(u)(M)$; so the integral over ∂M drops out, and we obtain

$$\left.\frac{d^2 V(\bar{\alpha}(u))}{du^2}\right|_{u=0} = \int_M [\mathrm{Ric}_M(\nu) - \Sigma_2]\, dV \ .$$

If N has sectional curvatures ≥ 0, then $\mathrm{Ric}_M(\nu) \leq 0$, so we obtain

$$\left.\frac{d^2 V(\bar{\alpha}(u))}{du^2}\right|_{u=0} \leq 0 \ .$$

Moreover, we have strict inequality unless $\Sigma_2 = 0$, which happens only when $s = 0$, so that our hypersurface is totally geodesic. Thus a non-totally geodesic minimal hypersurface in a space of non-negative sectional curvature always has <u>greater</u> volume than nearby parallel surfaces.

Now let us consider a minimal immersion $f\colon M \to \mathbb{R}^3$, where M is a compact surface-with-boundary. Let $\alpha\colon (-\varepsilon,\varepsilon) \times M \to \mathbb{R}^3$ be a variation of f keeping ∂M fixed, such that $W = hn$ for some function h vanishing on ∂M, where n is a unit normal field. Then our formula becomes

$$(1)\quad \left.\frac{d^2A(\bar{\alpha}(u))}{du^2}\right|_{u=0} = \int_M [-h^2({k_1}^2 + {k_2}^2) - h\Delta h]\, dA$$

where k_1 and $k_2 = -k_1$ are the principal curvatures

$$= \int_M [2h^2K - h\Delta h]\, dA$$

$$= \int_M [2h^2K + I_f(\text{grad } h, \text{grad } h)]\, dA\ ,$$

by Proposition 7-59.

In particular, consider a compact 2-dimensional manifold-with-boundary $D \subset \mathbb{R}^2$ and a minimal immersion $\Phi\colon D \to \mathbb{R}^3$ given by

$$(*)\qquad \begin{cases} \Phi^1 = \operatorname{Re} \int \frac{1}{2}F(w)(1-w^2)\, dw \\ \Phi^2 = \operatorname{Re} \int \frac{i}{2}F(w)(1+w^2)\, dw \\ \Phi^3 = \operatorname{Re} \int F(w)w\, dw \end{cases}$$

for a nowhere 0 complex analytic function $F\colon D \to \mathbb{C}$. For this immersion we have (p. 400)

$$(2)\quad I_f = \Phi^*\langle\ ,\ \rangle = \mu(dx\otimes dx + dy\otimes dy)\ ,$$

$$\text{where}\quad \mu(z) = \frac{|F(z)|^2(1+|z|^2)^2}{4}\ ,\qquad z = x + iy\ .$$

We can compute (Problem 5) that the curvature K for the metric $\Phi^*\langle\ ,\ \rangle$ on D is given by

$$(3)\quad K(z) = \frac{-16}{|F(z)|^2(1+|z|^2)^4}\ ,\qquad z = x + iy\ .$$

Suppose now that we have a variation α of Φ which keeps ∂D fixed, and such that $W(\Phi(z)) = h(z)\cdot n(z)$ for some function $h\colon D \to \mathbb{R}$ with $h = 0$ on ∂D. Using (2), we compute, from the last equation in the proof of Proposition 5, that

$$(4) \qquad I_f(\text{grad } h, \text{grad } h)(z) = \frac{(h_1{}^2 + h_2{}^2)(z)}{\left(\dfrac{|F(z)|^2(1+|z|^2)^2}{4}\right)} .$$

Substituting (4) and (3) into (1), and remembering that the volume element dA of I on D is

$$\sqrt{\det(g_{ij})}\, dx \wedge dy = \mu\, dx \wedge dy ,$$

we obtain

$$\left.\frac{d^2A(\bar{\alpha}(u))}{du^2}\right|_{u=0} = \int_D \left[h_1{}^2 + h_2{}^2 - \frac{8h^2}{(1+x^2+y^2)^2}\right] dxdy$$

[where h and h_i should really be $h(x,y)$ and $h_i(x,y)$, respectively]. Notice that this expression does not involve the original map (*) at all; it involves only the region D, and the function h. If we recall (p. 400) that $n = \sigma^{-1}$, we see that D contains the unit disc $B = \{(x,y)\colon x^2+y^2 \le 1\}$, if and only if the normal map n of image Φ covers the whole southern hemisphere of the unit sphere.

<u>39. THEOREM (SCHWARZ-RADO).</u> If the interior of D contains the unit disc $B = \{(x,y)\colon x^2+y^2 \le 1\}$, then there is a function $h\colon D \to \mathbb{R}$ with $h = 0$ on ∂D such that

(1) $$\int_D \left[{h_1}^2 + {h_2}^2 - \frac{8h^2}{(1+x^2+y^2)^2} \right] dxdy < 0 .$$

Consequently, for every nowhere 0 complex analytic function $F: D \to \mathbb{C}$, the minimal surface $\Phi(D)$ given by (*) does not have minimum area among all nearby surfaces with the same boundary.

(Since the solution to the Plateau problem tells us that there is some minimal disc with the same boundary as $\Phi(S^1)$, this proves that $\Phi(S^1)$ is the boundary of at least 2 different minimal surfaces.)

Proof. Let $B(r) = \{(x,y): x^2+y^2 \le r^2\}$, and define $h^r: B(r) \to \mathbb{R}$ by

(2) $$h^r(x,y) = \frac{x^2+y^2-r^2}{x^2+y^2+r^2} .$$

Set

(3) $$I(r) = \int_{B(r)} \left[(h_1^r)^2 + (h_2^r)^2 - \frac{8(h^r)^2}{(1+x^2+y^2)^2} \right] dxdy .$$

Substituting (2) into (3), we obtain the explicit formula

$$I(r) = \int_{B(r)} \frac{16(x^2+y^2)r^4}{(x^2+y^2+r^2)^4} dxdy - \int_{B(r)} \frac{8(x^2+y^2-r^2)^2}{(x^2+y^2+r^2)^2(x^2+y^2+1)^2} dxdy .$$

Making the substitution $x = u \cdot r$, $y = v \cdot r$, we get

$$I(r) = \int_B \frac{16(u^2+v^2)}{(u^2+v^2+1)^4} dudv - \int_B \frac{8(u^2+v^2-1)^2 r^2}{(u^2+v^2+1)^2(u^2r^2+v^2r^2+1)^2} dudv .$$

Finally, computing $I'(1)$ by Leibniz's Rule, we obtain

$$I'(1) = 16 \int_B \frac{(u^2+v^2-1)^3}{(u^2+v^2+1)^5}\, dudv$$

$$< 0 .$$

On the other hand, we claim that $I(1) = 0$. To prove this, we use Proposition 7-59 and the fact that $h^r = 0$ on $\partial B(r)$ to write (3) as

$$I(r) = - \int_{B(r)} h^r \left[h^r_{11} + h^r_{22} + \frac{8h^r}{(1+x^2+y^2)^2} \right] dxdy \ ;$$

then we just compute that the term in brackets is 0 for h^1.

Since $I(1) = 0$ and $I'(1) < 0$, there is a number $r_0 > 1$ such that $I(r) < 0$ for $1 < r < r_0$. Now there is some r with $1 < r < r_0$ such that $D \supset B(r)$. Define h on D by

$$h(x,y) = \begin{cases} h^r(x,y) \ , & (x,y) \in B(r) \\ 0 \ , & \text{otherwise.} \end{cases}$$

This h has all the desired properties, except that the first partial derivatives of h are discontinuous on $B(r)$. However, it is easy to see that we can round off h to a C^∞ function without changing the sign of the integral in (1). ■

PROBLEMS

1. Show that formula (*) on p. 400 gives

$$\text{a catenoid for } F(w) = \frac{1}{w^2}$$

$$\text{a helicoid for } F(w) = \frac{i}{w^2}$$

$$\text{Scherk's minimal surface for } F(w) = \frac{4}{1 - w^4} .$$

2. Let $f\colon \mathbb{R} \longrightarrow \mathbb{R}$ be convex.

(a) We have

$$f'(x+) = \inf_{h>0} \frac{f(x+h) - f(x)}{h}$$

$$f'(x-) = \sup_{h>0} \frac{f(x+h) - f(x)}{h} .$$

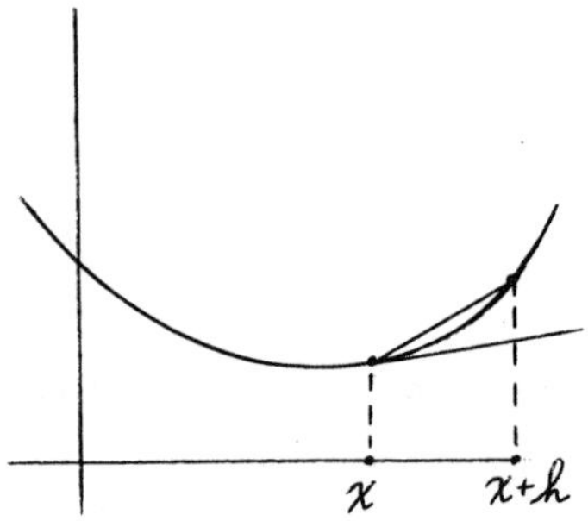

(b) If $f'(x)$ exists for all x, then f' is continuous. Hint: Consider $h > 0$, say, with $[f(x+h) - f(x)]/h < f'(x) + \varepsilon$.

3. Let $\gamma\colon [a,b] \to N$ be an arc-length parameterized geodesic, with unit tangent vector $V = d\gamma/dt$, and let $\alpha\colon (-\varepsilon,\varepsilon) \times [a,b] \to V$ be a variation, with variation vector field $\tilde{W}$.

(a) If Z is a vector field along γ with $\langle V,Z\rangle = 0$, then $\perp\nabla'_V Z = \nabla'_V Z$.

(b) $\Delta(\perp W) = D^2 \perp W/dt^2$.

(c) We have

$$\left.\frac{d^2 L(\bar{\alpha}(u))}{du^2}\right|_{u=0} = \int_a^b - \left\langle \frac{D^2 \perp W}{dt^2}, \perp W(t) \right\rangle - \langle R'(W,V)V,W\rangle(t)\, dt$$

$$+ \langle \nabla_V \mathbf{T} W, V\rangle \cdot \langle \mathbf{T} W, V\rangle + \langle [\perp W, \mathbf{T} W], V\rangle \Big|_b^a .$$

(d) Let

$$B = \langle \nabla_V \mathbf{T} W, V\rangle \cdot \langle \mathbf{T} W, V\rangle + \langle [\perp W, \mathbf{T} W], V\rangle$$

$$= \langle \nabla_V \mathbf{T} W, \mathbf{T} W\rangle + \langle \nabla_{\perp W} \mathbf{T} W, V\rangle - \langle \nabla_{\mathbf{T} W} \perp W, V\rangle .$$

Noting that $\mathbf{T} W$ is a multiple of V, say $\mathbf{T} W = hV$, show that

$$\langle \nabla_{\mathbf{T} W} \perp W, V\rangle = 0 \quad \text{and} \quad \langle \nabla_V \mathbf{T} W, \mathbf{T} W\rangle = \langle \nabla_{\mathbf{T} W} \mathbf{T} W, V\rangle .$$

Thus

$$B = \langle \nabla_{\mathbf{T} W} \mathbf{T} W, V\rangle + \langle \nabla_{\perp W} \mathbf{T} W, V\rangle = \langle \nabla_W \mathbf{T} W, V\rangle$$

$$= \langle \nabla_W W, V\rangle - \langle \nabla_W {\perp} W, V\rangle$$

$$= \langle \nabla_W W, V\rangle + \langle \nabla_V {\perp} W, {\perp} W\rangle \ .$$

(e) Conclude that

$$\left.\frac{d^2 L(\bar{\alpha}(u))}{du^2}\right|_{u=0} = \int_a^b \left\langle \frac{D{\perp}W}{dt}, \frac{D{\perp}W}{dt} \right\rangle - \langle R'(W,V)V, W\rangle(t)\, dt$$

$$+ \left.\langle \nabla_W, W, V\rangle\right|_a^b$$

4. If $M \subset \mathbb{R}^3$ is a minimal surface, then at any point $p \in M$ the Gaussian curvature $K(p)$ is given by

$$K(p) = -\frac{\langle \nu_* X, \nu_* X\rangle}{\langle X, X\rangle} \qquad \text{for any } X \in M_p .$$

<u>Hint</u>: The numerator is $\mathrm{III}(X,X)$.

5. Consider a minimal immersion $\Phi\colon V \to \mathbb{R}^3$ given by (*) on p. 400, so that $n = \sigma^{-1}$ and $g_{ij} = \mu\delta_{ij}$, where

$$\mu(z) = \frac{|F(z)|^2(1+|z|^2)^2}{4} \ .$$

Use Problems 4 and 7-20 to show that

$$K(z) = -\left(\frac{2}{1+|z|^2}\right)^2 \Big/ \mu(z)$$

$$= \frac{-16}{|F(z)|^2(1+|z|^2)^4} .$$

6. (a) Let $M \subset \mathbb{R}^3$ be a minimal surface with $K < 0$ everywhere, and consider an imbedding $f\colon U \to M$ whose parameter lines are lines of curvature. Using the formulas on p. III.323, show that if $k_1 > 0$ is the positive principal curvature, then

$$E(s,t) = S(s)/k_1(s,t) \qquad G(s,t) = T(t)/k_1(s,t)$$

for certain functions $S, T > 0$. Then show that there is a new imbedding with

$$E = G = \frac{1}{k_1} .$$

Conclude that if $\langle\ ,\ \rangle$ is the metric on M, then

$$\sqrt{-K}\,\langle\ ,\ \rangle$$

is a **flat** metric (Ricci).

(b) Let $\langle\ ,\ \rangle$ be a metric on a 2-dimensional manifold M such that $K < 0$ and $\sqrt{-K}\langle\ ,\ \rangle$ is flat. Thus there is a coordinate system (u,v) such that

$$\sqrt{-K}\langle\ ,\ \rangle = du \otimes du + dv \otimes dv \quad \Rightarrow \quad \langle\ ,\ \rangle = \frac{1}{\sqrt{-K}}(du \otimes du + dv \otimes dv)$$

$$= g(du \otimes du + dv \otimes dv), \text{ say.}$$

Using the formula on p. III.322, show that

$$K = -\frac{1}{2g}\left[\left(\frac{g_v}{g}\right)_v + \left(\frac{g_u}{g}\right)_u\right].$$

Then show that there is an imbedding $f\colon U \longrightarrow \mathbb{R}^3$ with

$$E = G = \frac{1}{\sqrt{-K}}, \qquad F = 0$$

$$\ell = 1, \quad n = -1, \quad m = 0.$$

Thus $f(U)$ is a minimal surface isometric to M.

Mini-Bibliography for Volume IV

Alexandrov, A.D.

[1] Uniqueness theorems for surfaces in the large. I. (Russian) Vestnik Leningrad. Univ. 11 (1956), No. 19, 5-17.

[2] A characteristic property of spheres, Ann. Mat. Pura Appl. 58 (1962), 303-315.

Bianchi, L.

[1] Sulle superficie a curvatura nulla in geometria ellittica, Ann. Mat. Pura Appl. 24 (1896), 93-129.

Blaschke, W.

{1} Vorlesungen über Differential Geometrie, Vols. I and II, third edition, Chelsea, New York, 1967.

{2} Kreis und Kugel, De Gruyter, Berlin, 1966.

Boys, C.V.

{1} Soap Bubbles. Their Colors and the Forces which Mold Them, third edition, Dover, New York, 1959.

Chern, S.S. and Lashof, R.K.

[1] On the total curvature of immersed manifolds, Amer. J. Math. 79 (1957), 306-318; II. Michigan Math. J. 5 (1958), 5-12.

Courant, R.

[1] Soap film experiments with minimal surfaces, Amer. Math. Monthly 47 (1940), 167-174.

Do Carmo, M. and Lima, E.

[1] Isometric immersions with semi-definite second quadratic forms, Arch. Math. (Basel) 20 (1969), 173-175.

[2] Immersions of manifolds with non-negative sectional curvatures, Bol. Soc. Brasil. Mat. 2 (1972), 9-22.

Do Carmo, M. and Warner, F.W.

[1] Rigidity and convexity of hypersurfaces in spheres, J. Differential Geometry 4 (1970), 133-144.

Eisenhart, L.P.

{1} Riemannian Geometry, Princeton Univ. Press, Princeton, N.J., 1925.

Fialkow, A.

[1] Hypersurfaces of a space of constant curvature, Ann. of Math. 39 (1938), 762-785.

Gromoll, D., Klingenberg, W., and Meyer, W.

{1} Riemannsche Geometrie im Grossen, Lecture Notes in Math. No. 55, Springer-Verlag, Berlin-Heidelberg, 1968.

Hartshorne, R.

{1} Foundations of Projective Geometry, W.A. Benjamin, New York, 1967.

Hopf, H.
[1] Über Flächen mit einer Relation Zwischen den Hauptkrümmungen, Math. Nachr. 4 (1951), 232-249.

Lawson, H.R.
[1] Complete minimal surfaces in S^3, Ann. of Math. 92 (1970), 335-374.

Milnor, J.
{2} Morse Theory, Ann. Math. Studies No. 51, Princeton Univ. Press, Princeton, N.J., 1963.

O'Neil, B.
[1] Isometric immersions which preserve curvature operators, Proc. Amer. Math. Soc. 13 (1962), 759-763.

Ryan, P.J.
[1] Homogeneity and some curvature conditions for hypersurfaces, Tohoku Math. J. 21 (1969), 363-388.

Sacksteder, R.
[1] On hypersurfaces with no negative sectional curvatures, Amer. J. Math. 82 (1960), 609-630.

Scheffers, G.W.
{1} *Anwendung der Differential- und Integral-Rechnung auf Geometrie*, 2 vols., Veit & Co., Leipzig, 1901-1902.

Stoker, J.J.
[1] Über die Gestalt der positiv gekrümmten offenen Flächen im dreidimensionalen Raume, Composito Math. 3 (1936), 55-89.

Warner, F.W.
{1} Foundations of Differentiable Manifolds and Lie groups, Scotts, Foresman and Co., Glenview, Illinois, 1971.

NOTATION INDEX

Chapter 7A

Chapter 7B

Chapter 7C

(cont.)

Chapter 7D

INDEX